METHODS IN MOLECULAR BIOLOGY™

Series Editor
John M. Walker
School of Life Sciences
University of Hertfordshire
Hatfield, Hertfordshire, AL10 9AB, UK

For other titles published in this series, go to
www.springer.com/series/7651

Mesenchymal Stem Cell Assays and Applications

Edited by

Mohan C. Vemuri

Stem Cells and Regenerative Medicine, Life Technologies, Frederick, MD, USA

Lucas G. Chase

Cellular Dynamics International, Inc., Madison, WI, USA

Mahendra S. Rao

Stem Cells and Regenerative Medicine, Life Technologies, Frederick, MD, USA

Humana Press

Editors
Mohan C. Vemuri, Ph.D.
Stem Cells and Regenerative Medicine
Life Technologies
Frederick, MD, USA
Mohan.Vemuri@lifetech.com

Lucas G. Chase, Ph.D.
Cellular Dynamics International, Inc.
525 Science Drive, Suite 200
Madison, WI, USA
lchase@cellulardynamics.com

Mahendra S. Rao
Stem Cells and Regenerative Medicine
Life Technologies
Frederick, MD, USA
mahendra.rao@lifetech.com

Please note that additional material for this book can be downloaded from http://extras.springer.com

ISSN 1064-3745 e-ISSN 1940-6029
ISBN 978-1-60761-998-7 e-ISBN 978-1-60761-999-4
DOI 10.1007/978-1-60761-999-4
Springer New York Dordrecht Heidelberg London

Library of Congress Control Number: 2011922525

Printed on acid-free paper

Humana Press is part of Springer Science+Business Media (www.springer.com)

Preface

Mesenchymal stem cells (marrow-derived stromal cells – MSC) were first harvested from the marrow well over two decades ago and were shown to likely play diverse functions in vitro and in vivo. MSC are thought to contribute to the stem cell niche in the marrow; contribute to smooth muscle, adipocyte, bone and cartilage development and repair; and in general contribute to the parenchyma of the most tissues and organs. MSC or MSC-like cells can therefore be isolated from a variety of tissues, and while subtle differences between these populations have been identified, their properties seem similar and different researchers have used different subpopulations of MSC and tested their efficacy in a variety of diverse models of disease.

This (MSC) class of somatic (adult) stem cells has seen an unprecedented level of interest in the last decade. They are the only stem cell type that has two annual meetings devoted solely to it and the only cell type that has spawned over a dozen companies and whose therapeutic efficacy is being tested in over 90 clinical trials. Perhaps the reason for this level of activity is the relative ease of isolation, the large numbers of cells present in the adult, and the ability to propagate these cells in culture (in contrast to hematopoietic stem cells or cord blood-derived hematopoietic stem cells).

Surprisingly, despite this level of activity, there is little consensus on the exact lineage of this population of cells and which set of markers defines these cells in vitro and in vivo. Given this ambiguity, we felt there was a need to compile a set of protocols and assays used by the leading investigators in this field that would enable others to standardize the population of cells they used in their experiments.

The end result of this effort is 35 chapters from leading experts from all over the world who have graciously shared their experience to describe how to isolate, propagate, characterize, and manipulate this special cell type. We hope the researchers will find the information in this book as useful as we and our laboratory groups have found it to be.

Frederick, MD — *Mohan C. Vemuri*
Madison, WI — *Lucas G. Chase*
Frederick, MD — *Mahendra S. Rao*

Contents

Contributors

BRIAN AGNEW • *Life Technologies, Eugene, OR, USA*

PEDRO Z. ANDRADE • *IBB-Institute for Biotechnology and Bioengineering, Centre for Biological and Chemical Engineering, Instituto Superior Técnico, Lisboa, Portugal*

AGNIESZKA ARTHUR • *Mesenchymal Stem Cell Group, Division of Haematology, Institute of Medical and Veterinary Science/Hanson Institute/CSCR, University of Adelaide, Adelaide, SA, Australia*

KERRY ATKINSON • *Mater Health Services, Brisbane, Queensland 4101, Australia; Mater Medical Research Institute, Brisbane, Queensland 4101, Australia; University of Queensland, Brisbane, Queensland 4072, Australia*

DENIZ BALCI • *Ankara University Biotechnology Institute, Ankara, Turkey*

SARAH BARLOW • *Mater Medical Research Institute, Brisbane, Queensland, Australia*

P. MARK BARTOLD • *Colgate Australian Clinical Dental Research Centre, Dental School, University of Adelaide, Adelaide, SA, Australia*

ROBERT H. BATCHELOR • *Primary and Stem Cell Systems, Life Technologies, Eugene, OR, USA*

SARINDR BHUMIRATANA • *Laboratory for Stem Cells and Tissue Engineering, Columbia University, New York, NY, USA*

SIMONE BORK • *Heidelberg Academy of Sciences and Humanities, Heidelberg, Germany; Department of Medicine V, University of Heidelberg, Heidelberg, Germany*

SHAYNE E. BOUCHER • *Life Technologies, Frederick, MD, USA*

JOLENE A. BRADFORD • *Life Technologies, Eugene, OR, USA*

GARY BROOKE • *Mater Medical Research Institute, Brisbane, Queensland, Australia*

SANDRA BRÜCKNER • *First Department of Medicine, Martin-Luther University of Halle-Wittenberg, Halle/Saale, Germany*

JOAQUIM M. S. CABRAL • *IBB-Institute for Biotechnology and Bioengineering, Centre for Biological and Chemical Engineering, Instituto Superior Técnico, Lisboa, Portugal*

CYNTHIA CAMARILLO • *Rutgers Stem Cell Research Center and W.M. Keck Center for Collaborative Neuroscience, Rutgers University, Piscataway, NJ, USA*

MELISSA CAMASSOLA • *Programa de Pós-Graduação em Diagnóstico Genético e Molecular, Universidade Luterana do Brasil, Canoas, RS, Brazil*

ALP CAN • *Laboratory for Stem Cell Science, Department of Histology and Embryology, Ankara University School of Medicine, Ankara University Stem Cell Institute, Ankara, Turkey*

CHRISTOPHER CANNIZZARO • *Laboratory for Stem Cells and Tissue Engineering, Columbia University, New York, NY, USA*

CHU-YAN CHAN • *Stanley Ho Centre for Emerging Infectious Diseases, The Chinese University of Hong Kong, Hong Kong, P. R. China*

LUCAS G. CHASE • *Cellular Dynamics International, 525 Science Dr # 200, Madison 53711-1018, WI, USA*

ANDRE CHOO • *Bioprocessing Technology Institute, Agency for Science Technology and Research, Singapore 138668, Singapore; Division of Bioengineering, National University of Singapore, 117574 Singapore*

BRUNO CHRIST • *First Department of Medicine, Martin-Luther University of Halle-Wittenberg, Halle/Saale, Germany*

SCOTT T. CLARKE • *Life Technologies, Willow Creek Road, Eugene, OR, USA*

DIMAS TADEU COVAS • *National Institute of Science and Technology for Stem Cells and Cell Therapy, Centro Regional de Hemoterapia de Ribeirão Preto – HCFMRP/Universidade de São Paulo, Rua Tenente Catão Roxo 2501, Ribeirão Preto, SP, Brazil; Department of Clinical Medicine, Universidade de São Paulo, Av. Bandeirantes 3900, Ribeirão Preto, SP, Brazil*

CLÁUDIA LOBATO DA SILVA • *IBB-Institute for Biotechnology and Bioengineering, Centre for Biological and Chemical Engineering, Instituto Superior Técnico, Lisboa, Portugal*

JOERY DE KOCK • *Department of Toxicology, Vrije Universiteit Brussel, Laarbeeklaan 103, 1090 Brussels, Belgium*

FRANCISCO DOS SANTOS • *IBB-Institute for Biotechnology and Bioengineering, Centre for Biological and Chemical Engineering, Instituto Superior Técnico, Lisboa, Portugal*

GEMMA EIBES • *Department of Chemical Engineering, School of Engineering, University of Santiago de Compostela, Santiago de Compostela, Spain*

ALANA ESHPETER • *Alberta Diabetes Institute, University of Alberta, Edmonton, AB, Canada*

DARIO O. FAUZA • *Department of Surgery, Children's Hospital Boston and Harvard Medical School, Boston, MA, USA*

TRINE FINK • *Laboratory for Stem Cell Research, Aalborg University, Aalborg, Denmark*

ZHAODI GONG • *Department of Anesthesiology, Yale University School of Medicine, New Haven, CT, USA*

WARREN L. GRAYSON • *Laboratory for Stem Cells and Tissue Engineering, Columbia University, New York, NY, USA*

CARL A. GREGORY • *Institute for Regenerative Medicine at Scott and White Hospital, Texas A and M Health Science Center, 5701 Airport Rd Module C, Temple 76502, TX, USA*

STAN GRONTHOS • *Mesenchymal Stem Cell Group, Division of Haematology, Institute of Medical and Veterinary Science/Hanson Institute/CSCR, University of Adelaide, Adelaide, SA, Australia*

ASHISH GUPTA • *Department of Medicine, Brigham and Women's Hospital, Harvard Medical School, Harvard Stem Cell Institute, Cambridge, MA, USA*

DEEPAK M. GUPTA • *Hagey Laboratory for Pediatric Regenerative Medicine, Division of Plastic and Reconstructive Surgery, Department of Surgery, Stanford University School of Medicine, Stanford, CA, USA*

MANDANA HAACK-SØRENSEN • *Cardiology Stem Cell Laboratory and Cardiac Catheterization Laboratory, The Heart Centre, Rigshospitalet Copenhagen University Hospital, Copenhagen, Denmark*

MAHBOD HAJIVANDI • *Life Technologies, Eugene, OR, USA*

SONIA HANCOCK • *Mater Medical Research Institute, Brisbane, Queensland, Australia*
GEORGE T. HANSON • *Life Technologies, Eugene, OR, USA*
COURTENAY HART • *Life Technologies, eugene, OR, USA*
RONALD P. HART • *Rutgers Stem Cell Research Center and W.M. Keck Center for Collaborative Neuroscience, Rutgers University, Piscataway, NJ, USA*
BORIS HINZ • *Laboratory of Tissue Repair and Regeneration, Matrix Dynamics Group, Faculty of Dentistry, University of Toronto, M5S 3E2, ON, Toronto, Canada*
PATRICK HORN • *Department of Medicine V, University of Heidelberg, Heidelberg, Germany*
SHIAW-MIN HWANG • *Bioresource Collection and Research Center (BCRC), Food Industry Research and Development Institute, Hsinchu, Taiwan*
NINA ILIC • *Mater Health Services, Brisbane, Queensland 4101, Australia*
CHRISTIAN JORGENSEN • *INSERM, U844 Montpellier, France*
ROHIT KARNIK • *Massachusetts Institute of Technology, Cambridge, MA, USA*
JEFFREY M. KARP • *Department of Medicine, Center for Regenerative Therapeutics Brigham and Women's Hospital, Harvard Medical School, Harvard Stem Cell Institute, Cambridge, MA, USA*
JENS KASTRUP • *Cardiology Stem Cell Laboratory and Cardiac Catheterization Laboratory, The Heart Centre, Rigshospitalet Copenhagen University Hospital, Copenhagen, Denmark*
CHRISTOPHER W. KEMP • *Kempbio, Inc, Frederick, MD, USA*
JUSTIN D. KLEIN • *Department of Surgery, Children's Hospital Boston and Harvard Medical School, Boston, MA, USA*
GREGORY S. KORBUTT • *Alberta Diabetes Institute, University of Alberta, Edmonton, AB, Canada*
ULF KRAUSE • *Institute for Regenerative Medicine at Scott and White Hospital, Texas A and M Health Science Center, 5701 Airport Rd Module C, Temple 76502, TX, USA*
HSIANG-FU KUNG • *Stanley Ho Centre for Emerging Infectious Diseases, The Chinese University of Hong Kong, Hong Kong, P. R. China*
RUENN CHAI LAI • *NUS Graduate School for Integrative Sciences & Engineering, National University of Singapore, Singapore; Institute of Medical Biology, Agency for Science Technology and Research, Singapore*
ANITA LAITINEN • *Finnish Red Cross Blood Service, Helsinki, Finland*
SAARA LAITINEN • *Finnish Red Cross Blood Service, Helsinki, Finland*
UMA LAKSHMIPATHY • *Primary and Stem Cell Systems, Life Technologies, Carlsbad, CA, USA*
ENG HIN LEE • *Department of Orthopaedic Surgery, Yong Loo Lin School of Medicine, NUS Tissue Engineering Program National University of Singapore, Singapore*
YUN-SHIEN LEE • *Genomic Medicine Research Core Laboratory (GMRCL), Chang Gung Memorial Hospital, Tao-Yuan, Taiwan; Department of Biotechnology, Ming Chuan University, Tao-Yuan 33302, Taiwan*
SYLVAIN LEHMANN • *CNRS, Institut de Génétique Humaine UPR1142, Montpellier, France; CHU Montpellier, Plateforme de Protéomique Clinique, Biochimie, Hôpital St. Eloi, Montpellier 34000, France; Université Montpellier 1, 34000 Montpellier, France*

GUO LI • *Stanley Ho Centre for Emerging Infectious Diseases, The Chinese University of Hong Kong, Hong Kong, P. R. China*
SAI KIANG LIM • *Institute of Medical Biology, Agency for Science Technology and Research, 138648 Singapore; Department of Surgery, Yong Loo Lin School of Medicine, National University of Singapore, 119074 Singapore*
WEI LI LOH • *Department of Medicine, Brigham and Women's Hospital, Harvard Medical School, Harvard Stem Cell Institute, Cambridge, MA, USA*
MICHAEL T. LONGAKER • *Hagey Laboratory for Pediatric Regenerative Medicine, Division of Plastic and Reconstructive Surgery, Department of Surgery, Stanford University School of Medicine, Stanford, CA, USA*
HICHAM MAJD • *Department of Surgery, Centre Medical Universitaire, University of Geneva, Geneva, Switzerland*
LINDOLFO DA SILVA MEIRELLES • *National Institute of Science and Technology for Stem Cells and Cell Therapy, Centro Regional de Hemoterapia de Ribeirão Preto – HCFMRP/Universidade de São Paulo, Ribeirão Preto, SP, Brazil*
PATRICIA MURRAY • *Mater Medical Research Institute, Brisbane, Queensland, Australia*
NANCE BEYER NARDI • *Programa de Pós-Graduação em Diagnóstico Genético e Molecular, Universidade Luterana do Brasil, Canoas, RS, Brazil*
LAURA E. NIKLASON • *Department of Anesthesia & Biomedical Engineering, Yale University School of Medicine, New Haven, CT, USA*
JOHANNA NYSTEDT • *Finnish Red Cross Blood Service, Helsinki, Finland*
MICHAEL O'GRADY • *Primary and Stem Cell Systems, Life Technologies, Eugene, OR, USA*
NICHOLAS J. PANETTA • *Hagey Laboratory for Pediatric Regenerative Medicine, Division of Plastic and Reconstructive Surgery, Department of Surgery, Stanford University School of Medicine, Stanford, CA, USA*
REBECCA PELEKANOS • *Mater Medical Research Institute, Brisbane, Queensland, Australia*
PATRICE PENFORNIS • *Gene Therapy Center, Tulane University Health Science Center, New Orleans, LA 70112, USA*
KITSIE J. PENICK • *Department of Biology, Skeletal Research Center, Case Western Reserve University, Cleveland, OH, USA*
RADHIKA POCHAMPALLY • *Gene Therapy Center, Department of Pharmacology, Tulane University Health Science Center, New Orleans, LA, USA*
MONIQUE PROVANSAL • *CNRS, Institut de Génétique Humaine UPR1142, Montpellier, France; Université Montpellier 1, 34000 Montpellier, France*
THOMAS M. QUINN • *Department of Chemical Engineering, McGill University, Montreal, QC, Canada*
PRANELA RAMESHWAR • *Department of Medicine Hematology/Oncology, University of Medicine and Dentistry of New Jersey – New Jersey Medical School, 185 South Orange Avenue, Newark, NJ, USA*
MAHENDRA S. RAO • *Stem Cells and Regenerative Medicine, Life Technologies, 7335 Executive Way, Frederick 21702, MD, USA*
JEPPE GRØNDAHL RASMUSSEN • *Laboratory for Stem Cell Research, Aalborg University, Aalborg, Denmark*

STÉPHANE ROCHE • *CNRS, Institut de Génétique Humaine UPR1142, Montpellier, France; CHU Montpellier, Plateforme de Protéomique Clinique, Biochimie, Hôpital St. Eloi, Montpellier 34000, France*

VERA ROGIERS • *Department of Toxicology, Vrije Universiteit Brussel, Laarbeeklaan 103, 1090 Brussels, Belgium*

TONY ROSSETTI • *Mater Medical Research Institute, Brisbane, Queensland, Australia*

DEBANJAN SARKAR • *Department of Medicine, Brigham and Women's Hospital, Harvard Medical School, Harvard Stem Cell Institute, Cambridge, MA, USA*

KELLY SCHEYHING • *Primary and Stem Cell Systems, Life Technologies, Carlsbad, CA, USA*

JACQUELINE FRIDA SCHMITT • *Department of Orthopaedic Surgery, Yong Loo Lin School of Medicine, NUS Tissue Engineering Program National University of Singapore, Singapore*

ANJA SECKINGER • *Section for Multiple Myeloma, Department of Internal Medicine V, Heidelberg University Hospital, Im Neuenheimer Feld 410, 69120, Heidelberg, Germany*

KAREN L. SEEBERGER • *Alberta Diabetes Institute, University of Alberta, Edmonton, AB, Canada*

SONGTAO SHI • *Center for Craniofacial Molecular Biology, University of Southern California School of Dentistry, Los Angeles, CA, USA*

SARAH SNYKERS • *Department of Toxicology, Vrije Universiteit Brussel, Brussels, Belgium*

LUIS A. SOLCHAGA • *Case Comprehensive Cancer Center, Case Western Reserve University, Cleveland, OH, USA*

PEGGY STOCK • *First Department of Medicine, Martin-Luther University of Halle-Wittenberg, Heinrich-Damerow-Strasse 1, 06120 Halle/Saale, Germany*

MAVIS SWERDEL • *Rutgers Stem Cell Research Center and W.M. Keck Center for Collaborative Neuroscience, Rutgers University, Piscataway, NJ, USA*

VANHAECKE TAMARA • *Department of Toxicology, Vrije Universiteit Brussel, Brussels, Belgium*

KATARZYNA A. TRZASKA • *Department of Medicine Hematology/Oncology, University of Medicine and Dentistry of New Jersey – New Jersey Medical School, Newark, NJ, USA*

MOHAN C. VEMURI • *Stem Cells and Regenerative Medicine, Life Technologies, 7335 Executive Way, Frederick 21704, MD, USA*

GORDANA VUNJAK-NOVAKOVIC • *Laboratory for Stem Cells and Tissue Engineering, Columbia University, New York, NY, USA*

WOLFGANG WAGNER • *Helmholtz Institute for Biomedical Engineering, Department of Cell Biology, RWTH Aachen University Medical School, Pauwelsstrasse 20, 52074 Aachen, Germany*

HUA WANG • *Stanley Ho Centre for Emerging Infectious Diseases, The Chinese University of Hong Kong, Hong Kong, P. R. China*

TZU-HAO WANG • *Department of Obstetrics and Gynecology, Lin-Kou Medical Center, Chang Gung Memorial Hospital, Chang Gung University, Tao-Yuan 33302, Taiwan; Genomic Medicine Research Core Laboratory (GMRCL), Chang Gung Memorial Hospital, Tao-Yuan 33302, Taiwan*

Jean F. Welter • *Department of Biology, Skeletal Research Center, Case Western Reserve University, Cleveland, OH, USA*
Pierre-Jean Wipff • *Laboratory of Cell Biophysics, Ecole Polytechnique Fédérale de Lausanne (EPFL), Station 15, 1015 Lausanne, Switzerland*
Zheng Yang • *Department of Orthopaedic Surgery, Yong Loo Lin School of Medicine, NUS Tissue Engineering Program National University of Singapore, 27 Medical Drive, Singapore 117510; Stem Cell Laboratory, National University of Singapore, 5 Lower Kent Ridge Road, Singapore 119074*
Vladimir Zachar • *Laboratory for Stem Cell Research, Aalborg University, Aalborg, Denmark*
Weian Zhao • *Department of Medicine, Brigham and Women's Hospital, Harvard Medical School, Harvard Stem Cell Institute, Cambridge, MA, USA*

Part I

Introduction

Chapter 1

Mesenchymal Stem Cell Assays and Applications

Mohan C. Vemuri, Lucas G. Chase, and Mahendra S. Rao

Abstract

Research on mesenchymal stem cells (MSC) is progressing with increasing popularity. Currently there are a significant number of clinical trials exploring the use of MSCs for the treatment of various diseases including graft-versus-host disease, Crohn's disease, myocardial infarction, stroke, bone defects, diabetes, and wound repair (www.clinicaltrails.gov). At the same time, there are questions associated with MSCs in terms of their isolation, culture expansion, phenotype, multipotential differentiation, and transplantation efficiency. This chapter outlines the current status of the field and emphasizes the need for clearly defined protocols to better define the function and use of MSCs in cell therapy.

Key words: MSC, Stem cell, Self renewal, Differentiation, Immunophenotype, Adipocyte, Osteocyte, Chondrocyte

1. Introduction

The term mesenchymal stromal cell or mesenchymal stem cell (MSC) refers to a population of adherent adult stem cells defined by their ability to give rise to differentiated mesenchymal cell types including osteoblasts, adipocytes, and chondrocytes (1). Originally identified by Friedenstein et al. (2) as fibroblast precursors within the bone marrow, this population was later defined as originating from the stromal framework (3), hence the term mesenchymal stromal cell. Since then, this population has been considered an adult stem cell population within the bone marrow (4), and called mesenchymal stem cell (MSC). Although questions have been raised (5, 6), about using the term stem cell it appears futile to suggest changing the nomenclature (7).

Mohan C. Vemuri et al. (eds.), *Mesenchymal Stem Cell Assays and Applications*, Methods in Molecular Biology, vol. 698, DOI 10.1007/978-1-60761-999-4_1, © Springer Science+Business Media, LLC 2011

2. How are Mesenchymal Stem Cells Defined?

Under standard in vitro culture conditions (i.e., 10% fetal bovine serum (FBS)-containing medium), MSCs are a heterogeneous population of plastic-adherent cells that exhibit a fibroblast-like morphology, form colonies when grown at clonal densities (referred to as colony-forming units – fibroblast or CFU-F) (2), and can differentiate into bone, cartilage, and fat cells (6). Human MSCs have been defined by the positive expression of the cell surface antigens CD73, CD90, CD105 and a lack of expression of hematopoietic antigens including CD11b or CD14, CD34, CD45, CD79 or CD19, and HLA-DR (5). A comprehensive list of surface antigen expression on MSC was recently published (8). Despite the exhaustive number of studies conducted to define MSCs by their surface antigen expression profile, variability still exists within MSC populations due to differences in the species, tissue source, and the in vitro culture conditions used. In an attempt to address this problem, the International Society for Cellular Therapy (ISCT) released a consensus position paper stating the minimal criteria required for defining MSCs (5). While this effort provides a necessary first step in standardization of the human MSC field, future efforts are necessary to better characterize these cells and identify markers capable of teasing apart this seemingly heterogeneous population and relating cellular phenotypes to biological function. Potential markers and uses of MSCs are outlined (Fig. 1).

3. MSC Isolation and Expansion

MSCs can be isolated from different organs or tissue compartments including the bone marrow (BM), umbilical cord blood (UCB), umbilical cord stroma (Wharton's jelly), placenta, adipose tissue (AT), and many others (8). Under standard growth conditions (i.e., 10% FBS-containing medium), human MSCs isolated from a tissue source (i.e., bone marrow mononuclear cells) and expanded in a culture dish gradually change their appearance from spherical to spindle in shape as they transition from suspension to adherent culture condition. Expanding MSCs undergo changes in their growth pattern from an initial lag phase to a rapid expansion phase. As adherent cultures are propagated, only a fraction of cells remains clonogenic (able to generate CFU-F under standard growth conditions), suggesting the presence of an in vitro niche for true MSCs within these cultures (1). Unlike other stem cell populations such as human embryonic stem cells, most human MSCs in culture display a limited expansion potential in vitro (i.e., ~5–10 passages)

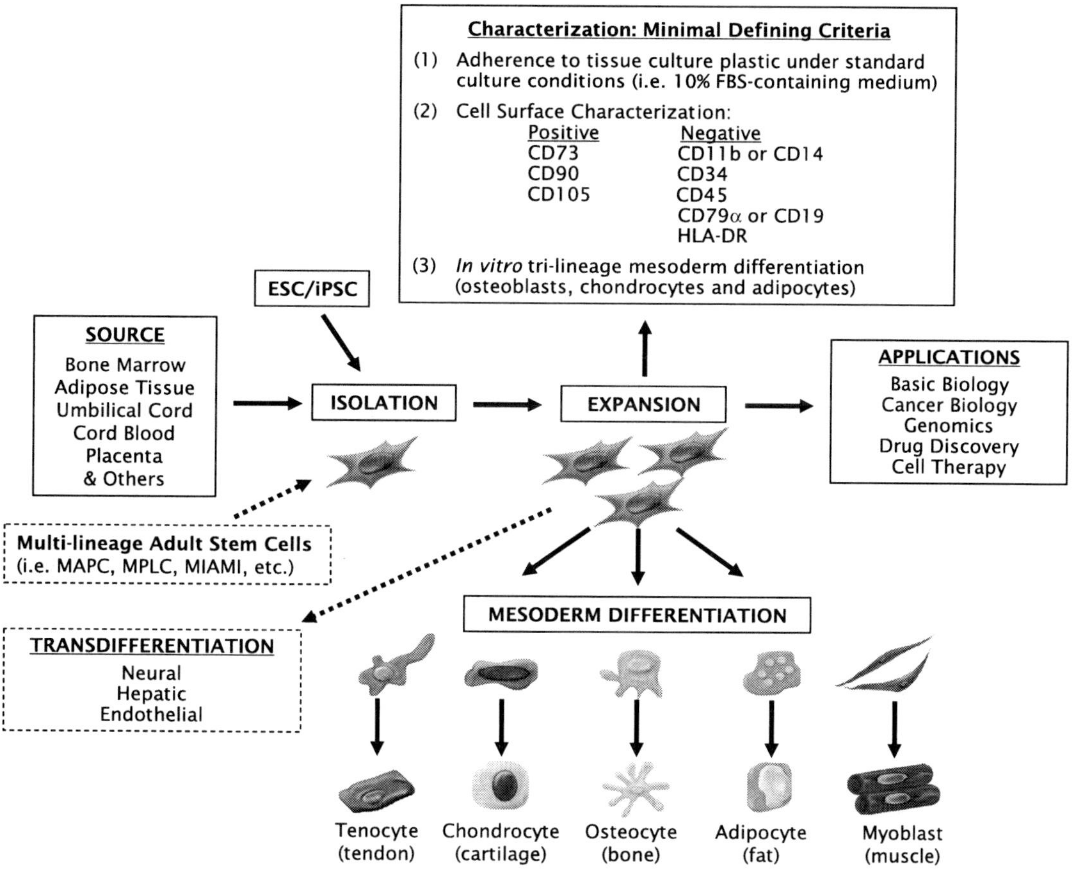

Fig. 1. Mesenchymal stem cells isolated and expanded from diverse sources are characterized by a minimal set of defining criteria and may be applied to various discovery and therapeutic applications.

and decreased differentiation potential with increased culture. Whether this lack of unlimited self-renewal is a result of insufficient or suboptimal isolation techniques, growth in cell culture conditions or is a true indicator of the stemness associated with this population is yet to be determined. Although some studies have indicated no significant differences in regard to the morphology, phenotype, and immunosuppressive properties of MSC like cells, differences have been observed in regard to the success rate of isolating MSCs from tissue sources (i.e., 100% for BM- and AT-MSC and 63% for UCB-MSC) (9). Also, while the MSC colony frequency was lowest in UCB and highest in AT, it was found that UCB-MSCs could be cultured longer and showed higher proliferative capacity while BM-MSCs displayed reduced culture longevity. Most strikingly, UCB MSCs showed limited adipogenic differentiation capacity, while BM- and AT-MSCs displayed a standard tri-lineage differentiation potential (9). By contrast, another report showed that applying identical culture conditions to MSCs from diverse sources resulted in major differences in MSC frequency and expansion potential, but the basic biological features of the expanded cells were comparable (10).

4. MSC Differentiation Potential

The large body of work that has accumulated since the discovery of human MSCs has convincingly shown that MSCs from diverse sources retain the ability to differentiate into various mesodermal lineages including osteoblasts (bone), chondrocytes (cartilage), and adipocytes (fat) (11). Differentiation into additional cell types of mesodermal origin (skeletal muscle, smooth muscle, cardiac muscle, endothelial cells, etc.) and cells derived from other germ layers (neurons, hepatocytes, beta cells, etc.) have been reported (Fig. 1), but thus far there is a lack of definitive evidence as to the functionality of these differentiated cells (6, 12). To elucidate the phenotype of differentiated MSCs, several classical assays have been employed including alkaline phosphatase and alizarin red S staining for osteoblasts, oil red O staining for adipocytes, and alcian blue staining for chondrogenesis. But these methods are limited in terms of defining the multipotential differentiation nature of MSC (12, 13). There is a need for new additional assays to define the multipotential nature of MSCs as well as to identify early stage precursors of differentiating lineages, such as preosteoblast, preadipocyte, and prechondrocyte populations.

5. MSC Properties and Functions

Establishing stem cell identity and function is defined by the prospective isolation of a single stem cell in order to prove (a) self renewal, (b) multipotential differentiation, and (c) repopulating capacity (14). These considerations are difficult to demonstrate with a single MSC cell. For human MSCs, intrinsic tissue turnover poses a barrier to assess the self-renewal rate of undifferentiated stem cells (i.e., the human skeleton turns over ~5 times during adulthood). Similarly, it is not clear whether the greater differentiation potential is acquired during cell culture of MSCs. Extensive transcription profiling and in vitro and in vivo assays have identified specific genes implicated in osteogenic differentiation (FHL2, ITGA5, Fgf18), chondrogenesis (FOXO1A), and tenogenesis (Smad8) (15). Further data is required to develop effective directed differentiation techniques to isolate a specific lineage. A decline in the frequency of potent MSCs has been implicated in aging and degenerative diseases. Thus, identifying the ideal cells for ex vivo expansion will form a major pursuit for an efficacious clinical application (16). Similarly, though it is difficult to demonstrate the efficiency of repopulation of MSCs per se, studies have confirmed the cotransplantation of MSCs and CD34+ cells can promote hematopoietic stem cell transplantation and hematopoietic recovery in vivo (17). Following transplantation, MSCs are expected to reduce the damage and activate the endogenous regenerative potential in recipient tissue (18).

The challenges in MSC field can be addressed by developing clearly defined conditions of MSC isolation, expansion in cell culture, low or high seeding density, different culture media used such as, serum-free and xeno-free culture media or with serum supplementation, extent of ex vivo expansion, that will significantly improve the understanding of basic properties of MSC and their utility in immune and therapeutical benefit. These observations suggest that the cell processing protocols can be modified to optimize the MSCs for a given clinical situation or disease. Chapters in this book have been selected to address these critical issues in the field.

References

1. Prockop, D.J., Repair of tissues by adult stem/progenitor cells (MSCs): controversies, myths, and changing paradigms. Mol Ther, 2009. **17**(6): 939–46.
2. Friedenstein, A.J., R.K. Chailakhjan, and K.S. Lalykina, The development of fibroblast colonies in monolayer cultures of guinea-pig bone marrow and spleen cells. Cell Tissue Kinet, 1970. **3**(4): 393–403.
3. Dexter, T.M., T.D. Allen, and L.G. Lajtha, Conditions controlling the proliferation of haemopoietic stem cells in vitro. J Cell Physiol, 1977. **91**(3): 335–44.
4. Caplan, A.I., Mesenchymal stem cells. J Orthop Res, 1991. **9**(5): 641–50.
5. Dominici, M., et al., Minimal criteria for defining multipotent mesenchymal stromal cells. The International Society for Cellular Therapy position statement. Cytotherapy, 2006. **8**(4): 315–7.
6. Horwitz, E.M., et al., Clarification of the nomenclature for MSC: The International Society for Cellular Therapy position statement. Cytotherapy, 2005. **7**(5): 393–5.
7. Bianco, P., P.G. Robey, and P.J. Simmons, Mesenchymal stem cells: revisiting history, concepts, and assays. Cell Stem Cell, 2008. **2**(4): 313–9.
8. Meirelles Lda, S. and N.B. Nardi, Methodology, biology and clinical applications of mesenchymal stem cells. Front Biosci, 2009. **14**: 4281–98.
9. Kern, S., et al., Comparative analysis of mesenchymal stem cells from bone marrow, umbilical cord blood, or adipose tissue. Stem Cells, 2006. **24**(5): 1294–301.
10. Bieback, K., et al., Comparing mesenchymal stromal cells from different human tissues: bone marrow, adipose tissue and umbilical cord blood. Biomed Mater Eng, 2008. **18**(1 Suppl): S71–6.
11. Bernardo, M.E., F. Locatelli, and W.E. Fibbe, Mesenchymal stromal cells. Ann N Y Acad Sci, 2009. **1176**: 101–17.
12. Bianco, P., et al., Postnatal skeletal stem cells. Methods Enzymol, 2006. **419**: 117–48.
13. Gronthos, S., et al., Stem cell properties of human dental pulp stem cells. J Dent Res, 2002. **81**(8): 531–5.
14. Verfaillie, C.M., Bony endothelium: tumor-mediated transdifferentiation? Cancer Cell, 2008. **14**(3): 193–4.
15. Augello, A. and C. De Bari, The regulation of differentiation in mesenchymal stem cells. Human Gene Ther, 2010. **10**: 1226–38.

16. Xiao, Y., et al., Stem cell related gene expression in clonal populations of mesenchymal stromal cells from bone marrow. Tissue Eng Part A, 2010. **16**(2): 749–58.

17. Hao, M., et al., [Study of influence of umbilical cord mesenchymal stem cells on CD34+ cells in vivo homing in NOD/SCID]. Zhonghua Xue Ye Xue Za Zhi, 2009. **30**(2): 103–6.

18. Noort, W.A., et al., Mesenchymal stromal cells to treat cardiovascular disease: strategies to improve survival and therapeutic results. Panminerva Med, 2010. **52**(1): 27–40.

Part II

Isolation and Expansion of MSCs from Various Sources

Chapter 2

Isolation and Expansion of Mesenchymal Stem Cells/Multipotential Stromal Cells from Human Bone Marrow

Patrice Penfornis and Radhika Pochampally

Abstract

In recent years, human mesenchymal stem cells (multipotential stromal cells) from bone marrow (hMSCs) have attracted enormous attention owing to their broad therapeutic potential. One of the problems in the overall therapeutic use of hMSCs has been the significant variability in the culture conditions used for their isolation and expansion. Since the seminal publications by Friedenstein and colleagues, the isolation and expansion of mesenchymal stromal cells (MSCs) from bone marrow have been of interest to several laboratories. As a result, numerous isolation protocols have been published. This chapter provides a simple protocol whereby a total of 80–100 million human MSCs, with an average viability greater than 90%, can be produced from a relatively small (1–3 mL) bone marrow aspirate in 14–20 days using double stacks culture chambers. MSCs were originally referred to as fibroblastoid colony forming cells because one of their characteristic features is adherence to tissue culture plastic and generation of colonies when plated at low densities. The efficiency with which they form colonies still remains an important assay for the quality of cell preparations. To assess the quality of cell preparations, two different colony forming unit (CFU) assays are also provided.

Key words: MSCs, Isolation, Expansion, Culture, Colony forming unit assay

1. Introduction

The recent explosion of interest in developing cell and gene therapies using adult stem/progenitors cells from human bone marrow can be partly attributed to the ease of isolation and expansion of cells from this source in vitro. In addition, the possibility of generating genetically manipulated bone marrow-derived stem cells to introduce specific genes of interest makes them attractive vehicles for gene therapy (1–5). In this review, the term human mesenchymal stem cell (hMSC) will be used to describe the plastic adherent cells from human bone marrow, first defined in the

Mohan C. Vemuri et al. (eds.), *Mesenchymal Stem Cell Assays and Applications*, Methods in Molecular Biology, vol. 698, DOI 10.1007/978-1-60761-999-4_2,

literature as fibroblastoid colony forming units (CFU), then as mesenchymal stem/progenitor cells, and most recently as multipotent mesenchymal stromal cells (MSCs) (6).

Human MSCs are readily isolated from bone marrow by their adherence to tissue culture plastic and can be expanded through multiple passages in medium containing high concentrations of fetal bovine serum (FBS) (4, 7–13). However, the proliferation rates and other properties of the cells gradually change during expansion, and therefore, it is advisable to not expand hMSCs beyond four or five passages (14, 15). As originally indicated by Friedenstein, the most prominent properties of MSCs are their ability to generate colonies after they are plated at a low density, but both the colonies and the cells within a colony are heterogeneous in morphology, rates of proliferation, and efficacy with which they differentiate (8). Also, cultures of expanded cells are heterogeneous in their content of cells possessing an early progenitor phenotype. Human MSCs are highly sensitive to plating density, and early progenitors are rapidly lost if the cultures are grown to confluence (14, 16). Although the most recent definition of MSCs includes the expression of CD105, CD90, and CD73 surface antigens as potential biomarkers for MSCs, they alone are not sufficient to isolate cells directly from human bone marrow (6). Therefore, it is important to devise standardized assays for isolating and characterizing MSCs.

For the primary isolation of bone marrow-derived MSCs, the critical steps include the isolation of mononucleated cells from a marrow aspirate by centrifugation on a density gradient followed by recovery and expansion of cells that adhere to tissue culture plastic in standard serum-containing medium (passage zero cells). Passage zero cells are subsequently expanded by plating at a low density, which enhances the percentage of rapidly proliferating spindle-shaped cells. These cells would be replaced by large, flat, and thereby more mature hMSCs if the passage zero cells were plated at higher density or continually passaged for more than four to six times (Fig. 1a, b). Mature hMSCs will expand more slowly and have less multilineage differentiation potential, but still retain the ability to differentiate into mineralizing osteoblasts and secrete factors that enhance the growth of hematopoietic stem cells and perhaps other cells (13).

The efficiency with which hMSCs form colonies still remains an important assay for the quality of cell preparations. This chapter also describes two methods used to assay the colony forming ability of MSCs: (a) a traditional assay for colony forming units – fibroblast assay (CFU-F, Fig. 1c, d) and (b) single-cell colony forming unit assay (sc-CFU, Figs. 2 and 3). In the traditional CFU assay, cells are plated at low density in large plates and discrete colonies counted after 2 or 3 weeks. When used for assay of human MSCs, a single cell generates each colony. Noted when

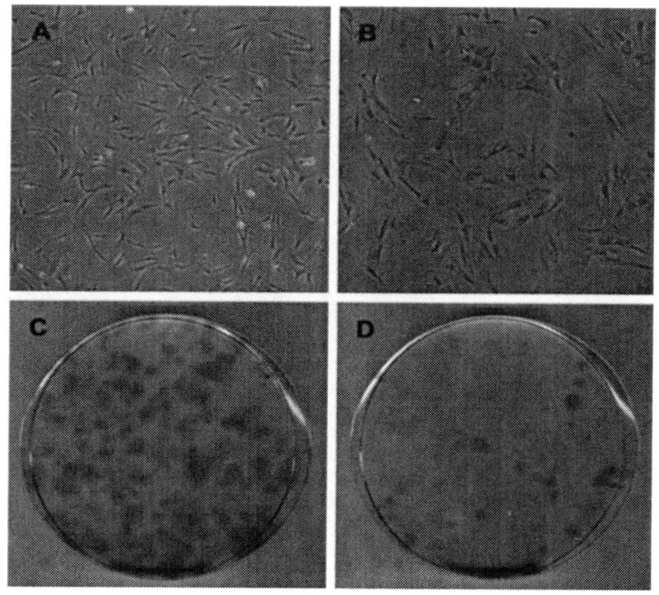

Fig. 1. Representative phase contrast microscopic images of cultures with (**a**) rapidly proliferating early passage mesenchymal stromal cells (MSCs) and (**b**) slowly proliferating late passage MSCs. Crystal violet stained plates of colony forming units – fibroblast assays (CFU-F) performed on (**c**) rapidly proliferating MSCs (as shown in (**a**)) and (**d**) slowly proliferating MSCs (as shown in (**b**)).

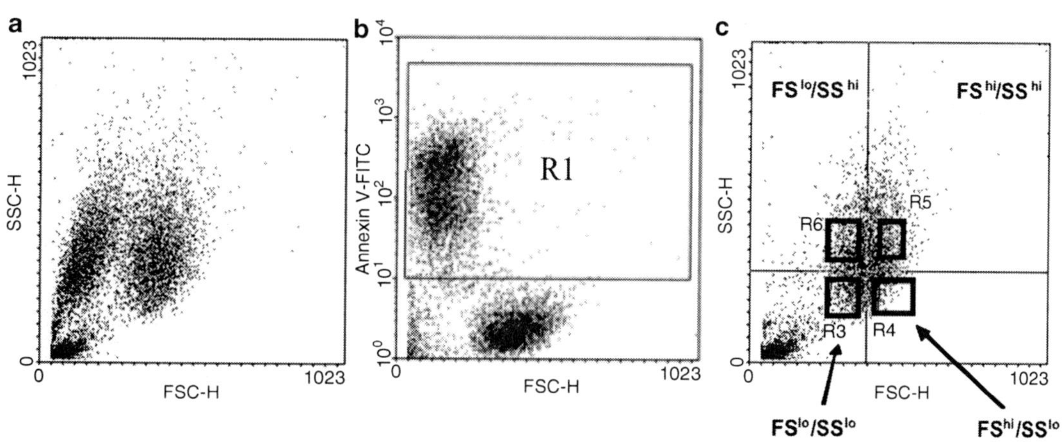

Fig. 2. Representative assay of FS/SS of passage two human mesenchymal stem cells (hMSCs) initially plated at 500 cells/cm^2 and incubated for 6 days to obtain high-density P3 cells. The cells were lifted with trypsin/EDTA and assayed on the open stream flow cytometer. (**a**) Uncorrected plot of FS/SS. (**b**) Same sample stained with Annexin V-FITC (R1). (**c**) Same sample after gating out Annexin V$^+$ events. A small fraction of very low FS/SS events was Annexin V$^-$ debris. Cell sorting based on FS/SS results in fractions that differ on subsequent analysis for FS/SS at the same instrument settings. *FITC* fluorescein isothiocyanate; *FS* forward scatter; *SS* side scatter.

used with rat or mouse MSCs, single cells can generate more than one colony because the cells can detach as they expand and reseed the plate (17, 18). In this chapter, we also describe a refined assay in which single MSCs are plated using a fluorescent flow cytometer

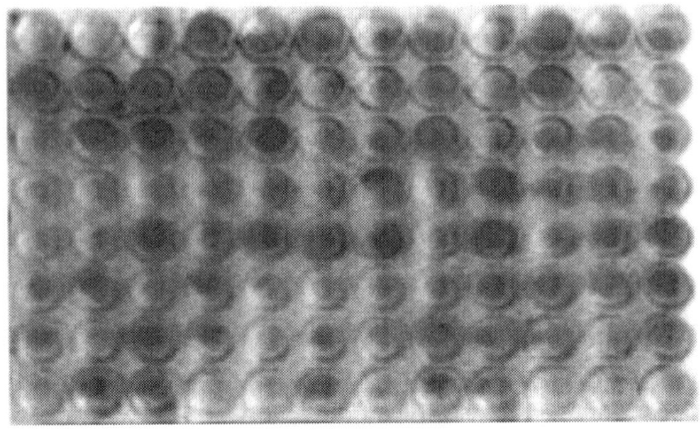

Fig. 3. Representative microtiter plate in which single FS^{lo}/SS^{lo} cells were deposited. The plate shown was stained with crystal violet to count colonies after incubation in standard serum-containing medium for 12–14 days.

with an automated cell sorter (FACSVantage SE with Clonesort accessory; Becton-Dickinson) to plate single cells into individual wells of a 96-well microtiter plate as described in Smith et al. (15). The cells are incubated in complete medium for 10–14 days and assayed visible colonies by staining the plates with Crystal Violet. With the sc-CFU assay, it is possible to distinguish the colony forming potential of two distinct kinds of MSCs present in early passage cultures: (1) spindle-shaped cells that are rapidly self-replicating are predominant in the first few days after plating the cells at low density, and (2) broader, slowly replicating cells that predominate as colonies or cultures become confluent. The proliferative spindle-shaped cells can be distinguished from larger, slower proliferating cells by their lower forward scatter (FS^{lo}) and lower side scatter (SS^{lo}) of light. As the delineation of subpopulations based upon FS/SS is somewhat difficult to standardize (Fig. 2), the sc-CFU assay is more useful in estimating the proportion of early progenitors in different preparations of MSCs.

2. Materials

2.1. Isolation and Culture of Bone Marrow-Derived hMSCs

1. Complete culture medium (CCM): α-MEM (Invitrogen, Carlsbad, CA) containing 16.5% (v/v) FBS (Atlanta Biologicals, Lawrenceville, GA) and 1% (v/v) Penicillin–Streptomycin (Invitrogen).
2. Hank's Balanced Salt Solution (HBSS) without Ca^{2+} and Mg^{2+} (Invitrogen).
3. Ficoll-Paque (GE Healthcare, Piscataway, NJ).

4. Phosphate buffered saline (PBS), pH 7.4 (Invitrogen).
5. Trypsin-EDTA in HBSS (Invitrogen).
6. Trypan blue, 0.4% (Invitrogen).
7. PBS, pH 7.4 (Invitrogen).
8. Cryopreservation medium: α-MEM (Invitrogen) containing 30% FBS (Atlanta Biologicals, Lawrenceville, GA) and 5% (v/v) DMSO (Sigma-Aldrich, St. Louis, MO).
9. T175 flasks or 150-mm tissue culture dish (Nunc, Thermo-Fisher Scientific, Rochester, NY).
10. Cryo 1°C Freezing Container (Nalgene, Thermo-Fisher Scientific).
11. Biological Safety cabinet Class II plugged to a vacuum system.
12. Water bath set at 37°C.
13. Water jacketed CO_2 incubator with HEPA filter system in humidified atmosphere and set at 37°C and 5% CO_2.
14. Bench centrifuge with swinging bucket rotor and brake ON/OFF option.
15. Inverted phase microscope.
16. Hemocytometer.
17. Sterile cell culture plastic pipets individually wrapped (2, 5, 10, 25 mL).
18. Pipet-Aid.
19. Sterile conical centrifuge tubes (15 and 50 mL).
20. 175 cm^2 Tissue culture flasks (Nunc).
21. Sterile plastic transfer pipets.
22. CellSTACK (2-stack) culture chambers (Corning, Corning, NY).
23. Solid cap, 33 mm threaded cap (Corning).

2.2. Colony Forming Unit Assay

1. Crystal violet (3%) (Sigma-Aldrich) in methanol. Filter through 25 μm filter paper and store at room temperature. Before use, dilute to 0.5% in PBS.
2. Annexin V-FITC apoptosis detection kit (Sigma-Aldrich).
3. Flow Cytometry Microbead Standards (Polysciences Inc., Warrington, PA).
4. 96-Well tissue culture plate (Nunc).
5. EPICS FC500 flow cytometer running with CXP software (Beckman-Coulter, Brea, CA).
6. FACSVantage SE with FACSDiva Option (BD Biosciences, San Jose, CA).

3. Methods

3.1. Isolation and Culture of Bone Marrow-Derived hMSCs

1. Source bone marrow aspirates from the iliac crest and placed in 10 mL heparinized tubes prefilled with 3 mL of plain α-MEM. Keep samples on ice until to be processed.
2. Transfer each aspirate into a 50 mL conical tube and dilute to 15 mL with HBSS.
3. Rinse aspirate tubes twice with 5 mL of HBSS and combine with the diluted aspirate (25 mL total volume).
4. For each aspirate, place 10 mL of prewarmed (37°C) Ficoll-Paque into a separate 50 mL conical tube.
5. Gently overlay each aspirate onto the Ficoll. Take care to angle the tube containing Ficoll and very slowly pipet on the diluted aspirate over the border of the Ficoll meniscus. Once done, gently replace the tube in a vertical position (see Note 1).
6. Centrifuge tubes at 1,800 ×*g* for 30 min at room temperature in a swinging bucket rotor with the brake OFF (see Note 2).
7. After centrifugation, carefully collect the buffy coat, located at the Ficoll-HBSS interface, with a sterile Pasteur transfer pipet and place the cells into a clean 50 mL conical tube.
8. Dilute each sample to 25 mL with HBSS and invert the tube three to five times to mix (see Note 3).
9. Centrifuge tubes at 1,000 ×*g* for 10 min in a swinging bucket rotor with the brake ON.
10. Remove the supernatant by vacuum aspiration and resuspend the cells with 30 mL of prewarmed CCM.
11. Count viable cells with a hemocytometer using Trypan blue and plate at a cell density of 50–100 cells/cm^2 in 175 cm^2 flasks or 150 mm dishes.
12. Incubate the cells at 37°C with 5% humidified CO_2 for 24 h to allow adherent cells to attach.
13. After 24 h, remove the media and nonadherent cells (see Note 4).
14. Add 10 mL of prewarmed PBS to the culture, rock gently to cover the entire surface area and aspirate. Repeat the wash two additional times (see Note 5).
15. Add 30 mL of fresh CCM to the flask and return flasks to the incubator.
16. Examine cultures daily by phase microscopy.
17. Every 3 days, remove the medium and rinse the flask with 10 mL of prewarmed PBS. Aspirate the wash and feed cultures with 30 mL of fresh CCM. Continue until the cells reach 70–80% confluence (see Note 6).

18. To harvest cultures, remove the media and rinse the flask with 30 mL PBS and aspirate.
19. Add 10 mL of prewarmed trypsin-EDTA solution to the flask. Distribute the trypsin across the surface area of the flask. Incubate the flask for 2–5 min at 37°C. Examine the cells by phase microscopy.
20. After 80–90% of the cells have rounded up or become detached, gently tap the sides of the flask to dislodge any remaining attached cells.
21. Add 10 mL CCM to the flask. Rock the flask back and forth to swirl the media around the flask and transfer the entire cell suspension into a clean 50-mL conical tube.
22. Rinse the flask with 30 mL of 1× PBS and combine with the cell suspension.
23. Centrifuge at 1,000 × g for 10 min in a swinging bucket rotor with the brake ON.
24. Remove the supernatant and resuspend the cells in 1–2 mL of prewarmed PBS.
25. Count the cells with a hemocytometer and trypan blue or preferred method (see Note 7).
26. Reseed harvested cells at a density of 50–100 viable cells/cm^2 in an appropriate culture vessel. The resultant hMSC cultures can usually be successfully expanded through passage three or four without significant loss of the stem cell phenotype.

    The remainder of this procedure will describe the expansion of hMSCs in a Corning CellSTACK (2-stack) culture chambers (total surface = 1,272 cm^2).
27. In order to obtain between 0.8 and 1 × 10^8 cells, we recommend using five 2-stack culture chambers. Add 300 mL of CCM/double stack culture chamber and place each chambers in incubator for at least 2 h before seeding (see Note 8).
28. Plate 6 × 10^4 cells/stack and carefully distribute cells evenly by gentle agitation using solid caps.
29. Grow cells for 2–3 weeks with complete medium changes made every 3–4 days. To remove medium, use a vacuum aspirator and a Pasteur pipet. Gently angle the culture chambers to avoid bubbling. The bottom stack can be monitored by using a regular inverted microscope.
30. To harvest cells, wash stacks with 100 mL PBS/stack and aspirate. Add 15 mL trypsin-EDTA/stack and incubate for 5 min. Follow cells lifting with the microscope, stop trypsinization when almost all cells are detached, reincubate an additional minute if needed but no more than 7 min.

31. Use the harvested cells for experimental purposes or reseed additional flasks or stacks at a density of 50–100 cells/cm^2. It is recommended to determine expanded MSC quality by using the CFU assays, see Subheadings 3.2.1 and 3.2.2.
32. Cryopreserve unused expanded hMSCs in Cryopreservation Medium (see Subheading 2.1) at 1×10^6 cells/mL (see Note 9).

3.2. Colony Forming Unit Assays

The efficiency with which MSCs form colonies still remains an important assay for the quality control of MSCs preparations. This section describes two methods to assay the colony forming ability of MSCs including (1) a traditional assay for CFU-F and (2) a sc-CFU.

3.2.1. Colony Forming Unit: Fibroblast Assay

1. Expand hMSC cultures to 70–80% confluence and harvest with trypsin-EDTA (see Subheading 3.1).
2. To ensure cell separation, a glass Pasteur pipet can be flamed to create a narrowed tip. Draw cells through the narrowed pipet several times (see Note 10).
3. Count the number of cells using a hemocytometer.
4. Dilute cells in CCM and plate at 100 cells/100-mm tissue culture dish or 10 cells/well in a 6-well plate.
5. Incubate for 10–14 days at 37°C in a humidified 5% CO_2 incubator.
6. Wash plates with PBS and stain with 0.5% (v/v) Crystal Violet solution for 5–10 min at room temperature.
7. Wash thoroughly with water and count visible colonies with a diameter greater than 5 mm. (Fig. 1).

3.2.2. Colony Forming Unit: Single Cell Assay

Rapidly self-renewing MSCs are characterized by low forward scatter (FS^{lo}) and low side scatter (SS^{lo}) of light. The following protocol describes the isolation of FS^{lo}/SS^{lo} MSCs that are rapidly self-renewing. It is also a rapid, standardized assay for FS/SS, a useful protocol to identify preparations of MSCs enriched for proliferative cells that will expand rapidly during subsequent passage in culture. The use of the assay should help to resolve discrepancies in data obtained by different laboratories with presumably similar preparations of hMSCs.

1. Standardize the closed stream flow cytometer (EPICS FC500 running CXP software) using microbeads with known uniform diameters (i.e., 6, 10, 15, and 20 μm).
2. Adjust the gains and voltages on the photomultiplier tubes so that the mean value of the FS peak for the 20 μm bead is about 650 and the peak of the SS for the 6 μM bead is about 450. With these settings, the standard deviation for FS of the largest bead should be less than ±0.4% ($n = 3$) of the mean and the slope of FS on a linear scale of 0–1,023 at least 41 (see Note 11).

3. For the assay, lift cells expanded as described in Subheading 3.1 with trypsin/EDTA and centrifuge in CCM at $450 \times g$ for 10 min (see Note 12).
4. Count cells on a hemocytometer and resuspend in cold PBS (4°C) at a concentration of about 5×10^5 cells/mL. The assay should be run shortly thereafter.
5. Stain cells with the Annexin V-FITC (using manufacturer recommended protocol) and maintain at 4°C to prevent aggregation due to the presence of calcium and reagent-induced toxicity. Staining with Annexin V-FITC demonstrates that the events in the upper left of the plot are cell debris and dead cells (Rl in Fig. 2b). To obtain subfractions of cells, the Annexin V^+ events are gated out and four sub-populations are defined on the basis of FS and SS (Fig. 2c).
6. Analyze cells using the above method and sort single FS^{lo}/SS^{lo} cells per well of a 96-well plate using the FACSVantage instrument. Divide the AnnexinV^- events into four quadrants on the basis of FS and SS (Fig. 2c). Offset the sort gates from the boundaries (see Note 13).
7. Incubate the microtiter plates with one hMSC per well in 0.15 mL CCM at 37°C and 5% CO_2.
8. Every 4–5 days, aspirate CCM from each well and replace with 0.15 mL fresh medium.
9. After 2 weeks in culture, remove the medium and wash the wells with PBS. Incubate samples with 0.5% crystal violet solution for 5–10 min wash with water and count colonies with diameters greater than 1 mm using an inverted phase contrast microscope with a 4× objective (Fig. 3).

4. Notes

1. If the Ficoll and HBSS-cell suspension layers are mixed, the mononuclear cells will not completely and efficiently separate during centrifugation.
2. The brake is left off to allow a slow deceleration that helps to avoid disturbance of the Ficoll-HBSS cell suspension interface.
3. It is recommended diluting the collected buffy coat with HBSS at a 3:1 volume ratio of diluent to sample.
4. If the nonadherent cells are not removed, hematopoietic cells may become attached and contaminate the hMSC culture.
5. There may not be many adherent cells seen at this point.

6. To preserve progenitor cell phenotype, do not allow the cells to become confluent. Because hMSCs are not evenly distributed in the marrow, some aspirates do not have enough hMSCs to obtain large cultures. If a sample does not grow well or do not have a good morphology by the eighth day, discard it.
7. A typical yield from a 175 cm^2 primary culture flask is between 1 and 3×10^6 total cells, with an average viability usually greater than 90%.
8. It is important to allow the chambers containing CCM to equilibrate to 37°C and 5% CO_2 before use.
9. Freeze at a rate of −1°C/min using a Nalgene Cryo 1°C Freezing container placed at −80°C. After 24 h, transfer vials to liquid nitrogen for long-term storage.
10. It is critical that the cells are well dissociated.
11. The variation in values for log (%G/%T) should be established against samples containing 0.5 or 1 million MSCs/mL when the following parameters are varied: (a) the flow rate was 250, 500 or 900 cells/s; (b) the FS was assayed with 67 or 122 V and a gain of 2 or with 353 V and a gain of 1; and (c) the peak for FS of the 20 μm bead was set at 550, 650, or 750; and (d) the peak for SS for the 7 μm bead was set at 350, 450, or 550.
12. Cell culture confluence is important, and cells should be harvested when they are less than 80% confluent.
13. The accuracy of sorting single cells into each well of a microtiter plate should be verified routinely by sorting fluorescent beads (i.e., Flowchek; Beckman-Coulter) into a test plate and examining the wells with an epifluorescence microscope.

References

1. Azizi, S. A., Stokes, D., Augelli, B. J., Digirolamo, C., and Prockop, D. J. (1998) Engraftment and migration of human bone marrow stromal cells implanted in the brains of albino rats – similarities to astrocyte grafts, *Proc. Natl. Acad. Sci. U S A 95*, 3908–3913.
2. Chopp, M., Zhang, X. H., Li, Y., Wang, L., Chen, J., Lu, D., Lu, M., and Rosenblum, M. (2000) Spinal cord injury in rat: treatment with bone marrow stromal cell transplantation, *Neuroreport 11*, 3001–3005.
3. Kopen, G. C., Prockop, D. J., and Phinney, D. G. (1999) Marrow stromal cells migrate throughout forebrain and cerebellum, and they differentiate into astrocytes after injection into neonatal mouse brains, *Proc. Natl. Acad. Sci. U S A 96*, 10711–10716.
4. Liechty, K. W., MacKenzie, T. C., Shaaban, A. F., Radu, A., Moseley, A. M., Deans, R., Marshak, D. R., and Flake, A. W. (2000) Human mesenchymal stem cells engraft and demonstrate site-specific differentiation after in utero transplantation in sheep, *Nat. Med. 6*, 1282–1286.
5. Pereira, R. F., Halford, K. W., O'Hara, M. D., Leeper, D. B., Sokolov, B. P., Pollard, M. D., Bagasra, O., and Prockop, D. J. (1995) Cultured adherent cells from marrow can serve as long-lasting precursor cells for bone, cartilage, and lung in irradiated mice, *Proc. Natl. Acad. Sci. U S A 92*, 4857–4861.
6. Dominici, M., Le, B. K., Mueller, I., Slaper-Cortenbach, I., Marini, F., Krause, D., Deans, R., Keating, A., Prockop, D., and Horwitz, E.

(2006) Minimal criteria for defining multipotent mesenchymal stromal cells. The International Society for Cellular Therapy position statement, *Cytotherapy 8*, 315–317.

7. Caplan, A. I. (1990) Stem cell delivery vehicle, *Biomaterials 11*, 44–46.
8. Friedenstein, A. J., Petrakova, K. V., Kurolesova, A. I., and Frolova, G. P. (1968) Heterotopic of bone marrow. Analysis of precursor cells for osteogenic and hematopoietic tissues, *Transplantation 6*, 230–247.
9. Krause, D. S., Theise, N. D., Collector, M. I., Henegariu, O., Hwang, S., Gardner, R., Neutzel, S., and Sharkis, S. J. (2001) Multi-organ, multi-lineage engraftment by a single bone marrow-derived stem cell, *Cell 105*, 369–377.
10. LaBarge, M. A. and Blau, H. M. (2002) Biological progression from adult bone marrow to mononucleate muscle stem cell to multinucleate muscle fiber in response to injury, *Cell 111*, 589–601.
11. Owen, M. and Friedenstein, A. J. (1988) Stromal stem cells: marrow-derived osteogenic precursors, *Ciba Found. Symp. 136*, 42–60.
12. Pittenger, M. F., Mackay, A. M., Beck, S. C., Jaiswal, R. K., Douglas, R., Mosca, J. D., Moorman, M. A., Simonetti, D. W., Craig, S., and Marshak, D. R. (1999) Multilineage potential of adult human mesenchymal stem cells, *Science 284*, 143–147.
13. Prockop, D. J. (1997) Marrow stromal cells as stem cells for nonhematopoietic tissues, *Science 276*, 71–74.
14. Colter, D. C., Class, R., DiGirolamo, C. M., and Prockop, D. J. (2000) Rapid expansion of recycling stem cells in cultures of plastic-adherent cells from human bone marrow, *Proc. Natl. Acad. Sci. U S A 97*, 3213–3218.
15. Smith, J. R., Pochampally, R., Perry, A., Hsu, S. C., and Prockop, D. J. (2004) Isolation of a highly clonogenic and multipotential subfraction of adult stem cells from bone marrow stroma, *Stem Cells 22*, 823–831.
16. Sekiya, I., Larson, B. L., Smith, J. R., Pochampally, R., Cui, J. G., and Prockop, D. J. (2002) Expansion of human adult stem cells from bone marrow stroma: conditions that maximize the yields of early progenitors and evaluate their quality, *Stem Cells 20*, 530–541.
17. Javazon, E. H., Colter, D. C., Schwarz, E. J., and Prockop, D. J. (2001) Rat marrow stromal cells are more sensitive to plating density and expand more rapidly from single-cell-derived colonies than human marrow stromal cells, *Stem Cells 19*, 219–225.
18. Peister, A., Zeitouni, S., Pfankuch, T., Reger, R. L., Prockop, D. J., and Raber, J. (2006) Novel object recognition in ApoE(−/−) mice improved by neonatal implantation of wild-type multipotential stromal cells, *Exp. Neurol. 201*, 266–269.

Chapter 3

Standardized Isolation of Human Mesenchymal Stromal Cells with Red Blood Cell Lysis

Patrick Horn, Simone Bork, and Wolfgang Wagner

Abstract

Human mesenchymal stromal cells (MSC) raise high hopes for tissue engineering and therapeutic applications. So far, it is not possible to isolate pure fractions from bone marrow and therefore MSC cell preparations notoriously represent heterogeneous mixtures of different cell types. The composition of subpopulations can already be affected by the initial steps of cell preparation. Usually, isolation of MSC involves density fractionation to separate the mononuclear cells (MNCs) from erythrocytes and granulocytes. However, this method is difficult to standardize especially under GMP conditions. Here, we describe an alternative approach for isolation of human MSC based on red blood cell (RBC) lysis with ammonium chloride. This results in a slightly higher number of fibroblastic colony forming units (CFU-F), whereas morphological analysis of the CFU-F reveals the same heterogeneous composition of MSC cultures indicating that the proportion of subpopulations is not affected by RBC lysis. Immunophenotype ($CD73^+$, $CD90^+$, $CD105^+$, $CD31^-$, $CD34^-$, $CD45^-$), adipogenic, and osteogenic differentiation potential of MSC were also similar with both methods. In conclusion, RBC lysis comprises an efficient method for the isolation of human MSC from bone marrow aspirate. This technique is faster and can be standardized more easily for clinical application of MSC.

Key words: Mesenchymal stromal cells, Mesenchymal stem cells, Isolation, Red blood cell lysis, Density gradient centrifugation, Ammonium chloride, CFU-F

1. Introduction

Mesenchymal stromal cells (MSC) comprise multipotent somatic cells. Given appropriate culture conditions they differentiate into different mesodermal cell lineages including adipocytes, osteocytes, and chondrocytes. Albeit controversial, there is evidence that MSC can also differentiate into myocytes and cardiomyocytes and even into cells of nonmesodermal origin including hepatocytes and neurons (1–4). Hence, they are alternatively termed "mesenchymal stem cells" using the same acronym. They can be

Mohan C. Vemuri et al. (eds.), *Mesenchymal Stem Cell Assays and Applications*, Methods in Molecular Biology, vol. 698, DOI 10.1007/978-1-60761-999-4_3,

isolated from bone marrow (BM) aspirates or other tissues by in vitro culturing. Specific molecular markers for quality control of the multipotent subsets are yet elusive. Therefore, MSC are characterized by: (a) plastic adherent growth, (b) a panel of surface markers that includes CD13, CD73, CD90, and CD105, whereas other markers such as hematopoietic antigens (e.g., CD31, CD34 or CD45) are not expressed, and (c) their differentiation potential towards osteocytes, adipocytes, and chondrocytes under specific in vitro differentiating conditions (5–7).

The possibility for large-scale expansion of MSC to high cell numbers, their differentiation potential and their immunomodulatory activity gave rise to high hopes for regenerative medicine. The spectrum of clinical applications includes treatment of steroid refractory graft *versus* host disease (G*v*HD), peridontitis, severe chronic myocardial ischemia, distal tibia fractures, decompensated liver cirrhosis, multiple sclerosis, tumor induced osteomalacia, and Crohn's disease (www.clinicaltrials.gov). So far there are hardly any reports on critical side effects by MSC treatment and preliminary results of some ongoing trials are very promising (8, 9). A major challenge of producing cells for clinical use is the need to implement GMP-compliant production processes to satisfy regulatory requirements. However, this is hampered by a tremendous variation of culture isolation methods for MSC throughout different laboratories. These culture methods have major impact on the composition of subpopulations and their differentiation potential. Usually, MNCs are separated by density fractionation to eliminate the vast amount of granulocytes and erythrocytes. However, the process of density gradient centrifugation is very time consuming and requires several steps and manipulations, which may lead to a higher risk of contamination. Additionally, the procedure is more difficult to standardize and may be influenced by the operator. It is possible to perform density gradient centrifugation under GMP conform conditions. Gastens et al. have even demonstrated a closed Ficoll isolation system for MSC preparation, although the manual technique appeared to be slightly more efficient (10). However, even in an automated closed system the density fractionation can hardly be standardized due to the heterogeneity of samples. BM aspirates can also be seeded without further treatment for isolation of MSC (11). However, the formation of a thick erythrocyte layer may inhibit settling and nutrition of initiating colonies. Therefore, the sample has to be further diluted and this requires a higher amount of medium and larger cell culture surfaces.

Here, we describe an alternative protocol for the isolation of human MSC through red blood cell lysis (RBC lysis) with ammonium chloride. Lysis with ammonium chloride is the most effective method for erythrocyte lysis that has relatively little toxic effects on MSC. A similar method has recently been described that is based on lysis under hypoxic conditions (12). RBC lysis requires fewer steps and appears to be more suitable for a closed isolation system. Furthermore, this technique can be applied to probes of very small volumes (<1 mL BM aspirate). Thus, RBC lysis provides a reliable and effective alternative to density gradient centrifugation for isolation of human MSC. Furthermore, it results in larger colonies compared to Ficoll procedure (13). This possibly results from the platelets that are maintained by this method. There is evidence that the platelet fraction greatly enhances the initial growth of MSC colonies. This might be attributed to growth factors such as PDGF. Other authors have demonstrated that platelet lysate or thrombin-activated platelet-rich plasma are very effective supplements for MSC growth (14, 15). Very large colonies with more than 10^4 cells were observed after 7 days only upon erythrocyte lysis but not after density gradient centrifugation. A single cell would have to divide more than 14 times to give rise to this cell number. Hence, these large colonies might rather result from small cell aggregates which might be lost in the pellet by Ficoll gradient centrifugation.

Taken together, RBC lysis provides an easier, more standardized, and efficient method for isolation of MSC for clinical application.

2. Materials

2.1. Biological Material

Human BM samples are taken after written consent using guidelines approved by an Ethic Committee of the respective institution. BM aspirates are combined with 5% (v/v) Heparin anticoagulant (Becton Dickinson, San Jose, CA) in the syringe to prevent coagulation.

2.2. Culture Medium

For this study we have used culture medium that was first described by Reyes and coworkers (1): Dulbecco's Modified Eagles Medium-Low Glucose (DMEM-LG) (PAA, Pasching) supplemented with 40% (v/v) MCDB201 (Sigma Aldrich, St. Louis, MO), 2% (v/v) FCS (HyClone, Bonn), 2 mM L-Glutamine (Sigma), 100 U/mL Pen/Strep (Lonza, Basel), 1% (v/v) insulin transferrin selenium (Sigma), 1% (v/v) linoleic acid-albumin from bovine serum albumin (Sigma), 10 nM dexamethasone (Sigma), 0.1 mM L-Ascorbic acid 2-phosphate (Sigma), PDGF-bb and EGF (10 ng/mL each) (PreproTech, Rocky Hill, NJ) (see Note 1).

2.3. Density Gradient Centrifugation

Ficoll-Paque® (density 1.077 g/mL) (Sigma)

2.4. Red Blood Cell Lysis

Lysis buffer: 10× stock solution: ultrapure water with 1.5 M ammonium chloride (NH_4Cl), 100 mM potassium bicarbonate ($KHCO_3$), 1 mM ethylenediamine tetraacetic acid (EDTA) (all from Sigma). Store in aliquots at –20°C. For sterile filtration, a 0.22 μm cellulose mixed ester filter device (Carl Roth, Karlsruhe) is used.

2.5. Passaging and Cell Culture

1. 10 ng/mL fibronectin solution (Sigma) in 1× phosphate-buffered saline (PBS) (PAA). Store at 4°C.
2. 0.25% Trypsin–EDTA solution (1×) (Gibco, Invitrogen, Carlsbad, CA).
3. Stop solution: PBS with 10% FCS (HyClone).
4. Türck solution (Merck, Darmstadt), kept in the dark until use.
5. Hemocytometer (Neubauer counting chamber; LO Laboroptik, Friedrichsdorf).

2.6. Preparation of Paraformaldehyde (4%) Fixative Solution

1. Paraformaldehyde (Sigma).
2. 1 M NaOH.
3. 10× PBS.
4. 1 M HCl.
5. 0.22 μm cellulose mixed ester filter device.

2.7. Adipogenic Differentiation and Staining

1. Adipogenic differentiation medium (16): DMEM low glucose (PAA), 10% (v/v) FCS (HyClone), 2 mM L-Glutamine (Sigma), 100 U/mL Pen/Strep (Lonza), 1 μM dexamethason (Sigma), 0.5 mM 1-methyl-3-isobutylxanthin (IBMX) (Sigma), 10 μg/mL Insulin (Sigma).
2. Oil red O (Sigma) for staining of lipid droplets.
3. 4% Paraformaldehyde.
4. 0.22 μm cellulose mixed ester filter device.

2.8. Osteogenic Differentiation and Staining

1. Osteogenic differentiation medium (1): DMEM low glucose (PAA), 10% (v/v) FCS (HyClone), 2 mM L-Glutamine (Sigma), 100 U/mL Pen/Strep (Lonza), 100 nM dexamethason (Sigma), 200 μM L-Ascorbic acid 2-phosphate (Sigma), 10 mM β-Glycerophosphate (Sigma).
2. Alizarin red S (Sigma) for staining of calcium-phosphate precipitates.
3. 4% Paraformaldehyde.

2.9. Immunophenotyping

The following murine monoclonal anti-human antibodies were used for flow cytometric characterization of MSC: CD31-fluorescein isothiocyanate (FITC, MBC 78.2, Caltag, Burlingame, CA), CD34-phycoerythrin (PE, 8G12, BD Biosciences, Franklin Lakes, NJ), CD45-FITC (2D1, BD Biosciences), CD73-PE (AD2, BD Biosciences), CD90-PE (G7, BD Biosciences), CD105-PE (MHCD10504, BD Biosciences).

3. Methods

3.1. Density Gradient Centrifugation

The most commonly used method for isolation of MSC from BM aspirate is separation of MNCs by density fractionation. Granulocytes and erythrocytes have a higher density and are thereby removed. The BM aspirate needs to be carefully layered over a solution such as Ficoll-Paque (density 1.077 g/mL) which represents a synthetic neutral, highly-branched hydrophilic polymer of sucrose. After centrifugation for 30 min, the sample is separated into distinct layers of erythrocytes and granulocytes, Ficoll solution, MNCs, and plasma with platelets. MNCs are collected in an interphase ring below the plasma fraction (Fig. 1). Depending on the operator there may be differences in harvesting of the interphase ring and the overlaying serum-fraction with platelets.

1. Mix BM aspirate in a 1:1 ratio with PBS.
2. For the separation of MNC from the residual BM compartments, add 15 mL of Ficoll-Paque in 50-mL tubes.
3. Carefully layer diluted BM on top of the Ficoll-Paque without mixing (see Note 2).
4. Centrifuge tube at 800 × *g* for 30 min at room temperature (RT), without brake to prevent blending of the MNC ring with the neighboring layers.
5. Harvest diffuse MNC interphase ring with a pipette and transfer into a fresh tube.
6. Fill tube with MNCs up to 50 mL with PBS and centrifuge at 400 × *g* for 7 min at RT.
7. Discard supernatant, wash the pellet with 30 mL PBS and centrifuge at 400 × *g* for 7 min at RT.
8. Resuspend pellet in culture medium.
9. Determine the total cell number and seed cells as described in Subheading 3.3.

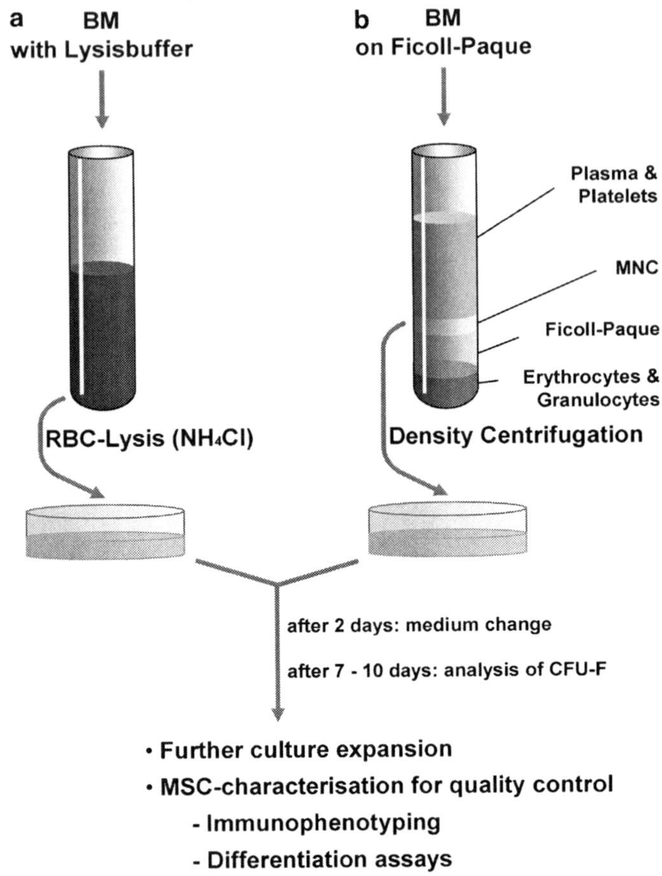

Fig. 1. Scheme of RBC-Lysis and density gradient centrifugation. (**a**) BM aspirate is mixed with lysis buffer. When the erythrocytes are lysed, the cell pellet is washed, cells are counted and seeded in culture medium. (**b**) BM aspirate is mixed with PBS and carefully layered on the Ficoll-Paque for density gradient centrifugation. Subsequently, the MNC interphase ring is harvested, the cells are counted and seeded in culture medium.

3.2. Red Blood Cell Lysis

Treatment with ammonium chloride provides an alternative method to remove erythrocytes by lysis. Thereby, this step facilitates settling and nutrition of the initial CFU-F. As mentioned above, RBC lysis requires less manipulation and gives rise to larger colonies than density gradient centrifugation. The cell lysis is driven by the passage of NH_3 and CO_2 through the membrane. The spontaneous conversion of NH_3 to NH_4^+ within the cell and of CO_2 to HCO_3^- by carbonic anhydrase, which is highly abundant in erythrocytes, results in a continuous flow of NH_3 and CO_2 entering the cell.

1. Thaw 10× lysis buffer stock solution and mix 1:10 with ultrapure water to make a 1× lysis buffer.
2. Sterile filter 1× lysis buffer through a 0.22-μm filter device.

3. Mix the BM 1:3 with 1× lysis buffer, invert several times for proper mixing (see Note 3).
4. Shake on a vertical rotary shaker at approximately 20 rpm for 10 min at RT (see Notes 4 and 5).
5. Centrifuge sample at 400 × *g* for 10 min at RT and discard the supernatant.
6. Wash the pellet once PBS, centrifuge at 400 × *g* for 10 min at RT and re-suspend in culture medium.
7. Determine the total cell number and seed cells as described in Subheading 3.3.

3.3. Cell Number Determination and Seeding

1. Before initial seeding, total cell number needs to be determined (e.g., by using a Neubauer cell counting chamber).
2. Coat culture flasks with 10 ng/mL fibronectin before use for at least 30 min at 37°C (see Note 6).
3. Seed MNCs at a density of 1×10^6 cells/cm^2 in fibronectin coated culture flasks (see Notes 7 and 8).
4. After incubation for 2–3 days, carefully wash the flask to remove nonadherent cells and replace with fresh culture medium.
5. Initial colony formation should be evident within 7–10 days and should be passaged before an overgrowth takes place (see Subheading 3.4) (see Note 9) (Fig. 2).

3.4. Passaging and Culture Expansion

To prevent overgrowth, reduced proliferation by contact inhibition, and a reduced differentiation capacity, initial colonies should be harvested at an optimal time and reseeded at a lower density (17). This passaging should be performed after ~7–10 days for primary cultures or when the cells reached ~70% confluence during subsequent passages.

1. Wash cells twice with PBS and treat with 60 μL/cm^2 0.25% Trypsin–EDTA solution for 3 min at 37°C (see Note 10).
2. Stop trypsination by adding 2 volumes of stop solution.
3. Collect the cell solution, wash the flask with PBS and centrifuge at 400 × *g* for 10 min at RT.
4. Discard the supernatant, resuspend the cell pellet in culture medium and count the total number of cells. Reseed cultures at a density of 1×10^4 cells/cm^2 for continual expansion or further process for characterization or differentiation.

3.5. Preparation of Paraformaldehyde (4%) Fixative Solution

1. For 100 mL, dissolve 4 g of paraformaldehyde in 50 mL ultrapure water.
2. Add 1 mL of 1 M NaOH to the solution.
3. Gently stir the mixture on a heating block at 65°C until the paraformaldehyde is completely dissolved.

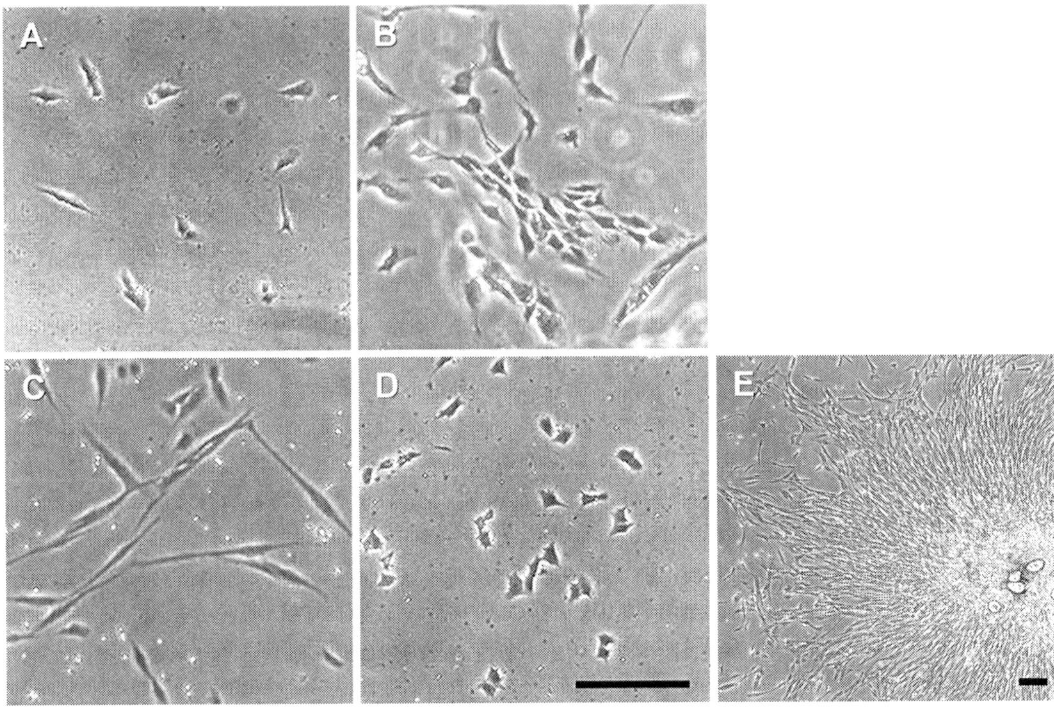

Fig. 2. Morphological variation between initial CFU-F. The morphology of the initial colonies can be discerned by several characteristics concerning density, cell shape, and colony size. The density of the colonies ranges from a relatively dispense distribution (**a**) to very tight packed epitheloid growth pattern (**b**), with rather elongated (**c**) or round cells (**d**). The colony size varies from 10 cells up to very large colonies with more than 10^4 cells (**e**). *Scale bar* = 100 μm.

4. Add 10 mL of 10× PBS and allow the solution to cool to RT.
5. Adjust the pH to 7.4 using 1 M HCl.
6. Adjust the volume to 100 mL with ultrapure water and filter the solution through a 0.22-μm filter device.
7. Aliquot and store at −20°C for several months. Avoid repeated freezing/thawing of the paraformaldehyde.

3.6. Adipogenic Differentiation and Staining

1. Seed expanded MSC at a density of 5×10^3 cells/cm^2 in adipogenic differentiation medium (16).
2. Maintain adipogenic cultures for 3 weeks with medium changes made twice per week (7).
3. Visualize adipogenic cultures under the microscope to check for the accumulation of lipid vesicles. To better visualize vesicle formation, stain cultures with oil red O dye.
4. Prepare oil red O stock-solution by dissolving 0.5% (w/v) oil red O in isopropanol.

5. Mix 6 volumes oil red O stock-solution with 4 volumes ultrapure water to prepare an oil red O work-solution. Incubated the work-solution for 24 h at RT and thereafter filter through a 0.22-μm filter device.
6. Source adipogenic cultures, remove culture medium, and carefully wash cells once with PBS.
7. Fix cells with 4% paraformaldehyde for 3 min at RT.
8. Discard paraformaldehyde and carefully wash cells twice with ultrapure water (see Note 11).
9. Add 60% isopropanol to fixed cells and incubate for 5 min (see Note 12).
10. Discard isopropanol and stain cells with oil red O work-solution for 15 min at RT.
11. Discard Oil red O work-solution and wash cells with 60% isopropanol.
12. Rinse cells twice with tap water followed by two additional wash steps with ultrapure water.
13. Layer the cells with glycerol and analyze by microscopy (Fig. 3).

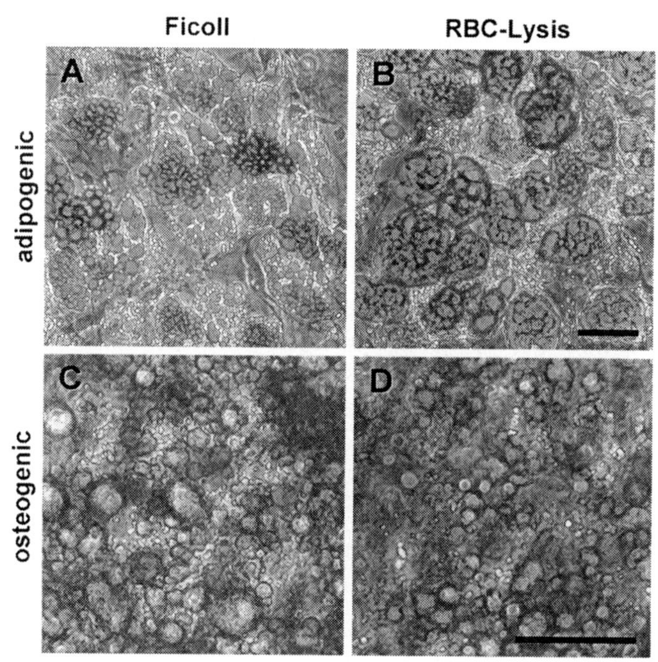

Fig. 3. Adipogenic and osteogenic in vitro differentiation. MSC were isolated either by RBC lysis or density gradient centrifugation. Upon differentiation for 3 weeks the cells were then stained with oil red O for adipogenic differentiation (**a**, **b**) and alizarin red S for osteogenic differentiation (**c**, **d**). *Scale bar* = 10 μm.

3.7. Osteogenic Differentiation and Staining

1. Seed expanded MSC at a density of 5×10^3 cells/cm^2 in osteogenic differentiation medium (1).
2. Maintain osteogenic cultures for 3 weeks with medium changes made twice per week (7).
3. Prepare alizarin red S solution by dissolving 0.5% (w/v) alizarin red S in ultra pure water.
4. Source osteogenic cultures, remove culture medium, and carefully wash cells once with PBS.
5. Fix cells with 4% paraformaldehyde for 3 min at RT.
6. Discard paraformaldehyde and carefully wash cells twice with ultra pure water (see Note 11).
7. Stain cells with the 0.5% alizarin red S solution for 10 min at RT.
8. Discard the staining solution and wash the cells 4–5 times with ultrapure water until the wash solution becomes clear.
9. Layer the cells with glycerol and analyze by microscopy (Fig. 3).

3.8. Immunophenotyping

Immunophenotypic analysis of MSC is performed as described previously in detail (6, 11) and elsewhere in this text. In brief, cells are trypsinized and washed twice with PBS. For each surface molecule, 1×10^5 cells are incubated with the respective concentration of antibody (according to manufacturer's instruction) for 30 min at 4°C and subsequently washed with PBS. The labeled cells are analyzed by flow cytometry in relation to proper isotype controls (Fig. 4).

4. Notes

1. Various different culture media can be used alternatively (18). For standardization purposes it might be easier to use DMEM-LG (PAA) with 10% (v/v) FCS (HyClone), 2 mM L-Glutamine (Sigma), 100 U/mL Pen/Strep (Lonza).
2. The tube with Ficoll solution should be inclined. Subsequently, the BM is pipetted very slowly on the edge of the tube to let it flow slowly over the Ficoll solution without mixing.
3. Proper mixing of the sample should be ensured to facilitate the lysis procedure.
4. A vertical rotary shaker guarantees suitable mixing (about 20 rpm). Alternatively a horizontal shaker can be used at 150–250 rpm (depending on applied tubes and the sample volume).

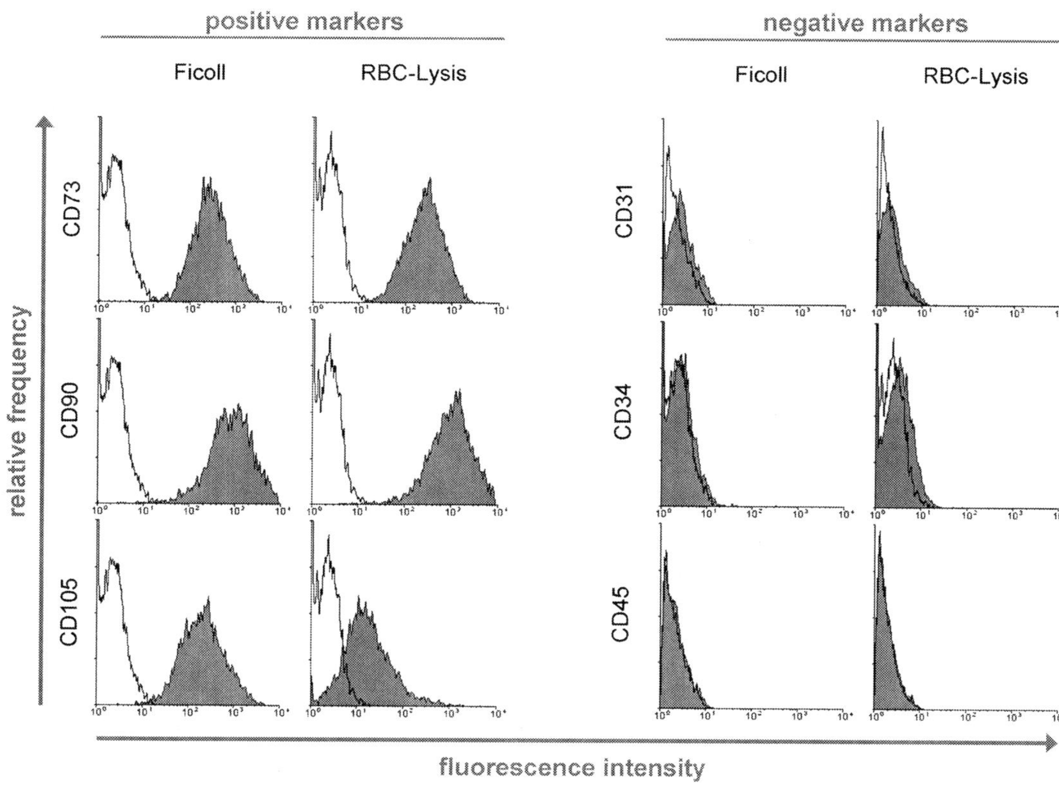

Fig. 4. Immunophenotypic characterization of isolated MSC. MSC preparations of either RBC lysis or density gradient centrifugation were analyzed by flow cytometry. MSC are, amongst others, positive for CD73, CD90, and CD105, whereas they do not express CD31, CD34, CD45. There were no differences between the two isolation methods and no subpopulations could be distinguished.

5. Depending on the source and cell number of the BM sample, the number of erythrocytes and BM-fragments varies. Therefore, it might be necessary to extend the lysis period. BM-fragments will not totally disintegrate, but the sample will appear uniformly as a reddish, diaphanous solution when lysed completely.
6. Culture flasks should be coated with fibronectin for at least 30 min in a humified atmosphere at 37°C. The fibronectin solution must not be allowed to evaporate and dry in the culture flasks. This may result in crystal formation and surface abnormalities which may impair the initial colony formation.
7. The red blood cell lysis method retains BM-fragments which are lost during density gradient centrifugation. The fragments contain high numbers of MSC and should be kept in the culture until the first washing step to increase the MSC yield. It needs to be further evaluated if small bone fragments remain throughout long-term culture and impair therapeutic intravenous application of MSC after expansion.

8. The yield of CFU-F varies between different donors and BM samples. Even within one aspiration procedure the number may vary. The initial aspirate will contain more BM fragments, whereas the percentage of peripheral blood increases during the aspiration procedure. Therefore, different MNC-concentrations should be seeded to evaluate the appropriate initial culture concentration (e.g., 5×10^5, 1×10^6 and 2×10^6 cells/cm^2).
9. MSC cultures comprise a very heterogeneous mixture of cells. This is reflected by the appearance of the initial colonies that vary in shape and size (13, 19).
10. Due to the variation in colony size, the trypsination may require a higher volume of Trypsin–EDTA or a longer incubation time until the colonies are separated. It may be difficult to separate the colonies and tapping of the culture flasks may be necessary to support cell separation and shorten trypsin-treatment.
11. If pipetting is performed too vigorously on differentiated cultures, the fixed cells may detach. Furthermore, the lipid droplets formed upon adipogenic differentiation may cause the cells to lift.
12. Isopropanol treatment should not be performed for more than 5 min. Otherwise, the fat droplets will be released from the cells.

Acknowledgments

We like to thank Anke Diehlmann for assistance in MSC culture. This work was supported by the German Ministry of Education and Research (BMBF) within the supporting program "cell based regenerative medicine" (START-MSC and CB-HERMES), the German Research Foundation DFG (HO 914/7-1), the Joachim Siebeneicher-Stiftung, Germany, and the Academy of Sciences and Humanities, Heidelberg (WIN-Kolleg) and the Stem Cell Network North Rhine Westphalia.

References

1. Reyes M., Lund T., Lenvik T., Aguiar D., Koodie L., Verfaillie C. M. (2001) Purification and ex vivo expansion of postnatal human marrow mesodermal progenitor cells. *Blood* **98**, 2615–25.
2. Jiang Y., Jahagirdar B. N., Reinhardt R. L. et al. (2002) Pluripotency of mesenchymal stem cells derived from adult marrow. *Nature* **418**, 41–9.
3. Petersen B. E., Bowen W. C., Patrene K. D. et al. (1999) Bone marrow as a potential source of hepatic oval cells. *Science* **284**, 1168–70.
4. Schwartz R. E., Reyes M., Koodie L. et al. (2002) Multipotent adult progenitor cells from bone marrow differentiate into functional hepatocyte-like cells. *J Clin Invest* **109**, 1291–302.

5. Dominici M., Le Blanc K., Mueller I. et al. (2006) Minimal criteria for defining multipotent mesenchymal stromal cells. The International Society for Cellular Therapy position statement. *Cytotherapy* **8**, 315–7.
6. Wagner W., Wein F., Seckinger A. et al. (2005) Comparative characteristics of mesenchymal stem cells from human bone marrow, adipose tissue, and umbilical cord blood. *Exp Hematol* **33**, 1402–16.
7. Wagner W., Horn P., Castoldi M. et al. (2008) Replicative senescence of mesenchymal stem cells – a continuous and organized process. *PLoS ONE* **5**, e2213.
8. Le Blanc K., Frassoni F., Ball L. et al. (2008) Mesenchymal stem cells for treatment of steroid-resistant, severe, acute graft-versus-host disease: a phase II study. *Lancet* **371**, 1579–86.
9. Nagaya N., Kangawa K., Itoh T. et al. (2005) Transplantation of mesenchymal stem cells improves cardiac function in a rat model of dilated cardiomyopathy. *Circulation* **112**, 1128–35.
10. Gastens M. H., Goltry K., Prohaska W. et al. (2007) Good manufacturing practice-compliant expansion of marrow-derived stem and progenitor cells for cell therapy. *Cell Transplant* **16**, 685–96.
11. Bartmann C., Rohde E., Schallmoser K. et al. (2007) Two steps to functional mesenchymal stromal cells for clinical application. *Transfusion* **47**, 1426–35.
12. Parekkadan B., Sethu P., van P. D., Yarmush M. L., Toner M. (2007) Osmotic selection of human mesenchymal stem/progenitor cells from umbilical cord blood. *Tissue Eng* **13**, 2465–73.
13. Horn P., Bork S., Diehlmann A. et al. (2008) Isolation of human mesenchymal stromal cells is more efficient by red blood cell lysis. *Cytotherapy* **10**, 676–85.
14. Schallmoser K., Bartmann C., Rohde E. et al. (2007) Human platelet lysate can replace fetal bovine serum for clinical-scale expansion of functional mesenchymal stromal cells. *Transfusion* **47**, 1436–46.
15. Kocaoemer A., Kern S., Kluter H., Bieback K. (2007) Human AB serum and thrombin-activated platelet-rich plasma are suitable alternatives to fetal calf serum for the expansion of mesenchymal stem cells from adipose tissue. *Stem Cells* **25**, 1270–8.
16. Pittenger M. F., Mackay A. M., Beck S. C. et al. (1999) Multilineage potential of adult human mesenchymal stem cells. *Science* **284**, 143–7.
17. Colter D. C., Class R., DiGirolamo C. M., Prockop D. J. (2000) Rapid expansion of recycling stem cells in cultures of plastic-adherent cells from human bone marrow. *Proc Natl Acad Sci USA* **97**, 3213–8.
18. Wagner W., Ho A. D. (2007) Mesenchymal stem cell preparations-comparing apples and oranges. *Stem Cell Rev* **3**, 239–48.
19. Ho A. D., Wagner W., Franke W. (2008) Heterogeneity of mesenchymal stromal cell preparations. *Cytotherapy* **10**, 320–30.

Chapter 4

Isolation and Growth of Adipose Tissue-Derived Stem Cells

Vladimir Zachar, Jeppe Grøndahl Rasmussen, and Trine Fink

Abstract

Fat tissue provides a rich and easily accessible supply of stem cells that feature the general differentiation potential and plasticity of mesenchymal stem cells. These stem cells, variably designated adipose-derived stem cells (ASCs), hold a great promise for regeneration in vivo and tissue engineering. In order to be able to obtain ASC preparations suitable for basic investigations as well as development of future therapeutic protocols, it is important that the critical isolation steps are properly carried out. Here, we describe in detail enzyme-based procedure for isolation and propagation of ASCs from the human subcutaneous fat tissue that is also adoptable to several animal species. The critical steps and impact of possible modifications on the final outcome are discussed, and useful hints are provided to streamline the troubleshooting.

Key words: Adipose stem cells, Isolation, Growth, Fat tissue, Collagenase, Hypoxia

1. Introduction

Since the seminal report by Zuk et al. (1) on the isolation of pluripotent mesenchymal stem cells (MSCs) from adipose tissue, the interest in adipose tissue-derived stem cells grew enormously and laid foundation to a dedicated field (http://www.ifats.org/). Several terms have been adopted to designate this heterogenous stem cell pool. Adipose stem cells, adipose tissue-derived stem cells, adipose stromal cells, or stromal vascular fraction cells are the most frequent terms that have been coined to refer to the stem cell entity within the fat tissue lysate. Recent data indicate that the stem cell type of the fat tissue is associated with the vascular bed (2, 3). For the purpose of the current paper, the designation adipose-derived stem cells (ASCs) will be adopted.

ASCs are indeed a very attractive type of somatic stem cells given the relative ease of accessibility and availability in abundant supply. The possibility of obtaining ASCs in therapeutically

Mohan C. Vemuri et al. (eds.), *Mesenchymal Stem Cell Assays and Applications,* Methods in Molecular Biology, vol. 698,
DOI 10.1007/978-1-60761-999-4_4,

relevant quantities without the need for an extended in vitro amplification, along with a broad differentiation potential, makes these cells appealable for cell-based therapies and tissue engineering applications. The areas of application range from osteo-chondral repair and wound healing, where the regenerative capacity has convincingly been documented (4, 5), to areas, such as functional cardiac repair, where clear benefits have yet to be established (6). In addition, substantial plasticity as exemplified by the capacity to differentiate along neuronal (7) or liver (8) specifications lends support to the notion that in the future these cells might become a preferred source for regenerative applications.

There are variables, however, that come into play during the isolation and propagation of ASCs (9, 10). Awareness of the quality of input reagents and the critical steps as well as rigorous adherence to standardized procedures is essential to provide for necessary reproducibility. In the following, we will describe the isolation of MSCs from lipoaspirates and from tissue biopsies of human and animal origin.

2. Materials

2.1. Tissue Preparation

1. Lipoaspirate can be stored overnight at room temperature and fat biopsies can be stored up to 36 h from harvest (see Note 1).
2. Dulbecco's phosphate-buffered saline (D-PBS) (GIBCO/Invitrogen, Leiden, The Netherlands).
3. Sterile scalpel or surgical scissors, forceps, and petri dishes are needed for biopsy processing.

2.2. Isolation of Adipose Tissue Stromal Stem Cells

1. Liberase Blendzyme 2 (Roche Applied Science, Hvidovre, Denmark) is reconstituted as recommended by the manufacturer, divided in single-use aliquots of 5.6 Wünsch units for lipoaspirates or 2.8 Wünsch units for biopsies, and stored at −80°C. In our experience, the stored enzymes retain their activity during storage at −80°C for more than a year. For the use of other collagenases, please refer to Note 2.
2. Milli-Q sterile-filtered water for lysis of red blood cells.
3. 10× D-PBS (GIBCO/Invitrogen).
4. Growth medium is composed of Minimum Essential Medium Alpha (A-MEM) (GIBCO/Invitrogen) supplemented with 10% fetal bovine serum (FBS), and penicillin (100 U/mL), streptomycin (100 μg/mL), and gentamicin (50 μg/mL) (all from GIBCO/Invitrogen). For the use of other types of media, please refer to Note 3.

5. Falcon 100- and 70-μm mesh cell strainers (BD Biosciences, Franklin Lakes, NJ).
6. 15- and 50-mL Conical plastic tubes (Corning, Schiphol-Rijk, The Netherlands).
7. Methyl violet in acetic acid for leukocyte counting (Bie&Berntsen, Rødovre, Denmark).
8. See-saw rocker SSM4 (Bibby Scientific Limited, Stone, United Kingdom).
9. Hemocytometer.

2.3. Determination of the Yield of Fibroblast Colony-Forming Units (CFU-Fs)

1. Growth medium (see Note 3).
2. PBS without divalent cations.
3. Methylene blue hydrate (Sigma-Aldrich, Brøndby, Denmark) is dissolved to saturation in absolute ethanol and diluted fourfold in milli-Q water to obtain a working solution.
4. Milli-Q water.
5. 4% Phosphate-buffered formaldehyde (Bie&Berntsen).
6. 96-Well flat-bottom cell culture grade plates (Corning).

2.4. Propagation of ASCs and Preparation for Immunophenotyping

1. PBS.
2. Trypsin/EDTA mixture: Trypsin (GIBCO/Invitrogen) is obtained as a solution of 2.5% in PBS. It is diluted 10 times with PBS and sterile-filtered. Aliquots of 10 mL are stored at −20°C. EDTA (anhydrous, crystalline, cell culture tested) (Sigma-Aldrich) is dissolved in PBS at a concentration of 0.02%, sterile-filtered, and stored in 10 mL aliquots at 4°C. Prior to use, the trypsin and EDTA solutions are mixed 1:1 to produce a ready-to-use 0.125% trypsin/0.01% EDTA blend, which is stable for up to a week at 4°C.
3. Growth medium (see Note 3).
4. 0.4% Trypan blue solution, cell culture tested, and sterile-filtered (Sigma-Aldrich).
5. Hemocytometer.
6. Steri-Cycle CO_2 incubator (Thermo Scientific, Marietta, OH) or workstation Xvivo System (BioSpherix, Redfield, NY) enabling incubation and handling in an uninterrupted hypoxic environment buffered with 5% CO_2 and nitrogen.

2.5. Determination of Proliferative Capacity

1. 48-Well cell culture grade plates (Corning).
2. Growth medium.
3. Milli-Q water, DNase-free.
4. 100× TE buffer: 1.0 M Tris–HCl, 0.1 M EDTA, pH 8.0 (Sigma-Aldrich) is diluted in DNase-free water to achieve 1× working concentration of TE buffer.

5. The 10% sodium dodecyl sulfate (SDS) (Sigma-Aldrich) stock solution is prepared by dissolving 1 g in 10 mL of DNase-free water and stored at room temperature. If SDS forms a precipitate, it is gently warmed by heating in a microwave oven for 5 s. The working solution is made ready by diluting to a final 0.02% concentration in DNase-free water.
6. Quant-iT™ PicoGreen® dsDNA reagent (Molecular Probes/Invitrogen, Leiden, The Netherlands) is stored at −20°C. Just prior to use, it is diluted 200-fold in TE buffer to obtain the working solution. The working solution is stored at room temperature in dark until use and is stable for up to 5 h. To prepare 6.4 mL of working solution, 32 μL of PicoGreen dye is mixed with 6.37 mL of TE buffer.
7. The 0.5 mg/mL lambda DNA (New England Biolabs, Ipswich, MA) stock solution is diluted 250-fold with TE buffer to obtain a working concentration of 2 μg/mL. It is stored in 1 mL aliquots at −20°C.
8. 96-Well black microtiter plates (Corning).
9. Fluorescence reader.

3. Methods

3.1. Processing of Lipoaspirate Samples (see Note 4)

1. Lipoaspirate is transferred to a sterile bottle or large Erlenmeyer flask.
2. After the blood has settled below the adipose tissue, the blood is removed with a 50-mL pipet.
3. An equal volume of D-PBS is added to the adipose tissue.
4. The flask is swirled for 10 s to mix the lipoaspirate and D-PBS.
5. The bottle is placed in the laminar flow bench for 2–3 min to allow the D-PBS and adipose tissue to separate.
6. The D-PBS is removed with a 50-mL pipet.
7. Steps 3–6 are repeated until the wash is clear (usually 3–5 times).
8. The lipoaspirate is divided into 20 mL aliquots in 50-mL tubes.

3.2. Processing of Adipose Tissue Biopsies

1. This procedure will be suitable for tissue biopsies between 1 and 4 mL.
2. 7 mL of D-PBS is added to a sterile petri dish (10 cm diameter).
3. The tissue biopsy is transferred with forceps to the petri dish.
4. The tissue is minced with a scalpel or scissors, while holding on with a forceps, and it is important that it is continuously kept moist with the D-PBS. The tissue should be minced to pieces not bigger than 2 mm (see Note 5).

5. With the aid of forceps and pipette, the minced tissue is transferred to a 50-mL tube.

3.3. Enzymatic Digestion (see Note 6)

1. For the lipoaspirates, 20 mL of collagenase buffer (or 10 mL of collagenase buffer for each biopsy) is added and the samples are incubated on a see-saw rocker at 37°C for 120 min. The tubes should be positioned horizontally, such that maximum mixing is ensured (see Note 7).
2. The digests are centrifuged at 400 × g for 10 min and the top fluid and fat layers are discarded (see Note 8).
3. To lyse erythrocytes, 9 mL of sterile milli-Q water is added to the pellet and pipetted up and down briskly for 7 times. 1 mL of 10× D-PBS is then promptly added to adjust the salt concentration (see Note 9).
4. The cells are filtered through a 100 μm cell strainer (see Note 10).
5. The filtrate is centrifuged at 400 × g for 10 min and the supernatant is discarded.
6. Each pellet is resuspended in 5 mL of growth medium and filtered through a 70-μm mesh cell strainer.
7. For lipoaspirate preparations, an additional 20 mL of growth medium is added to each tube, before the cells are seeded in T175 flasks (25 mL cell solution per flask). At this point, the yield of nucleated cells can be determined by staining with methyl violet and counting in hemocytometer. Also, at this point, the cells can be analyzed for total yield of CFU-Fs.
8. The cells from tissue biopsies are seeded in T25 flasks in 5 mL of growth medium. Due to impurities, such as pieces of plastic from the petri dishes after the mincing of tissue, the determination of cell number is difficult at this point.
9. Cells are transferred to CO_2 incubator overnight.
10. After 24 h, medium is changed and nonadherent cells are removed (see Note 11).

3.4. Determining Total Yield of CFU-Fs (see Note 12)

1. The cells are counted and resuspended in growth medium at a concentration of 50,000 cells/mL in 4 mL in a 15-mL centrifuge tube (see Note 13).
2. Eleven 15-mL centrifuge tubes are prepared, each with 2 mL of growth medium.
3. The cell suspension is diluted serially twofold across the 11 tubes.
4. 200 μL is dispensed from each dilution in a flat-bottom 96-well plate, such that each column will be seeded with 8 replicates of the corresponding dilution. Thus, the first column will contain 10,000 cells per well, and in the following

columns, it will be 5,000, 2,500, 1,250, 625, 313, 156, 78, 39, 20, 10, and 5 cells per well.

5. The cells are cultured for 11 days with medium changed twice a week.
6. At day 11, the medium is removed and the cells are washed twice with PBS.
7. The cells are fixed with 50 μL of 4% formaldehyde at room temperature for 5 min.
8. The formaldehyde is removed, the cells are washed once with PBS, and 50 μL of methylene blue is added and incubated at room temperature for 10 min.
9. The wells are gently washed with milli-Q water until all excess dye is removed.
10. The plate is left to dry.
11. For each dilution, the number of wells with colonies is determined. A colony is defined as having no less than 20 cells.
12. The number of CFU-Fs per well (U) is calculated by the equation U = –lnF0, where *F*0 is the number of empty wells divided by 8. This can naturally only be done for the dilution(s) where the number of empty wells is between 1 and 7. In the example shown below, in the column with 40 cells seeded per well, three wells scored positive for the presence of colonies. The U then equals –ln(5/8) = 0.47.

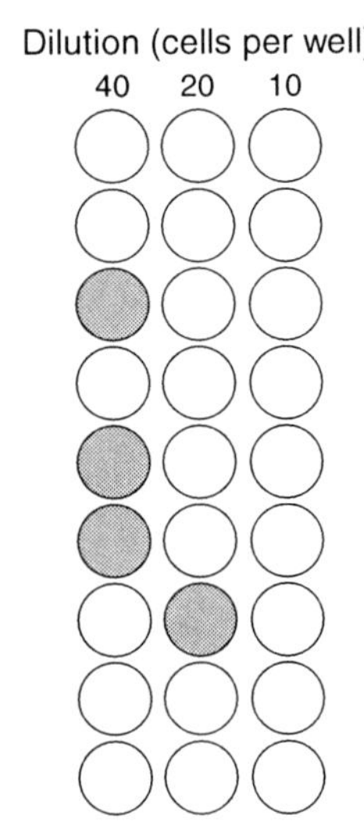

13. The fraction of CFU-Fs from the total cell yield is calculated as U divided by the corresponding cell seeding density. In our example, 0.47 U/(40 cells/well) yields 0.01 CFU-F per cell, which also corresponds to 1 CFU-F per 100 cells.

3.5. Preparing Cells for Immunostaining (see Note 14)

1. The medium is removed and the cells in the T25 culture flask are washed twice with 5 mL of PBS (see Note 15).
2. 1 mL of Trypsin/EDTA is added and incubated at 37°C.
3. In 2-min intervals, the flask is knocked sharply to loosen the cells. The progress of detachment is monitored visually using a microscope.
4. As soon as sufficient detachment has been achieved, the reaction is stopped by adding 4 mL of growth medium and the cell suspension is homogenized by repetitive pipetting with a 5-mL pipette.
5. The yield of live cells is determined by counting in a hemocytometer chamber using trypan blue. Cells are now ready for immune analysis.

3.6. Determining Cell Proliferation Rate

1. A suspension of cells in growth medium is prepared as detailed in Subheadings 3.5 or 3.7.
2. The concentration is adjusted to 600 cells/mL, and the total volume should be at least 25 mL.
3. The cells are seeded in a 48-well plate at 0.5 mL/well (300 cells/cm^2) and transferred to a CO_2 incubator.
4. 3–4 h after the seeding, when the attachment is complete, the cells from four replicate wells are lysed and the cell lysates collected as follows:
   (a) Media is removed from the wells, after which the wells are rinsed twice with 200 μL of PBS.
   (b) The cells are lysed with 200 μL of 0.02% SDS.
   (c) The plate is incubated in the CO_2 incubator for 20–30 min.
   (d) The lysates are collected and frozen at −20°C.
5. Cell lysates from four replicate wells are collected each subsequent day until all wells are processed. The growth medium is replaced with fresh one every 3–4 days (see Note 16).
6. At the day of analysis, lysates are thawed and 0.3 mL of TE buffer added to each sample.
7. DNA standards are prepared according to Table 1 below.
8. An aliquot of 100 μL of each sample and duplicate aliquots of the standards are transferred to a black 96-well microtiter plate and mixed with 100 μL of PicoGreen working solution.
9. The lysates and PicoGreen reagent are mixed well on a microplate shaker and incubated protected from light at room temperature for 10 min.

Table 1 Dilution series of standard DNA

Volume of 2 μg/mL λDNA stock (μL)	Volume of TE buffer (μL)	Final DNA concentration (ng/mL)
500	500	500
100	900	100
50	950	50
10	990	10
1	999	1
0	1,000	0

10. The fluorescence is measured in reader with excitation at 485 nm and emission at 535 nm.
11. Based on the values from the lambda DNA standards, a standard curve is constructed, and the DNA content of the samples is calculated.
12. The Microsoft Office Excel package is then used to calculate the doubling time based on the DNA measurements as follows:
    (a) The amount of DNA at a given time is calculated as a mean of the four replicates.
    (b) An *x*–*y* scatter plot is constructed, where the amount of DNA is plotted as a function of time, and logarithmic scale is applied to the *y*-axis.
    (c) The linear part of the resulting plot is a reflection of an exponential growth period. Usually, we find that the linear part/exponential phase is between days 2 and 7.
    (d) A second *x*–*y* scatter plot is constructed, depicting DNA vs. time again; however, this plot is solely based on the data from the exponential phase and both axes are linear.
    (e) The Excel program is used to plot a regression line using the second scatter plot. This can be done by right-clicking on one of the data points, then adding an exponential trendline. At this point, in the "options" window, the display of the equation of the regression line is prompted. It will be of the format $y = a * e^{kx}$. The *k*-value is then used to calculate the doubling time T_d as $T_d = \ln(2/k)$. An example is given below, where the proliferation rate has been calculated as 1.6 days per cell division.

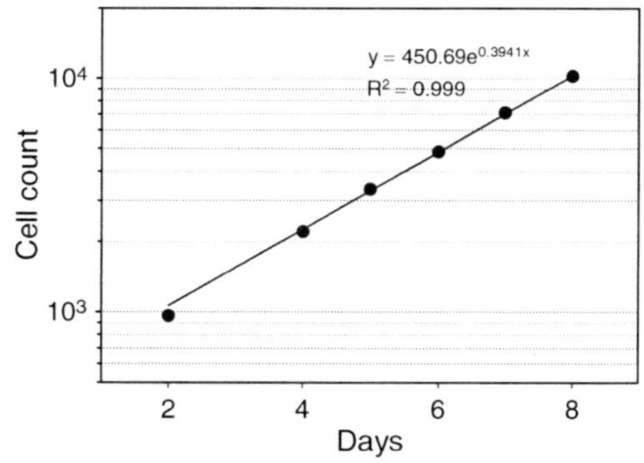

3.7. Propagation of ASCs (see Note 17)

1. The growth medium is removed, and the cells in the T25 culture flask are washed twice with 5 mL of PBS (see Note 15).
2. 1 mL of 0.125% Trypsin/EDTA is added and incubated at room temperature for 10 min. The detachment is assessed visually and under the microscope and can be aided by manually tapping the flask.
3. The trypsinization is stopped through addition of 5 mL growth medium.
4. An aliquot of the cell suspension is mixed with an equal volume of trypan blue, and the yield of live cells is determined by counting in a hemocytometer.
5. The cells are reseeded at 5,000 cells/cm^2 in a total volume of 5 mL per T25. Further incubation proceeds at ambient air oxygen concentration or in hypoxic conditions.

4. Notes

1. Prior to performing the tissue biopsy, a 15-mL centrifuge tube should be prepared with D-PBS supplemented with penicillin (100 U/mL), streptomycin (100 μg/mL), and gentamicin (50 μg/mL) (all from Invitrogen, Leiden, The Netherlands). The biopsy material is then transferred to the tube and stored at room temperature until processed. We have successfully harvested stem cells from material 48 h postharvest; however, the yield of stem cells was low.
2. In our experience, Liberase Blendzyme 1 and H1 (Roche Applied Science) and crude collagenase mix (CCM) (Waco, Neuss, Germany) can also be used. If the CCM is used, it should be dissolved in D-PBS with 2% bovine serum albumin just prior to use.
3. We have assayed different types of media, sera, and sera supplements for their capacity to support mesenchymal cell

growth and to maintain their differentiation potential (9). We cannot recommend using sera replacers, and although A-MEM is recommended, DMEM/F12 and F12 media also appear to be suited for MSC propagation and differentiation.

4. GLP should strictly be observed, especially in all steps requiring sterile handling. All media and solutions prepared for ASC isolation and growth in-house should be sterile-filtered using a 0.2-μm exclusion cell culture grade filters (TPP, Trasadingen, Switzerland). We have successfully isolated ASCs from fat biopsies from sheep, rat, rabbit, and brown bear using the described procedure. In general, when using this approach, the yield of stem cells from rabbit is quite high, average for rat and brown bear samples, and moderately low for sheep.
5. It is quite critical for the yield that the tissue is finely minced.
6. Differences between tissue samples from different donors are such that it is difficult, if not impossible, to identify a universal best type of enzyme (10). However, underperforming types should be avoided, and highly defined enzymes (e.g., Liberase Blendzymes; Roche Applied Science) are preferable due to minimal batch-to-batch variability. As these defined enzymes are rather expensive, in order to keep experimental cost under control, tested batches of crude collagenase can be used. As for the digestion time, longer periods tend on average to result in higher yields; however, we have found a 2 h digestion to provide the best trade-off between yield and assay duration.
7. Though we have found that digestion time of 2 h in general gives the best yield, the digestion may be stopped sooner. After 45 min, the digests are inspected every 15 min, and when the consistency appears homogenous, digestion may be stopped.
8. For the biopsy material, it may be necessary to use forceps to remove the top fat layer after centrifugation. Also, the pellet may be virtually invisible due to low cell numbers.
9. As an alternative to lysis of erythrocytes in distilled water, the lysis can be accomplished by resuspending the pellet in 15 mL of erythrocyte lysis buffer for 10 min. The composition of the buffer is 155 mM NH_4Cl (8.3 g/L) and 5.7 mM K_2HPO_4 (0.993 g/L) in 0.1 mM EDTA-Na (0.04 g/L). The solution is sterile-filtered and stored at room temperature.
10. For material from the same patients, the 100 or 70 μm strainers may be used. However, the 100 μm cell strainers, in particular, become clogged soon.
11. In our experience, approximately 25% of the nucleated cells attach to the cell culture flasks.

12. The phenotypical properties of cells in crude lysate can be assessed directly by identification of cell surface or intracellular markers by immunostaining. Such analysis can also be carried out after the initial attachment. The selection of cells which adhere to plastic, in combination with a short period of growth after the seeding, provides for expansion of ASCs, thus the profile will differ from that of crude lysate. Frequently used general functional parameters include determination of the frequency of precursor cells by the CFU-F assay and the proliferation rate. Functional assessment routinely also involves differentiation assays that are dealt with in other chapters of this book.
13. This protocol is optimized for analysis of crude stromal vascular fraction. For analysis of the more purified cell population obtained after selection of plastic adherence, during overnight culture, the cells are first gently trypsinized as detailed in the Subheading 3.3, after which the cells are resuspended in growth medium at a concentration of 5,000 cells/mL. Apart from using tenfold fewer cells throughout the procedure, the protocol for determining the frequency of CFU-Fs is identical to the one described in Subheading 3.4.
14. This gentle trypsinization procedure may be used to identify major MSCs markers, such as CD29, CD44, CD90, and CD105 (11). Immunostaining protocol is not detailed in this section, since it is covered in other chapters of this book.
15. With different growth areas, the volumes are changed proportionally.
16. When performing this assay, we also harvest cells during weekends. If that is not feasible, we recommend starting the experiment on a Monday (day 0) and harvesting the cells Monday through Friday for the following 2 weeks.
17. Conditions for expansion are determined by the intended purpose. Two crucial factors are represented by the growth medium and gaseous conditions. A selection of several types of media and serum supplements is available and different concentrations of oxygen can be used in the culture atmosphere. Thus, a variety of different schemes based on their combinations are possible. A case for synthetic serum replacements is pushing more and more in the forefront with the prospect of introducing stem cells and their products into clinical practice. Nevertheless, in general, their capacity to support expansion of ASCs, at this point in time, appears inferior to that of FCS (9). As for the oxygen concentration, it is now well established that mild hypoxia is beneficial for the proliferative rate of MSCs (Fig. 1) (12, 13). The seeding density has an impact on the dynamic of the culture, and very low densities, in the proximity of 400 cells/cm^2, should be chosen if rapid

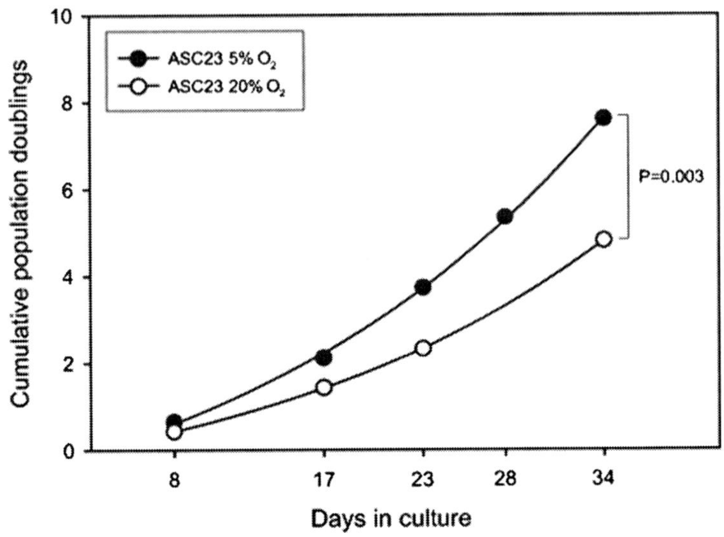

Fig. 1. Effect of hypoxia on the proliferation of human ASCs. The culture of the cell line ASC23 was initiated at passage P0 and subcultured at 5,000 cells/cm^2 for overall duration of 34 days.

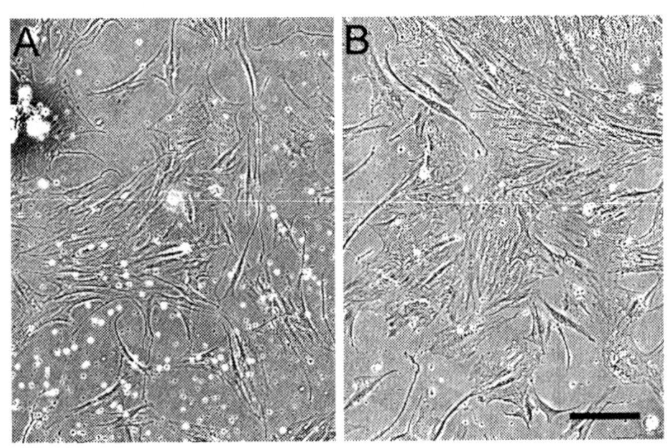

Fig. 2. Phase contrast images of human ASCs after different periods in culture. (**a**) The cells were maintained in culture for one passage, or (**b**) eight passages after isolation. The cultures were initiated at 5,000 cells/cm^2 and incubated in normoxic conditions for 7 days. The cells have a characteristic fibroblastic appearance and, with higher passages, the cytoplasm to nucleus ratio typically increases. The scale bar indicates 200 μm.

proliferation is desirable. This procedure describes ASCs propagation in an optimized environment of A-MEM supplemented with 10% FCS in an atmosphere of 5% oxygen. With subculturing on a tissue culture-treated polystyrene surface, early passage cells undergo an adaptation reflected by a slightly faster replication rate and a change in morphology whereby the cells increase their contact surface with the polystyrene and the cytoplasm to nucleus ratio increases (Fig. 2).

Regarding long-term propagation, in our hands, the different ASC lines display undiminished replication rate for over 30 passages, corresponding to over 60 population doublings (approximately half a year). Even after such extensive culturing, these cell lines maintain their differentiation potential when tested in standard assays for adipogenesis, chondrogenesis, and osteogenesis.

References

1. Zuk, P. A., Zhu, M., Mizuno, H., Huang, J., Futrell, J. W., Katz, A. J., Benhaim, P., Lorenz, H. P., and Hedrick, M. H. (2001) Multilineage cells from human adipose tissue: implications for cell-based therapies, *Tissue Eng* 7, 211–228.
2. Cossu, G., and Bianco, P. (2003) Mesoangioblasts--vascular progenitors for extravascular mesodermal tissues, *Curr Opin Genet Dev* **13**, 537–542.
3. Kahn, C. R. (2008) Medicine. Can we nip obesity in its vascular bud? *Science* **322**, 542–543.
4. Rada, T., Reis, R. L., and Gomes, M. E. (2009) Adipose tissue-derived stem cells and their application in bone and cartilage tissue engineering, *Tissue Eng Part B Rev* **15**, 113–125.
5. Wu, Y., Chen, L., Scott, P. G., and Tredget, E. E. (2007) Mesenchymal stem cells enhance wound healing through differentiation and angiogenesis, *Stem Cells* **25**, 2648–2659.
6. Mollmann, H., Nef, H., Elsasser, A., and Hamm, C. (2009) Stem cells in myocardial infarction: from bench to bedside, *Heart* **95**, 508–514.
7. Anghileri, E., Marconi, S., Pignatelli, A., Cifelli, P., Galie, M., Sbarbati, A., Krampera, M., Belluzzi, O., and Bonetti, B. (2008) Neuronal differentiation potential of human adipose-derived mesenchymal stem cells, *Stem Cells Dev* **17**, 909–916.
8. Aurich, H., Sgodda, M., Kaltwasser, P., Vetter, M., Weise, A., Liehr, T., Brulport, M., Hengstler, J. G., Dollinger, M. M., Fleig, W. E., and Christ, B. (2009) Hepatocyte differentiation of mesenchymal stem cells from human adipose tissue in vitro promotes hepatic integration in vivo, *Gut* **58**, 570–581.
9. Lund, P., Pilgaard, L., Duroux, M., Fink, T., and Zachar, V. (2009) Effect of growth media and serum replacements on the proliferation and differentiation of adipose-derived stem cells, *Cytotherapy* **11**, 189–197.
10. Pilgaard, L., Lund, P., Rasmussen, J. G., Fink, T., and Zachar, V. (2008) Comparative analysis of highly defined proteases for the isolation of adipose tissue-derived stem cells, *Regen Med* **3**, 705–715.
11. Lee, R. H., Kim, B., Choi, I., Kim, H., Choi, H. S., Suh, K., Bae, Y. C., and Jung, J. S. (2004) Characterization and expression analysis of mesenchymal stem cells from human bone marrow and adipose tissue, *Cell Physiol Biochem* **14**, 311–324.
12. Grayson, W. L., Zhao, F., Bunnell, B., and Ma, T. (2007) Hypoxia enhances proliferation and tissue formation of human mesenchymal stem cells, *Biochem Biophys Res Commun* **358**, 948–953.
13. Fink, T., Rasmussen, J. G., Lund, P., Pilgaard, L., Søballe, K., and Zachar, V. (2011) Isolation and expansion of adipose-derived stem cells for tissue engineering, *Front Biosci* **3**, 256–263.

Chapter 5

Isolation, Culture, and Characterization of Human Umbilical Cord Stroma-derived Mesenchymal Stem Cells

Alp Can and Deniz Balci

Abstract

As the collection and isolation of human bone marrow-derived mesenchymal stem cells (MSCs) require invasive and often undesirable procurement procedures, investigators have begun to seek alternative sources of human MSC including the umbilical cord stroma. Here we describe the noninvasive isolation, culture, and basic characterization of human umbilical cord stroma-derived mesenchymal stem cells (hUCS-MSCs). Although some technical and observational variations exist between laboratories, there has been a relatively common consensus regarding the immunological and functional characteristics of hUCS-MSCs. Successful in vitro and in vivo differentiation to several lineages makes these cells an invaluable stem cell source, deserving further testing as a cellular therapy or other applications in regenerative medicine. Therefore, the isolation and culture of hUCS-MSCs still need better clarification to ultimately build an optimal standard procedure among laboratories, tissue banks, and clinics.

Key words: Cell culture, Cell isolation, Mesenchymal stem cell, Stem cell, Umbilical cord stroma

1. Introduction

Human umbilical cord (UC) consists of three tissue components including the (1) amniotic membrane, (2) stroma (namely, Wharton's jelly), and (3) blood vessels (two arteries and one vein). The stroma is also further divided into three histological zones including the (1) subamniotic zone, (2) intervascular zone, and (3) perivascular zone (1) (Fig. 1). Over the last few years, human umbilical cord stroma-derived mesenchymal stem cells (hUCS-MSCs) have been isolated to examine their stem cell nature. The expression of certain embryonic and mesenchymal stem cell markers on the naïve form of these cells brings suggestive evidence as to their multipotency. This potential has ultimately led many groups to initiate the evaluation of these cells in

Mohan C. Vemuri et al. (eds.), *Mesenchymal Stem Cell Assays and Applications*, Methods in Molecular Biology, vol. 698, DOI 10.1007/978-1-60761-999-4_5, © Springer Science+Business Media, LLC 2011

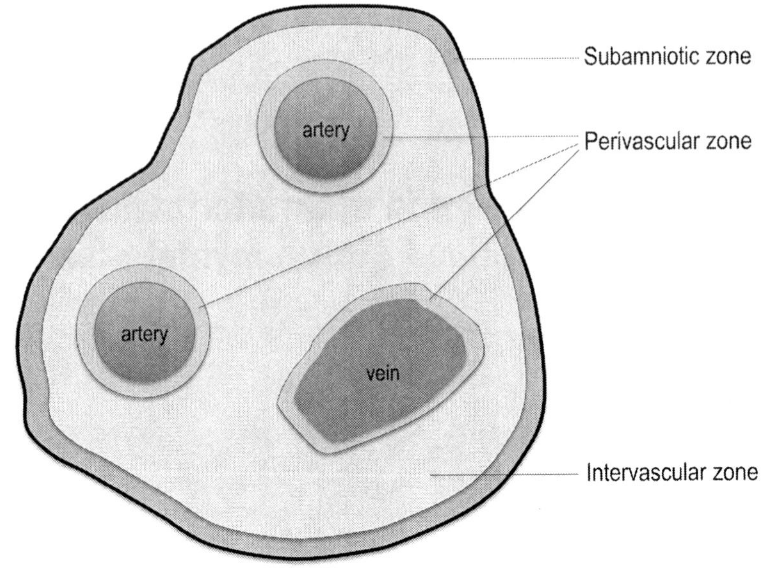

Fig. 1. In a given human umbilical cord cross section, three histological zones are recognized as subamniotic (*narrow area* beneath the amniotic membrane), intervascular (*light grey* area), and perivascular (*wide areas* surrounding the vessels) zones. These zones are distinctive by the morphological characteristics of stromal cells, the amount and the distribution of extracellular matrix proteins.

therapeutic regimens. Even though UC stroma-derived cells still need further characterization, both intervascular and perivascular stroma-derived cells have been successfully differentiated into various cell types including adipocytes (2), chondrocytes (3–5), osteocytes (3–8), cardiomyocytes (5, 9), skeletal myocytes (6), hepatocytes (10), insulin-producing cells (11) as well as neuron-like cells (3, 7, 12–16). These studies indicate that hUCS-MSCs stand as a promising source of stem cells, especially for allogeneic therapeutic applications in regenerative medicine.

This chapter aims to describe the step-wise isolation, culture, and characterization of hUCS-MSCs. Despite this seemingly standard approach, readers must be aware that there are still significant variations that exist between laboratories, which must be taken into account when comparing results using the described methodology. Also, we have noted that there is a wide range of individual differences among donor tissues. While most UC samples will provide a reasonable number of hUCS-MSCs using the provided protocols, some samples may result in suboptimal cell isolation and expansion. The reasons behind this phenomenon still remain to be elucidated.

2. Materials

2.1. Transport of Human UC Samples to the Laboratory

See Note 1 prior to starting the procedure.

1. Transport medium: Leibovitz L-15 medium (Sigma, St. Louis, MO) supplemented with 2.05 mM L-Glutamine (HyClone, Ogden, UT), 3% (w/v) penicillin and streptomycin (Sigma), 5 µg/mL amphotericin B (Sigma) in a 100–200 mL glass or polystyrene capped-bottle.

2.2. Isolation of hUCS-MSCs

1. Digestion medium: Dulbecco's Modified Eagle's Medium (DMEM)/Ham's Nutrient Mixture F12 (1:1) (Sigma) supplemented with 10% heat-inactivated fetal bovine serum (FBS) (Invitrogen, Carlsbad, CA), 1% (w/v) penicillin and streptomycin (Sigma), 2.5 µg/mL amphotericin B (Sigma), 0.1% (w/v) collagenase type B (La Roche Ltd., Basel) or 0.1% (w/v) collagenase type I (Sigma).
2. 2% (w/v) dispase (Sigma).
3. Dulbecco's phosphate buffered saline (D-PBS) solution containing 0.6 mM calcium chloride and 0.5 mM magnesium chloride (Sigma).
4. 50 mL conical polypropylene centrifuge tubes (Biochrom, Berlin, Germany).
5. Glass petri dishes (100 mm diameter).
6. Hemocytometer (American Optical, Buffalo, NY) or automated cell counting device (Vi-CELL, Automated Cell Viability Analyzer, Beckman Coulter, Fullerton, CA).
7. Shaking water bath (set to 37°C) (Memmert, Schwabach).
8. Refrigerated centrifuge with 50 mL adaptor (Hettich, Tuttlingen).
9. Chopping pad (5–8 mm thick Teflon pad).
10. Stainless steel dissecting scissor (Electron Microscopy Sciences, Hatfield, PA).
11. Surgical scalpel handle and blade (no: 20, 21, or 22) (Electron Microscopy Sciences).
12. Two fine-tipped stainless steel tissue forceps with 1 × 2 teeth (Electron Microscopy Sciences).
13. Serological pipettes (5, 10, 25 mL) and pipette holder (Biochrom).

2.3. Primary and Subculture of hUCS-MSCs

1. Culture medium: DMEM/Ham's Nutrient Mixture F12 (1:1) (Sigma) supplemented with 10% heat-inactivated FBS (Invitrogen), 1% (w/v) penicillin and streptomycin (Sigma), 2.5 µg/mL amphotericin B (Sigma).

2. Trypsin (0.25%) (w/v) and ethylenediamine tetraacetic acid (EDTA) (1 mM) solution (Invitrogen).
3. Phosphate buffered saline (PBS) solution containing 0.6 mM calcium chloride and 0.5 mM magnesium chloride (Sigma).
4. Polystyrene sterile tissue culture flasks (T25, T75, 24-well, or 100 mm in diameter) (Biochrom).
5. 15-mL polypropylene sterile conical tubes (Biochrom).
6. Circular glass coverslips (12-mm diameter) (Fisher Scientific, Pittsburgh, PA).

2.4. Cryopreservation and Recovery

1. Freezing medium: DMEM/Ham's Nutrient Mixture F12 (1:1) (Sigma) supplemented with 20% FBS (Invitrogen).
2. Dimethyl sulfoxide (DMSO) (Sigma).
3. Freezing polycarbonate container (Mr. Frosty®, Nalgene Labware, Rochester, NY).
4. 2-mL polypropylene capped cryo vials (Biochrom).
5. Shaking water bath (set to 37°C) (Memmert).
6. Computer-controlled rate freezer (Sy-Lab, Neupurkersdorf).

2.5. Immunocyto-chemical Characterization of hUCS-MSCs

1. Paraformaldehyde (4%) (freshly prepared from 10% stock solution) (Merck, San Diego, CA).
2. PBS (Sigma).
3. PBS/glycerol mixture (1:1) containing 25 mg/mL sodium azide (Merck).
4. Antibodies against cytoskeletal proteins; (1) Anti-vimentin monoclonal IgG produced in mouse (clone V9) (Sigma). 1:100 dilution in blocking and aldehyde reducing solution of PBS containing 2% bovine serum albumin (Sigma), 2% powdered milk, 2% normal goat serum (Sigma), 0.1 mol/L glycine (Merck), and 0.01% Triton X-100 (Sigma). (2) Anti-cytokeratin (types 1, 4, 5, 6, 8, 10, 13, 18, and 19) monoclonal IgG produced in mouse (Sigma). 1:100 dilution in blocking buffer. (3) Anti-desmin monoclonal IgG produced in mouse (clone DE-U-10) (Sigma). 1:50 dilution in blocking buffer. (4) Anti-α smooth muscle actin IgG produced in mouse (Sigma). 1:100 in blocking buffer.
5. Antibody against CD 105; mouse monoclonal IgG (Becton, Dickinson and Company, NJ). 1:5 dilution in PBS.
6. Fluorescent dye-conjugated secondary antibodies; (1) FITC-conjugated goat anti-mouse IgG (Jackson Immunoresearch Laboratories, West Grove, PA). 1:100 dilution in PBS. (2) Cy3-conjugated goat anti-mouse IgG (Jackson Immunoresearch Laboratories). 1:100 dilution in PBS.

7. Tetramethylrhodamine B isothiocyanate (TRITC)-phalloidin (specific to F-actin filaments) (Sigma). 35 μg/mL in PBS.
8. Hoechst 33242 stain (1 μg/mL in mounting medium).

3. Methods

3.1. Transport of Human UC Samples to the Laboratory

1. All instruments should be sterile. As a precaution, uncomplicated cesarean deliveries at full term are preferred versus vaginal deliveries due to the risk of microbial contamination.
2. Cut a 15 cm-long UC piece close to the placenta; drain the excess blood by simply squeezing the cord with two gloved fingers. Place the cord into the transport medium (see Note 2).
3. Transfer the sample to the laboratory as quickly as possible in a cooled environment (see Note 3).

3.2. Isolation of hUCS-MSCs

A number of methods are available for the isolation and/or culture of hUCS-MSCs (see (1) for review). The technique described below is a slight modification of a method described previously (2, 3). All of the below described procedure should be conducted in a class II laminar airflow hood.

1. Take out the cord from the bottle using a long forceps and transfer the tissue into a glass petri dish.
2. Cut the cord into 5 cm-long pieces, take one piece at a time onto the chopping Teflon pad (Fig. 2a), and retain other pieces in the transport medium.
3. Cut the cord longitudinally using two teethed forceps by rigid movements. Vessels will be distinguished by their typical appearance from the surrounding tissue. Arteries resemble white rigid cords while vein has a slightly yellowish appearance (Fig. 2b). As the vessels are not very fragile, using a teethed forceps, pinch off the vessels by coarse movements along the long axis of the cord (Fig. 2c, d). At the end of this procedure, an approximately 5 cm-long cord will become a relatively flat pinkish-white tissue weighing around 3,000 mg (Fig. 2e).
4. Thoroughly chop the cord using a dissecting scissor in a 50 mL centrifuge tube (Fig. 2f) until approximately 2–3 mm tissue fragments are obtained.
5. Add 20 mL of culture medium to the chopped cord pieces in the 50 mL conical tube.
6. Centrifuge at 250 × *g* for 5 min at 4°C. Carefully pour off the supernatant and repeat this washing process twice by adding 20 mL culture medium to the tube.

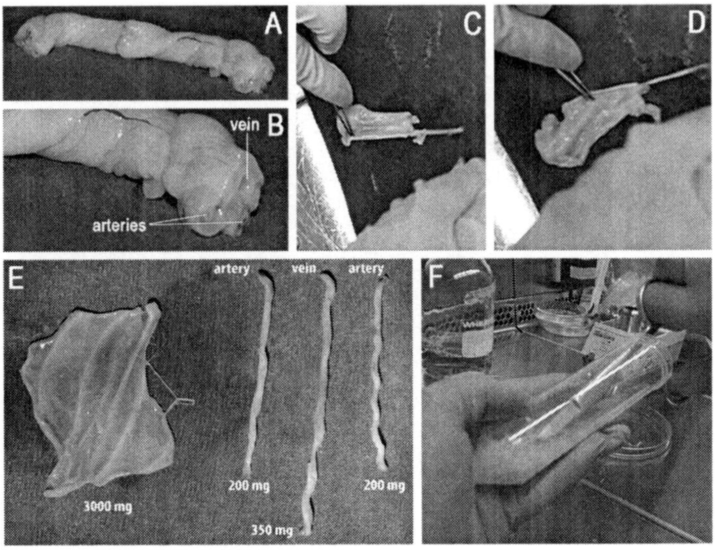

Fig. 2. Isolation of hUCS-MSCs (**a**) requires a delicate removal of blood vessels (two arteries and one vein) (**b**) by pinching off the vessels using two teethed forceps (**c**, **d**) resulting a pinkish-white cord material (**e**) containing subamniotic, intervascular, and perivascular stromal elements. Before digestion, cord must be chopped into 3–4 mm^3 pieces by fast movements of a dissecting scissor (**f**).

7. Resuspend the pellet in culture medium (see Note 4) containing collagenase type B or type I. Digest the tissue pieces for 3 h in a shaking water bath at 37°C (see Note 5).
8. Add dispase (2 mg/mL working concentration) and incubate tissue pieces for another 30 min at 37°C.
9. Split the digested material into three 50 mL tubes and add D-PBS to terminate the digestion. The amount of D-PBS must be at least threefold of the volume of digestion fluid (see Note 6).
10. Centrifuge at 350 × *g* for 20 min at 4°C. The supernatant will still be very viscous and contains many unsettled cells. So pipette off the supernatant carefully and do not discard.
11. Resuspend the cell pellets in 3–4 mL of culture medium; gently pipette the pellet up and down until a homogeneous cell suspension is achieved. If some undigested tissue pieces still exist, allow them to settle down in the bottom of the tube.
12. Collect the cell suspension into one 15 mL tube and count the cells using a hemocytometer or an automated cell counting device to determine the total number of cells in the suspension. The total number of cells obtained varies greatly; however, typically $4.5 \times 10^6 - 5 \times 10^6$ per 15 cm cord are obtained; therefore, total yield is plated into two T75 flask.
13. If low number of cells is collected, plate the supernatant directly onto T150 flask.

3.3. Primary and Subculture of hUCS-MSCs

1. Plate approximately 2×10^5 cells into each T75 flask in 10 mL culture medium.
2. Allow cells to attach and grow for 3 days in a humidified, 5% CO_2 culture incubator. As a few explants may still exist in the flask, the 3–4 day incubation will also allow the migration of cells from the explants.
3. To remove the nonadherent cells or fragments, gently wash the flasks using prewarmed D-PBS and gently add 10 mL prewarmed culture medium.
4. Change culture medium every third day (or twice per week).
5. Upon reaching 80–90% confluence at day 12–13, remove the cells with prewarmed trypsin-EDTA solution (4 mL per flask) for 10 min at 37°C. If undetached cells still exist, firmly tap the flask to detach.
6. Collect detached cells in a 15-mL conical centrifuge tube and count. Approximately $1–1.2 \times 10^6$ cells are typically obtained at the end of this primary culture period (Passage 0; P_0).
7. Centrifuge the tubes at $200 \times g$ for 10 min, carefully pour off the supernatant.
8. Resuspend the cell pellet in culture medium. Adjust the amount of medium so as to plate 5,000 cells/cm^2 in the next passage (Passage 1; P_1). Typically one flask of P_0 is subcultured into four flasks.
9. Alternatively, plate the cells onto poly-D lysine coated glass coverslips (in 6- or 24-well tissue culture plates) when an advance microscopic evaluation of the cells (immunostaining etc.) is desired (see Subheading 3.5).
10. Replace culture medium twice a week and examine the cells under a phase contrast inverted microscope.
11. Repeat the harvest procedure (see steps 5–10) as needed for desired applications. The maximum number of passages varies among laboratories, but in our experience cells can be cultured up to 13 passages (40–60 population doublings), whereby approximately 1.3×10^9 total cells can be obtained.

3.4. Cryopreservation and Recovery

1. After centrifugation of cells following trypsin-EDTA treatment, resuspend the cell pellet in freezing medium.
2. Adjust the number of cells to 1×10^6 total cells in 1.5 mL freezing medium.
3. Transfer 1,350 μL of cell suspension into a 2 mL cryo vial and gently add 150 μL DMSO dropwise to result in a total volume of 1.5 mL (10% DMSO).
4. Transfer vials to 4°C for 10 min allowing cells to accommodate to the freezing medium.

a

Step	Temperature (°C)	Rate (°C/min)	Time (min)
start	5.00	0.00	0.00
1	0.00	2.00	2.50
2	0.00	0.00	10.00
3	-6.00	1.00	6.00
4	-45.00	60.00	0.65
5	-26.00	10.96	1.73
6	-15.00	3.00	3.67
7	-35.00	1.00	22.00
8	-80.00	5.00	9.00
9	-120.00	10.00	4.00

b

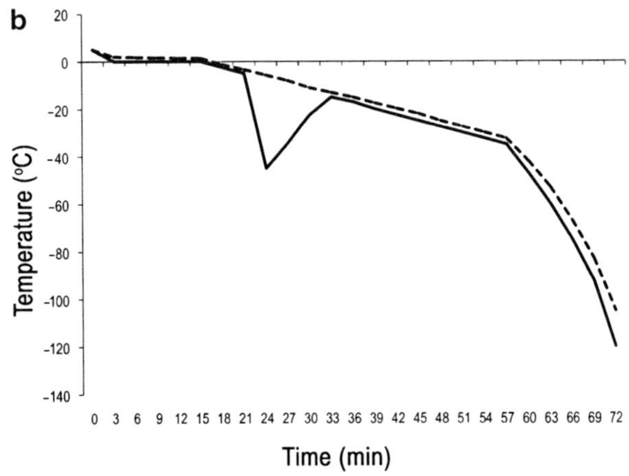

Fig. 3. (**a**) A nine-step freezing program is used to set up the time, temperature, and rates in a computer-controlled freezer specifically optimized for the hUCS-MSCs. (**b**) Note that some steps require a relatively high freezing rate (60°C/min) to compensate the *latent heat of fusion* that normally emerges from the specimen. The *solid line* refers to the chamber temperature whereas the *dashed line* refers to the sample temperature.

5. Transfer vials to −80°C in the freezing container and keep them overnight. Then, transfer to liquid nitrogen tanks for permanent storage.
6. If a computer-controlled slow rate freezer is accessible, follow the program optimized for hUCS-MSCs (Fig. 3). After controlled freezing, transfer the vials to liquid nitrogen tanks (see Note 7).
7. To thaw frozen cells, transfer the cryo vials directly to a 37°C water bath. Upon thawing in less than a minute, centrifuge cell suspension at 150×*g* for 10 min, gently remove the supernatant and resuspend the cell pellet in culture medium (see Subheading 2.3, item 1).

3.5. Immunocyto-chemical Characterization of hUCS-MSCs

The scope of this section is limited to certain techniques to emphasize the basic features of isolated hUCS-MSCs. Here, in vitro morphological features, in situ detection of certain cytoskeletal proteins, and CD 105 as one of the most recognized cell surface markers are described. Flow cytometric detection of cell surface markers and further characterization techniques (i.e., in vitro differentiation to mesenchymal lineages) could also be implemented, all of which are described elsewhere in the literature (3, 10, 17–19).

1. For direct morphological evaluation of cells during different time points of expansion, use an inverted microscope equipped with phase contrast or a differential interference contrast viewing system (Fig. 4).
2. For advanced microscopy and higher optical clarity, use subcultures plated onto round poly-D lysine coated glass coverslips in 6- or 24-well plates. We have noted no difference in morphology or structural protein expression levels between cells grown on poly-D lysine coated glass or polystyrene surfaces.
3. After reaching near confluence, fix the cells on the coverslips with 4% PFA (250–300 μL of PFA is enough per well of a 24-well plate) for 20 min at room temperature.
4. Remove fixative and add PBS to each well (0.5 mL/well) and store the plates at 4°C until use. If an airtight insulation is provided, plates can be stored for several months at 4°C.
5. For immunolabeling of cells (Fig. 5), take the coverslips off the plate; place them in a humidified chamber with the cell-attached side face-up. Add 20 μL primary antibody (see Subheading 2.5, item 4) onto the coverslip and incubate the cells for 2 h at 37°C.

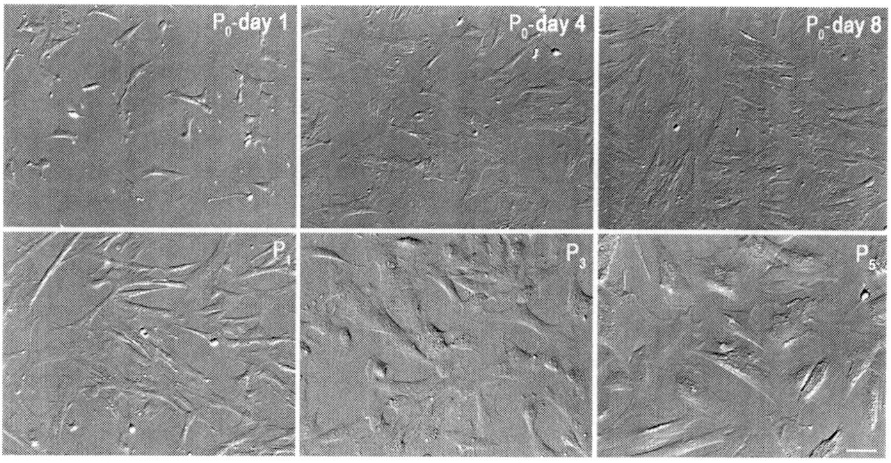

Fig. 4. Following isolation (Passage 0; P_0-day 1), hUCS-MSCs display a fibroblast-like phenotype. On the following days (as early as day 4; P_0-day 4), however, these cells gain their distinct cell morphology displaying an extremely flat and fiber-rich morphology with a wide cytoplasm that lasts through multiple passages (P_5). All images were obtained using a Nomarsky differential interference contrast microscope system. *Scale bar*, 50 μm.

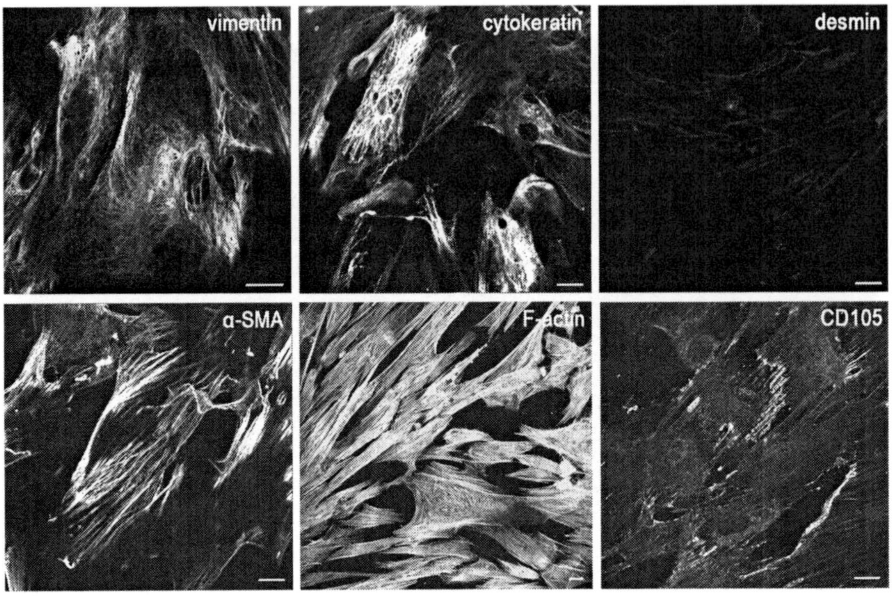

Fig. 5. Confocal microscopic analysis of certain cytoskeletal proteins such as *vimentin, cytokeratin, desmin, α-smooth muscle actin (α-SMA), F-actin,* and *CD105* (endoglin, an integral membrane protein) are used to positively or negatively confirm and characterize the hUCS-MSCs. Desmin is consistently lacking in cultured hUCS-MSCs while α-smooth muscle actin and cytokeratin filaments are expressed in only some of the cultured expanded cells. *Scale bars*; 20 μm.

6. Wash the coverslip surface by twice change of 25 μL PBS, add 20 μL secondary antibody (see Subheading 2.5, item 6) for 90 min at 37°C.
7. For F-actin labeling, incubate coverslips with TRITC-phalloidin dye (see Subheading 2.5, item 7) for 30 min at 37°C.
8. Wash the coverslip surface by twice change of 25 μL PBS.
9. Finally, drop 3–4 μL mounting medium containing Hoechst 33242 DNA dye onto a glass slide and then mount the coverslip onto the drop with the cell-attached side facing down. Apply nail polish circumferentially to permanently stick the coverslip on the slide.
10. Store the stained slides at –20°C until observation.

4. Notes

1. All media, instruments, and glassware should be sterilized prior to use.
2. Freshly prepared medium is preferred (less than 2-day-old) and should be stored at 4°C until use.

3. Our observations show that resting of unprocessed UC pieces even in 4°C for more than 6–8 h do not give good yield after isolation.
4. The volume of digestion solution should be around two- to threefold of the pellet volume (approximately 15 mL of pellet is obtained from a 15 cm-long cord).
5. Check occasionally during digestion to make sure that the tissue clumps gradually disappear while digestion fluid gains an intense cloudy appearance.
6. Since the UC tissues are highly rich in extracellular matrix proteins such as hyaluronic acid, even the digested material will be very viscous. So, in order to free the undigested tissue pieces from the isolated cells, shake the tubes thoroughly after adding PBS.
7. We have found that cell survival rate dramatically increases from 59 to 86% using the controlled freezing protocol (manuscript in preparation) compared to uncontrolled freezing (see Subheadings 3.4, steps 4 and 5).

Acknowledgments

The authors would like to thank Drs. Serçin Karahüseyinoğlu and Özgür Çinar who worked as Ph.D. students to improve the described techniques with great pleasure and care. This work was partly supported by The Turkish Ministry of Industry STZ-139-2007-2.

References

1. Can, A., and Karahuseyinoglu, S. (2007) Concise review: human umbilical cord stroma with regard to the source of fetus-derived stem cells. *Stem Cells* **25**, 2886–2895.
2. Karahuseyinoglu, S., Kocaefe, C., Balci, D., Erdemli, E., and Can, A. (2008) Functional structure of adipocytes differentiated from human umbilical cord stroma-derived stem cells. *Stem Cells* **26**, 682–691.
3. Karahuseyinoglu, S., Cinar, O., Kilic, E., Kara, F., Akay, G. G., Demiralp, D. O., Tukun, A., Uckan, D., and Can, A. (2007) Biology of stem cells in human umbilical cord stroma: in situ and in vitro surveys. *Stem Cells* **25**, 319–331.
4. Baksh, D., Yao, R., and Tuan, R. S. (2007) Comparison of proliferative and multilineage differentiation potential of human mesenchymal stem cells derived from umbilical cord and bone marrow. *Stem Cells* **25**, 1384–1392.
5. Wang, H. S., Hung, S. C., Peng, S. T., Huang, C. C., Wei, H. M., Guo, Y. J., Fu, Y. S., Lai, M. C., and Chen, C. C. (2004) Mesenchymal stem cells in the Wharton's jelly of the human umbilical cord. *Stem Cells* **22**, 1330–1337.
6. Conconi, M. T., Burra, P., Di Liddo, R., Calore, C., Turetta, M., Bellini, S., Bo, P., Nussdorfer, G. G., and Parnigotto, P. P. (2006) CD105(+) cells from Wharton's jelly show in vitro and in vivo myogenic differentiative potential. *Int J Mol Med* **18**, 1089–1096.
7. Lu, L. L., Liu, Y. J., Yang, S. G., Zhao, Q. J., Wang, X., Gong, W., Han, Z. B., Xu, Z. S., Lu, Y. X., Liu, D., Chen, Z. Z., and Han, Z. C. (2006) Isolation and characterization of

human umbilical cord mesenchymal stem cells with hematopoiesis-supportive function and other potentials. *Haematologica* **91**, 1017–1026.

8. Sarugaser, R., Lickorish, D., Baksh, D., Hosseini, M. M., and Davies, J. E. (2005) Human umbilical cord perivascular (HUCPV) cells: a source of mesenchymal progenitors. *Stem Cells* **23**, 220–229.
9. Kadivar, M., Khatami, S., Mortazavi, Y., Shokrgozar, M. A., Taghikhani, M., and Soleimani, M. (2006) In vitro cardiomyogenic potential of human umbilical vein-derived mesenchymal stem cells. *Biochem Biophys Res Commun* **340**, 639–647.
10. Campard, D., Lysy, P. A., Najimi, M., and Sokal, E. M. (2008) Native umbilical cord matrix stem cells express hepatic markers and differentiate into hepatocyte-like cells. *Gastroenterology* **134**, 833–848.
11. Wu, L. F., Wang, N. N., Liu, Y. S., and Wei, X. (2009) Differentiation of Wharton's jelly primitive stromal cells into insulin-producing cells in comparison with bone marrow mesenchymal stem cells. *Tissue Eng Part A* **15**, 2865–2873.
12. Fu, Y. S., Cheng, Y. C., Lin, M. Y., Cheng, H., Chu, P. M., Chou, S. C., Shih, Y. H., Ko, M. H., and Sung, M. S. (2006) Conversion of human umbilical cord mesenchymal stem cells in Wharton's jelly to dopaminergic neurons in vitro: potential therapeutic application for Parkinsonism. *Stem Cells* **24**, 115–124.
13. Fu, Y. S., Shih, Y. T., Cheng, Y. C., and Min, M. Y. (2004) Transformation of human umbilical mesenchymal cells into neurons in vitro. *J Biomed Sci* **11**, 652–660.
14. Ma, L., Feng, X. Y., Cui, B. L., Law, F., Jiang, X. W., Yang, L. Y., Xie, Q. D., and Huang, T. H. (2005) Human umbilical cord Wharton's Jelly-derived mesenchymal stem cells differentiation into nerve-like cells. *Chin Med J (Engl)* **118**, 1987–1993.
15. Weiss, M. L., Medicetty, S., Bledsoe, A. R., Rachakatla, R. S., Choi, M., Merchav, S., Luo, Y., Rao, M. S., Velagaleti, G., and Troyer, D. (2006) Human umbilical cord matrix stem cells: preliminary characterization and effect of transplantation in a rodent model of Parkinson's disease. *Stem Cells* **24**, 781–792.
16. Mitchell, K. E., Weiss, M. L., Mitchell, B. M., Martin, P., Davis, D., Morales, L., Helwig, B., Beerenstrauch, M., Abou-Easa, K., Hildreth, T., Troyer, D., and Medicetty, S. (2003) Matrix cells from Wharton's jelly form neurons and glia. *Stem Cells* **21**, 50–60.
17. Friedman, R., Betancur, M., Boissel, L., Tuncer, H., Cetrulo, C., and Klingemann, H. (2007) Umbilical cord mesenchymal stem cells: adjuvants for human cell transplantation. *Biol Blood Marrow Transplant* **13**, 1477–1486.
18. Fong, C. Y., Richards, M., Manasi, N., Biswas, A., and Bongso, A. (2007) Comparative growth behaviour and characterization of stem cells from human Wharton's jelly. *Reprod Biomed Online* **15**, 708–718.
19. Li, G., Zhang, X. A., Wang, H., Wang, X., Meng, C. L., Chan, C. Y., Yew, D. T., Tsang, K. S., Li, K., Tsai, S. N., Ngai, S. M., Han, Z. C., Lin, M. C., He, M. L., and Kung, H. F. (2009) Comparative proteomic analysis of mesenchymal stem cells derived from human bone marrow, umbilical cord, and placenta: implication in the migration. *Proteomics* **9**, 20–30.

Chapter 6

The Isolation and Culture of Human Cord Blood-Derived Mesenchymal Stem Cells Under Low Oxygen Conditions

Anita Laitinen, Johanna Nystedt, and Saara Laitinen

Abstract

There is growing evidence that low oxygen conditions are beneficial for in vitro stem cell culturing. Mimicking the physiological oxygen tension of the placental stem cell niche in cell expansion can ultimately result in more robust cell expansion. Growing evidence also suggests that hypoxic preconditioning of cells may improve therapeutic outcomes. Here we describe a scalable method that enables mesenchymal stromal cell expansion from virtually every cord blood unit, including those that would normally be disqualified from banking. In addition, the cells obtained by the described method fulfill exclusively the mesenchymal stromal cell characteristics defined by the *International Society for Cellular Therapy*.

Key words: Low-oxygen, Hypoxia, Normoxia, Mesenchymal stem cell

1. Introduction

Mesenchymal stem or stromal cells (MSCs) are adult stem cells that can be derived from many human tissues including the bone marrow, adipose tissue, and cord blood. Both stem cell research and therapy require a large amount of cells and therefore efficient expansion methods are being sought after by many laboratories. Since the frequency of cord blood stromal cells is very low, improved isolation and expansion methods are needed. Stem cell cultures are traditionally performed in 20–21% oxygen, a condition far from their true physiological environment. Most of the cells in our bodies are never exposed to such high levels of oxygen, and several types of human cells in fact produce their energy in a near anoxic environment (1, 2).

Excess supply of oxygen in isolated stem cells has been shown to alter energy metabolism, change differentiation potential, and

Mohan C. Vemuri et al. (eds.), *Mesenchymal Stem Cell Assays and Applications*, Methods in Molecular Biology, vol. 698, DOI 10.1007/978-1-60761-999-4_6, © Springer Science+Business Media, LLC 2011

influence surface protein properties (3, 4). By contrast, hypoxia has been shown to increase the life span and proliferation of stem cells (5, 6). A widely accepted statement is that an enhanced life span of cells cultured in low rather than atmospheric oxygen tension is due to decreased ROS production (7). As a consequence, reduced oxygen tension leads to reduced DNA damage and lipid peroxidation, both of which are harmful for cells (8). Surprisingly, in contrast to the described results, it has been also shown that hypoxia increases ROS production (9, 10).

Importantly, culture conditions more closely mimicking physiologic conditions have shown to enhance survival of therapeutic cells when injected intravenously (11), and hypoxic preconditioning of stem cells enhances migration to biologically relevant niches (12). In both cases, the proposed mechanism is the change of surface receptor expression important in chemotaxis.

Growth stimulating factors, like platelet-derived growth factor (PDGF) (13), epidermal growth factor (EGF) (14), dexamethasone (DEXA) (15), and other cytokines (16), have been shown to enhance the proliferation of MSCs in vitro. These and other factors are under extensive studies as well as the development of better defined and serum-free media (17).

It has been previously concluded that the success rate for cord blood-derived MSCs lies between 20 and 30% (18). Here we describe a method that increases the success rate to above 90%, and the cord blood-derived MSCs obtained by this method fulfill the ISCT minimum characterization criteria for MSCs (19). Moreover, the described protocol enables efficient use of small units typically not acceptable for banking. We have established that the key to acquiring MSCs from cord blood is the use of optimal oxygen conditions combined with the knowledge of proper MSC cell surface structure characteristics.

2. Materials

2.1. Fibronectin Coating of 6-Well Plates

1. Sterile phosphate-buffered saline (PBS): 1.9 mM NaH_2PO_4, 8.1 mM Na_2HPO_4, 154 mM NaCl, pH 7.2. Sterile filter or autoclave. Store at 4°C.
2. Fibronectin (FN, Sigma-Aldrich, St. Louis, MO) is dissolved at 1 mg/ml in sterile water. Store in aliquots at –20°C. Prepare always a fresh working solution (10 ng/ml) by further diluting in PBS.
3. 6-Well plates (Nunc, Roskilde, Denmark).

2.2. Isolation of Mononuclear Cells

1. Anticoagulated cord blood collected ex-utero by CB-Collect, double blood bag 300/20 ml for cord blood

(collection methods accredited by FACT-Netcord) (see Note 1).

2. Autoclaved 500 ml glass bottle (e.g., Schott Duran, Duran Group GmbH, Meinz, Germany).
3. 18G Needle (e.g., Terumo, Leuven, Belgium).
4. 50 ml Syringe (e.g., Terumo, Leuven, Belgium).
5. 50 ml Centrifuge tubes (e.g., Greiner Bio-One, Frickenhausen, Germany).
6. 25 ml Pipettes (e.g., Falcon, BD, Franklin Lakes, NJ).
7. PBS/EDTA: 2 mM disodium EDTA in PBS. Store at 4°C.
8. Ficoll-Paque Plus (GE Healthcare, Uppsala, Sweden).
9. Sterile pasteur pipettes (Scherf Praezision Europa, Kimberle & Chase Life Science, Germany).
10. Bürker chamber.

2.3. Cell Culture

1. FN-coated 6-well plates. Store at 4°C for up to 2 weeks.
2. EGF (Sigma-Aldrich) is dissolved at 10 μg/ml in 10 mM acetic acid containing 0.1% BSA (Sigma-Aldrich). Store in aliquots at −20°C.
3. PDGF-BB (R&D Systems, Minneapolis, MN) is dissolved at 10 μg/ml in 4 mM HCl containing 0.1% BSA (Sigma-Aldrich). Store in aliquots at −20°C.
4. DEXA (Sigma-Aldrich) is dissolved at 1 mg/ml in absolute ethanol and then further diluted to 20 μg/ml in sterile medium (e.g., Minimum Essential Medium (MEM)-α, Invitrogen, Paisley, UK). Store in aliquots at −20°C.
5. Proliferation medium: MEM-α Medium with Glutamax (Invitrogen) supplemented with 10% fetal bovine serum (FBS, Invitrogen, Grand Island, NY) (see Note 2), 50 nM DEXA (Sigma-Aldrich), 10 ng/ml EGF (Sigma-Aldrich), 10 ng/ml PDGF (R&D Systems), and 1× Penicillin–Streptomycin (Invitrogen). Store at 4°C for up to 1 week (see Table 1).
6. PBS. Store at 4°C.
7. Trypsin-EDTA solution (0.25%, 1 mM, Invitrogen). Store in aliquots at −20°C.
8. Tissue culture-treated plates (Nunc).
9. Freezing medium: 50% proliferation medium, 40% FBS (Invitrogen), 10% dimethyl sulfoxide (DMSO, Sigma-Aldrich). Make fresh before each use.
10. Cryovials (Nunc).
11. Cell freezing container (Nalgene, USA).

Table 1
Medium composition

Reagent	Volume/100 ml	Stock concentration	Final concentration	Manufacturer/ cat. no.
MEM-α with Glutamax	88.7 ml	1×	1×	Invitrogen 32571
MSC-qualified FBS	10 ml	100%	10%	Invitrogen 16000-044
Dexamethasone (DEXA)	100 μl	50 μM	50 nM	Sigma D8893
Epidermal growth factor (EGF)	100 μl	10 μg/ml	10 ng/ml	Sigma E9644
rhPDGF-BB	100 μl	10 μg/ml	10 ng/ml	R&D Systems 220-BB
Penicillin–Streptomycin	1 ml	Pen 1,000 U/ml, Strep 1,000 μg/ml	Pen 100 U/ml, Strep 100 μg/ml	Invitrogen 15140-122

3. Methods

3.1. Fibronectin Coating of Culture Plates

1. Source at least 6 × 6-well plates (Nunc) (see Note 3).
2. Prepare at least 36 ml of 10 ng/ml FN (Sigma-Aldrich) in PBS solution.
3. Pipet 1 ml of the diluted FN solution to each well of the 6-well plates and place plates into a 37°C incubator for at least 2 h.
4. Aspirate the FN solution before use or before storing.
5. Store excess FN-coated plates in an air tight container for up to 2 weeks at 4°C.

3.2. Isolation of Mononuclear Cells

1. Transfer cord blood from the plastic collection bag into a glass bottle. Using an 18G needle attached to a 50 ml syringe, transfer blood from the bag to the glass bottle. Remove the needle from the syringe before emptying into the glass bottle to avoid excess shear stress to the cells.
2. Determine the blood volume and dilute the blood 1:2–1:4 with PBS/EDTA (see Note 4).
3. Place 15 ml Ficoll-Paque (GE Healthcare) solution into 50 ml centrifuge tubes. Prepare 1 tube for every 33 ml or less of diluted cord blood.

4. Using 25 ml pipet, carefully layer the diluted cord blood on top of the Ficoll-Paque solution. Drain the diluted blood slowly down the wall of the tube to avoid mixing with the Ficoll-Paque layer.
5. Place the tubes in a centrifuge (with swinging-bucket rotor) and spin for 40 min at 400 × *g*, 18–20°C with the brake off.
6. Collect the mononuclear cells from the interface of plasma and Ficoll-Paque layers using a Pasteur pipet. Take special care to harvest cells from the walls of the tubes as cells may adhere to the plastic tube. Transfer and pool harvested cells from two to three mononuclear cell layers into a new 50 ml centrifuge tube (see Note 5).
7. Wash the cells with PBS/EDTA solution. Fill up the tubes with PBS/EDTA and centrifuge for 10 min at 300 × *g*. Remove the supernatant by aspiration. The supernatant should not be aspirated too closely as the cells may have formed a loose pellet at the time of washing.
8. Resuspend the cell pellet(s) into a small volume of PBS/EDTA and transfer all suspension into a single 50 ml centrifuge tube. Fill the tube up to 50 ml with PBS/EDTA and evenly resuspend the solution by pipetting up and down.
9. Count resuspended cells in a Bürker chamber or using a preferred counting method.
10. Centrifuge the cells at 300 × *g* for 10 min.

3.3. Plating and Culturing MSCs

1. Prepare the FN-coated culture plates and proliferation medium before starting. Prewarm proliferation medium to 37°C in a water bath.
2. Take the washed mononuclear cells (see Subheading 3.2, step 10) and resuspend the cells in an appropriate volume of proliferation medium (see Note 6).
3. Plate the cells in the culture plates. Pipet 3 ml of the cell suspension evenly into each well of the 6-well plate. Try not to shake the plates as the cells in the suspension may clump together and settle unevenly.
4. Place the culture plates into a humidified incubator under the following conditions: 5% CO_2, 3% O_2 at 37°C (see Note 7).
5. The next day, remove the entire medium and replace with fresh, prewarmed proliferation medium (see Note 8). Take special care to remove the reddish sediment. This is easily achieved by gently whirling the plates in a circular motion, causing the nonadherent settled cells to float in the medium (see Note 9).
6. Monitor the wells every second day and change the medium twice per week by aspirating the entire medium and replacing with fresh, prewarmed proliferation medium.

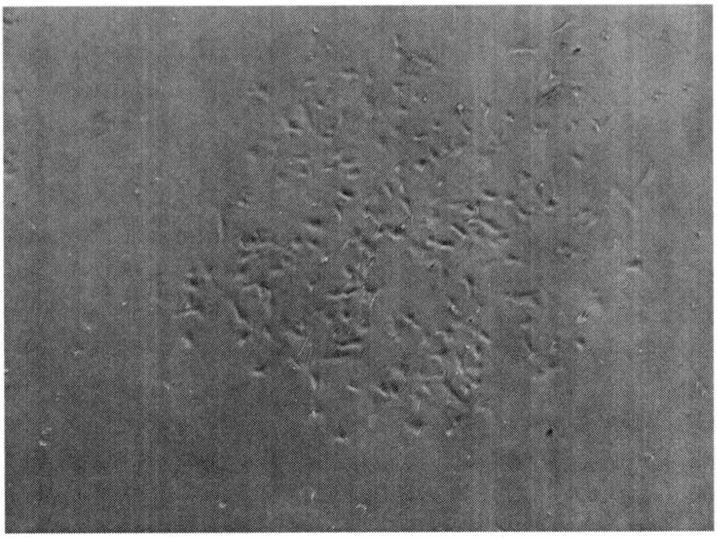

Fig. 1. Example of an UCB-MSC colony visualized by 40-fold magnification.

7. When the adherent cells form a small colony as seen in Fig. 1, it will be necessary to harvest and replate the culture into a new plate. Do not let colonies grow too dense (see Note 10).
8. Aspirate growth medium and wash the desired wells with 1 ml of PBS.
9. Aspirate the PBS and add 150 μl trypsin-EDTA solution to each well being harvested and incubate at room temperature for approximately 5 min. Observe the wells under a microscope while detaching the cells to monitor when the cells are completely detached. The plate can be shaken gently to facilitate cell detachment. Do not allow cells to remain in trypsin solution too long as this may harm the cells (see Note 11).
10. Add 1 ml of proliferation medium into the well containing trypsin-EDTA (Invitrogen) and pipet the medium gently up and down.
11. Transfer the cells into a new culture plate (Nunc) without FN coating (see Note 12). Reseed the cells at density of 1,000 cells/cm^2 if possible to count them. It is not important to know the exact number of the cells at this time. Therefore, if one small colony is replated, the cells can be transferred into one new well of a 6-well plate or smaller. It is important to ensure that the cells will not be too close to each other and will have enough space to proliferate. Return primary culture plates with remaining unharvested wells to the incubator (see Notes 13 and 14).
12. Add an appropriate amount of fresh medium into the reseeded wells and place the cells back into the incubator.

13. Monitor growth of the reseeded cells (see Note 15). Take care also to monitor the remaining primary culture wells as some colonies may appear later.
14. Change medium every 3–4 days with prewarmed complete proliferation medium.
15. Once cultures reach ≤80% confluence, wash the well(s) with PBS and harvest by adding an appropriate amount of trypsin-EDTA (Invitrogen) solution (1.5 ml for 10 cm plates, 150 μl for 6-well, 100 μl for 12-well, and 50 μl for 24-well plate/well). Incubate at room temperature until the cells are completely detached (usually less than 5 min).
16. Add appropriate amount of fresh medium to the well(s) (3 ml for 10 cm plates, 1.5 ml for 6-well, 1 ml for 12-well, and 0.5 ml for 24-well plate wells).
17. Transfer the cells into a 50 ml centrifuge tube and centrifuge at 300 × *g* for 4 min.
18. Resuspend the cells in small volume of proliferation medium and count using a Bürker chamber or a preferred counting method.
19. Determine the surface area required for reseeded cells at a density of 1,000 cells/cm^2 and source appropriate noncoated tissue culture plates (Nunc).
20. Dilute resuspended cells in an appropriate volume of prewarmed proliferation medium and add to culture plates.
21. Return cultures to the incubator and provide fresh medium twice per week.
22. Continue propagating cultures until you have enough cells for cryopreservation or other downstream applications (see Note 16).
23. The ability of expanded MSCs to differentiate into chondrocytes, adipocytes, and osteoblasts as well as their surface phenotype should be tested as these are part of the defining minimum criteria established for human mesenchymal stromal cells (19). Please refer to other chapters within this book for the necessary characterization and differentiation protocols.

3.4. Cryopreservation and Thawing of MSCs

1. Prepare fresh freezing medium and keep in a fridge or on ice until needed. Source an appropriate number of cryovials (Nunc) and place in the fridge or on ice to chill them before use.
2. Harvest expanded MSCs and count using a Bürker chamber or a preferred counting method (see Subheading 3.3, steps 15–18).
3. Centrifuge the cells at 300 × *g* for 4 min and aspirate the supernatant.

4. Gently resuspend cells in prechilled freezing medium at a density of 10^6 cells/ml.
5. Pipet aliquots of resuspended cells into each prechilled cryovial. The number of cells per cryovial depends on the desired downstream use of cryopreserved cells. A generally recommended number of cells per cryovial is 5×10^5 (0.5 ml), but $0.3–1.5 \times 10^6$ cells/vial may be used.
6. Place the cryovials into a cell freezing container and place it in −70°C freezer.
7. The next day, transfer the cryovials into liquid nitrogen tank for long-term storage.
8. When thawing the cells, source cryovials from the liquid nitrogen and let the vials sit at room temperature for about 1 min. Immediately place the vial into a 37°C water bath and allow the cells to thaw until the frozen cell pellet is almost completely melted.
9. Transfer the vial to the flow hood and place the entire volume of cells into a centrifuge tube containing 5 ml of prewarmed proliferation medium.
10. Centrifuge the tube at $300 \times g$ for 4 min and aspirate the supernatant (see Note 17).
11. Resuspend the cells into appropriate amount of medium and plate the cells on tissue culture-treated plates (see Note 18).

4. Notes

1. The volume of the cord blood should be at least 40 ml. We have not down-scaled this method for smaller units.
2. Use serum that is tested for MSCs since all sera do not support the growth of these cells.
3. The number of needed plates depends on the number of mononuclear cells. Since the cell content in cord blood units varies, the number of mononuclear cells cannot be predicted from the volume of the blood. It is better to prepare an excess number of plates and store them for later use as the preparation of the plates takes over 2 h.
4. More dilute blood will provide better result as the dilution of the blood reduces aggregate formation and thus ensures better mononuclear fraction collection from the density gradient. Remember to prewarm all needed reagents to room temperature before use since the temperature range of 18–20°C is optimal for mononuclear cell isolation.

5. As cord blood contains a significant population of nucleated erythroid progenitors, the mononuclear cell layer will contain these red cells among other mononuclear cells. These cells will not affect the culture of MSCs.

6. The amount of medium depends on the cell number and the type of culture plates that are used. The cells should be plated at a density of 10^6 cells/cm^2. If 6-well plates are used, the cell density in the suspension should be 3.2×10^6 cells/ml as the area of each well of a 6-well plate is 9.6 cm^2 and the culturing volume is 3 ml.

7. It is important to use an incubator that has a capacity to regulate the oxygen concentration. The low oxygen condition is more close to the physiological oxygen conditions and thus supports the growth of MSCs. There is evidence that MSCs grow and form colonies better in low oxygen conditions (20). Low oxygen has also been shown to induce expression of fibronectin and integrin $\alpha 5$ $\beta 1$ (21), the cell surface receptor that is known to bind fibronectin. Open the doors of the hypoxic incubator as little as possible since the gases mix very quickly and will cause the level of O_2 to rise.

8. The cultures may look very reddish at this time as the erythroid progenitors are floating in the medium. These progenitors as well as the other nonadherent cells will be removed by subsequent medium changes.

9. Do not wash the plates at this time as the desired cells may have adhered only lightly and are easily washed away.

10. The growth of MSCs should be seen within 3 weeks. Screen the wells of the 6-well plates under a microscope systematically for adherent cells. When adherent spindle-shaped cells are seen, mark the wells and keep monitoring these cells every day.

11. Some cell colonies formed in culture may not detach from the plates with trypsin-EDTA solution. This may be indicative that the cells are not MSCs and may be cells of a monocyte lineage. In addition, some expanded cell colonies will indeed detach similar to MSCs but will stop proliferating after propagation. These may represent endothelial cell colonies or differentiated MSC populations.

12. The fibronectin coating of the culture plates is not necessary after primary isolation of the cells.

13. If not all of the cells appeared to detached during trypsin-EDTA incubation, it is advisable to add fresh, prewarmed proliferation medium into the well where the cells were detached and place back into the incubator. These wells will serve as backup cultures.

14. If there are several small colonies growing in one well, they might need a larger growth area. Also if there are several colonies growing in different wells, the harvested cells can be combined and the cell number can be counted as in the following passages.
15. When the passaged cells have started to proliferate, take care that the cells do not grow too dense as this may affect the cells' capacity to proliferate and may cause spontaneous differentiation.
16. Cells seeded at 1,000 cells/cm^2 will typically require propagation twice per week. Cells will proliferate quicker in early passages and will gradually slow in later passages.
17. This centrifugation step is necessary to remove the residual DMSO that is harmful for the recovered cells.
18. It may be beneficial for the recovered cells to be seeded at a slightly higher density after thawing (e.g., 2,000 cells/cm^2).

References

1. Erecinska M, Silver IA. Tissue oxygen tension and brain sensitivity to hypoxia. Respir Physiol 2001;128:263–76.
2. Papandreou I, Powell A, Lim AL, Denko N. Cellular reaction to hypoxia: sensing and responding to an adverse environment. Mutat Res 2005;569:87–100.
3. Ivanovic Z. Hypoxia or in situ normoxia: the stem cell paradigm. J Cell Physiol 2009; 219:271–5.
4. Hung SC, Pochampally RR, Hsu SC, et al. Short-term exposure of multipotent stromal cells to low oxygen increases their expression of CX3CR1 and CXCR4 and their engraftment in vivo. PLoS One 2007;2:e416.
5. Bell EL, Klimova TA, Eisenbart J, Schumacker PT, Chandel NS. Mitochondrial reactive oxygen species trigger hypoxia-inducible factor-dependent extension of the replicative life span during hypoxia. Mol Cell Biol 2007; 27:5737–45.
6. Carrancio S, Lopez-Holgado N, Sanchez-Guijo FM, et al. Optimization of mesenchymal stem cell expansion procedures by cell separation and culture conditions modification. Exp Hematol 2008;36:1014–21.
7. Michelakis ED, Rebeyka I, Wu X, et al. O_2 sensing in the human ductus arteriosus: regulation of voltage-gated K+ channels in smooth muscle cells by a mitochondrial redox sensor. Circ Res 2002;91:478–86.
8. Bertram C, Hass R. Cellular responses to reactive oxygen species-induced DNA damage and aging. Biol Chem 2008;389:211–20.
9. Duranteau J, Chandel NS, Kulisz A, Shao Z, Schumacker PT. Intracellular signaling by reactive oxygen species during hypoxia in cardiomyocytes. J Biol Chem 1998;273:11619–24.
10. Guzy RD, Schumacker PT. Oxygen sensing by mitochondria at complex III: the paradox of increased reactive oxygen species during hypoxia. Exp Physiol 2006;91:807–19.
11. Tang YL, Zhu W, Cheng M, et al. Hypoxic preconditioning enhances the benefit of cardiac progenitor cell therapy for treatment of myocardial infarction by inducing CXCR4 expression. Circ Res 2009;104:1209–16.
12. Rosova I, Dao M, Capoccia B, Link D, Nolta JA. Hypoxic preconditioning results in increased motility and improved therapeutic potential of human mesenchymal stem cells. Stem Cells 2008;26:2173–82.
13. Muller I, Kordowich S, Holzwarth C, et al. Animal serum-free culture conditions for isolation and expansion of multipotent mesenchymal stromal cells from human BM. Cytotherapy 2006;8:437–44.
14. Krampera M, Pasini A, Rigo A, et al. HB-EGF/HER-1 signaling in bone marrow mesenchymal stem cells: inducing cell expansion and reversibly preventing multilineage differentiation. Blood 2005;106:59–66.

15. Hong L, Sultana H, Paulius K, Zhang G. Steroid regulation of proliferation and osteogenic differentiation of bone marrow stromal cells: a gender difference. J Steroid Biochem Mol Biol 2009;114:180–5.
16. Fan X, Liu T, Liu Y, Ma X, Cui Z. Optimization of primary culture condition for mesenchymal stem cells derived from umbilical cord blood with factorial design. Biotechnol Prog 2009;25:499–507.
17. Chase LG, Firpo MT. Development of serum-free culture systems for human embryonic stem cells. Curr Opin Chem Biol 2007;11:367–72.
18. Reddy NP, Vemuri MC, Pallu R. Isolation of stem cells from human umbilical cord blood. Methods Mol Biol 2007;407:149–63.
19. Dominici M, Le Blanc K, Mueller I, et al. Minimal criteria for defining multipotent mesenchymal stromal cells. The International Society for Cellular Therapy position statement. Cytotherapy 2006;8:315–7.
20. Grayson WL, Zhao F, Izadpanah R, Bunnell B, Ma T. Effects of hypoxia on human mesenchymal stem cell expansion and plasticity in 3D constructs. J Cell Physiol 2006;207:331–9.
21. Milner R, Hung S, Erokwu B, Dore-Duffy P, LaManna JC, del Zoppo GJ. Increased expression of fibronectin and the alpha 5 beta 1 integrin in angiogenic cerebral blood vessels of mice subject to hypobaric hypoxia. Mol Cell Neurosci 2008;38:43–52.

Chapter 7

Amniotic and Placental Mesenchymal Stem Cell Isolation and Culture

Justin D. Klein and Dario O. Fauza

Abstract

The amniotic fluid and placenta are sources of diverse progenitor cell populations, including mesenchymal, hematopoietic, trophoblastic, and possibly more primitive stem cells. Given that much of the amniotic cavity and placenta share a common origin, namely the inner cell mass of the morula, perhaps it is not surprising that most types of progenitor cells that can be isolated from these two sources also share many characteristics. This chapter focuses solely on the most abundant and easy to isolate progenitor cell population found therein, the mesenchymal stem cells (MSCs). Unlike some of the other stem cell types, MSCs are present throughout gestation. Methods of isolation, expansion, freezing, and thawing of these cells will be presented with preference given to the simplest methods available for any given procedure.

Key words: Mesenchymal stem cells, Stem cells, Amniotic stem cells, Amniocytes, Mesenchymal amniocytes, Amniotic fluid, Placenta, Fetus, Perinatal, Tissue engineering, Fetal tissue engineering

1. Introduction

Mesenchymal stem cells (MSCs) have been thoroughly studied for over 40 years, ever since Friedenstein's initial description in 1966 of colony-forming, spindle-shaped stromal cells with multipotent differentiation potential isolated from bone marrow (1–3). The plasticity, self-renewal, and multi-lineage potential of MSCs have generated growing interest in their use in a constantly expanding variety of experimental regenerative therapy applications (4, 5).

A number of reports have identified diverse sources of MSCs; however, most would not be compatible with perinatal cell-based therapies (6). In the perinatal setting, the amniotic fluid and placenta have emerged as the prime sources of MSCs. Ethical objections to the isolation of amniotic and placental MSCs should be obviated by the fact that amniocentesis and/or chorionic villus

Mohan C. Vemuri et al. (eds.), *Mesenchymal Stem Cell Assays and Applications*, Methods in Molecular Biology, vol. 698, DOI 10.1007/978-1-60761-999-4_7, © Springer Science+Business Media, LLC 2011

sampling are the widely performed screening diagnostic procedures. Amniocentesis is the safest of any invasive prenatal diagnostic method, as it is associated with a spontaneous abortion rate of less than 0.5% (7). Additionally, when a congenital anomaly is detected on prenatal imaging, the mother is routinely offered a diagnostic amniocentesis to rule out the presence of chromosomal abnormalities. The procurement of a small extra aliquot of fluid during the procedure, for the isolation of amniotic MSCs (aMSCs) to be used in tissue engineering, gene therapy, cell transplantation, or other cell-based purposes, would essentially pose no additional risk to the mother or fetus, hence it would be ethically unobjectionable.

The amniotic cavity and placenta share a common origin, namely the inner cell mass of the morula, which gives rise to the embryo, yolk sac, mesenchymal core of the chorionic villi, chorion, and amnion. Likely because of their shared origin, both the amniotic fluid and the placenta contain a similarly heterogeneous population of progenitor cells, which includes mesenchymal, hematopoietic, trophoblastic, and, perhaps, more primitive stem cells.

Cells from all three germ layers have been identified in the amniotic fluid for over 30 years (8–11). Yet, the isolation of progenitor cells from the amniotic fluid was first reported only in 1993, when small, nucleated, round cells identified as hematopoietic progenitor cells were found therein before week 12 of gestation (12). The multi-lineage potential of nonhematopoietic stem cells was first suggested in 1996 by the demonstration of myogenic conversion of select amniocytes (13). However, the proper characterization, differentiation potential, and foreseeable therapeutic applications of mesenchymal amniocytes have started to be determined only quite recently (14–20).

Throughout gestation, the cellular profile of the amniotic fluid changes predictably, as the amniotic cavity receives cells from the fetus and possibly from the placenta as well (21). The cell types present in the amniotic cavity are also influenced by the mechanisms responsible for the production and turnover of the amniotic fluid itself. Certain fetal pathologic states (e.g., neural tube and body wall defects) may lead to changes in the amniotic cell population profile. Although the chemical and cellular composition of the amniotic fluid varies with gestational age and fetal health status, MSCs can be consistently isolated from it at any time during pregnancy.

Placental and amniotic fluid MSCs have been shown to differentiate into most cell types of mesodermal lineage, as well as a few cells of ectodermal and endodermal lineages (22). It is likely that their full spectrum of differentiation remains to be entirely defined. Because of their unique differentiation potential and ease of access, much interest has been generated of late on the prospective clinical value of amniotic fluid and placental MSCs in regenerative therapies. One example of such an application would be the use of these cells for fetal tissue engineering strategies aimed at the surgical repair of congenital anomalies in the perinatal period. Translated into clinical context, from a simple amniocentesis, grafts could be engineered in parallel to the remainder of gestation so that a newborn, or even a fetus, could benefit from having autologous tissue promptly available for surgical reconstruction, either before or at birth. Thus far, we have validated this concept in large animal models of diaphragmatic, tracheal, and sternal repairs (14, 18, 19, 23). As a prerequisite to eventual clinical trials of amniotic MSC-based therapies, we have also shown that human aMSCs can be isolated and expanded in the absence of animal products and that they can be submitted to staged cell manufacturing (24, 25).

This chapter describes deliberately simple and easily reproducible methodologies for isolation, expansion, and storage of amniotic fluid and placental MSCs. Isolation of these MSCs is achieved through a basic combination of natural selection by the culture medium and mechanical separation. Two means of implementing the mechanical separation are presented for both amniotic fluid and placental samples, namely either the use of multiple wells (see Subheadings 3.2 and 3.4) or of coverslips (see Subheadings 3.3 and 3.5). The basic methods for passaging (see Subheading 3.6) and freezing/thawing (see Subheading 3.7) these cells are independent to whether they were isolated from a placental or amniotic fluid source. Finally, it must be noted that, in addition to the straightforward methodologies described here, other methods that rely on oxygen tension, two-stage cultures, and alternative media formulations have also been reported (22, 26).

2. Materials

2.1. Preparation of Mesenchymal Stem Cell Medium

1. Dulbecco's Modified Eagle's Medium (DMEM) containing L-glutamine (Sigma, St. Louis, MO) (see Note 1).
2. Mesenchymal Stem Cell (MSC) – qualified fetal bovine serum (FBS) (Invitrogen, Carlsbad, CA), stored in 50 mL single use aliquots at –20°C (see Note 2).
3. 10,000 U/mL penicillin G sodium/10 mg/mL streptomycin sulfate antibiotic solution (Sigma).
4. rhFGF-basic stock (Promega, Madison, WI), 25 μg recombinant human-basic-fibroblast growth factor; reconstituted in 500 μL sterile water. Split into 50-μL aliquots and stored at –20°C for up to 1 year (see Note 3).

2.2. Isolation of Mesenchymal Stem Cells from Amniotic Fluid

1. Amniotic fluid.
2. Mesenchymal-20 medium (see Subheading 3.1).
3. Dulbecco's phosphate-buffered saline (cation-free; CMF-DPBS; Invitrogen).

4. 0.025% (w/v) trypsin/0.04% (w/v) EDTA (Invitrogen).
5. Conical centrifuge tubes (15- and 50-mL) (BD Biosciences, San Jose, CA).
6. Centrifuge.
7. 6-Well culture plate precoated with collagen type I (BD Biosciences).
8. Tissue culture dishes (10-cm) (BD Biosciences).
9. Inverted microscope.

2.3. Isolation of Mesenchymal Stem Cells from Amniotic Fluid Using Coverslips

1. Amniotic fluid.
2. Mesenchymal-20 medium (see Subheading 3.1).
3. Dulbecco's phosphate-buffered saline (cation-free; CMF-DPBS; Invitrogen).
4. 0.025% (w/v) trypsin/0.04% (w/v) EDTA (Invitrogen).
5. Conical centrifuge tubes (15- and 50-mL) (BD Biosciences).
6. Centrifuge.
7. 6-Well culture plate precoated with collagen type I (BD Biosciences).
8. Tissue culture dishes (10-cm) (BD Biosciences).
9. Inverted microscope.
10. Coverslips (5-mm^2) (Allegiance, McGaw Park, IL).

2.4. Isolation of Mesenchymal Stem Cells from Placenta

1. Placental sample.
2. 10% Type 2 collagenase (Worthington Biochemical, Lakewood, NJ).
3. Dispase II (Roche Applied Science, Indianapolis, IN).
4. $CaCl_2$ (Sigma).
5. Mesenchymal-20 medium (see Subheading 3.1).
6. Dulbecco's phosphate-buffered saline (cation-free; CMF-DPBS; GIBCO).
7. 0.025% (w/v) trypsin/0.04% (w/v) EDTA (Invitrogen).
8. Serum-free DMEM (Sigma).
9. Conical centrifuge tubes (15- or 50-mL) (BD Biosciences).
10. 100-μm Mesh (Thermo Fisher Scientific, Waltham, MA).
11. Centrifuge.
12. 6-Well culture plate precoated with collagen type I (BD Biosciences).
13. Inverted microscope.
14. Tissue culture dishes (10-cm) (BD Biosciences).

2.5. Isolation of Mesenchymal Stem Cells from Placenta Using Coverslips

1. Placental sample.
2. 10% Type 2 collagenase (Worthington Biochemical).
3. Dispase II (Roche Applied Science).
4. $CaCl_2$ (Sigma).
5. Mesenchymal-20 medium (see Subheading 3.1).
6. Dulbecco's phosphate-buffered saline (cation-free; CMF-DPBS; GIBCO).
7. 0.025% (w/v) trypsin/0.04% (w/v) EDTA (Invitrogen).
8. Serum-free DMEM (Sigma).
9. Conical centrifuge tubes (15- or 50-mL) (BD Biosciences).
10. 100-μm Mesh (Thermo Fisher Scientific).
11. Centrifuge.
12. 6-Well culture plate precoated with collagen type I (BD Biosciences).
13. Inverted microscope.
14. Tissue culture dishes (10-cm) (BD Biosciences).
15. Coverslips (5-mm^2) (Allegiance).

2.6. Expansion of Mesenchymal Stem Cells from Amniotic Fluid of Placenta

1. MSC in culture at 80–90% confluence.
2. Dulbecco's phosphate-buffered saline (cation-free; CMF-DPBS; Invitrogen).
3. 0.025% (w/v) trypsin/0.04% (w/v) EDTA (Invitrogen).
4. Mesenchymal-10 medium (see Subheading 3.1).
5. Inverted microscope.
6. 10-cm Tissue culture dishes (BD Biosciences).

2.7. Freezing and Thawing Mesenchymal Stem Cells

1. MSC in culture, ~50% confluent.
2. Dulbecco's phosphate-buffered saline (cation-free; CMF-DPBS; Invitrogen).
3. 0.025% (w/v) trypsin/0.04% (w/v) EDTA (Invitrogen).
4. Mesenchymal-10 medium (see Subheading 3.1).
5. Freezing solution: Mesenchymal-10 medium supplemented with 20% DMSO (Sigma).
6. Pasteur pipette.
7. Inverted microscope.
8. Conical centrifuge tube (15-mL) (BD Biosciences).
9. Centrifuge.
10. Cryogenic vials (2.0 mL) (Fisher Scientific).
11. –80°C freezer.

12. 37°C water bath.
13. Tissue culture dishes (10-cm) (BD Biosciences).

3. Methods

Ethics approval for the described protocols is usually required from the appropriate institutional research office, typically an Institutional Review Board (IRB) for human cells or Institutional Animal Care and Use Committee (IACUC) for animal cells. All experiments should be performed under sterile conditions in either Class II biological hazard flow hoods or laminar flow horizontal draft hoods. When working with human cells, Class II biological hazard flow hoods are recommended. Appropriate sterilization measures for reagents and containers are a necessity. On the spectrum of cell culture, although amniotic MSCs are fairly hardy cell lines, our experience has shown that strict adherence to sterile techniques is a must.

When first performing MSC isolation, it is prudent to confirm MSC identity. Such steps in MSC characterization are outside the scope of this particular chapter and are covered elsewhere within this text. Human MSCs display a typical pattern of cell surface markers compatible with a multipotent mesenchymal progenitor lineage including CD73 (SH3), CD105 (SH2), CD44, CD29, CD90, CD13, CD10, and CD71. These cells are also positive for human leukocyte antigens A–C and negative for CD45, CD34, CD14, CD19, CD8, CD56, and CD31 (4, 16, 17, 24).

MSC isolation via the protocols in this unit is relatively straightforward and consistently reliable. However, as stated in the introduction, the cellular and chemical context of the amniotic fluid varies with gestational age, which may lead to variations in initial cell density between samples. Thus, there can be no "expected" or "typical" yield of MSCs from the initial isolation to be used as a guide. Regardless, the initial yield is not the rate limiting step because amniotic fluid samples procured during pregnancy can generate sufficient numbers of MSCs within just a few generations. Typically, it takes 2–3 weeks to generate ample amounts of MSCs from a small sample of amniotic fluid or placenta. The amounts of FBS and FGF added to the culture media have a direct impact on cell kinetics.

3.1. Preparation of Mesenchymal Stem Cell Media

1. Thaw one or two 50-mL aliquot(s) of FBS in a 37°C water bath, respective to mesenchymal-10 or mesenchymal-20 recipe (see Note 4). The complete media formulations will hereafter be referred to as mesenchymal-10 or mesenchymal-20 media.
2. Add thawed FBS to 500 mL high-glucose DMEM containing L-glutamine resulting in 10% (mesenchymal-10 medium) or 20% (mesenchymal-20 medium) (v/v) solution.

3. Add 5 mL 10,000 U/mL penicillin G sodium/10 mg/mL streptomycin sulfate antibiotic solution (see Note 5).
4. Thaw and add an entire 50 μL aliquot of prepared rhFGF-basic stock (final concentration 5 ng/mL).
5. Store complete medium up to 3 weeks at 4°C.

3.2. Isolation of Mesenchymal Stem Cells from Amniotic Fluid

1. Human amniotic fluid is typically obtained from diagnostic amniocentesis and occasionally from therapeutic amnioreductions. Conceivably, one could also procure it during delivery. Place 40 mL amniotic fluid in a 50-mL conical centrifuge tube (see Note 6).
2. Proceed with MSC isolation or store amniotic fluid at 4°C for up to 48 h until needed (see Note 7).
3. Centrifuge the amniotic fluid sample for 15 min at 500 ×*g* at room temperature.
4. Aspirate the supernatant and discard. Resuspend the pellet in 6 mL mesenchymal-20 medium.
5. Plate 1 mL of the cell suspension into each well of a 6-well collagen-coated culture plate and incubate for 48 h (see Note 8).
6. After 48 h, aspirate medium and replace with 1 mL of fresh medium per well (see Note 9).
7. Aspirate medium and replace with 1 mL of fresh medium per well every 3 days thereafter.
8. After 7–14 days, inspect individual wells for the predominance of amniocytes with the characteristic "mesenchymal" morphology and the absence of contamination. Select only wells with mesenchymal-like cells for continual expansion (see Note 10).
9. When selected wells reach near confluence, aspirate medium and wash by adding 2 mL CMF-DPBS to each well and aspirate.
10. Add 2 mL of 0.025% trypsin/0.04% EDTA, incubate 2–4 min to facilitate cell detachment. Confirm detachment visually or under an inverted microscope and transfer each cell suspension into a separate 15-mL conical centrifuge tube.
11. Add 8 mL mesenchymal-20 medium to each tube.
12. Centrifuge each tube for 5 min at 400 ×*g* at room temperature.
13. Aspirate the supernatant carefully and resuspend the pellet in 6 mL mesenchymal-20 medium.
14. Dispense 1-mL aliquots of cell suspension into 10-cm tissue-culture dishes containing 9 mL mesenchymal-20 medium.
15. Grow cells to 80–90% confluence (replacing 9–10 mL of the medium every 3 days) and continually expand cultures

(see Subheading 3.6) or grow cells to ~50% confluence and freeze (see Subheading 3.7).

3.3. Isolation of Mesenchymal Stem Cells from Amniotic Fluid Using Coverslips

It is noteworthy that Subheadings 3.2 and 3.4 describe the isolation of MSCs by choosing only wells containing predominantly the characteristic mesenchymal-like cells. Subheadings 3.3 and 3.5 describe an alternative protocol where the mechanical separation of coverslips is used instead of selecting distinct wells.

1. Process the amniotic fluid as in Subheading 3.2, steps 1–4.
2. Distribute as many 5-mm^2 coverslips (other sizes may also be used) as can fit on the surface of a single 10-cm collagen-coated plate.
3. Seed cells on the plate at a density of two to three million cells per 150 cm^2 (see Note 11) (27).
4. After 48 h, inspect each individual coverslip and select those containing the morphologically distinct, mesenchymal-like cells.
5. Place the coverslips into separate 30-cm^2 culture plates containing 6 mL fresh mesenchymal-20 medium.
6. Replace the medium every 3 days and proceed with Subheading 3.2, steps 9–15.

3.4. Isolation of Mesenchymal Stem Cells from Placenta

1. Mechanically remove the maternal decidua by peeling it away from the remainder of the placenta, which contains the chorionic villi (each with a mesenchymal core) and discard (see Note 12).
2. Mince the residual specimen (chorionic villi with mesenchymal core) and transfer into a 15- or 50-mL conical tube (depending on the volume of the sample).
3. Add a solution containing 10% (w/v) type 2 collagenase and 4.0 U dispase II/2.5 mM $CaCl_2$ per liter to cover the tissue (see Note 13) (28).
4. Filter the mixture through 100-μm mesh into a 15-mL conical centrifuge tube (see Note 14).
5. Centrifuge for 15 min at $500 \times g$ at room temperature.
6. Resuspend the pellet in 10 mL mesenchymal-20 medium (see Note 15).
7. Plate 1 mL of the cell suspension into each well of a 6-well, collagen-coated culture plate and place in an incubator (see Note 16).
8. After 48 h, aspirate medium and replace with 1 mL of fresh medium per well (see Note 17).
9. Aspirate and replace with 1 mL of fresh medium per well every 3 days thereafter.

10. After 7–14 days, inspect individual wells for the predominance of cells with the characteristic "mesenchymal" morphology and the absence of contamination. Select only wells with mesenchymal-like cells for continual expansion (see Note 18).
11. When selected wells reach near confluence, aspirate medium and wash by adding 2 mL CMF-DPBS to each well and aspirate.
12. Add 2 mL of 0.025% trypsin/0.04% EDTA and incubate for 2–4 min to facilitate cell detachment. Confirm detachment visually or under an inverted microscope and transfer each cell suspension into a separate 15-mL conical centrifuge tube.
13. Add 8 mL of mesenchymal-20 medium to each tube.
14. Centrifuge each tube for 5 min at $400 \times g$ at room temperature.
15. Aspirate the supernatant carefully and resuspend the pellet in 6 mL mesenchymal-20 medium.
16. Dispense 1-mL aliquots of cell suspension into 10-cm tissue-culture dishes containing 9 mL mesenchymal-20 medium.
17. Grow cells to 80–90% confluence (replacing 9–10 mL of the medium every 3 days) and continually expand cultures (see Subheading 3.6) or grow cells to ~50% confluence and freeze (see Subheading 3.7) (28).

3.5. Isolation of Mesenchymal Stem Cells from Placenta Using Coverslips

1. Process the placenta as in Subheading 3.4, steps 1–6.
2. Distribute as many 5-mm^2 coverslips (other sizes may also be used) as can fit on the surface of a single 10-cm collagen-coated plate.
3. Seed cells on the plate at a density of two to three million cells per 150 cm^2 (see Note 19) (27).
4. After 48 h, inspect each individual coverslip and select those containing the morphologically distinct, mesenchymal-like cells.
5. Place the coverslips into separate 30-cm^2 culture plates containing 6 mL fresh mesenchymal-20 medium.
6. Replace the medium every 3 days and proceed with Subheading 3.4, steps 11–17.

3.6. Expansion of Mesenchymal Stem Cells from Amniotic Fluid or Placenta

1. Upon reaching 80–90% confluence, aspirate medium from cultured cells, using a Pasteur pipette (see Note 20).
2. Wash the plate with 2 mL of either CMF-DPBS or 0.025% trypsin/0.04% EDTA. Apply the solution gently along the side of the dish, swirl, immediately aspirate and discard.

3. Add 3 mL of 0.025% trypsin/0.04% EDTA and return the plate to the incubator for ~2–4 min to allow detachment of the cells.
4. Confirm cell detachment by observing under an inverted microscope.
5. *If splitting cells 2:1*: Load 10-cm tissue culture dishes with 6 mL mesenchymal-10 medium. Add 5 mL mesenchymal-10 medium to the trypsinized plate; mix with gentle repeated pipetting for ~30 s and aliquot 4 mL into each of the two prepared dishes.
6. *If splitting cells 3:1*: Load 10-cm tissue culture dishes with 7 mL mesenchymal-10 medium. Add 6 mL mesenchymal-10 medium to the trypsinized plate, mix with gentle repeated pipetting for ~30 s and dispense 3 mL into each of the three prepared dishes.
7. *If splitting cells 4:1*: Load 10-cm tissue culture dishes with 8 mL mesenchymal-10 medium. Add 5 mL mesenchymal-10 medium to the trypsinized plate, mix with gentle repeated pipetting for ~30 s and dispense 2 mL into each of the four prepared dishes.
8. Return dishes to the incubator. Grow cells to 80–90% confluence (replacing 9–10 mL of the medium every 3 days) and passage by repeating steps 1–7 or grow cells to ~50% confluence and freeze (see Subheading 3.7).

3.7. Freezing Mesenchymal Stem Cells

1. Aspirate medium from a ~50% confluent MSC culture using a Pasteur pipette (see Note 21).
2. Wash the plate with 2 mL of either CMF-DPBS or 0.025% trypsin/0.04% EDTA by applying the solution gently along the side of dish, swirl, and immediately aspirate.
3. Add 3 mL of 0.025% trypsin/0.04% EDTA and return the plate to the incubator for ~2–4 min to allow detachment of the cells.
4. Confirm cell detachment by observing under an inverted microscope.
5. Add 7 mL of mesenchymal-10 medium to neutralize trypsin, bringing total volume to 10 mL, mix with gentle repeated pipetting for ~30 s, and transfer the cell suspension to a 15-mL conical centrifuge tube.
6. Centrifuge for 5 min at 200 × *g* at room temperature.
7. Aspirate the supernatant and discard. Resuspend the pellet in 1.5 mL mesenchymal-10 medium.
8. Add 1.5 mL of freezing solution and mix with gentle repeated pipetting for ~30 s (see Note 22).

9. Dispense 1-mL aliquots into each of three cryogenic vials.
10. Place cryovials in a –80°C freezer and store for up to 1 year (see Note 23).

3.8. Thawing Mesenchymal Stem Cells

1. Remove a cryovial from the –80°C freezer and roll the vial between the hands for 10–15 s until the outside of vial is frost free.
2. Hold the cryogenic vial in a 37°C water bath until the contents are visibly thawed (see Note 24).
3. Immerse the vial in 95% ethanol bath or spray with 70% ethanol to kill microorganisms from the water bath. Air dry the cryovial in a sterile hood.
4. Dispense the vial contents into a 10-cm tissue culture plate containing 9 mL of mesenchymal-10 medium.
5. Incubate at 37°C, 5% CO_2.
6. See Subheading 3.6 for continual cell propagation.

4. Notes

1. Prepared can typically be stored up to 2 or 3 weeks and should be inspected prior to each use. The color of the medium, as a reflection of pH, will evolve from pink-amber (fresh) to pink-magenta (old). In addition, evidence of crystallization or flocculation should prompt discarding.
2. FBS purchased from Hyclone has previously been used in our laboratory for MSC cultures, but recently it has been observed that MSC-qualified serum from Invitrogen leads to more robust cell growth. Batch testing of all serum is imperative to ensure optimal expansion of MSCs. Batch testing of all serum is also a requirement to validate the purity of the batch and its ability to maintain undifferentiated MSCs in culture. FBS should only be thawed once because multiple freeze/thaw cycles impede its efficacy for cell culture.
3. For the remainder of the culture recipes and steps, use sterile tissue culture-grade water unless otherwise specified. Avoid multiple freeze–thaw cycles.
4. The recipe name reflects the FBS concentration in the medium (10 or 20%; v/v). Preparation of media typically takes 30 min and needs to be performed every 2–3 weeks. If sterility of medium components is a concern, it is possible to sterilize medium by passing through 0.22-µm polyethersulfone sterilizing, low protein-binding membrane (Corning) using a vacuum driven filter system.

5. Final antibiotic concentration should be 100 U/mL penicillin G and 0.1 mg/mL streptomycin.
6. A sample size greater than 10 mL is recommended, particularly when initially implementing these protocols, although a minimal sample size of as little as 2 mL is feasible. For these small amounts of amniotic fluid, a 15-mL conical tube should be used instead. The initial seeding of the amniotic or placental sample onto the precoated plates takes ~45 min.
7. Regardless of source, the amniotic fluid sample should be processed within 48 h and preferably within 24 h of harvest.
8. Upon inspection, mostly dead cells and cellular debris will be evident; typically, fewer than 2% of the cells are viable. Instead of collagen, other options for coating the wells are fibronectin and laminin. All incubations are performed in a humidified, 37°C, 5% CO_2 incubator unless otherwise specified.
9. Mesenchymal amniocytes will adhere to collagen-coated plates by 48 h; this step removes the nonadherent cells.
10. This time period can be relatively variable. Accordingly, daily plate inspections are recommended. The typical mesenchymal morphology is that of attached, spindle-shaped cells.
11. The coverslips can also be precoated with collagen. As for the well-based method, other options for coating the culture plates and coverslips are fibronectin and laminin.
12. Human placental tissue is typically procured from the whole organ, obtained at delivery or from diagnostic chorionic villus sampling (CVS) specimens. This method will refer to whole organ processing (i.e., from specimens obtained at delivery), but much of the protocol can be adapted to the processing of CVS specimens, in accordance with the peculiarities of each individual CVS procurement. The mechanical separation of the decidual layer from the remainder of the placental specimen, which contains the chorionic villi, generally takes about 1 h but will depend on the familiarity of the operator with basic tissue dissection technique and placental anatomy. Dissolving, filtering, and resuspending the tissue into single cells can take up to an additional 1 h.
13. The optimal amount of enzyme mixture per weight of tissue has not yet been established.
14. Once the placental tissue is cellularized (dissociated into a cell suspension), this protocol is identical to the one for amniotic fluid (see Subheading 3.2).
15. The placental samples that the authors have processed ranged in weight between 12 and 19 g and the resulting pellet could be resuspended in 10 mL medium. The optimal volume of medium per weight of tissue has not yet been established.

16. Instead of collagen, other options for coating the wells are fibronectin and laminin.
17. MSCs will adhere to collagen-coated plates by 48 h; this step removes the nonadherent cells.
18. This time period can be fairly variable. Accordingly, daily plate inspections are recommended. The typical mesenchymal morphology is of attached, spindle-shaped cells.
19. The coverslips can also be precoated with collagen. (As for the well-based method, other options for coating the culture plates and coverslips are fibronectin and laminin.)
20. After isolation, cells will typically reach optimal confluence (80–90%) in 3–5 days and at that point require passage. Beyond passage 2, the authors typically lower the FBS content of the medium from 20 to 10%, solely for cost savings because these cells expand quite rapidly even with 10% FBS. The transfer of selected MSCs from the precoated primary isolation plates to the tissue culture dishes requires ~45 min.
21. Once derived, MSCs have a relatively long lifespan in culture and have been carried in the authors' laboratory for up to 45 passages. Nonetheless, siphoning off early-passage MSCs for freezing is routinely performed. From early passage through any point during MSC expansion, cells can be frozen for future use. Freezing of MSCs is optimally performed with exponentially growing cells, typically near 50% confluence.
22. CAUTION: DMSO is hazardous. When preparing the freezing solution, work in a fume hood and wear gloves.
23. The time span for MSC viability at −80°C is yet to be reliably determined, but it is certainly possible to store them for at least 1 year.
24. To protect sterility do not submerge the cap.

References

1. Friedenstein, A. J., Chailakhjan, R. K., and Lalykina, K. S. (1970) The development of fibroblast colonies in monolayer cultures of guinea-pig bone marrow and spleen cells. *Cell Tissue Kinet* **3**, 393–403.
2. Friedenstein, A. J., Petrakova, K. V., Kurolesova, A. I., and Frolova, G. P. (1968) Heterotopic of bone marrow. Analysis of precursor cells for osteogenic and hematopoietic tissues. *Transplantation* **6**, 230–47.
3. Friedenstein, A. J., Piatetzky, S., II, and Petrakova, K. V. (1966) Osteogenesis in transplants of bone marrow cells. *J Embryol Exp Morphol* **16**, 381–90.
4. Pittenger, M. F., Mackay, A. M., Beck, S. C., Jaiswal, R. K., Douglas, R., Mosca, J. D., Moorman, M. A., Simonetti, D. W., Craig, S., and Marshak, D. R. (1999) Multilineage potential of adult human mesenchymal stem cells. *Science* **284**, 143–7.
5. Prockop, D. J. (1997) Marrow stromal cells as stem cells for nonhematopoietic tissues. *Science* **276**, 71–4.
6. Vaananen, H. K. (2005) Mesenchymal stem cells. *Ann Med* **37**, 469–79.
7. Jauniaux, E., and Rodeck, C. (1995) Use, risks and complications of amniocentesis and chorionic villous sampling for prenatal

diagnosis in early pregnancy. *Early Pregnancy* **1**, 245–52.

8. Gosden, C. M. (1983) Amniotic fluid cell types and culture. *Br Med Bull* **39**, 348–54.
9. Hoehn, H., and Salk, D. (1982) Morphological and biochemical heterogeneity of amniotic fluid cells in culture. *Methods Cell Biol* **26**, 11–34.
10. Milunsky, A. (1979) Genetic disorders and the fetus: diagnosis, prevention, and treatment, Plenum, New York.
11. Prusa, A. R., Marton, E., Rosner, M., Bernaschek, G., and Hengstschlager, M. (2003) Oct-4-expressing cells in human amniotic fluid: a new source for stem cell research? *Hum Reprod* **18**, 1489–93.
12. Torricelli, F., Brizzi, L., Bernabei, P. A., Gheri, G., Di Lollo, S., Nutini, L., Lisi, E., Di Tommaso, M., and Cariati, E. (1993) Identification of hematopoietic progenitor cells in human amniotic fluid before the 12th week of gestation. *Ital J Anat Embryol* **98**, 119–26.
13. Streubel, B., Martucci-Ivessa, G., Fleck, T., and Bittner, R. E. (1996) [In vitro transformation of amniotic cells to muscle cells–background and outlook]. *Wien Med Wochenschr* **146**, 216–7.
14. Fuchs, J. R., Kaviani, A., Oh, J. T., LaVan, D., Udagawa, T., Jennings, R. W., Wilson, J. M., and Fauza, D. O. (2004) Diaphragmatic reconstruction with autologous tendon engineered from mesenchymal amniocytes. *J Pediatr Surg* **39**, 834–8; discussion 34–8.
15. In 't Anker, P. S., Scherjon, S. A., Kleijburg-van der Keur, C., Noort, W. A., Claas, F. H., Willemze, R., Fibbe, W. E., and Kanhai, H. H. (2003) Amniotic fluid as a novel source of mesenchymal stem cells for therapeutic transplantation. *Blood* **102**, 1548–9.
16. Kaviani, A., Guleserian, K., Perry, T. E., Jennings, R. W., Ziegler, M. M., and Fauza, D. O. (2003) Fetal tissue engineering from amniotic fluid. *J Am Coll Surg* **196**, 592–7.
17. Kaviani, A., Perry, T. E., Dzakovic, A., Jennings, R. W., Ziegler, M. M., and Fauza, D. O. (2001) The amniotic fluid as a source of cells for fetal tissue engineering. *J Pediatr Surg* **36**, 1662–5.
18. Kunisaki, S. M., Freedman, D. A., and Fauza, D. O. (2006) Fetal tracheal reconstruction with cartilaginous grafts engineered from mesenchymal amniocytes. *J Pediatr Surg* **41**, 675–82; discussion 75–82.
19. Kunisaki, S. M., Fuchs, J. R., Kaviani, A., Oh, J. T., LaVan, D. A., Vacanti, J. P., Wilson, J. M., and Fauza, D. O. (2006) Diaphragmatic repair through fetal tissue engineering: a comparison between mesenchymal amniocyte- and myoblast-based constructs. *J Pediatr Surg* **41**, 34–9; discussion 34–9.
20. Kunisaki, S. M., Jennings, R. W., and Fauza, D. O. (2006) Fetal cartilage engineering from amniotic mesenchymal progenitor cells. *Stem Cells Dev* **15**, 245–53.
21. Fauza, D. (2004) Amniotic fluid and placental stem cells. *Best Pract Res Clin Obstet Gynaecol* **18**, 877–91.
22. Tsai, M. S., Lee, J. L., Chang, Y. J., and Hwang, S. M. (2004) Isolation of human multipotent mesenchymal stem cells from second-trimester amniotic fluid using a novel two-stage culture protocol. *Hum Reprod* **19**, 1450–6.
23. Steigman, S. A., Ahmed, A., Shanti, R. M., Tuan, R. S., Valim, C., and Fauza, D. O. (2009) Sternal Repair with Bone Grafts Engineered from Amniotic Mesenchymal Stem Cells. *J Pediatr Surg* **44**, 1120–6; discussion 1126.
24. Kunisaki, S. M., Armant, M., Kao, G. S., Stevenson, K., Kim, H., and Fauza, D. O. (2007) Tissue engineering from human mesenchymal amniocytes: a prelude to clinical trials. *J Pediatr Surg* **42**, 974–9; discussion 79–80.
25. Steigman, S. A., Armant, M., Bayer-Zwirello, L., Kao, G. S., Silberstein, L., Ritz, J., and Fauza, D. O. (2008) Preclinical regulatory validation of a 3-stage amniotic mesenchymal stem cell manufacturing protocol. *J Pediatr Surg* **43**, 1164–9.
26. In 't Anker, P. S., Scherjon, S. A., Kleijburg-van der Keur, C., de Groot-Swings, G. M., Claas, F. H., Fibbe, W. E., and Kanhai, H. H. (2004) Isolation of mesenchymal stem cells of fetal or maternal origin from human placenta. *Stem Cells* **22**, 1338–45.
27. Phelan, M. C. (2006) Techniques for mammalian cell tissue culture. *Curr Protoc Mol Biol* **Appendix 3**, Appendix 3F.
28. Kaviani, A., Perry, T. E., Barnes, C. M., Oh, J. T., Ziegler, M. M., Fishman, S. J., and Fauza, D. O. (2002) The placenta as a cell source in fetal tissue engineering. *J Pediatr Surg* **37**, 995–9; discussion 95–9.

Chapter 8

Manufacture of Clinical Grade Human Placenta-Derived Multipotent Mesenchymal Stromal Cells

Nina Ilic, Gary Brooke, Patricia Murray, Sarah Barlow, Tony Rossetti, Rebecca Pelekanos, Sonia Hancock, and Kerry Atkinson

Abstract

Clinical grade human mesenchymal stem cells (MSC) are manufactured and used in clinical trials for a range of regenerative and inflammatory diseases. Human MSC have now been derived from tissues other than bone marrow, such as placenta, as described in this laboratory protocol. It provides instructions for clinical grade MSC manufacturing according to the Code of Good Manufacturing Practice (cGMP) principles and according to policies and procedures of our internal Quality Management System (QMS), which is based on the International Organization for Standardization (ISO) standard requirements. Relevant organizational structure and QMS elements are presented.

Key words: Mesenchymal stem cells/multipotent progenitor cells, Placenta, Isolation, Expansion, Tissue culture, cGMP, Clinical grade, Quality assurance

1. Introduction

Clinical grade human mesenchymal stem/stromal cells (MSC) are currently being manufactured and used in clinical trials for a range of regenerative and inflammatory diseases (1). Human MSC have now been derived from tissues other than bone marrow, such as placenta (2). Placentas are a natural "waste product" at birth and at the Mater Mother's Hospital, Brisbane, there are approximately 9,000 births/year, thus providing ample donor tissue. Bone marrow-derived MSC and placenta-derived MSC share many properties (3) and we have opted to use placental-derived MSC for our clinical trial program (2).

We apply a Code of Good Manufacturing Practice (cGMP) standard (or its close equivalent) across the entirety of our cell manipulation process, including acquisition of the starting cell

Mohan C. Vemuri et al. (eds.), *Mesenchymal Stem Cell Assays and Applications*, Methods in Molecular Biology, vol. 698, DOI 10.1007/978-1-60761-999-4_8, © Springer Science+Business Media, LLC 2011

population (in this case, placental tissue), isolation of cells, processing, storage, and transport (4). We have also devised a Quality Management System (QMS) to provide support for the manufacturing process based on International Organization for Standardization (ISO) standard requirements (5). This includes an extensive documentation system for all relevant procedures (including laboratory protocols) to comply with cGMP requirements (Table 1). It covers all stages of MSC manufacturing. It also applies, but is not limited to the following: facilities, equipment, materials, staff, monitoring, validation, process change, record generating, and record keeping (2).

Production scientists work closely with the quality assurance (QA) staff to ensure safe cellular therapy product. Education of staff is of ultimate importance and adherence to QMS principles is required from all staff, including scientific support and laboratory management. Implementation of the QMS often presents significant change in laboratory as well as in many supportive activities. The conventional research scientists' mindset for performing discovery-motivated laboratory-based scientific experiments is not appropriate for the provision of a cellular therapeutic product that is safe for the recipient, staff, and the institution. We have established a formal training system, which includes a comprehensive training manual, on-line learning modules and assessments, task-based training, as well as continuing education for staff (6).

An efficient and robust system of reporting and addressing of nonconformances (noncompliances with external or internal requirements), improvement suggestions, and change control has been established (5). The QMS staff conduct regular internal audits, while providing relevant reporting to stakeholders. In areas where the QMS interacts with hospital departments and infrastructure, relevant service level agreements are established. The entire process is founded on the following elements: quality program, organization and personnel, standard operating procedures, facilities, environmental control, equipment monitoring, supplies and reagents, process controls and changes, process validation, labeling design and control, storage requirements, records, tracking, nonconformances and complaints management, risk assessment, and reporting and reviewing (4).

Staffing requirements are presented on the schematic organizational structure (Fig. 1) which complies with cGMP requirements and the complex organizational structure inevitably involved in an activity such as cell manufacturing for therapeutic purposes.

An overview of the elements of the QMS is shown in Fig. 2. The decision making process is based on risk management principles and specific risk management framework elements (5) (Fig. 3). A flow diagram of MSC production and the time requirements involved, including a period of preparation prior to

Table 1
Current elements of cGMP practice utilized in manufacturing of clinical grade human placenta-derived MSC (with assigned priority and with estimation of time and resources allocation)

cGMP code requirements	QMS requirements			Assigned priority	Time allocation	Increase of resources required
Level of requirements	Policies	Procedures	Records	H, M, L[a]	I, C, L, O[b]	Yes/No
Quality management system	Quality objectives, quality system requirements			H	I	No
	Organizational structure: director, quality nominee and production nominee(s)			H	I	No
	Resources and personnel			H	C	Yes
	Monitoring systems, internal audits, corrective actions and management review			M	C	No
	Change control			L	C	No
Personnel and training	General personnel objectives			M	I	No
	Initial training and induction			H	I	No
	Laboratory training			H	L	No
	QA and QMS training			M	L	No
	Annual training, task-based training			M	O	No
Buildings and facilities	General building and facilities objectives, floors, walls and fittings			H	L	Yes
	Environmental control			H	C	Yes
	Cleaning and pest control			H	C	Yes
	Goods receipt and storage areas			M	L	No
	Manufacturing areas including general, donor areas, despatch			M	L	No
Equipment	General equipment objectives, preventative maintenance			H	C	Yes
	Qualification and validation including IQ, OQ, PQ[c]			M	I	Yes
	Calibration			H	O	Yes
	Specific equipment including refrigeration equipment, irradiation equipment, barcode equipment			H	I	Yes
Documentation	General document control objectives			H	C	No
	Control of documentation			H	C	No
	Storage and retention			M	O	No
Records	General record keeping objectives, donor sessions, donor deferral			H	C	No
	Collection, retrieval and processing records			H	C	No

(continued)

Table 1 (continued)

cGMP code requirements	QMS requirements			Assigned priority	Time allocation	Increase of resources required
Level of requirements	Policies	Procedures	Records	H, M, L[a]	I, C, L, O[b]	Yes/No
	Final product and batch record			H	C	No
	Storage, retention and archiving			H	O	Yes
	Product complaints and recall, corrective action			H	C	No
	Computer records including general computer records requirements, documentation, validation data, data control			M	L	Yes
Control of materiel	General control of materiel objectives			H	C	No
	Starting materiel including general, reagents, collection packs, labels			H	L	Yes
	In-process materiel			H	L	Yes
	Suppliers and subcontractors review			M	O	No
	Nonconforming materiel, recall of products			H	L	No
Donor selection, donation, and testing[d]	General donor selection requirements including predonation, donation, postdonation			H	L	No
	Tissue donors including predonation, donation, postdonation			H	L	No
	Testing of donor samples including general, contracted laboratories			H	L	Yes
Process control	General process control requirements			H	C	No
	Validation			M	I	Yes
	Quality control, monitoring			M	C	No
	Product release			H	L	No
Storage, packaging and transport	Storage including general, labeling, quarantine, discard			H	L	No
	Packaging including general, labeling, records			H	O	Yes
	Transport			M	O	Yes

[a]*H* High; *M* medium; *L* low

[b]*I* Initial stage, required in the first 1–3 months only; *O* once off or less than 3 months at any stage; *L* long-term, duration 6–12 months; *C* continuous, duration over 12 months

[c]*IQ* Installation qualification; *OQ* operational qualification; *PQ* process qualification

[d]Includes additional requirements of both Australian Red Cross and Auscord (The Australian Cord Blood Registry) guidelines for donor selection and testing

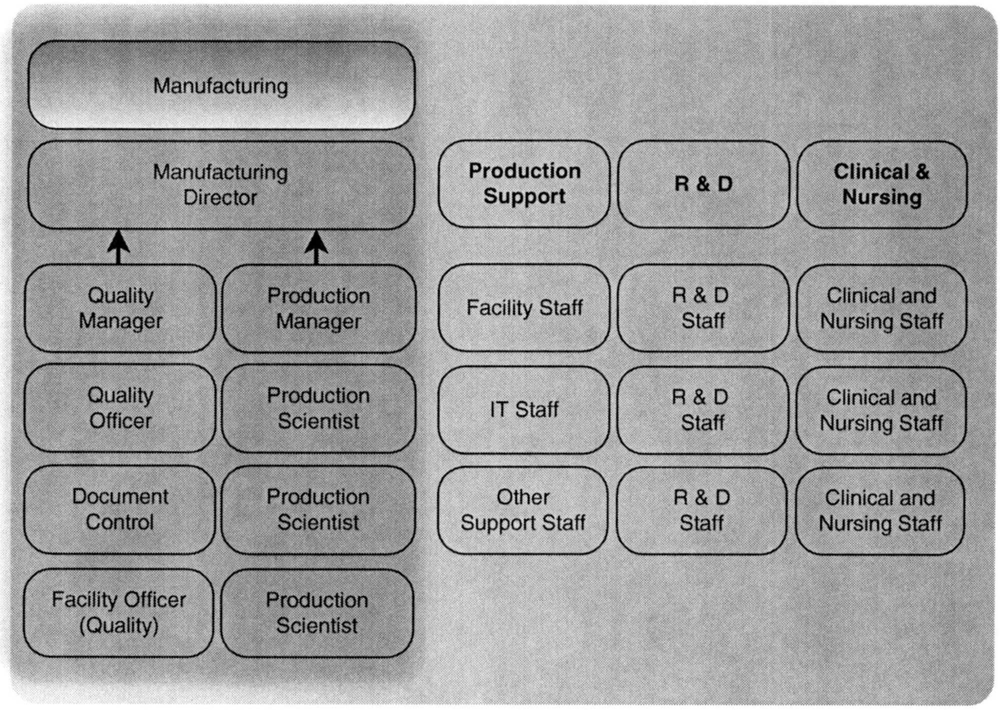

Fig. 1. Organizational structure which meets requirements of Code of Good Manufacturing Practice (cGMP), hospital and research institute organizational structures.

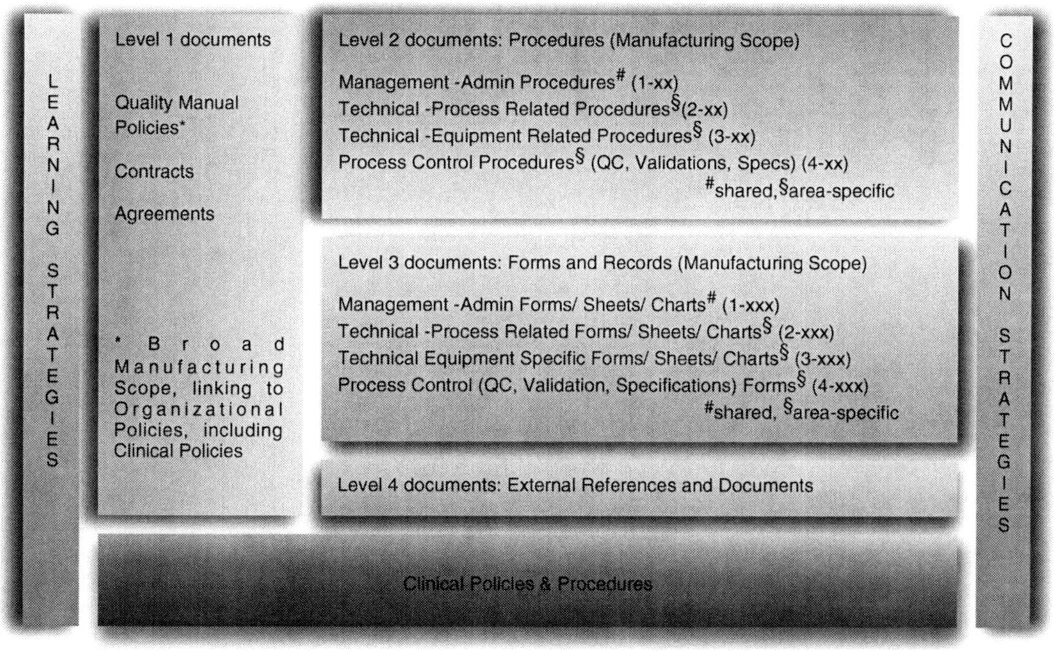

Fig. 2. Elements of the Quality Management System (QMS) stretching across complex organizational structure (hospital and research institute).

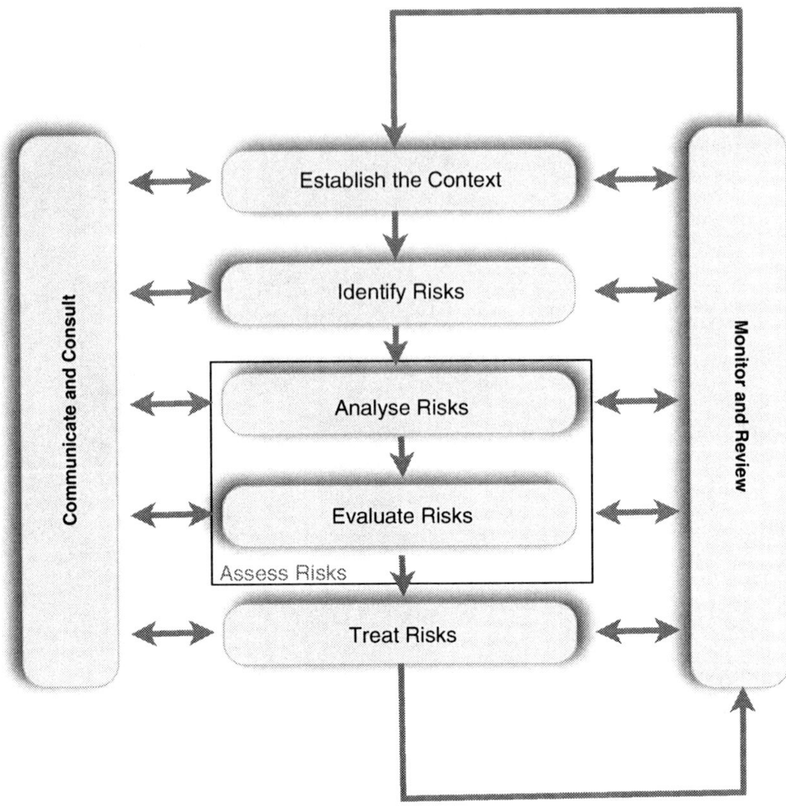

Fig. 3. Risk management framework overview according to AS/NZS 4360:2004 and internal organizational requirements (adapted from AS/NZS 4360:2004).

production commencement and the mandated wait after production for safety testing requirements, is shown in Fig. 4. Ultimately, the project brings together a collaborative team with a unique skill-set to deliver safe cellular therapy manufacturing of human clinical grade MSC (4).

2. Materials

2.1. Cell Isolation, Culture, and Cryopreservation

1. On-demand label printer (Birch).
2. Weighing balance (Sartorius).
3. Heat Sealer CRG (Ljungberg and Kogel AB).
4. Bench centrifuge (Sigma).
5. Sterile 70% ethanol (Pharmacia).
6. Sterile surgical gowns, face masks, overshoes and hair nets (Kimberley-Clark).

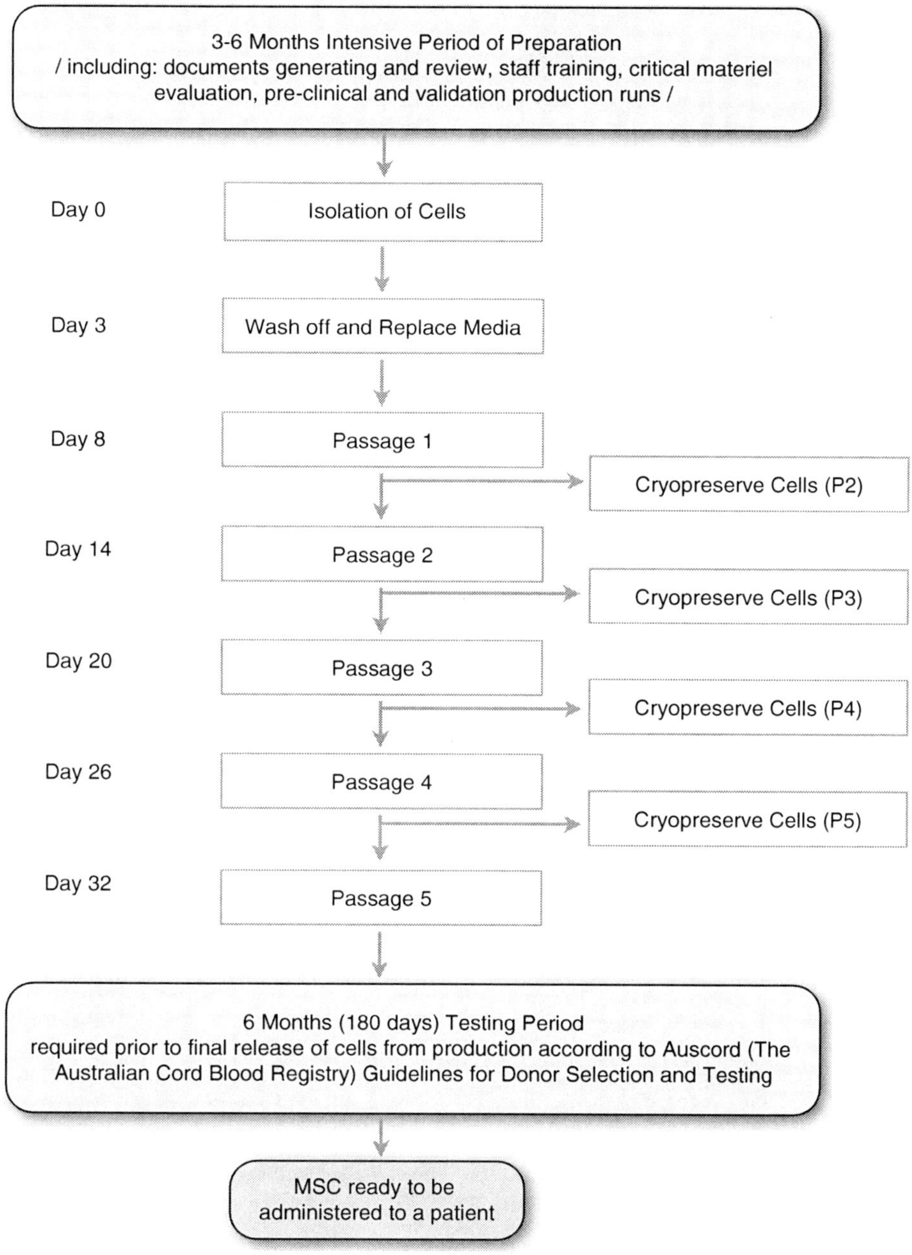

Fig. 4. Flow diagram of manufacturing schedule of human placental-derived mesenchymal stem cells (MSC).

7. Hanks Balanced Salt Solution (HBSS) (Invitrogen).
8. TrypLE Select (recombinant trypsin-like substitute) (Invitrogen).
9. Dulbecco's Modified Eagle Medium – low glucose 1 g/100 ml (DMEM-LG, Invitrogen).

10. Fetal calf serum, Australian sourced (Invitrogen).
11. Trypan blue (Invitrogen).
12. Sterile water (Baxter).
13. Gentamicin, 40 mg/ml (Pharmacia).
14. Tissue culture (TC) medium (TCM): DMEM-LG supplemented with 20% fetal bovine serum (Invitrogen) and 50 μg/ml gentamicin (40 mg/ml stock).
15. Ficoll-Paque Premium (GE Healthcare) (see Note 1).
16. Digest Medium: 50 ml Collagenase I stock (Worthington, stock is 0.2 μm filter sterilized 2,000 U/ml collagenases in sterile RO-grade water (see Note 2).
17. 0.2 μm sterile filter (Becton Dickinson).
18. Incubated orbital shaker (Bioline).
19. Pulmozyme (recombinant cGMP grade DNase I) (Genentech).
20. 70 μm cell strainer (BD Pharmingen).
21. Centrifuge tubes, 50 ml, conical bottom (Nunc).
22. Syringes (60 ml) (Terumo).
23. Mixing cannula (Unomedical).
24. Seriological pipettes; 5, 10 and 25 ml (Sarstedt) and motorized pipetter.
25. Tissue culture flasks (175 and 25 cm^2) (Nunc).
26. Albumex 40 (40% human serum albumin, Australian Red Cross Blood Service).
27. Bovine serum albumin (BSA) (Sigma).
28. Plasma-Lyte 148 replacement intravenous infusion (Baxter).
29. CryoSure-Dimethyl-sulphoxide (DMSO) (Wak Chemie).
30. Cryopreservation media: 65 ml Plasmalyte, 25 ml Albumex-20 and 10 ml DMSO. To avoid protein precipitation always add Albumex-20 to the Plasmalyte and mix before adding DMSO.
31. Cryocyte freezing bags, 50 ml (Baxter).
32. Controlled-rate freezer (Planar).
33. 7-Amino-actinomycin D (7-AAD) (BD Pharmingen).
34. mAbs: $CD45^{APC-Cy7}$, $CD73^{PE}$, $CD105^{FITC}$ $CD45^{APC-Cy7}$, $CD73^{PE}$, $CD105^{FITC}$ (BD Pharmingen).
35. Calibrite fluorescent beads (BD Pharmingen).
36. Sterile phosphate buffered saline (PBS) (Invitrogen).
37. FACS tubes (BD Pharmingen).

3. Methods

3.1. Preparation Prior to Production

1. Documents: Print a Batch Record Form and all documents required for the production day, as outlined in the Batch Record Form (see Note 3).
2. Equipment Records: Check that equipment records show that all equipment servicing is up-to-date for the entire production run (see Note 4). Record equipment status in Batch Record Form (see Note 5).
3. Environmental Monitoring Records: Check that current environmental monitoring data is their within defined limits (see Note 6). If any values are out of the limits, inform Production Managers, Quality Manager, and the Director before deciding whether to proceed with production (see Note 7).

3.2. Equipment Sterilization

Pipettes, dissection equipment, and incubator fixtures must be sterilized no more than 4 weeks prior to use. Once sterilized, equipment is to be stored in the production facility storage cupboard prior to use (see Note 8). If equipment is stored for longer than 4 weeks sterilization must be repeated prior to use (see Note 9).

3.3. Production Authorization

To proceed with the production run notify the *Director* of the status of the cleaning, environmental monitoring, equipment status and any other building, and facility-related issues, wait for confirmation that it is satisfactory to proceed with manufacturing (see Note 10).

3.4. Labels

Print all labels for production day according to standard operating procedure (SOP) (see Note 11).

3.5. Cleaning

Clean equipment required and clean Production Laboratory as described in SOP (see Note 12). Ensure that incubator is thoroughly cleaned according to SOP. Sterile water for irrigation is used for the humidifying pan (see Note 13). During a single placenta production run the incubator is cleaned weekly when the cells are removed for passaging. Sterile water for irrigation is replaced after each cleaning (see Note 13).

3.6. Production Day Preparation

1. Clean biological safety cabinet (BSC) according to SOP.
2. Prepare Batch Record Form.
3. Place sterile drapes on the base of the BSC. Use several layers, so that one layer at a time can be removed if spills occur.
4. Spray the packaged, sterilized waste containers with 70% ethanol and place in BSC (see Note 14).
5. Prepare all media and label as appropriate (see Note 15).

3.7. Mononuclear Cell Isolation from Placenta

1. Obtain a placenta (with prior patient consent and prior ethics committee approval, as per SOP) and place in BSC (see Note 16).
2. Spray 70% ethanol on gloved hands and the outer plastic bag containing the placenta and place into BSC. Open the two sterile bags which contain the placenta, keeping placenta within the bags.
3. Using tweezers and scissors, remove, and discard the umbilical cord (see Note 17).
4. Cut the placenta into approximately 5–10 g pieces and temporarily store in 30 cm Petri dishes (see Note 18).
5. Divide placental pieces evenly into 1 l wash containers (3×) using tweezers. Add approximately 100 g of placenta (300 g total) to each container (~100 ml mark).
6. Add 500 ml HBSS to each beaker. Discard any unused placenta.
7. Wash placental pieces to remove excess blood by stirring contents of the beaker with a 5 ml serological pipette for approximately 10 s. Using the pipette, gently squeeze the placental pieces against the sides of the beaker to aid process.
8. With care, pour red blood cell-containing HBSS into waste container.
9. Repeat steps 6–8 twice.
10. Using tweezers, transfer all placenta pieces from wash containers to fresh Petri dishes.
11. Cut each piece of placenta into very small (approximately 5–10 mm pieces) with scissors.
12. Transfer approximately 10 g of placenta pieces using tweezers (this will fill the 50 ml conical centrifuge tubes up to 10 ml mark). Repeat for 24 × 50 ml tubes. Approximately 240 g of placenta will be used in this process. This figure was chosen due to centrifuge capacity.
13. Add 25 ml Digest Medium to each tube.
14. Transfer tubes to incubated shaker and incubate tubes at 250 rpm, 37°C for 1 h. Fix tube racks onto shaker at 20–40° angle to aid digestion.
15. Place 70 μm nylon cell strainers on top of 16× fresh 50 ml tubes, positioning cell strainers on an angle so that there is a small air gap between the bottom of each cell strainer and the top of each tube.
16. When incubation is complete, place tubes into centrifuge, and pulse spin at 350 × *g*. This requires allowing the centrifuge to reach full speed, then stopping immediately. This process

pellets large clumps of undigested tissue leaving the cell suspension on top.

17. In BSC, being careful not to dislodge the loosely pelleted solid tissue; remove 15 ml of the overlaying cell suspension, and transfer through the 70 μm cell strainers.
18. In the original tubes containing undigested pelleted tissue, add a further 15 ml HBSS. Attach lids to tubes, tighten firmly and shake vigorously for 5 s.
19. Repeat steps 16–17 and discard tubes containing remaining pelleted undigested tissue.
20. Transfer tubes containing the 70 μm filtered cell suspensions to centrifuge and spin at 540×*g* for 5 min.
21. Carefully discard the supernatant using a pipette (supernatant will appear cloudy due to subcellular debris still being present).
22. Attach lids to all tubes and flick tube base to resuspend cells.
23. Add 28 ml HBSS to resuspended cells and ensure that they are evenly mixed.
24. Take up 15 ml Ficoll-Paque Premium in a 60 ml syringe with mixing cannula attached. Place the mixing cannula into the tube with the cells so that its tip reaches the bottom of the tube. Gently push on the syringe plunger to obtain a slow gradual flow trying to ensure that the Ficoll-Paque does not mix with the cell suspension.
25. Centrifuge tubes at 540×*g* for 20 min at 20°C with the brake off. It is important that the brake is left off.
26. Transfer tubes back to BSC. Carefully remove cells from interface and transfer into a fresh 50 ml tube (up to a maximum volume of 10 ml/tube). Repeat for all density gradient tubes.
27. Add HBSS to each tube to bring the total volume up to 50 ml.
28. Centrifuge tubes at 540×*g* for 8 min with the brake on.
29. Transfer TCM from 4°C storage in production laboratory to BSC and spray bottle with 70% ethanol and allow to air dry. It is not necessary to prewarm the media before the next step.
30. On completion of centrifugation, transfer tubes back to BSC and using a pipette, discard supernatant from tubes in the waste container.
31. Reattach lids to tubes and resuspend cells by gently flicking bottom of tubes.
32. Add 3 ml TCM to each tube. Pool the contents of all tubes and place lids back on. Thoroughly mix the cell suspension by inversion.

33. Determine total cell viable cell numbers using a trypan blue count. Total cell number should exceed 100×10^6. If the number is below this, the production manager should be informed to determine if the production run should continue as incubation times may become excessive.

3.8. Cell Culture

1. Place 6×175 cm^2 flasks into BSC.
2. Add 22 ml TCM to each flask using a 25 ml pipette.
3. Add 8 ml of the pooled cell suspension to each flask.
4. Gently swirl flask contents to disperse cells evenly.
5. Transfer flasks to tissue culture (TC) incubator (37°C in 5% CO_2, humidified). Ensure that the entire flask base is covered with media.
6. Lock TC incubator and place label on incubator door.
7. After 48 h, remove and discard all media from flasks using a pipette.
8. Add 20 ml of HBSS to the flask. Gently swirl the flask to dislodge nonadherent cells. Remove HBSS containing the nonadherent cells and discard HBSS wash.
9. Add 60 ml TCM to each flask using a pipette and place back in TC incubator.

3.9. Passaging Cells

1. When cells are approximately 90–95% confluent, check morphology using an inverted microscope with ×20 objective. Cells should have a fibroblast (elongated) morphology. Move T175 flasks containing cells into BSC.
2. Remove all medium from flasks using pipettes and discard.
3. Add 25 ml HBSS to each flask and swirl contents.
4. Remove HBSS from each flask using serological pipette and discard in waste container.
5. Add 5 ml TrypLE Select to each flask.
6. Incubate flasks horizontally for 15 min in the BSC. Agitate flasks using gentle swirling at least once during this period.
7. Add 5 ml HBSS to each flask. Pipette flow should be directed against the monolayer to ensure that all cells are mechanically forced off the surface of the flask.
8. Transfer the cell suspension to 50 ml conical tubes.
9. Centrifuge tubes at $540 \times g$ for 5 min at 4°C.
10. In the BSC, remove supernatant using a pipette, discard to waste container.
11. Resuspend cell pellets by gently flicking tubes.
12. Add enough TCM to resuspend cell pellets in 10 ml.

13. Remove an aliquot of cells and mix with trypan blue at 1:1 dilution for counting on a hemocytometer.
14. Calculate total number of cells for all flasks (see Note 19).
15. Aliquot 1×10^6 cells in media into the FACS cryo-tube for flow cytometry testing and 2.5×10^5 cells into a labeled T25 flask for karyotype analysis. We recommend that this is achieved by trained personnel in a pathology laboratory.
16. Aliquot cells for further propagation into a labeled 50 ml falcon tube. This will require 4.4×10^5 cells for each T175 flask. The numbers of flasks used will depend upon the total number of cells required. Store cells for propagation in the fridge.
17. The remaining cells (from passage 2 onwards) are prepared for cryopreservation.

3.10. Cryopreservation of Cells

1. Label the correct number of cryocyte freezing bags according (see Note 20).
2. In BSC, heat seal 2 of the 3 input tubes three times in a row leaving one tube available for cell input.
3. Cut both the sealed tubes at the third seal (i.e. the seal that is furthest from the cryocyte bag) with scissors.
4. In BSC, attach a mixing cannula to a 60 ml syringe.
5. Take up cryopreservation media containing cell suspension in the 60 ml syringe and insert the mixing cannula into the one remaining input tube. Have a second person check that the correct cell dose is added to the appropriate cryocyte freezing bag. Add the cells/cryopreservation mixture to the bag.
6. Use heat-sealing machine to seal each open tube on the cryocyte bag three times, taking care to seal each tube as close to the bag as possible.
7. Cut the third seal (the seal furthest from the cryocyte bag) with scissors.
8. Place one cryocyte bag per cryocyte bag cassette in the controlled-rate freezer.
9. Controlled-rate freezer program: hold at 4°C for 10 min. Drop 1°C/min to –90°C.
10. Transfer cryobags to vapor phase liquid N_2 tank. We use custom made aluminium racks to store our cryobags.

3.11. Phenotyping by Flow Cytometry

1. Spin cells aliquoted for flow cytometry in two FACS tubes at $350 \times g$ for 5 min. Discard the supernatant and resuspend cells.
2. Add 100 µl of mAb mixtures to each tube (see Note 21).
3. Transfer tubes to a fridge and incubate for 15 min at 4°C.

Table 2
Production release criteria for manufacturing of clinical grade human placenta-derived MSC

Cellular product testing	Release criteria
14 Days microbiology culture	Sterile
Viability	>70% by trypan blue exclusion
Purity by flow cytometry	>90% $CD73^+$, >90% $CD105^+$, <1% $CD45^+$
Karyotype	Cytogenetic analysis normal
Gram stain	Negative
Mycoplasma test of final product	Negative
Endotoxin test of final product	<2 EU/ml
Day 180, donor serology for infectious disease markers	Negative
Day 180, repeat infectious disease testing on mother and baby health questionnaire	Normal

4. Resuspend cells with 3 ml PBS and spin 350×*g*, 5 min, 4°C. Discard supernatant.
5. Resuspend cells in 200 μl PBS and 7-AAD (1:40 dilution) for dead cell discrimination.
6. Perform flow cytometry analysis. A defined threshold should be set; we have chosen our release criteria as >90% $CD73^+$/$CD105^+$ (double positive), <1% $CD45^+$ (Table 2).

3.12. Storage, Release, and Dispatch of Cells

1. Store cells in cryobags in the vapor phase of liquid N_2 tank; maintain optimal storage conditions (see Note 22).
2. Release cells from production for use and/or dispatch based on the *production release criteria* (see Note 23).
3. Dispatch cells for use to an internal or external clinical site for future patient administration (see Note 24).

4. Notes

1. A major obstacle has been a lack of clinical grade cGMP reagents in the cell isolation steps. If clinical grade consumables are available, these should given priority (e.g. Ficoll-Paque is produced as a cGMP reagent whereas Percoll is not). Any change in reagent should be validated, e.g. we have found

that Ficoll-Paque produces equivalent results for isolation of placental cells in our preclinical data.

2. Consumables with a sterility testing certificate and/or a Certificate of Analysis (COA) stating the minimal level of "contaminants" should also be considered (e.g. collagenase I remains a relatively crude isolation fraction containing a range of proteins and enzymes such as collagenase, trypsin, and dispase).
3. The Batch Record Form and procedural documents (e.g. SOP) are subject to strict document control; each version is reviewed, approved and released. Archived versions are stored and made available, if required.
4. Equipment servicing and calibration is obtained on a regular basis and along with the cleaning records kept on the premises; it is required that all equipment used in the production run is up-to-date with its maintenance and calibration (including items such as thermometers, timers, automatic pipettes, data-loggers monitoring temperature, and humidity).
5. Equipment status can be recorded on separate documents linked to the Batch Record.
6. Environmental monitoring shall at a minimum include temperature, humidity, and microbial contamination. Acceptable limits (alert vs. action limits) are set according to historical data and/or literature data.
7. A system of recording nonconformances (noncompliances) should be in place. Environmental monitoring data outside the acceptable limits are considered to be a nonconformance with the QMS requirements and addressed as such (including root cause analysis, corrective, and preventative action implementation).
8. Production facility storage cupboards are temperature mapped, temperature and humidity monitored, and access-controlled.
9. Requirements for each equipment item maintenance, cleaning, and sterilization are outlined in equipment-specific procedures (SOP).
10. Requirements for production approval are outlined in a separate procedure; cGMP requirements are followed in terms of responsibilities and Quality Assurance staff and production staff act independently ("arm's length" – production nominee vs. quality nominee). The ultimate decision-maker is the Director (both production and quality staff independently report to the Director). This also provides a relevant Risk Management framework and decision making structure.
11. Labels and related procedural documents are document controlled. A label reconciliation process is implemented for

each production run. Unique identifiers are devised and no numbers can be repeated in any other production process. Patient confidentiality is maintained throughout the process.

12. Production facility maintenance requirements are outlined and recorded in separate documents that may be directly linked to the Batch Record Form or cross-referenced (by using date, production ID, unique product identifier or similar).
13. It is important to use sterile water and replace it on a regular basis in order to prevent fungal contamination.
14. Control of critical material (that directly or indirectly may influence the quality of the cell product) is obtained according to specific requirements outlined in procedural documents. This includes *all* items used in the production process (such as media, enzymes, flasks and tubes), but also items such as 70% ethanol, sterile drapes, sterile gloves, and disposable laboratory gowns.
15. It is useful to have a brief check-list of tasks to be completed prior to and on the day of production. This list enables staff to focus and perform their tasks according to procedural requirements and record any nonconformances occurring in the manufacturing process.
16. We currently use elective cesarean sections as our placenta source as, although the operating procedures are open, they are performed in relatively sterile environment. The placenta is subsequently double bagged in plastic draw string bags. After weighing, the placenta is placed in a cool box and transferred to our manufacturing facilities on the same campus for processing. Appropriate documentation is signed (donor details and ID check, date, time, witnessing process). The placenta must never be left unattended; it must be given directly to the manufacturing staff to be placed into the laboratory and attended immediately by the manufacturing team. Although class II safety cabinets are used in our production of MSC, this still represents an "open" system of manufacture. Even with a quality system in place and numerous safety precautions, there remains a greater chance of microbial contamination in an open system versus a closed system. Our hospital manufacturing laboratory is not classified as a clean room and is therefore not required to meet defined levels for air particles or viable organisms as set out by appropriate regulatory authorities for clean room standards. However, in order to provide a highly controlled environment and to ensure processing is carried out in an environment where microbial contamination levels are known, environmental monitoring is conducted on a regular basis.
17. If cord blood has not been drained prior to receiving placenta, blood in the placenta will still be at fetal arterial pressure.

Care should be taken when taking the cord blood and placental anatomy should be known in order to find the cord vein. Relevant safety equipment such as eye protection, mask, and gloves should always be used.

18. Multilayer high surface area flasks have been considered. However, efficient washing and removal of adherent cells after trypsinisation becomes much more difficult.
19. We generally obtain approximately 9×10^6 cells per T175 flask and have cultured up to 90 flasks per passage, which equates to approximately 850×10^6 cells.
20. Accurate labeling of products is an important component of Good Clinical Practice (GCP) and is required to ensure compliance with the local QMS Process. Labels must be compatible with the use/handling and storage conditions of the labeled sample and must not detach when stored in incubators or in the vapor phase of liquid nitrogen, or if immersed in water (e.g. use cryovials labels that can be stored at −180°C and remain attached to the cryovial).

    Labels should be affixed in such a way as to allow all information to be read clearly. Labels must detail that it is label X of Y e.g. 4 of 12, to allow traceability; labels must be used in sequential number order. Extra copies of labels must be destroyed or defaced. Copies of labels should be attached to the label reconciliation form and where possible to batch records, storage logs etc as a confirmation of the identity of the product that was being handled.
21. Add 20 µl of each of phenotyping mAbs; $CD45^{PerCP}$, $CD14^{FITC}$, $CD73^{PE}$ and $CD105^{APC}$ (BD Pharmingen) to tube 1 together with in 20 µl PBS/2% BSA. In tube 2, add 20 µl each of negative control mAb; $IgG1^{FITC}$, $IgG1^{PE}$, $IgG1^{APC}$ and $IgG1^{PerCP}$ and 20 µl PBS/2% BSA. In tube 3 add 1 ml PBS/2% BSA and one drop each of Calibrite beads (FITC, PE, APC and PerCP beads). Tube 3 should be used for compensation settings on the flow cytometer.
22. Conditions of the cryostorage should be clearly defined and action limits set. Cryostorage conditions should be continuously monitored and alarmed. Cellular product specifications have been devised in order to provide a defined set of parameters for their storage, transport and testing prior to use. Maintenance of the vapor phase liquid N_2 tank is required, including regular refill of the liquid N_2 by an approved supplier.
23. The cell release process consists of two phases. The very first step is production release according to production release criteria (Table 2). These tests are obtained in an accredited Pathology laboratory and it is recommended to outsource them rather than doing own (in-house) testing for the release

criteria purposes. The second step is release prior to infusion obtained at the clinical site, externally or internally, with the testing of sterility 14-day microbiology culture and viability >70% by trypan blue exclusion. Cells have to pass both of these sets of testing in order to be administered to a patient.

24. Dispatch of cells is by an approved courier only, in a validated and approved dry shipper (that maintains required temperature during the entire transport process). Accompanying documentation (instruction for handling, storage and testing of the cells) is required and regular liaison with clinical site staff, externally or internally, occurs.

Acknowledgments

The authors would like to acknowledge funding from the Inner Wheel Australia, Mater Foundation and Mater Health Services for making this work possible.

References

1. Brooke G, Cook M, Blair C, Han R, Heazlewood C, Jones B, Kambouris M, Kollar K, McTaggart S, Pelekanos R, Rice A, Rossetti T, Atkinson K (2007) Therapeutic applications of mesenchymal stromal cells *Semin Cell Dev Biol* **18**, 846–858.
2. Brooke G, Rossetti T, Pelekanos R, Ilic N, Murray P, Hancock S, Antonenas V, Huang G, Gottlieb D, Bradstock K, Atkinson K (2008) Manufacturing of human placenta-derived mesenchymal stem cells for clinical trials *Br J Haematol* **144**, 571–579.
3. Barlow S, Brooke G, Chatterjee K, Price G, Pelekanos R, Rossetti T, Doody M, Venter D, Pain S, Gilshenan K, Atkinson K (2008) Comparison of human placenta- and bone marrow-derived multipotent mesenchymal stem cells *Stem Cells Dev* **17(6)**, 1095–1108.
4. Brooke G, Rossetti T, Ilic N, Murray P, Hancock S, Hancock S, Atkinson K (2008) Points to consider in designing mesenchymal stem cell-based clinical trials *Transfus Med Hemother* **35**, 279–285.
5. Internal policies and procedures (2009) Mater Cell Therapy, Mater Health Services, Brisbane, Australia.
6. Ilic N (2008) Mater Cell Therapy cGMP Training Manual, Mater Health Services, Brisbane, Australia.

Chapter 9

A Method to Isolate and Culture Expand Human Dental Pulp Stem Cells

Stan Gronthos, Agnieszka Arthur, P. Mark Bartold, and Songtao Shi

Abstract

Dentinal repair in teeth occurs through the activity of specialized cells known as odontoblasts that are thought to be maintained by a precursor population associated with the perivascular cells within dental pulp tissue. We have previously isolated candidate dental pulp stem cells (DPSC) from adult human third molars, with the ability to generate clonogenic cell clusters (CFU-F: colony-forming units-fibroblastic), a high proliferation rate, and multi-potential differentiation in vitro. When cultured DPSC are transplanted into immunocompromised mice, they generated a dentin-like structure lined with human odontoblast-like cells that surrounded a pulp-like interstitial tissue, composed of collagen and a vascular network. The present protocol describes a methodology to generate highly purified preparations of human DPSC. This process involves the enzymatic digestion of fresh samples of human dental pulp tissue followed by the isolation of DPSC using magnetic bead cell separation, based on their expression of mesenchymal stem cell associated markers.

Key words: Dental pulp stem cells, Dentin, Teeth, Mesenchymal stem cells, Magnetic bead cell sorting

1. Introduction

During tooth growth and development, ameloblasts form the enamel of the crown and odontoblasts generate the underlying primary dentin (1). After tooth eruption, ameloblasts disappear from the surface of the enamel where enamel formation ceases to occur. In contrast, odontoblasts, along the inner surface of the dentin inside the pulp chamber, continue to deposit dentin matrix to form secondary dentin throughout life (2, 3). In addition to secondary dentin, odontoblasts can form tertiary (reactionary/reparative) dentin in response to several stimuli, such as mechanical, chemical, and/or bacterial stimulation (4–7). Even when odontoblasts have been damaged, the reparative dentin can be

Mohan C. Vemuri et al. (eds.), *Mesenchymal Stem Cell Assays and Applications*, Methods in Molecular Biology, vol. 698, DOI 10.1007/978-1-60761-999-4_9, © Springer Science+Business Media, LLC 2011

formed in the dental pulp to protect against further disruption of the pulp tissue. This reparative dentinogenesis has been thought to be mediated by newly generated odontoblasts that seem to arise from dental pulp tissue (8, 9). These findings have led to the speculation that odontogenic progenitor cells or stem cells may exist in dental pulp tissue (7, 9–12). Several reports have demonstrated that pulp tissue contains proliferating odontoblast-like cells and that these cells are capable of forming mineralized nodules in vitro (13–15). However, these cells isolated from dental pulp appear to have a limited capacity to differentiate into odontoblast-like cells and an inability to differentiate into other cell types such as adipocytes or neurons (14–18).

Putative stem cell populations from the dental pulp of human third molars (dental pulp stem cells; DPSC) and deciduous teeth (stem cells from human exfoliated deciduous teeth; SHED) which exhibit similar properties to bone marrow derived mesenchymal stem cells have been identified (19–21). Unlike odontoblast-like cells, these clonogenic DPSCs have the capacity to form dentin, fibrous pulp, adipose, and neural cells under in vitro inductive conditions and following implantation in vivo (19–23). These stem cell populations harboured in dental pulp tissue were found to reside in a specific perivascular microenvironment (24) or "stem cell niche" similar to that described for other mesenchymal stem cell-like populations from bone marrow and peripheral adipose tissues (24–30). Recent studies have confirmed that proliferating odontogenic precursor cells appear to be mobilized from blood vessels, to sites of damaged pulp or dentine tissue (7, 12). Collectively, these studies describe the presence of pre-odontoblast cells present in dental pulp tissue, leading to speculation of the existence of putative dental stem cell populations. In this chapter, we describe procedures for the isolation, expansion, and tissue regeneration potential of purified preparations of DPSC.

2. Materials

1. Collagenase/Dispase solution; 3 mg/ml Collagenase I (Worthington Biochemical, Lakewood, NJ) and 4 mg/ml Dispase II in phosphate-buffered saline (PBS; Roche Diagnostics, Indianapolis, IN).
2. 0.5% Trypsin/EDTA solution (JRH Biosciences, Lenexa, KS, USA).
3. Hanks balanced salt solution (HBSS; JRH Biosciences).
4. PBS solution, pH 7.4 (JRH Biosciences).
5. Clonogenic Growth Medium; Alpha-Modification of Eagle's Medium (α-MEM; JRH Biosciences) supplemented with 20%

(v/v) Foetal Bovine Serum (FBS; JRH Biosciences), 2 mM L-glutamine (JRH Biosciences), 100 μM L-ascorbate-2-phosphate (WAKO Pure Chemical Industries, Japan), 50 U/ml penicillin, 50 μg/ml streptomycin (JRH Biosciences), and 5×10^{-5} M β-mercaptoethanol (BDH Chemicals, Poole, UK).

6. DPSC Growth Medium; α-MEM supplemented with 10% (v/v) FBS, 2 mM L-glutamine, 100 μM L-ascorbate-2-phosphate, 50 U/ml penicillin, 50 μg/ml streptomycin, and 5×10^{-5} M β-mercaptoethanol.
7. Osteogenic inductive medium; α-MEM supplemented with 5% (v/v) FBS, 2 mM L-glutamine, 10 mM HEPES, pH 7.35 (JRH Biosciences), 100 μM L-ascorbate-2-phosphate, 10^{-7} M dexamethasone sodium phosphate (David Bull Laboratory, Sydney, Australia), 1.8 mM KH_2PO_4 (Sigma-Aldrich), 50 U/ml penicillin, 50 μg/ml streptomycin, and 5×10^{-5} M β-mercaptoethanol.
8. Adipogenic inductive medium; α-MEM supplemented with 10% (v/v) FBS, 2 mM L-glutamine, 100 μM L-ascorbate-2-phosphate, 0.5 mM isobutylmethylxanthine (Sigma-Aldrich), 60 μM indomethacin (Sigma-Aldrich), 0.5 μM hydrocortisone (Sigma-Aldrich), 1 mM sodium pyruvate, 50 U/ml penicillin and 50 μg/ml streptomycin, and 5×10^{-5} M β-mercaptoethanol.
9. Neural inductive medium A; Neurobasal A Media, 1× B27 Supplement (Invitrogen), 500 U/ml penicillin, 50 μg/ml streptomycin, 20 ng/ml epidermal growth factor (CytoLab/PeproTech, Rehovot, Israel), 40 ng/ml basic fibroblast growth factor (CytoLab/PeproTech).
10. Neural inductive medium B: 1:1 ratio of DMEM and F12 Media (Invitrogen), Insulin-transferrin-sodium selenite supplement (ITSS: Roche Diagnostics), 50 U/ml penicillin and 50 μg/ml streptomycin, 40 ng/ml basic fibroblast growth factor.
11. Retinoic acid (Sigma-Aldrich).
12. HHF: HBSS supplemented with 5% (v/v) FBS.
13. 70 μm Falcon cell strainer (Becton Dickinson Biosciences, San Jose, CA, USA).
14. 50 ml Conical centrifuge tubes.
15. 4 ml Polystyrene centrifuge tubes.
16. 4 ml Polypropylene centrifuge tubes.
17. FACS Fix solution; 1% (v/v) formalin, 0.1 M D-glucose, 0.02% sodium azide in PBS (Sigma-Aldrich).
18. Cell counting fluid: 0.4% trypan blue in PBS (Sigma-Aldrich).
19. Blocking buffer; HHF supplemented with 5% (v/v) normal human serum.

20. 2% $CaCl_2$ in distilled water (Sigma-Aldrich).
21. Polyornithine (Sigma-Aldrich).
22. Fibrinogen and thrombin (Sigma-Aldrich).
23. Recombinant human laminin (Millipore, Billerica, MA, USA).
24. 0.5% Trypsin/EDTA solution (JRH Biosciences).
25. T-25 and T-75 culture flasks (CellStar, Greiner Bio-One, Frickenhausen, GmbH).
26. 6-Well culture plates and 8-well chamber slides (NUNC, Intermed, Roskilde, Denmark).
27. 0.22 μm Ministart low protein-binding filters (Sartorius, Goettingen, Germany).
28. 1.8 ml Cryotubes (Nalge Nunc International, Rochester, NY, USA).
29. 100 mm Petri dishes (CellStar).
30. Periodontal scaler (Henry Schein Halas, Waterloo, Sydney, NSW).
31. Haemostat (Henry Schein Halas).
32. Fissure bur, #556 (Henry Schein Halas).
33. Periodontal curettes (Henry Schein Halas).
34. Endodontic file and broach (Henry Schein Halas).
35. Dimethyl sulphoxide (DMSO; BDH Chemicals).
36. 0.1% (w/v) Toluidine blue in 1% paraformaldehyde solution (BDH Chemicals).
37. 10% Neutral-buffered formalin (BDH Chemicals).
38. 4% Paraformaldehyde in PBS (BDH Chemicals).
39. 0.5% w/v Oil Red O in isopropyl alcohol (Sigma-Aldrich).
40. 1% w/v Alizarin Red S in distilled water (Sigma-Aldrich).
41. Hematoxylin and eosin solution (DAKO, Carpinteria, CA).
42. 10% EDTA in PBS (BDH Chemicals).
43. Mouse IgM anti-human MSC (STRO-1 clone; Developmental Studies Hybridoma Bank, University of Iowa, Iowa City, IA, USA).
44. Mouse IgG_{2a} anti-human CD146/MUC-18 (CC9 clone; A/Prof. S. Gronthos, IMVS, Adelaide, SA, Australia).
45. Mouse IgM anti-human pericyte marker (3G5 clone; American Type Culture Collection Manassas, VA).
46. Mouse IgG_1 anti-human heat shock protein-β (STRO-4 clone; A/Prof. S. Gronthos, IMVS, Adelaide, SA, Australia).
47. Murine anti-β-III tubulin IgG (TUJ1 clone; Covance).

48. Murine anti-neural filament-medium chain IgG (13-0700 clone; Zymed).
49. Rabbit anti-neural filament-heavy chain (AB1991 clone; Chemicon).
50. Isotype-matched murine IgM, IgG_1, and IgG_{2a} and rabbit (Ig) controls (Caltag Laboratories CA).
51. Goat anti-mouse Alexa 488 (Jackson Immunoresearch, A11029).
52. Goat anti-rabbit Alexa 488 (Molecular Probes, A11034).
53. Prolong gold anti-fade with DAPI (Invitogen).
54. (Caltag Laboratories, Burlingame, CA).
55. Sheep anti-mouse IgG-conjugated or rat anti-mouse IgM-conjugated magnetic Dynabeads (Dynal, Oslo, Norway).
56. MPC-1 magnetic particle concentrator (Dynal).
57. 10% w/v Povidine Iodine antiseptic solution (Professional Disposables Inc., Orangeburg, NY).
58. 9 mm Wound clips (Becton Dickinson).
59. TRIzol® (Invitrogen).
60. Hydroxyapatite/tricalcium phosphate (HA/TCP) ceramic particles (Zimmer, Warsaw, IN).

3. Methods

The methods described below outline: (1) the isolation and preparation of the DPSC; (2) the enrichment of clonogenic DPSC or CFU-F by magnetic bead cell separation; (3) ex vivo expansion and cryopreservation of human DPSC; (4) immunophenotypic analysis of cultured human DPSC; (5) in vitro and in vivo assays for the differentiation of human DPSC.

3.1. Preparation of Single Cell Suspensions from Human Dental Pulp Tissue

1. Following informed consent, extracted third molars are collected from patients and placed into α-MEM medium containing double strength penicillin (100 U/ml) and streptomycin (100 μg/ml) and stored at 4°C (see Note 1).
2. Aseptic techniques are used throughout the procedure. Working within a biohazard laminar flow hood, set up 2–3 100 mm petri dishes with approximately 10 ml HBSS for each tooth to be processed. While working in one of the petri dishes, use a periodontal scaler to clean off debris and periodontal ligament on the roots. Scrape in the direction of the root tips away from the dento-enamel junction (DEJ) to avoid contamination from periodontal ligament cells.

Wipe teeth on the outside surfaces with 10% w/v povidone iodine or 70% alcohol swabs.

3. Hold tooth by the root with a haemostat. Using a sterile dental handpiece fitted with a #556 fissure bur, split tooth circumferentially into two halves at the level of the DEJ. Cut intermittently and keep surface wet with HBSS in the petri dish to reduce overheating of the dental tissue. Cut at the level of cementum to avoid the hard enamel layer. Cut a grove circumferentially around the tooth at the level of the DEJ, but do not cut into the pulp tissue (see Note 2).
4. Transfer the split tooth halves into another petri dish containing HBSS. Use sterile periodontal curettes and an endodontic file or a broach to recover the pulp tissue cells.
5. Cut pulp tissue into small pieces with a scalpel blade and transfer into ice cold 5 ml HFF. Wash once by adding 5 ml HFF and centrifuge at $400 \times g$ for 10 min. Remove supernatant and digest pellet with 5 ml collagenase/dispase solution for 1 h in a 37°C water bath (see Note 3). Vortex every 15 min to help break up tissue. Wash twice with HFF with centrifugation at $400 \times g$ for 10 min. After the second wash, carefully pipette off supernatant and re-suspend cell pellet in 1 ml clonogenic growth medium. Pass cell suspensions through a 70 μm strainer to remove large cell aggregates.
6. Add 20 μl of cell suspension to 80 μl 0.4% trypan blue solution and count the number of cells using a hemocytometer.

3.2. Magnetic Bead Isolation and Expansion of DPSC

The use of magnetic bead sorting allows for purification of the DPSC population. Following dental pulp tissue harvest and enzymatic digestion, we routinely recover approximately $1–2 \times 10^5$ cells from one adult third molar. Pool cells from multiple teeth collected from the same patient.

1. Prior to immunolabelling, dental pulp cells are re-suspended in 0.5 ml blocking buffer and incubated on ice for approximately 30 min to reduce the possibility of Fc receptor-mediated binding of antibodies.
2. The cells are then incubated with primary antibodies reactive with DPSC using either antibodies specific to STRO-1, CD146/MUC-18, 3G5, or STRO-4 (heat shock protein-β) for 1 h on ice at a concentration of 10 μg/ml for purified antibodies and 1:2 dilution for hybridoma supernatants (24, 31, 32) (see Note 4).
3. Labelled cells are washed twice with PBS + 1% BSA with centrifugation at $400 \times g$ for 10 min, then incubated with either sheep anti-mouse IgG-conjugated or rat anti-mouse IgM-conjugated magnetic Dynabeads (4 beads/cell: Dynal, Oslo, Norway) for 40 min on a rotary mixer at 4°C. Cells binding

to beads are removed using the MPC-1 magnetic particle concentrator (Dynal) according to the manufacturer's recommended protocol.

4. The STRO-1, STRO-4, CD146, or 3G5 bead-positive cells are then seeded at 3×10^4 cells/cm^2 in growth medium to generate primary cultures in clonogenic growth medium.
5. After reaching 90% confluency, primary DPSC can be subcultured by washing once with HBSS and liberated by enzymatic digestion through the addition of 3 ml of trypsin/EDTA solution per T-75 culture flask for 2–5 min at 37°C. The single cell suspension is then washed twice in growth medium containing as above 10% FBS with centrifugation at $400 \times g$ for 10 min and re-seeded into fresh culture flasks at a plating density of $0.5–1 \times 10^4$ cells/cm^2.

3.3. Assessment of Dental Pulp Colony-Forming Efficiency in Dental Pulp Tissue

The expected incidence of CFU-F colony in human dental pulp tissue is approximately 20–60 CFU-F/10^4 unfractionated cells plated (20).

1. Single cell suspension of dental pulp tissue is seeded into 6-well culture plates at 0.3, 1.0, and 3.0×10^4 cells/well in clonogenic growth medium. Cultures are set up in triplicate and incubated at 37°C in 5% CO_2 and >90% humidity for 12 days (see Note 5).
2. For enumeration, day 12 colonies are washed twice with PBS and then fixed for 20 min in 1% (w/v) paraformaldehyde in PBS.
3. The fixed cultures can then be stained with 0.1% (w/v) toluidine blue for 1 h, then rinsed with tap water and allowed to dry. Aggregates of greater than 50 cells are scored as CFU-F using a dissecting light microscope. Colonies should be visually checked at day 10 to ensure that there is no overgrowth of cells which would make it difficult to enumerate individual colonies, particularly at the higher cell densities.

3.4. Flow Cytometric Analysis of DPSC

To characterize the immunophenotype of ex vivo expanded DPSC, flow cytometric analysis can be used to measure the expression of mesenchymal and non-mesenchymal stem cell associated surface markers at different cell passages. The relatively low number of cells initially harvested from the digestion of pulp tissue limits the use of flow cytometric analysis without ex vivo expansion.

1. Adherent, expanded DPSC are washed once with HBSS and liberated by enzymatic digestion through the addition of 3 ml trypsin/EDTA solution per T-75 culture flask for 5 min at 37°C. The single cell suspension is then washed twice in HHF with centrifugation at $400 \times g$ for 10 min.

2. Cell count and assessment of viability are performed as described above in Subheading 3.1, step 6.
3. Re-suspend DPSC for immunolabelling in 0.5 ml blocking buffer and incubate on ice for approximately 30 min to reduce the possibility of Fc receptor-mediated binding of antibodies.
4. Individual 4 ml round bottom polypropelyne tubes containing 1×10^5 culture-expanded DPSC are incubated with appropriate primary murine monoclonal IgG antibodies or isotype-matched controls at a concentration of 10 μg/ml for purified antibodies or 1:2 diluted hybridoma supernatant for 1 h on ice. Wash the cells twice in 2 ml HHF as described above.
5. Incubate cells with the appropriate secondary detection reagent goat anti-mouse IgG_1-fluorescein isothiocyanate (FITC)-conjugated antibody or anti-mouse IgM-FITC (1:50, Southern Biotechnology, Birmingham, AL) for 45 min on ice (see Note 6). Wash the cells twice in 2 ml HHF with centrifugation at $400 \times g$ for 10 min.
6. Re-suspend cell pellet in 0.5 ml FACS Fix solution to each tube for 10 min.
7. Analyze the cells on any fluorescence-activated cell sorter fitted with a 250 MW argon laser emitting light at a wavelength of 488 nm and therefore able to detect FITC (Fig. 1).

3.5. Differentiation Potential of DPSC In Vitro

The capacity of DPSC to generate mesodermal and ectodermal tissues including those similar to which they were derived from is recognized as a hallmark feature of these cells. The ability of DPSC to differentiate into different cell lineages in vitro can be investigated by culturing under inductive conditions or in vivo following transplantation into immunocompromised mice.

3.5.1. In Vitro Differentiation into Osteoblasts

1. Seed 5×10^4 in vitro expanded DPSCs in triplicate 24-wells in DPSC growth medium at 37°C in 5% CO_2 and >90% humidity.
2. After 24 h, aspirate the culture media and add an equivalent volume of osteogenic inductive medium. Replace the osteogenic medium twice a week for 4 weeks.
3. Aspirate the medium and gently rinse the mineralized cultures with PBS five times, fix with 10% neutral-buffered formalin for 1 h at room temperature (RT), and then rinse three times with distilled H_2O.
4. Stain the mineral with Alizarin Red S staining solution for 1 h at room temperature. Mineralized deposits of calcium will appear red (Fig. 2a). Rinse flasks with distilled water until excess Alizarin Red stain is removed. Store at 4°C.

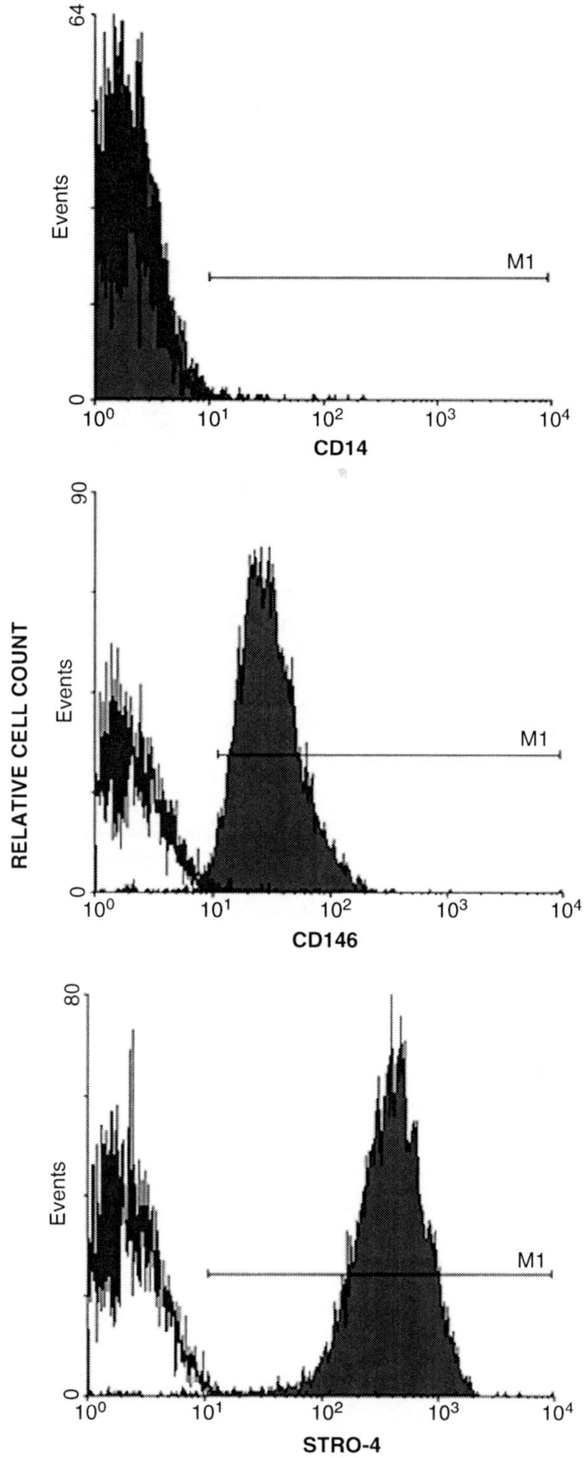

Fig. 1. Immunophenotype of cultured dental pulp stem cells (DPSC) by flow cytometric analysis. The representative flow cytometric histograms show the lack of cell surface expression for the monocyte/macrophage marker, CD14, on single cell suspensions of culture-expanded human DPSC and high cell surface expression of the mesenchymal stem cell markers, CD146 and STRO-4 (*black*) relative to the isotype (IgG_1) negative control (*open*) detected using goat anti-murine IgG_1-conjugated FITC secondary antibodies. Levels of fluorescence greater than 1% compared to the isotype control signify positivity.

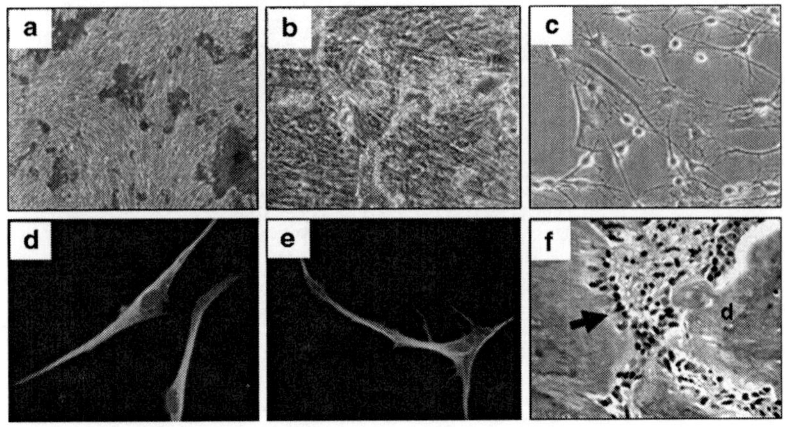

Fig. 2. Differentiation potential of DPSC. The primary cultures of cells derived from STRO-1⁺ selected DPSC were induced under (**a**) osteogenic, (**b**) adipogenic, or (**c**) neural inductive conditions. (**a**) Mineralized deposits stained positively with the Alizarin Red reagent formed within 4 weeks of culture under osteoinductive conditions (200×). (**b**) The presence of clusters of lipid containing adipocytes was also detected by Oil Red O staining within 2 weeks of adipogenic induction (200×). (**c**) Neural-like cellular morphology of DPSC (200×) following 3 weeks of neural induction as demonstrated by up-regulation of (**d**) β-III tubulin (*pale grey*) and (**e**) neural filament-medium chain (*pale grey*) protein expression with DAPI counterstained nuclei (*grey*) (400×). (**f**) Human dental pulp STRO-1⁺ DPSC was expanded in vitro, then implanted subcutaneously into NOD/SCID mice using hydroxyapatite/tricalcium phosphate (HA/TCP) carrier. Implants were harvested 8 weeks after the transplant. New dentin formation (**f**) can be seen with a distinct layer of odontoblasts (*arrow*) together with surrounding fibrous tissue. The sections were counterstained with haematoxylin and eosin (200×). (For the color version, please visit http://extras.springer.com).

3.5.2. In Vitro Differentiation into Adipocytes

1. Seed 5×10^4 in vitro expanded DPSCs in triplicate 24-wells in 500 μl DPSC growth medium. Cultures are incubated at 37°C in 5% CO_2 and >90% humidity.
2. After 24 h, aspirate the culture media and add an equivalent volume of adipogenic inductive media. Replace the adipogenic inductive medium twice a week for 4 weeks.
3. Aspirate the media and gently rinse the adipogenic culture once with PBS and fix with 10% neutral-buffered formalin for 10 min at room temperature.
4. Aspirate the formalin and stain the adipogenic culture with a solution of 3 parts 0.5% w/v Oil Red O staining solution for at least 2 h at room temperature. Lipid-laden droplets within fat cells will appear red (Fig. 2b). Rinse flasks with distilled water until excess Oil Red O stain is removed, do not allow to air dry, store in distilled water at 4°C.

3.5.3. In Vitro Differentiation into Neural Cells

1. Tissue culture 6-wells or 8-well chamber slides are coated with polyornithine (10 μg/ml) overnight at room temperature in laminar flow hood. The wells are washed twice with

sterile water and then coated with Laminin (5 μg/ml) overnight at 37°C in a humid incubator, then washed three times with PBS prior to use.

2. Ex vivo expanded DPSC are re-seeded for 3 days in coated 6-well plates at a density of 2×10^3 cells/well in coated 8-well chamber slides. Cells are then cultured in neural inductive medium A for 7 days, followed by a change in neural inductive medium B for 7 days with the addition of 0.5 μM retinoic acid over the last 7 days of culture. All media should be changed twice weekly. Following the final incubation, lyse cells with TRIzol® and store at −80°C for RNA isolation.
3. For immunohistochemical staining, replicate wells are fixed with 4% PFA for 30 min at room temperature. After washing three times in PBS, block cultures with 5% goat serum in PBS for 30 min at room temperature and then incubate with primary antibody (1:500 β-III tubulin clone, TUJ1 (MMS-435P); 1:200 neural filament-medium chain (13-0700); 1:500 neural filament-heavy chain (AB1991)) in blocking solution overnight at 4°C. Test mouse and rabbit (Ig) controls (Caltag Laboratories, CA) under the same conditions. After washing in PBS, add the secondary antibodies (1:200 Goat anti-mouse Alexa 488; 1:200 Goat anti-rabbit Alexa 488) in PBS for 2 h at room temperature in the dark.
4. Wash the slides with PBS and add a cover slip along with Prolong gold anti-fade containing the nuclear stain, DAPI. Visualize slides under a fluorescence microscope as shown in Fig. 2c–e (see Note 7).

3.6. In Vivo Differentiation Potential of DPSC

In order to demonstrate that ex vivo expanded DPSC can differentiate into functional odontoblast-like cells, cultured DPSC are mixed with osteoconductive HA/TCP ceramic carrier particles followed by subcutaneous transplantation into immunocompromised mice (see Note 8).

1. Prepare a single cell suspension of ex vivo expanded DPSC using trypsin/EDTA digestion and assess cell viability using 0.4% trypan blue as described above.
2. Re-suspend approximately 5×10^6 ex vivo expanded DPSCs in 1 ml DPSC culture medium and transfer to a 1.8 ml cryovial containing 40 mg HA/TCP carrier particles (Zimmer, Warsaw, IN) (see Note 9). Gently mix the cell suspension and HA/TCP particles using a rotator and incubate at 37°C for 1 h to enhance cell attachment to the particles.
3. Gently pellet the mix at $300 \times g$ for 2 min and discard the supernatant.
4. Approximately 10 min prior to implantation, add 15 μl mouse fibrinogen (30 mg/ml in PBS), followed by 15 μl mouse

thrombin (100 U/ml in 2% $CaCl_2$) to the cell pellets and gently mix with a pipette tip to form a fibrin-clotted plug.

5. Following an appropriate level of anaesthesia of mice (see Note 10) to 6–8-week-old immunocompromised NOD/SCID mice, perform a 1 cm mid-longitudinal skin incision with a sharp scissor or a scalpel on the dorsal surface of the NOD/SCID mice, then insert blunt-ended curved scissors beneath the skin and gently tease the skin from the underlining muscle layer to create subcutaneous pockets on both flanks. Place one transplant into each of the subcutaneous pockets and close the incision with surgical wound clips (see Note 11).
6. Recover the transplants 8 weeks after transplantation, cut into two pieces using a surgical blade and fix in 4% paraformaldehyde for 2 days.
7. Decalcify transplant for 10 days in 10% EDTA solution with daily changes prior to paraffin embedding.
8. Unstained 5 μM sections of paraffin-embedded 8-week-old DPSC transplants are deparaffinized in xylene 2× 5 min, then re-hydrated in 100% ethanol 2× 5 min, 90% ethanol 1× 5 min, 70% ethanol 1× 5 min, 50% ethanol 1× 5 min, water 2× 5 min. Finally, stain with hematoxylin and eosin (Fig. 2f) or use for immunohistochemistry.

3.7. Cryopreservation of Ex Vivo Expanded DPSC

1. Prepare single cell suspensions of culture-expanded DPSC with 0.5% trypsin/EDTA as described above and then wash the cells in cold HFF followed by centrifugation at 400 × *g* for 10 min.
2. Re-suspend the cell pellet in FBS at a concentration of 1×10^7 cells/ml and maintained on ice. Gradually add an equal volume of freeze mix (20% DMSO in cold FBS), while gently mixing the cells to give a final concentration of $2–5 \times 10^6$ cells/ml (final medium composition is 10% DMSO/90% FBS). Distribute 1 ml aliquots into 1.8 ml cryovials that have been pre-cooled on ice. Freeze vial at a rate of −1°C/min using a rate control freezer (see Note 12).
3. The frozen vials are then transferred to liquid nitrogen for long-term storage. Recovery of the frozen stocks is achieved by rapidly thawing the cells in a 37°C water bath. The cells are then re-suspended in cold HFF and spun at 400 × *g* for 10 min.
4. To assess viability of the cells, prepare a 1:5 dilution in 0.4% trypan blue and determine the number of cells using a haemocytometer. Typically, this procedure gives viabilities between 80 and 90%.

4. Notes

1. Teeth can be stored for 24 h at 4°C without any significant reduction in DPSC viability.
2. Dental pulp tissue can also be extracted by wrapping the teeth in several layers of sterile gauze and carefully cracking the teeth in a vice.
3. Collagenase stock (6 mg/ml in PBS) and Dispase stock (8 mg/ml in PBS) are mixed together 1:1 for working solution. Store at −20°C until required. Pre-warm to 37°C before use.
4. The STRO-1 antibody can also be purchased commercially from R&D Systems Inc., Minneapolis, MN, USA. Commercially available antibodies that identify human CD146 and HSP90 can also be sourced from Santa Cruz Biotechnology Inc. and BD Biosciences.
5. Variations in FBS batches can severely hamper establishment of CFU-F colonies and growth. Batch testing of FBS is highly recommended to ensure optimal growth conditions.
6. Other conjugates for secondary detection antibodies like phycoerythrin can also be used.
7. Alternatively, for colorimetric analysis, incubate with secondary goat anti-rabbit or anti-mouse Ig-biotinylated antibody (1:200; Caltag Laboratories) for 60 min, followed by washing three times with PBS. Antigen expression can be detected by peroxidase activity using a peroxidase Vectorstain ABC kit and peroxidase substrate AEC kit (Vector Laboratories). After staining, wash the cells three times with PBS and counterstain for 2 min with Mayer's hematoxylin. Rinse with water and mount with a glass cover slip using Gurr aqueous solution (Univert, BDH).
8. This procedure requires animal ethics approval from the appropriate body and should be performed in accordance with an institutional approved small-animal protocol.
9. Pre-wash HA/TCP particles in growth medium for 30 min at room temperature.
10. The anaesthetic Halothane delivered as vapour with oxygen allows quick recovery of mice. If using intraperitoneal injection of barbiturates such as Ketamine and Xylazine, place animals on heated pad or under heat lamp until the animals have recovered from the anaesthesia.
11. Other suitable strains for the xenogeneic transplants include Nude and NIH-bg-nu-xid mice.

12. The DPSC can also be frozen using a Cryo 1°C freezing container "Mr. Frosty" (Nalge Nunc International, Rochester, NY, USA) by placing the container holding the cryotubes at -80°C overnight before transferring the cells into liquid nitrogen. For serum-free applications, ProFreeze solution (CAMBREX Bio Science, Walkersville, MD, USA) containing a final concentration of 7.5% DMSO can be substituted for the 10% FBS/DMSO freeze mix.

Acknowledgements

This work was supported by National Health and Medical Council of Australia Project Grant #453599 and #453497.

References

1. Thesleff, I., and Aberg, T. (1999) Molecular regulation of tooth development, *Bone 25*, 123–125.
2. Baume, L. J. (1980) The biology of pulp and dentine. A historic, terminologic-taxonomic, histologic-biochemical, embryonic and clinical survey, *Monogr Oral Sci 8*, 1–220.
3. Smith, A. J., Tobias, R. S., Cassidy, N., Begue-Kirn, C., Ruch, J. V., and Lesot, H. (1995) Influence of substrate nature and immobilization of implanted dentin matrix components during induction of reparative dentinogenesis, *Connect Tissue Res 32*, 291–296.
4. Cox, C. F., White, K. C., Ramus, D. L., Farmer, J. B., and Snuggs, H. M. (1992) Reparative dentin: factors affecting its deposition, *Quintessence Int 23*, 257–270.
5. Murray, P. E., About, I., Lumley, P. J., Franquin, J. C., Remusat, M., and Smith, A. J. (2002) Cavity remaining dentin thickness and pulpal activity, *Am J Dent 15*, 41–46.
6. About, I., Murray, P. E., Franquin, J. C., Remusat, M., and Smith, A. J. (2001) The effect of cavity restoration variables on odontoblast cell numbers and dental repair, *J Dent 29*, 109–117.
7. Tecles, O., Laurent, P., Zygouritsas, S., Burger, A. S., Camps, J., Dejou, J., and About, I. (2005) Activation of human dental pulp progenitor/stem cells in response to odontoblast injury, *Arch Oral Biol 50*, 103–108.
8. Heikinheimo, K., Begue-Kirn, C., Ritvos, O., Tuuri, T., and Ruch, J. V. (1998) Activin and bone morphogenetic protein (BMP) signalling during tooth development, *Eur J Oral Sci 106 Suppl 1*, 167–173.
9. Sveen, O. B., and Hawes, R. R. (1968) Differentiation of new odontoblasts and dentine bridge formation in rat molar teeth after tooth grinding, *Arch Oral Biol 13*, 1399–1409.
10. Butler, W. T., Ritchie, H. H., and Bronckers, A. L. (1997) Extracellular matrix proteins of dentine, *Ciba Found Symp 205*, 107–115.
11. Ruch, J. V. (1998) Odontoblast commitment and differentiation, *Biochem Cell Biol 76*, 923–938.
12. Tecles, O., Laurent, P., Aubut, V., and About, I. (2008) Human tooth culture: a study model for reparative dentinogenesis and direct pulp capping materials biocompatibility, *J Biomed Mater Res B Appl Biomater 85*, 180–187.
13. About, I., Bottero, M. J., de Denato, P., Camps, J., Franquin, J. C., and Mitsiadis, T. A. (2000) Human dentin production in vitro, *Exp Cell Res 258*, 33–41.
14. Couble, M. L., Farges, J. C., Bleicher, F., Perrat-Mabillon, B., Boudeulle, M., and Magloire, H. (2000) Odontoblast differentiation of human dental pulp cells in explant cultures, *Calcif Tissue Int 66*, 129–138.
15. Tsukamoto, Y., Fukutani, S., Shin-Ike, T., Kubota, T., Sato, S., Suzuki, Y., and Mori, M. (1992) Mineralized nodule formation by cultures of human dental pulp-derived fibroblasts, *Arch Oral Biol 37*, 1045–1055.
16. Buurma, B., Gu, K., and Rutherford, R. B. (1999) Transplantation of human pulpal and

gingival fibroblasts attached to synthetic scaffolds, *Eur J Oral Sci 107*, 282–289.

17. Kuo, M. Y., Lan, W. H., Lin, S. K., Tsai, K. S., and Hahn, L. J. (1992) Collagen gene expression in human dental pulp cell cultures, *Arch Oral Biol 37*, 945–952.
18. Shiba, H., Fujita, T., Doi, N., Nakamura, S., Nakanishi, K., Takemoto, T., Hino, T., Noshiro, M., Kawamoto, T., Kurihara, H., and Kato, Y. (1998) Differential effects of various growth factors and cytokines on the syntheses of DNA, type I collagen, laminin, fibronectin, osteonectin/secreted protein, acidic and rich in cysteine (SPARC), and alkaline phosphatase by human pulp cells in culture, *J Cell Physiol 174*, 194–205.
19. Gronthos, S., Brahim, J., Li, W., Fisher, L. W., Cherman, N., Boyde, A., DenBesten, P., Robey, P. G., and Shi, S. (2002) Stem cell properties of human dental pulp stem cells, *J Dent Res 81*, 531–535.
20. Gronthos, S., Mankani, M., Brahim, J., Robey, P. G., and Shi, S. (2000) Postnatal human dental pulp stem cells (DPSCs) in vitro and in vivo, *Proc Natl Acad Sci U S A 97*, 13625–13630.
21. Miura, M., Gronthos, S., Zhao, M., Lu, B., Fisher, L. W., Robey, P. G., and Shi, S. (2003) SHED: stem cells from human exfoliated deciduous teeth, *Proc Natl Acad Sci U S A 100*, 5807–5812.
22. Arthur, A., Rychkov, G., Shi, S., Koblar, S. A., and Gronthos, S. (2008) Adult human dental pulp stem cells differentiate toward functionally active neurons under appropriate environmental cues, *Stem Cells 26*, 1787–1795.
23. Nosrat, I. V., Smith, C. A., Mullally, P., Olson, L., and Nosrat, C. A. (2004) Dental pulp cells provide neurotrophic support for dopaminergic neurons and differentiate into neurons in vitro; implications for tissue engineering and repair in the nervous system, *Eur J Neurosci 19*, 2388–2398.
24. Shi, S., and Gronthos, S. (2003) Perivascular niche of postnatal mesenchymal stem cells in human bone marrow and dental pulp, *J Bone Miner Res 18*, 696–704.
25. Bianco, P., Riminucci, M., Gronthos, S., and Robey, P. G. (2001) Bone marrow stromal stem cells: nature, biology, and potential applications, *Stem Cells 19*, 180–192.
26. Doherty, M. J., Ashton, B. A., Walsh, S., Beresford, J. N., Grant, M. E., and Canfield, A. E. (1998) Vascular pericytes express osteogenic potential in vitro and in vivo, *J Bone Miner Res 13*, 828–838.
27. Fuchs, E., Tumbar, T., and Guasch, G. (2004) Socializing with the neighbors: stem cells and their niche, *Cell 116*, 769–778.
28. Moore, K. A., and Lemischka, I. R. (2006) Stem cells and their niches, *Science 311*, 1880–1885.
29. Sacchetti, B., Funari, A., Michienzi, S., Di Cesare, S., Piersanti, S., Saggio, I., Tagliafico, E., Ferrari, S., Robey, P. G., Riminucci, M., and Bianco, P. (2007) Self-renewing osteoprogenitors in bone marrow sinusoids can organize a hematopoietic microenvironment, *Cell 131*, 324–336.
30. Gronthos, S., Zannettino, A. C., Hay, S. J., Shi, S., Graves, S. E., Kortesidis, A., and Simmons, P. J. (2003) Molecular and cellular characterisation of highly purified stromal stem cells derived from human bone marrow, *J Cell Sci 116*, 1827–1835.
31. Gronthos, S., McCarty, R., Mrozik, K., Fitter, S., Paton, S., Menicanin, D., Itescu, S., Bartold, P. M., Xian, C., and Zannettino, A. C. (2009) Heat shock protein-90 beta (Hsp90β) is expressed at the surface of multipotential mesenchymal precursor cells (MPC): generation of a novel monoclonal antibody, STRO-4, with specificity for MPC from human and ovine tissues, *Stem Cells Dev 18(9)*, 1253–1262.
32. Gronthos, S., and Zannettino, A. C. (2008) A method to isolate and purify human bone marrow stromal stem cells, *Methods Mol Biol 449*, 45–57.

Chapter 10

Isolation and Culture of Human Multipotent Stromal Cells from the Pancreas

Karen L. Seeberger, Alana Eshpeter, and Gregory S. Korbutt

Abstract

Mesenchymal stem cells, also termed multipotent mesenchymal stromal cells (MSCs), can be isolated from most adult tissues. Although the exact origin of MSCs expanded from the human pancreas has not been resolved, we have developed protocols to isolate and expand MSCs from human pancreatic tissue that remains after islet procurement. Similar to techniques used to isolate MSCs from bone marrow, pancreatic MSCs are isolated based on their cell adherence, expression of several cell surface antigens, and multilineage differentiation. The protocols for isolating, characterizing, and differentiating MSCs from the pancreas are presented in this chapter.

Key words: Multipotent mesenchymal stromal cells, Mesenchymal stem cells, Isolation, Culture, Differentiation, Nonendocrine pancreas

1. Introduction

An attractive alternative to daily insulin injections is to transplant insulin-producing tissue to achieve a more physiological means for restoring glucose homeostasis, thereby potentially reversing the metabolic and neurovascular complications of diabetes. However, the current shortage of human pancreases limits the availability of donor tissue. Therefore, to allow islet transplantation to become a more applicable form of therapy for more patients with type 1 diabetes, we need to identify an unlimited source of β-cells and develop safer antirejection strategies that will induce transplantation tolerance without chronic administration of immunosuppressive drugs.

The expansion of self-duplicating β-cells (1, 2) or the generation of islets from progenitor/stem cells is a practical source to address

Mohan C. Vemuri et al. (eds.), *Mesenchymal Stem Cell Assays and Applications*, Methods in Molecular Biology, vol. 698, DOI 10.1007/978-1-60761-999-4_10, © Springer Science+Business Media, LLC 2011

the shortage of donor islets. Several studies have reported that β-cell progenitors reside within the pancreas itself. Possible sources of these progenitors are the pancreatic ductal epithelium (3–11), acinar cells (12–15), islets (1, 2, 16–19), and mesenchymal cells (20–22). Several studies have indicated that mesenchymal stem cells, also referred to as multipotent mesenchymal stromal cells (MSCs; (23)), are capable of deriving functional endoderm (24–27). Although MSCs isolated from adipose tissue (26) and bone marrow (28, 29) can be differentiated into insulin producing cells, these newly derived β-cells secrete significantly lower amounts of insulin as compared to native adult β-cells. Therefore, since islet β-cells could be derived from cells of mesenchymal origin (26, 28, 29), we hypothesized that MSCs from the pancreas would possess a greater potential to differentiate into functional β-cells as compared to those isolated from nonpancreatic tissues. We have previously reported that MSCs could be isolated and expanded (12-fold) from human pancreatic ductal epithelium/nonendocrine pancreas (21). The pancreatic MSCs that were isolated were phenotypically similar to bone marrow-derived MSCs as they expressed CD13, CD29, CD44, CD49b, CD54, CD90, and CD105 cell surface antigens and could be differentiated in vitro into mesoderm (osteocytes, adipocytes, and chondrocytes).

In addition to MSCs expanded from the exocrine pancreas (20–22), several citations have demonstrated that pancreatic MSCs may originate from β-cells, which undergo a reversible epithelial-mesenchymal transition (17, 19, 30) or may preexist within the adult islet (31–34). Although the cell origin of pancreatic MSCs remains controversial (17–19, 30–34), we routinely isolate and expand MSCs from the nonendocrine pancreas that is left over following islet procurement, and in this chapter we describe the method for isolating and characterizing MSCs from this tissue.

2. Materials

2.1. General Lab Supplies

1. Sterile 100 and 150 mm tissue and nontissue culture-treated plates (Fisher Scientific, Nepean, Canada).
2. Sterile 1 and 2 well tissue culture-treated chamber slides (BD Biosciences, Bedford, MA).
3. Sterile 50 mL conical centrifuge tubes (BD Biosciences).
4. 1.5 mL centrifuge tubes (Fisher Scientific).
5. 0.2 μm bottle top filters, 250–1,000 mL size (Corning Life Sciences, Corning NY).

6. Sterile 50 mL disposable vacuum 0.2 μm filters (Millipore, Mississauga, Canada).
7. 2.0 mL cryovials (Corning Life Sciences).
8. V-bottom 96-well plates (Thermoscientific, Ashville, NC).
9. Sterile polystyrene round bottom 5 mL tubes (BD Biosciences).
10. Sterile Partec CellTrics 50 μm mesh filters (Partec, Swedesboro, NJ).
11. Tissue cassettes, microscope slides, coverslips, and cryomolds (Fisher Scientific).
12. Hemocytometer with coverslip (Fisher Scientific).
13. Glass coplin jars (Fisher Scientific).
14. Automatic pipettor, micropipettors, disposable sterile serological pipettes, and pipette tips (Fisher Scientific).
15. Vacuum source, vacuum bottle, and tubing.

2.2. Reagents, Media, and Solutions for Tissue Culture

1. Expansion medium: RPMI 1640 (Invitrogen, Burlington, Canada) supplemented with 10% FBS (Invitrogen), 1 mM sodium pyruvate (Invitrogen), 10 mM HEPES (Invitrogen), 100 U penicillin, 100 μg/mL streptomycin (Invitrogen), 71.4 μM beta-mercaptoethanol (Sigma-Aldrich), 20 ng/mL basic fibroblast growth factor, and 20 ng/mL epidermal growth factor (Invitrogen). Medium can be stored for 2 weeks at 4°C.
2. 1× phosphate buffered saline (PBS).
3. EDTA DNase solution: PBS supplemented with 0.5 mM EDTA (Invitrogen) and 0.01 mg/mL DNase I (Roche Applied Science, Laval, Canada).
4. Dissociation solution: EDTA DNase solution supplemented with 0.05% trypsin (Invitrogen).
5. Trypan blue dye (Sigma-Aldrich): Prepare 0.4% w/v in PBS.
6. Freezing medium: Undiluted FBS supplemented with 20% dimethyl sulfoxide (DMSO; Fisher Scientific).
7. Mesencult MSC Basal Medium (Stem Cell Technologies, Vancouver, Canada).
8. Osteogenic Stimulatory Kit (Stem Cell Technologies): contains osteogenic stimulatory supplements, 1 M β-glycerophosphate, dexamethasone, and ascorbic acid.
9. Adipogenic Stimulatory Supplements (Stem Cell Technologies).

2.3. Histochemistry and Flow Cytometry

1. Primary and secondary antibodies (see Table 1).
2. Formaldehyde (BDH Laboratory Supplies, UK): Prepare 1 and 4% v/v in PBS.

Table 1
Primary and secondary antibodies for cell characterization

Antibody	Dilution	Fluorochrome	Source
CD13[a]	3/50	PE-Cy5	Invitrogen, Burlington, Canada
CD29[a]	3/50	PE-Cy5	Invitrogen, Burlington, Canada
CD34[a]	2/5	PerCP-Cy5.5	BD Biosciences, Mississauga, CA
CD44[a]	1/10	Fluorescein isothiocyanate (FITC)	Santa Cruz Biotechnology, CA
CD45[a]	3/50	Phycoerythrin (PE)	Caltag Laboratories, Burlingame, CA
CD49b[a]	1/10	FITC	Sanquin, Amsterdam, The Netherlands
CD54[a]	1/50	FITC	Caltag Laboratories, Burlingame, CA
CD90[a]	1/50	PE	BD Biosciences, Mississauga, Canada
CD105[a]	3/50	PE	Caltag Laboratories, Burlingame, CA
CD117[a]	1/5	FITC	Millipore, Mississauga, Canada
E-Cadherin[a]	1/100	Unconjugated	R&D Systems Inc., Minneapolis, MN
EpCAM[a]	1/10	FITC	Stem Cell Technologies, Vancouver, Canada
Vimentin[b]	1/10	FITC	Progen, Heidelberg, Germany
CK19[b]	1/50	Unconjugated	Dako Cytomation Inc, Mississauga Canada
Insulin[b]	1/500	Unconjugated	Dako Cytomation Inc, Mississauga, Canada
Glucagon[b]	1/5,000	Unconjugated	Millipore, Mississauga, Canada
Amylase[b]	1/300	Unconjugated	Sigma-Aldrich, Oakville, Canada
Goat anti-mouse IgG[c]	1/200	PE	Southern Biotech, Birmingham, AL
Goat anti-mouse IgG[c]	1/200	FITC	Jackson Immuno Research Laboratories Inc., Westgrove PA
Goat anti-rabbit IgG[c]	1/200	FITC	Jackson Immuno Research Laboratories Inc., Westgrove PA
Goat anti-guinea pig IgG[c]	1/500	FITC	Southern Biotech, Birmingham, AL

[a]Antibody directed to cell surface marker
[b]Antibody directed to intracellular marker
[c]Secondary antibody

3. Permeabilization solution: Prepare 0.3% w/v saponin (Sigma-Aldrich) in PBS.
4. Alkaline phosphatase staining kit (Vector Red Alkaline Phosphatase Substrate Kit I, Vector Laboratories, Burlingame, CA).
5. 1 M Tris–HCl pH 8.0 (Fisher Scientific).
6. Oil red O (Sigma-Aldrich) staining solution: Prepare 0.4% w/v of oil red O in 95 mL of 80% ethanol and 5 mL of acetone (Fisher Scientific).
7. Acetone (Fisher Scientific).
8. Silver nitrate (Sigma-Aldrich): Prepare 2% w/v in distilled water.
9. Sodium thiosulfate (Fisher Scientific): Prepare 5% w/v in distilled water.
10. Ultra violet light or 60 W lamp.
11. Anhydrous ethanol.
12. Aqueous mounting medium (Immu-Mount, Thermo Shandon, Pittsburgh, PA).
13. Permanent mounting medium (Permount, Fisher Scientific).
14. Shandon Cryomatrix (Thermo Shandon).

3. Methods

Pancreatic MSCs can be expanded from nonendocrine human pancreatic tissue that remains following islet procurement. This exocrine tissue is usually devoid of islets (≤5% insulin positive cells) and is primarily composed of acinar tissue which consists of 40% amylase staining positive cells and ductal epithelium that consists of 60% CK19 staining positive cells (21, 35). Briefly, human donor pancreas are removed from cadaveric donors who had previously given informed consent, and islets are harvested via enzyme perfusion via the duct and purified on continuous Ficoll gradients (36). Following islet purification, the exocrine cell fraction is shipped in CMRL medium and upon receipt by our laboratory is washed with buffered salt solution and cultured in suspension for 1–2 days (21). These cell aggregates are then plated onto tissue culture-treated dishes, and it is from these adherent aggregates that fibroblast-like cells (multipotent stromal cells) grow out from. The methods below describe the isolation and expansion, cryopreservation, cell characterization via flow cytometry, and the differentiation of pancreatic multipotent stromal cells (MSCs; see Notes 1–3).

3.1. Isolation of Human MSCs from Nonendocrine Pancreas

1. Transfer the exocrine cell fraction to a 50-mL centrifuge tube. Centrifuge briefly (450×*g* for 1 min) and discard the supernatant. Resuspend the cell pellet in PBS and centrifuge again. Repeat this PBS wash once more for a total of two washes.
2. After the second wash, discard the supernatant and transfer 500 μL of the packed cell volume to a nontissue culture-treated plate and culture at 37°C in 5% CO_2, 95% air in 35 mL of RPMI supplemented with 10% FBS for 1–2 days (see Note 4).
3. Collect the cell aggregates from a single 150 mm plate and transfer to a 50-mL centrifuge tube. Wash the plate with RPMI 1640 supplemented with 10% FBS and add to the centrifuge tube. Centrifuge briefly (450×*g* for 1 min) and discard the supernatant. Cell debris, which includes single cells and small fragmented aggregates, are removed via this wash.
4. Resuspend the cell pellet in 10 mL RPMI 1640 supplemented with 10% FBS. Transfer 1 mL of the cell aggregate suspension to a 150-mm tissue culture-treated plate and culture at 37°C in 5% CO_2, 95% air in 35 mL expansion medium.
5. Allow cellular aggregates to attach and adherent stromal cells will grow out from these adhered aggregates (see Fig. 1a). Perform complete media changes every 2–3 days until cells are 80% confluent, which will be in approximately 7–14 days (see Fig. 1b).
6. After the cells reach 80% confluence, the cells can be subcultured, cryopreserved (see Subheading 3.2), and characterized via flow cytometry and cell differentiation (see Subheadings 3.3 and 3.4).
7. To detach adherent cells, rinse the cell monolayer with serum free RPMI 1640 medium or EDTA DNase solution (10–15 mL/150 mm dish). Add dissociation solution and return the plate to the incubator for a maximum of 5 min. Monitor the plate for cells lifting off using an inverted phase

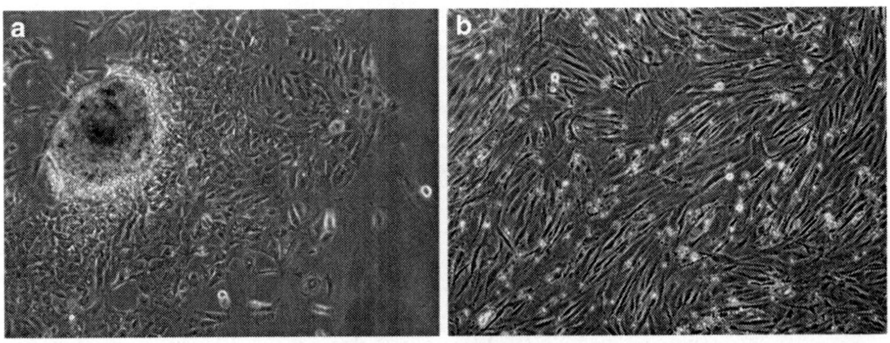

Fig. 1. Outgrowth of pancreatic MSCs from adhered cell aggregates (**a**). Pancreatic MSCs at 80% confluency (**b**). Magnification is 100×.

microscope. Tap the sides of the plate to detach any remaining cells and inactivate trypsin by adding 10–15 mL of medium containing 10% FBS.

8. Transfer the cell suspension to a 50-mL conical tube and centrifuge at $450 \times g$ for 10 min.
9. Remove the supernatant and resuspend the cell pellet in 10 mL of medium containing 10% FBS.
10. Mix the cell suspension and pipette 100 µL of the cell suspension into a 1.5-mL microcentrifuge tube. Add 100 µL of trypan blue dye to the 100 µL of cells and mix. This is a 1:1 ratio and a dilution factor of 2.
11. Determine total cell number using a hemocytometer by placing 10 µL of the diluted cells on both sides of a hemocytometer and count the total number of cells in the four outer squares (1 mm^2) of both sides of the hemocytometer.
12. Calculate the number of cells/mL as follows:

$$\frac{\text{Total cells counted}}{8} \times 2(\text{dilution factor}) \times 10^4 = \text{cells / mL}$$

13. Calculate the total number of cells by multiplying the cells/mL with the cell volume:

    Cells/mL × 10 mL = total number of cells

14. Seed $0.5–1 \times 10^6$ cells into a new 150 mm tissue culture-treated plate. At this seed density, the cells should become 80% confluent within approximately 7–10 days and are ready to be passaged again following steps 5–12 (see Note 5).

3.2. Cryopreservation and Thawing of Pancreatic MSCs

3.2.1. Freezing

1. Pancreatic MSCs can be successfully stored in undiluted FBS or culture medium containing 10% DMSO at a concentration of 2×10^6 cells/mL.
2. Detach cells from the growth surface with trypsin/EDTA as described in the previous section and count the cells using a hemocytometer.
3. Prepare freezing medium (undiluted FBS containing 20% DMSO).
4. Resuspend the cells in undiluted FBS at a concentration of 4×10^6 cells/mL.
5. Label cryovials and aliquot 0.5 mL of the cell suspension per vial (2×10^6 total cells).
6. Add 0.5 mL of freezing medium per vial, cap and invert to mix (The final concentration of DMSO is 10%).
7. Freeze vials overnight at −80°C and transfer to liquid nitrogen storage the next day.

3.2.2. Thawing

1. For each vial to be thawed, label a 50 mL centrifuge tube and dispense 10 mL of medium that contains 10% FBS.
2. Thaw cells by warming the cryovial quickly by passing it through 37°C water. Do not thaw the cells completely, but thaw just until some ice remains.
3. Wipe the outside of the cryovial with ethanol and quickly transfer the semi-frozen cell suspension by pouring the contents of the cryovial into the 50 mL centrifuge tube that contains 10 mL of medium (see Note 6).
4. Centrifuge tube at 400×*g* for 5–10 min, remove the supernatant, and resuspend the cell pellet in expansion medium. The number of viable cells can be determined by performing a cell count with trypan blue and using a hemocytometer.
5. Transfer resuspended cells into a tissue culture-treated dish and the medium should be changed every 2–3 days.

3.3. Flow Cytometry

As no single antigen can be used to positively identify this stem cell population, MSCs are characterized based on their expression of a number of cell surface antigens and their ability to differentiate into several different cell lineages (37–39). This section describes MSC characterization based on cell antigen expression and differentiation into mesoderm (adipocytes and osteocytes).

Flow cytometry analysis can be used to determine the cell surface and intracellular antigen expression of the starting cellular aggregates (exocrine tissue from donor pancreas) and pancreatic MSCs expanded from donor pancreas (21, 35). A panel of antibodies directed toward several cell surface molecules is used to determine the cell phenotype of pancreatic MSCs and to determine changes to the cell phenotype that may occur when MSCs are cultured for long periods of time (21, 40). The list of monoclonal antibodies that we use to characterize pancreatic MSCs is listed in Table 1, as are the antibodies to intracellular molecules used to quantify the endocrine and exocrine cells in the starting tissue used to expand MSCs. To prepare cell samples for flow cytometry, cell aggregates are dissociated into a single cell suspension (see method below) or cells are detached from the growth surface (see Subheading 3.1, steps 5–11) prior to cell labeling. The antigens expressed by nonendocrine pancreatic cells and expanded pancreatic MSCs are listed in Table 2.

3.3.1. Dissociation of Cell Aggregates

1. Dispense 250 μL of packed cell volume (nonendocrine pancreas) into a 50-mL centrifuge tube.
2. Wash the cell aggregates twice with 50 mL of EDTA DNase solution.
3. Centrifuge at 450×*g* for 1 min, discard the supernatant, and repeat for a total of two washes.

Table 2
Antigen expression of nonendocrine pancreatic cells and expanded pancreatic MSCs

Antigen	Nonendocrine pancreas	Pancreatic MSCs
CD13	+*	+
CD29	+	+
CD34	–	–
CD44	+	+
CD45	–	–
CD49b	+	+
CD54	±	+
CD90	+*	+
CD105	+*	+
CD117	–	–
Vimentin	+	+
E-Cadherin	+	–
EpCAM	+	–
CK19	+	–
Insulin	±*	–
Glucagon	±*	–
Amylase	+	–

Antigens that are expressed less than 10% (*), positive (+), negative (–), and varied (±)

4. Add 25 mL of prewarmed (37°C) dissociation solution (see Note 7).
5. Place the centrifuge tube in a 37°C shaking water bath until the cell suspension is turbid and cell aggregates dissociate (maximum 5 min).
6. Remove the tube from the water bath and pipet the cell suspension up and down using a 25-mL serological pipette to complete the dissociation.
7. Inactivate trypsin by adding 25 mL of medium containing 10% FBS.
8. Centrifuge the cell suspension at 450×*g* for 5–10 min, discard the supernatant, and resuspend the cell pellet in 10 mL of EDTA/DNAse I solution.
9. Remove any undissociated cell aggregates or clumps by filtering the cell suspension through a sterile 50 μm disposable

mesh filter placed on top of a 15-mL centrifuge tube (see Note 8).

10. Correct the volume (final volume of 10 mL) and perform a cell count using a hemocytometer and trypan blue dye (see Subheading 3.1 steps 8–11).

3.3.2. Cell Characterization by Flow Cytometry

It is important to plan in advance the number of wells required to allow for the analysis of a variety of cell antigens. Multiple antibodies can be combined per well, as determined by the fluorophore to which they are conjugated and the number of lasers your flow cytometer is equipped with. However, if multiple antibodies and colors are being analyzed, appropriate single color controls must be included in order to properly compensate the flow cytometer and account for the bleeding of fluorophores into adjacent channels. It is also imperative that proper controls are used to determine background fluorescence, including both directly conjugated and unconjugated isotype antibody controls (mouse IgG1 and mouse IgG2a) and secondary isotype controls, in order to properly analyze the data.

This description covers four different staining protocols including the following: (1) one-step directly conjugated antibodies to cell surface antigens, (2) one-step directly conjugated antibodies to intracellular cell antigens, and two-step methods to either (3) cell surface or (4) intracellular cell antigens involving unconjugated primary antibodies followed by conjugated secondary antibodies. An unstained sample should also be kept for the purpose of setting up the flow cytometer (see Note 9).

1. Plate cells at $0.5–1.0 \times 10^6$ per well of a 96-well V-bottom plate. Centrifuge at $450 \times g$ at 4°C for 2 min using 96-well plate buckets. The cell pellet should be visible at the bottom of the well, aspirate the supernatant from each well using a 100-μL pipette tip affixed to rubber tubing and a vacuum bottle attached to a vacuum source. Fix the cells by adding 100 μL of cold 1% formaldehyde in PBS to each well and incubate at 4°C for 30 min. Add 100 μL of PBS to each well and centrifuge again as described above, followed by aspiration of supernatant.
2. If cell permeabilization is required for intracellular antigen exposure, proceed to step 3. For staining cell surface antigens with directly conjugated antibodies, proceed to step 7. For one or two-step unconjugated antibody staining of cell surface antigens, proceed to step 5.
3. Permeabilize the cells by resuspending the cell pellet in 50 μL/well of permeabilization solution. Gently tap the side of the plate against the countertop to mix. Cover the plate with a lid and incubate at room temperature for 10 min in the dark.
4. Centrifuge at $450 \times g$ at 4°C for 2 min. Aspirate the supernatant from the wells. If using a directly conjugated antibody,

proceed to step 7. If the primary antibody is unconjugated, proceed to step 5.

5. Add the unconjugated primary antibody to the appropriate wells at the concentration required (see Table 1). For staining intracellular antigens, dilute the antibody in permeabilization solution, for cell surface antigens, dilute the antibody in PBS/1% FBS for a final volume of 50 μL/well. Mix several times using a multichannel pipettor and incubate at 4°C for 30 min.
6. Add 100 μL of PBS to each well, followed by centrifugation at 450×*g* for 2 min. Aspirate and discard the supernatant, then add the appropriate secondary antibody to each well (see Table 1). Dilute secondary antibody in applicable buffer as indicated in step 5 (50 μL/well). Mix as previously described, cover and incubate at 4°C for 30 min in the dark. Proceed to step 7 or 8.
7. Add the directly conjugated antibody to the appropriate wells at the concentration required (see Table 1). For staining intracellular antigens, dilute the antibody in permeabilization solution, for cell surface antigens, dilute the antibody in PBS/1% FBS for a final volume of 50 μL/well. Mix several times using a multichannel pipettor and incubate at 4°C for 30 min in the dark.
8. Add 100 μL of PBS to each well, followed by centrifugation as described previously. Aspirate and discard the supernatant from the wells, add another 100 μL PBS per well, resuspend the cell pellet, and centrifuge again for a second wash. Label tubes recommended for use with your flow cytometer and add 250 μL of PBS to each tube. After the second wash, resuspend the samples in 50 μL PBS and transfer the cells to the appropriate tubes for final volume of 300 μL. If analysis is not possible on the day of completion, samples can instead be transferred to 250 μL of PBS/1% formaldehyde for longer storage (up to 5 days).
9. Run samples on a flow cytometer as per the manufacturer's instructions and analyze the data with applicable software.

3.4. Differentiation

3.4.1. Adipogenic Differentiation

1. Plate single MSCs from an early passage onto one or two well tissue culture-treated chamber slides at a density of 2,500 cells/cm^2.
2. Expand the MSCs in expansion medium until 70% confluency is reached, replacing the medium every 2–3 days. Chamber slide volume capacity is 6 mL for the single well and 2.5 mL for the two well slide.
3. Remove the expansion medium and wash the cell monolayer twice with buffered salt solution (PBS).
4. Following the manufacturer's protocol, add adipogenic medium that contains Adipogenic Stimulatory Supplements.

5. Replace the medium after 3 days and proceed with half media changes every 2–3 days for 14 days.
6. Within 14 days, lipid droplets should be visible by phase microscopy and lipid droplets can be stained with oil red O (41).

3.4.2. Osteogenic Differentiation

1. Plate single MSCs from an early passage onto one or two-well tissue culture-treated chamber slides at a density of 2,500 cells/cm^2.
2. Expand cells in expansion medium until 70% confluency is reached, replacing medium every 2–3 days. Chamber slide volume capacity is 6 mL for the single well and 2.5 mL for the two well.
3. Remove expansion medium and wash the cell monolayer twice with buffered salt solution (PBS).
4. Following the manufacturer's protocol, add osteogenic medium that contains Osteogenic Supplements (15% final volume), 10^{-8} M dexamethasone and 50 μg/mL ascorbic acid.
5. After 5 days, replace two thirds of the medium every 2–3 days. Gently pipet medium down the sides of the chambers as the monolayer may lift off the growth surface during differentiation.
6. Observe changes to the cell monolayer using an inverted phase microscope. The cell monolayer should thicken, and dense foci should form within 14 days.
7. Once the cell monolayer has thickened, ß-glycerophosphate (3.5 mM) can be added to the osteogenic medium as per the manufacturer's instructions.
8. Continue with differentiation for 21–35 days with two third medium changes every 2–3 days.
9. Alkaline phosphatase activity can be measured by staining using a commercial antibody kit from Vector Laboratories. Mineralization and calcium deposition can be assessed via Von Kossa staining (41).
10. If the monolayer detaches during differentiation, continue with differentiation and collect the cell aggregate for histology (see Note 10).

3.5. Histology (see Note 11)

3.5.1. Sample Preparation

1. Upon completing differentiation, remove the medium from the chambers and wash the cell layer twice with PBS.
2. Add 4% formalin to the wells and fix for 30–60 min at 4°C.
3. Remove the fixative and wash the cell layer twice with PBS.
4. Proceed with staining or slides maybe stored for up to 2 weeks in PBS at 4°C.
5. If cells have detached during the differentiation process, collect the cell aggregate for histology (see Note 10; (21)).

3.5.2. Oil Red O Staining for Lipids

1. Prepare the oil red O staining solution and transfer to a clean glass coplin jar.
2. Remove PBS from the chambers that have been previously fixed (see Subheading 3.5.1), and remove the chambers and adhesive from the slide surface using the tool provided. Alternately, fixed frozen sections may be stained (see Note 10).
3. In a glass coplin jar, wash the slide in PBS.
4. Repeat wash step in PBS twice.
5. Briefly immerse the slides in 70% ethanol.
6. Transfer the slides to the oil red O staining solution for 3 min.
7. Wash with distilled water for 3 min.
8. Counterstain with hematoxylin and coverslip with aqueous mounting medium (see Note 12).
9. Lipid droplets will stain red and nuclei will be counterstained blue (see Fig. 2a).

3.5.3. Alkaline Phosphatase Stain

1. Utilize the commercial staining kit for the detection of alkaline phosphatase activity in developing osteocytes (Vector Red Alkaline Phosphatase Substrate Kit I).
2. Prepare working substrate staining solution according to manufacturer's instructions. To 2.5 mL of 100 mM of Tris–HCl (pH 8.2), add the following:
   (a) Add 1 drop of Reagent 1 and mix.
   (b) Add 1 drop of Reagent 2 and mix.
   (c) Add 1 drop of Reagent 3 and mix.
3. Remove PBS from the chambers that have been previously fixed (see Subheading 3.5.1) and wash adherent cells twice with PBS. Alternately, fixed frozen sections or paraffin sections may be stained (see Note 10).
4. Add enough working stain solution to chambers to cover the entire cell layer.

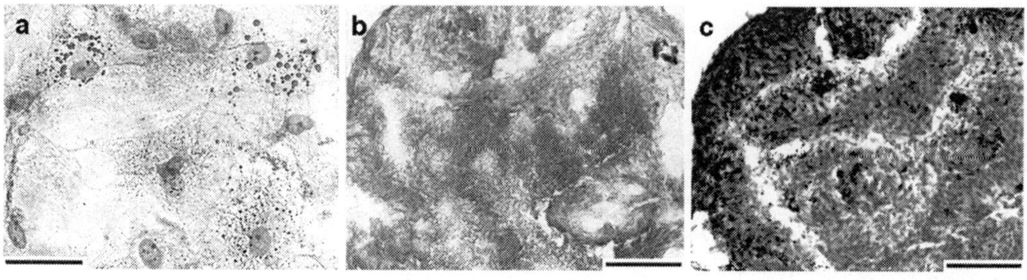

Fig. 2. Pancreatic MSCs differentiated into adipocytes (**a**) and osteocytes (**b**, **c**). Oil red O stains the lipid droplets in adipocytes (**a**). Alkaline phosphatase activity is stained (**b**) and Von Kossa staining stains the calcium deposits black in osteocytes (**c**). Scale bar represents 100 μm.

5. Incubate for up to 1 h in the dark.
6. Remove the stain and wash 3 times with distilled water, 2 min for each wash.
7. Separate the chambers from the slide and remove any remaining adhesive with the tool provided.
8. Coverslip with an aqueous mounting medium or dehydrate and coverslip with a permanent mounting medium.
9. Alkaline phosphatase activity should stain bright pink/red using brightfield microscopy or will fluoresce using fluorescent microscopy using the Texas Red filter (see Fig. 2b).

3.5.4. Von Kossa Stain

1. Remove PBS from the chambers that have been previously fixed (see Subheading 3.5.1), separate the chambers from the slide, and remove any remaining adhesive with the tool provided. Alternately, fixed frozen sections or paraffin sections may be stained (see Note 10).
2. In a coplin jar, wash the slides with distilled water 3 times, 2 min for each wash.
3. In a clean clear glass coplin jar, immerse the slides in silver nitrate and place the coplin jar in front of a 60-W lamp for 1 h or UV lamp for 30–45 min.
4. Rinse with distilled water.
5. In a new clean clear glass coplin jar, immerse the slides in sodium thiosulfate for 3 min at room temperature (see Note 13).
6. Rinse slides in running tap water.
7. Counterstain with hematoxylin and coverslip with an aqueous mounting medium or counterstain with hematoxylin, dehydrate, and coverslip with a permanent mounting medium.
8. Calcium deposits will appear as black deposits if the silver nitrate is reduced using a UV lamp or brown deposits if reduced with a 60-W lamp. Nuclei will appear blue if counterstained (see Fig. 2c).

4. Notes

1. Protocols outlined in this chapter involve the handling of human tissue/cells. As a general rule, *Universal Precautions* must be adhered to and guidelines set out by the users' institution for handling and disposing of biohazardous and chemical waste must be strictly followed.
2. All cell culture techniques described in this chapter must be carried out using aseptic technique. All glassware and disposables

must be sterile (autoclaved or gas sterilized). Solutions, reagents, and media listed should be filter sterilized (0.2 μm filter) prior to use, and all cell manipulations should be done in a Class II biological safety cabinet equipped with a vacuum source and vacuum bottle for removing medium.

3. This protocol outlines the expansion of MSCs from nonendocrine pancreatic tissue; however, several citations also describe the expansion of MSCs from islets (17–19, 30).
4. Our lab routinely cultures the exocrine tissue for 1–2 days in suspension to allow cells to recover after pancreatic dissociation. We then plate 1/10 of the original packed cell volume that was cultured in suspension. Similarly, one could plate the exocrine tissue at a concentration of 50 μL packed cell volume per 150 mm tissue culture-treated plate.
5. We have observed that similar to MSCs isolated from other tissues, pancreatic MSCs from later passages (6 and up) lose some of their stem cell-like properties (21, 40). Cells from late passages may have altered expression of cell surface antigens and lose their ability to differentiate. Therefore, it is advisable to work with cells from earlier passages and to cryopreserve cells or bank cells from these passages. In addition, donor patient variability may also impact MSC expansion, phenotype, and differentiation.
6. We prefer to pour the semi-thawed cell suspension, as we find that pipetting cells during the thawing procedure impacts cell viability.
7. The dissociation solution should be brought to 37°C for no longer than 5 min. If left at 37°C, trypsin will lose its activity.
8. Prior to any flow cytometry analysis, it is desirable to filter out any leftover cell aggregates and or clumps.
9. The flow cytometry techniques supplied in this chapter are brief and are covered in detail elsewhere within this book. The space provided does not allow for a detailed description of flow cytometry analysis. However, as most institutions have flow cytometry facilities staffed with experienced operators, the authors urge individuals interested in this technique to consult with experienced users when designing new assays.
10. If the cell monolayer has aggregated and detached from the growth surface, the aggregate can be fixed in formalin, embedded in paraffin and 5 μm sections cut. Sections need to be hydrated prior to staining. Alternately, the cell aggregate can be placed into a plastic mold and preserved for frozen sectioning by overlaying with cryomatrix embedding compound. After frozen sections are cut, fix slides in either 4% ice-cold formalin for 30–60 min or ice-cold acetone for 2 min prior to staining.

11. As with all techniques, positive and negative controls should be included. Cells that have been expanded but not differentiated should be used as negative controls to ensure that differentiation and staining are specific. In addition, tissue sections of bone and fat would be appropriate positive staining controls.
12. Aqueous mounting medium must be used with oil red O stain. If the slides are dehydrated for permanent cover slipping, the lipids will dissolve in the organic solvents and the stain will be lost.
13. Ensure all glassware is clean and rinsed sufficiently with deionized water. Place sodium thiosulfate into a clean coplin jar prior to transferring slides. If any residue is left on the glass, the sodium thiosulfate will precipitate and cannot be used.

References

1. Dor, Y., Brown, J., Martinez, O. I., and Melton, D. A. (2004) Adult pancreatic beta-cells are formed by self-duplication rather than stem-cell differentiation. *Nature* **429**, 41–46.
2. Khalaileh, A., Gonen-Gross, T., Magenheim, J., Nir, T., Porat, S., Salpeter, S., Stolovich-Rain, M., Swisa, A., Weinberg, N., and Dor, Y. (2008) Determinants of pancreatic β-cell regeneration. *Diabetes Obes. Metab.* **10**, 128–135.
3. Wang, R. N., Kloppel, G., and Bouwens, L. (1995) Duct- to islet-cell differentiation and islet growth in the pancreas of duct-ligated adult rats. *Diabetologia* **38**, 1405–1411.
4. D'Alessandro, J. S., Lu, K., Fung, B. P., Colman, A., and Clarke, D. L. (2007) Rapid and efficient in vitro generation of pancreatic islet progenitor cells from nonendocrine epithelial cells in the adult human pancreas. *Stem Cells Dev.* **16**, 75–89.
5. Bonner-Weir, S., Taneja, M., Weir, G. C., Tatarkiewicz, K., Song, K. H., Sharma, A., and O'Neil, J. J. (2000) In vitro cultivation of human islets from expanded ductal tissue. *PNAS* **97**, 7999–8004.
6. Heremans, Y., Van de Casteele, C. M., in't Veld, P., Gradwohl, G., Serup, P., Madsen, O., Pipeleers, D., and Heimberg, H. (2002) Recapitulation of embryonic neuroendocrine differentiation in adult human pancreatic duct cells expressing neurogenin 3. *J. Cell Biol.* **159**, 303–312.
7. Xu, X., D'Hoker, J., Stange, G., Bonne, S., De Leu, N., Xiao, X., Van de Casteele, M., Mellitzer, G., Ling, Z., Pipeleers, D., Bouwens, L., Scharfmann, R., Gradwohl, G., and Heimberg, H. (2008) Beta cells can be generated from endogenous progenitors in injured adult mouse pancreas. *Cell* **132**, 197–207.
8. Yato, S., Dodge, R., Akashi, T., Omer, A., Sharma, A., Weir, G. C., and Bonner-Weir, S. (2007) Differentiation of affinity-purified human pancreatic duct cells to β-cells. *Diabetes* **56**, 1802–1809.
9. Bonner-Weir, S., Inada, A., Yatoh, S., Li, W., Aye, T., Toschi, E., and Sharma, A. (2008) Transdifferentiation of pancreatic ductal cells to endocrine β-cells. *Biochem. Soc. Trans.* **36**, 353–356.
10. Inada, A., Nienaber, C., Katsuta, H., Fugitani, Y., Levine, J., Morita, R., Sharma, A., and Bonner-Weir, S. (2008) Carbonic anhydrase II-positive pancreatic cells are progenitors for both endocrine and exocrine pancreas after birth. *PNAS* **105**, 19915–19919.
11. Hao, E., Tyrberg, B., Itkin-Ansari, P., Lakey, J. R., Geron, I., Monosov, E. Z., Barcova, M., Mercola, M., and Levine, F. (2006) Beta-cell differentiation from nonendocrine epithelial cells of the adult human pancreas. *Nature Med.* **12**, 310–316.
12. Rooman, I., Lardon, J., and Bouwens, L. (2002) Gastrin stimulates beta-cell neogenesis and increases islet mass from transdifferentiated but not from normal exocrine pancreas tissue. *Diabetes* **51**, 686–690.
13. Baeyens, L., De Breuck, S., Lardon, J., Mfopou, J. K., Rooman, I., and Bouwens, L.

(2005) In vitro generation of insulin-producing beta cells from adult exocrine pancreatic cells. *Diabetologia* **48**, 49–57.

14. Baeyens, L., and Boewens, L. (2008) Can β-cells be derived from exocrine pancreas? *Diabetes Obes. Metab.* **10**, 170–178.
15. Lipsett, M. A., Castellarin, M. L., and Rosenberg, L. (2007) Acinar plasticity development of a novel in vitro model to study human acinar-to-duct-to-islet differentiation. *Pancreas* **34**, 452–457.
16. Teitelman, G. (2004) Islet-derived multipotential cells/progenitor cells. *Cell Biochem. Biophys.* **40**, 89–102.
17. Gallo, R., Gambelli, F., Gava, B., Sasdelli, F., Tellone, V., Masini, M., Marchetti, P., Dotta, F., and Sorrentino, V. (2007) Generation and expansion of multipotent mesenchymal progenitor cells from cultured human pancreatic islets. *Cell Death. Differ.* **14**, 1860–1871.
18. Eberhardt, M., Salmon, P., von Mach, M. A., Hengstler, J. G., Brulport, M., Linscheid, P., Seboek, D., Oberholzer, J., Barbero, A., Martin, I., Muller, B., Trono, D., and Zulewski, H. (2006) Multipotential nestin and Isl-1 positive mesenchymal stem cells isolated from human pancreatic islets. *Biochem. Biophys. Res. Commun.* **345**, 1167–1176.
19. Gershengorn, M. C., Hardikar, A. A., Wei, C., Geras-Raaka, E., Marcus-Samuels, B., and Raaka, B. M. (2004) Epithelial-to-mesenchymal transition generates proliferative human islet precursor cells. *Science* **306**, 2261–2264.
20. Todorov, I., Omori, K., Pascual, M., Rawson, J., Nair, I., Valiente, L., Vuong, T., Matsuda, T., Orr, C., Ferreri, K., Smith, C. V., Kandeel, F., and Mullen, Y. (2006) Generation of human islets through expansion and differentiation of non-islet pancreatic cells discarded (pancreatic discard) after islet isolation. *Pancreas* **32**, 130–138.
21. Seeberger, K. L., Dufour, J. M., Shapiro, A. M., Lakey, J. R., Rajotte, R. V., Korbutt, G. S. (2006) Expansion of mesenchymal stem cells from human pancreatic ductal epithelium. *Lab. Invest.* **86**, 141–153.
22. Baertschiger, R. M., Bosco, D., Morel, P., Serre-Beinier, V., Berney, T., Buhler, L. H., and Gonelle-Gispert, C. (2008) Mesenchymal stem cells derived from human exocrine pancreas express transcription factors implicated in beta-cell development. *Pancreas* **37**, 75–83.
23. Horwitz, E. M., Le Blanc, K., Dominici, M., Mueller I., Slaper-Cortenbach, I., Marini F. C., Deans, R. J., Krause, D. S., and Keating, A. (2005) Clarification of the nomenclature for MSC: The international society for cellular therapy position statement. *Cytotherapy* 7, 393–395.
24. Lee, K. D., Kuo, T. K., Whang-Peng, J., Chung, Y. F., Lin, C. T., Chou, S. H., Chen, J.R., Chen, Y. P., and Lee, O. K. (2004) In vitro hepatic differentiation of human mesenchymal stem cells. *Hepatology* **40**, 1275–1284.
25. Seo, M. J., Suh, S. Y., Bae, Y. C., and Jung, J. S. (2005) Differentiation of human adipose stromal cells into hepatic lineage in vitro and in vivo. *Biochem. Biophys. Res. Commun.* **328**, 258–264.
26. Timper, K., Seboek, D., Eberhardt, M., Linscheid, P., Christ-Crain, M., Keller, U., Muller, B., and Zulewski, H. (2006) Human adipose tissue-derived mesenchymal stem cells differentiate into insulin, somatostatin, and glucagon expressing cells. *Biochem. Biophys. Res. Commun.* **341**, 1135–1140.
27. Battula, V. L., Bareiss, P. M., Treml, S., Conrad, S., Albert, I., Hojak, S., Abele, H., Schewe, B., Just, L., Skutella, T., and Buhring, H. (2007) Human placenta and bone marrow derived MSC cultured in serum-free, b-FGF-containing medium express cell surface frizzled-9 and SSEA-4 and give rise to multilineage differentiation. *Differentiation* **75**, 279–291.
28. Tang, D. Q., Cao, L. Z., Burkhardt, B. R., Xia, C. Q., Litherland, S. A., Atkinson, M. A., and Yang, L. J. (2004) In vivo and in vitro characterization of insulin-producing cells obtained from murine bone marrow. *Diabetes* **53**, 1721–1732.
29. Oh, S. H., Muzzonigro, T. M., Bae, S. H., LaPlante, J. M., Hatch, H. M., and Petersen, B. E. (2004) Adult bone marrow-derived cells trans-differentiating into insulin-producing cells for the treatment of type I diabetes. *Lab. Invest.* **84**, 607–617.
30. Russ, H. A., Bary, Y., Ravassard, P., and Efrat, S. (2008) In vitro proliferation of cells derived from adult human ß-cells revealed by cell-lineage tracing. *Diabetes* **57**, 1575–1583.
31. Atouf, F., Park, C. H., Pechhold, K., Ta, M., Choi,Y., and Lumelsky, N. L. (2007) No evidence for mouse pancreatic beta-cell epithelial-mesenchymal transition in vitro. *Diabetes* **56**, 699–702.
32. Chase, L. G., Ulloa-Montoya, F., Kidder, B. L., and Verfaillie, C. M. (2007) Islet-derived fibroblast-like cells are not derived via epithelial-mesenchymal transition from Pdx-1 or insulin-positive cells. *Diabetes* **56**, 3–7.
33. Morton, R. A., Geras-Raaka, E., Wilson, L. M., Raaka, B. M., and Gershengorn, M. C. (2007)

Endocrine precursor cells from mouse islets are not generated by epithelial-to-mesenchymal transition of mature beta cells. *Mol. Cell Endocrinol.* **270**, 87–93.

34. Kayali, A. G., Flores, L. E., Lopez, A. D., Kutlu, B., Baetge, E., Kitamura, R., Hao, E., Beattie, G. M., and Hayek, A. (2007) Limited capacity of human adult islets expanded in vitro to redifferentiate into insulin-producing beta-cells. *Diabetes* **56**, 703–708.
35. Seeberger, K. L., Eshpeter, A., Rajotte, R. V., and Korbutt, G. S., (2009) Epithelial cells within the human pancreas do not coexpress mesenchymal antigens: epithelial-mesenchymal transition is an artifact of cell culture. *Lab. Invest.* **89**, 110–121.
36. Lakey, J. R. T., Warnock, G. L., Shapiro, A. M. J., Korbutt, G. S., Ao, Z., Kneteman, N. M., and Rajotte, R. V. (1999) Intraductal collagenase delivery into the human pancreas using syringe loading or controlled perfusion. *Cell Transplant.* **8**, 285–292.
37. Pittenger, M. F., Mackay, A. M., Beck, S. C., Jaiswal, R., Douglas, R., Mosca, J. D., Moorman, M. A., Simonetti, D. W., Craig, S., and Marshak, D. R. (1999) Multilineage potential of adult human mesenchymal stem cells. *Science* **284**, 143–147.
38. Pittenger, M. F., and Martin, B. J. (2004) Mesenchymal stem cells and their potential as cardiac therapeutics. *Circ. Res.* **9**, 9–20.
39. Prockop, D. J., Phinney, D. J., and Bunnell, B. A. (2008) Mesenchymal stem cells: methods and protocols. *Methods in Molecular Biology* **449**, Humana Press, Totowa.
40. Baxter, M. A., Wynn, R. F., Jowitt, S. N., Wraith, J. E., Fairbairn, L. J., and Bellantuono, I. (2004) Study of telomere length reveals rapid aging of human marrow stromal cells following in vitro expansion. *Stem Cells* **22**, 675–682.
41. McManus, J.F.A., and Mowry R. (1960) Staining Methods Histological and Histochemical. Paul B Hoeber, New York.

Chapter 11

Derivation and Characterization of Human ESC-Derived Mesenchymal Stem Cells

Ruenn Chai Lai, Andre Choo, and Sai Kiang Lim

Abstract

Mesenchymal stem cells (MSCs) are multipotent stem cells that have been isolated from numerous sources including human embryonic stem cells (hES). Derivation from hES is unique in that hES must be differentiated. In our hands, trypsinizing hES into single cells and plating them on gelatin coated plates in a DMEM medium supplemented with serum replacement media and FGF2 with either PDGF AB or EGF will induce differentiation of hES and selectively enhance the survival of MSCs over hES. Repeated passaging by trypsinization results in a highly enriched MSC culture. Enriched MSC cultures can be further purified to homogeneity by limiting dilution or FACS sorting for a CD105+ or CD73+ and CD24− cell population. The resulting hES-MSCs fulfill the ISCT minimal defining criteria for human MSCs, namely adherence to plastic, a surface antigen expression profile of CD29+, CD44+, CD49a+ CD49e+, CD73+, CD105+, CD166+, CD34−, CD45−, and a differentiation potential that includes adipogenesis, osteogenesis, and chondrogenesis. Finally, hES-MSCs can be extensively and stably propagated. This method of deriving hES-MSCs without the need for a xenogeneic feeder and use of animal serum could be used to derive clinically compliant MSCs from hESCs.

Key words: Mesenchymal stem cells, Human embryonic stem cells, CD105, CD73, FACS

1. Introduction

Mesenchymal stem cells (MSCs) are nonhematopoietic multipotential cells and have been reported to differentiate into an amazing array of cell types, e.g., osteocytes, chondrocytes, adipocytes, endothelial cells, etc (1). In addition to this extensive differentiation potential, MSCs are also one of the easiest adult stem cells to isolate and propagate in culture. These characteristics make them highly attractive for cell-based therapies, allowing easy expansion and differentiation ex vivo for autologous cell therapy for a myriad of diseases. To date, these cells have been reportedly isolated

Mohan C. Vemuri et al. (eds.), *Mesenchymal Stem Cell Assays and Applications*, Methods in Molecular Biology, vol. 698, DOI 10.1007/978-1-60761-999-4_11, © Springer Science+Business Media, LLC 2011

from many adult, fetal, and embryonic tissues such as bone marrow, fats, skin, etc (2). However, the use of different methods for isolation and the criteria for defining MSCs makes a comparison between different studies difficult. To better facilitate such comparison, the International Society for Cellular Therapy has issued a position statement for a minimal criterion to define multipotent MSCs (3). First, MSCs must be plastic-adherent when maintained in standard culture conditions. Second, they must express CD105, CD73, and CD90, and lack expression of CD45, CD34, CD14 or CD11b, CD79α or CD19, and HLA-DR surface molecules. Third, MSCs must differentiate to osteoblasts, adipocytes, and chondroblasts in vitro. On the basis of these criteria, several groups including ourselves have reported the derivation of MSCs from human ESCs (4–6). Unlike other reports, our derivation process does not require the use of viral vectors, DNA transformation, or the use of feeders (5). Instead, we first induce spontaneous differentiation of hESCs by trypsinizing hESCs into single cells then culture the trypsinized cells onto gelatinized plates without feeders or medium conditioned by feeders. Alternatively, hESCs can be first induced to form embryoid bodies for 14 days before trypsinization. The trypsinized cells are cultured in a medium supplemented with FGF2 and PDGF or EGF to promote MSC growth. By repeated trypsinization of hESCs into single cells followed by plating on gelatinized plates in culture medium supplemented with FGF2 and PDGF, highly enriched hES-MSCs can be obtained. The enriched hES-MSC cultures can be further purified by sorting for cells bearing MSC-associated markers such as CD105 and against cells bearing hESC-but not MSC associated markers such as CD24. Although hES-MSCs exhibit many characteristics of MSCs derived from adult bone marrow, they also exhibit some distinct biological differences. For example, they express higher levels of genes associated with early embryonic processes and, unlike bone marrow-derived MSCs, differentiate more efficiently into adipocytes than osteocytes or chondrocytes (5).

Recently, it was reported that MSCs mediate tissue repair by secreting paracrine factors that promote cell growth and reduce tissue injury (7). Our group has also demonstrated that hES-MSCs secrete more than 201 unique gene products (8) and this secretion reduces reperfusion injury in a pig model of myocardial ischemia/reperfusion (9). We also have demonstrated that hES-MSCs or their secreted products can be used as autogeneic feeders to support the propagation of hESCs in an undifferentiated state (10). This paracrine effect of MSCs may extend the clinical application of MSCs from cell-based to secretion-based therapies. For secretion-based applications, hES-MSCs offer several advantages over bone marrow-derived MSCs. The most distinct advantage is the almost infinitely expandable source of hESC for generating MSCs. This virtually ensures that highly uniformed batches of MSCs can be consistently generated from the same hESC source and therefore minimizes batch-to-batch variations in large-scale production of secreted products. In contrast, large-scale production of secreted products form adult tissue-derived MSCs such as bone marrow or adipose tissues will likely require multiple donors and expensive testing for each individual donor.

This chapter provides the detailed information for the derivation and characterization of hES-MSCs.

2. Materials

2.1. Cell Culture

1. Trypsin, 0.05% (1×) with EDTA 4Na, liquid (Invitrogen, Carlsbad, CA)
2. 0.1% gelatin in water (Stem Cell Technologies Inc, Vancouver, BC)
3. Dimethyl Sulphoxide (DMSO) Hybri-Max (Sigma-Aldrich, St. Louis, MO)
4. Phosphate buffered saline (PBS), pH 7.4 (1×) (Invitrogen)
5. Cryo 1°C Freezing Container, "Mr. Frosty" (Nalgene, Rochester, NY)
6. 15/10/6 cm tissue culture plate (BD Biosciences, San Jose, CA)
7. Six-well low adhesion culture dish (Corning, New York City, NY)

2.2. hES-MSC Medium

Knockout Dulbecco's modified eagele's medium (DMEM, Invitrogen Corporation, Carlsbad, CA) supplemented with 10%(v/v) knockout serum replacement (Invitrogen), nonessential amino acids (Invitrogen), penicillin–streptomycin–glutamine (PSG) (Invitrogen), β-mercaptoethanol (Invitrogen), 20 ng/ml basic fibroblast growth factor (bFGF) (Invitrogen), and 20 ng/ml recombinant human EGF (Invitrogen). Store at 4°C for up to 4 weeks (see Note 1).

2.3. Cell Sorting and Surface Antigen Profiling

1. Antibodies: CD24-PE (clone ML5), CD29-PE (clone MAR4), CD44-FITC (clone G44-26), CD49a-PE (clone SR84), CD49e-PE (clone IIA1), CD73-PE (clone AD2), CD166-PE (clone 3A6), CD34-FITC (clone 581), CD45-FITC (clone HI30), mouse IgG1κ-FITC Isotype Control, mouse IgG1κ-PE Isotype Control (all from BD Biosciences, San Jose, CA); CD105-FITC (clone SN6) (AbD Serotec, Oxford, UK)
2. FACSAria Cell Sorter (BD Biosciences)

3. FACSCalibur Flow Cytometer (BD Biosciences)
4. FACSDiva Software (BD Biosciences)
5. CELLQuest software (BD Biosciences)

2.4. Differentiation Assays

1. STEMPRO® Adipogenic, Chondrogenesis, Osteogenesis Differentiation Kit (Invitrogen) or Adipogenic, Chondrogenesis, Osteogenesis Differentiation Media BulletKits (Lonza, *Walkersville, MD*)

3. Methods

3.1. Derivation of MSCs from hESCs (hES-MSCs)

3.1.1. hES-MSC Derivation: Protocol 1 (see Note 2)

1. HuES9 human ESCs are maintained on mouse embryonic fibroblast feeder as previously described (11).
2. To generate MSCs, a starting confluent 6-cm plate of HuES9 hESCs is recommended.
3. Aspirate the culture medium and rinse cells with 5 ml of PBS.
4. Add 1 ml of trypsin/EDTA and return the plate to the 37°C CO_2 incubator for 8 min (see Note 3).
5. After incubation, remove the plate and tilt the plate from side to side to dislodge and disperse the cells into a cell suspension.
6. Neutralize trypsin with 5 ml of hES-MSC medium. Pipet cell suspension up and down gently (~5 times) to break up any cell clumps.
7. Centrifuge the cell suspension for 5 min at 800 × *g*, 4°C.
8. Discard the supernatant and loosen the cell pellet by flicking the side of tube with your finger.
9. Add 5 ml of hES-MSC medium to resuspend the cells (see Note 4).
10. Plate cell suspension on a 6-cm gelatinized plate (see Note 5). No feeder is required from this point.
11. After plating, return the cell suspension to the CO_2 incubator for 48 h to allow cells to adhere to the plate.
12. After 48 h, remove culture medium and cell debris, wash the cell culture with 5 ml of PBS, and replenish cells with fresh hES-MSC medium (see Note 6).
13. Repeat trypsinization when the culture is 90~100% confluent.
14. Seeding density is recommended at 500 cells per cm^2.
15. After the third or fourth trypsinization, the characteristic fingerprint whorl of confluent MSC cultures starts to become evident (see Notes 7, 8). Figure 1 shows differentiated hES-MSCs, HuES9.E1.

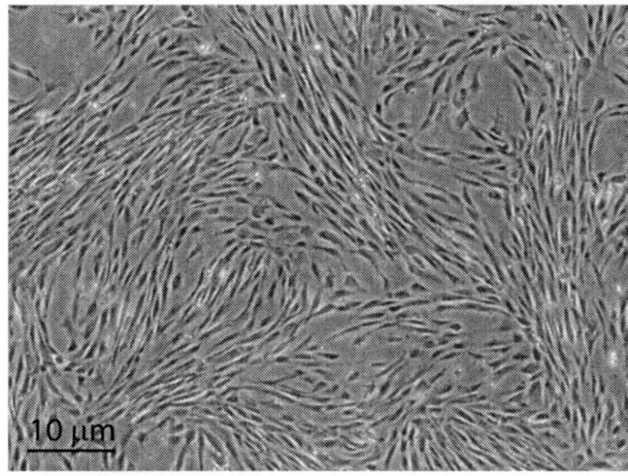

Fig. 1. hES-MSCs. Cellular morphology of huES9.E1 (hES-MSCs) under phase contrast.

16. When the fingerprint whorl of confluent MSC cultures forms, trypsinize the cell culture and plate cells on a gelatinized 10 cm plate. MSC cultures should not be expanded at more than a 1:3–1:4 split. When cells have reached a population size of 10^7 the culture is designated P1 (see Note 9).

3.1.2. hES-MSC Derivation: Protocol 2

1. After Subheading 3.1.1 step 3, add 2 ml hES-MSC medium and mechanically dissociate hESC into clumps using a pipette tip.
2. Detach the cells with a cell scrapper and transfer the clumps into two wells of a six-well low adhesion culture dish containing an additional 3 ml hES-MSC medium per well to allow EB formation.
3. Culture the EBs in suspension for 7 days with medium change every alternate day. Collect the EBs in a 15-ml conical centrifuge tube and centrifuged at 500×*g*, 4°C for 3 min. The supernatant is discarded and 3 ml of fresh hES-MSC medium (AC: typically we allow the EBs to settle in a 15-ml tube, remove most of the media – leaving ~0.5 ml and resuspend in fresh media. Rationale for this is that it is easier to resuspend the EBs when compared with centrifugation).
4. After 7 days, transfer the EBs into two gelatinized wells of a six-well tissue culture dish containing hES-MSC medium and allow the EBs to adhere to the dish.
5. Differentiate the attached EBs for an additional 7 days with medium change every alternate day.
6. Resume from step 13 of Subheading 3.1.1.

3.2. hES-MSC Cell Sorting

1. Trypsinize the differentiating hESCs with 1 ml of trypsin for 8 min at 37°C. Tilt plate from side to side to ensure that all cells are lifted off the plates and there are no visible cell clumps (see Notes 10, 11).
2. Neutralize the trypsin with an equal volume of hES-MSC medium. Gently pipet the cells up and down about 5 times to break up any cell clumps.
3. Centrifuge the cell suspension for 5 min at 800 × *g*, 4°C.
4. Discard the supernatant and resuspend the cell pellet in 10 ml culture medium.
5. Plate the cell suspension on 10 cm bacterial culture dish and place the plate on an orbital shaker with gentle shaking for 2 h in a CO_2 incubator at 37°C.
6. After 2 h shaking, harvest the cells and centrifuge the cell suspension for 5 min at 800 × *g*, 4°C.
7. Wash cells twice with PBS by resuspending the cell pellet in 10 ml PBS each time followed by centrifuging the cell suspension for 5 min at 800 × *g*, 4°C.
8. Resuspend 1×10^6 cells in 0.5 ml 2% FBS or KSR in PBS (v/v).
9. Add 10 μl of FITC-conjugated anti-human CD105 and 10 μl of PE-conjugated anti-human CD24 or the corresponding FITC- or PE-conjugated mouse IgG1 isotype controls to the cell samples.
10. Incubate with gentle shaking for 40 min at room temperature in the dark.
11. Centrifuge the cell suspension for 5 min at 800 × *g*, 4°C.
12. Repeat step 7.
13. Resuspend cells in 500 μl medium and sort for CD105+, CD24− cells on a FACS Aria using FACS Diva software.
14. After sorting, plate cells at a density of about 500 cells per cm^2 on a 15-cm gelatinized tissue culture plate with hES-MSC medium. Incubate the cellsat 37°C in a CO_2 incubator.

3.3. Passaging hES-MSCs

1. Source expanded hES-MSCs from the CO_2 incubator (see Note 12).
2. Aspirate medium and rinse cells with PBS.
3. For a 15-cm plate, add 5 ml trypsin to cells and place in the incubator for 8 min.
4. Remove plates from incubator, tilt plate from side to side to ensure that all cells are lifted off the plates and there are no visible cell clumps.
5. Add 5 ml hES-MSC medium to the plate to neutralize trypsin.
6. Gently pipette the cells up and down about 5 times to break up any remaining cell clumps.

7. Transfer cell suspension into a 50-ml falcon tube.
8. Wash the plate with another 10 ml culture medium. Transfer wash into the same 50 ml Falcon tube.
9. Centrifuge at 800 × *g* for 3 min, 4°C.
10. Aspirate supernatant, dislodge cell pellet by tapping on the outside of the tube and resuspend cells in 40 ml hES-MSC medium. Pipette up and down until cells are evenly dispersed.
11. Distribute 10 ml of the cell suspension onto new gelatinized 15-cm plates.
12. Add an additional 10 ml of hES-MSC medium to each plate.
13. Feed cells every 48 h with fresh medium until the culture is ~60–75% confluent and thereafter every 24 h. Cells reach 90–100% confluency in about 7–9 days. The cells should ideally be split before they are 100% confluent, preferably at 80% confluent.

3.4. Freezing and Thawing

1. Freezing Medium (2×), contains 80% (v/v) FBS and 20% (v/v) DMSO . Mix well. Keep medium at 4°C until use. Store complete freezing medium for ≤24 h.
2. Trypsinize a 15-cm MSC plate (~90% confluence) with 5 ml trypsin for 8 min at 37°C.
3. Remove plate from incubator, tilt plate from side to side to ensure that all cells are lifted off the plates and there are no visible cell clumps.
4. Neutralize the trypsin with an equal volume of hES-MSC medium. Gently pipette the cells up and down (~5 times) to break up any cell clumps.
5. Centrifuge the cell suspension for 5 min at 800 × *g*, 4°C.
6. Discard the supernatant and resuspend the cell pellet in 2 ml hES-MSC medium.
7. Add 2 ml of 2× Freezing Medium slowly while gently shaking the tube to mix the freezing medium and the cell suspension.
8. When addition is complete, pipette the cell suspension up and down a few times to ensure complete and even distribution of the freezing medium.
9. Quickly aliquot 1 ml of the cell suspension into prelabeled cyrovials.
10. Transfer the cryovials to a Cryo 1°C Freezing Container and place at −80°C for 48 h before transferring to long-term storage in liquid nitrogen or at −150°C.

3.5. Surface Antigen Profiling

1. Source hES-MSCs from the incubator and place in the tissue culture hood.
2. Aspirate the medium and rinse cells with PBS.

3. For a 15-cm plate, add 5 ml trypsin to cells and place in the incubator for 8 min. Tilt plate from side to side to ensure that all cells are lifted off the plates and there are no visible cell clumps.
4. Neutralize the trypsin with an equal volume of hES-MSC medium. Gently pipet the cells up and down about 5 times to break up any cell clumps.
5. Centrifuge the cell suspension for 5 min at 800 × *g*, 4°C.
6. Discard the supernatant and resuspend the cell pellet in 10 ml culture medium.
7. Plate the cell suspension in a 10-cm bacterial culture dish and place the plate on an orbital shaker with gentle shaking for 2 h in a CO_2 incubator at 37°C (see Note 13).
8. After 2 h shaking, harvest the cells and centrifuge the cell suspension for 5 min at 800 × *g*, 4°C.
9. Wash cells twice with PBS by resuspending the cell pellet in 10 ml PBS each time followed by centrifuging the cell suspension for 5 min at 800 × *g*, 4°C.
10. Resuspend 2.5×10^5 cells in 0.5 ml 2% FBS or KSR in PBS (v/v).
11. Add 10 μl of FITC- or PE-conjugated anti-human surface antigen antibodies or the corresponding FITC- or PE-conjugated mouse IgG1 isotype controls to the cell samples. Surface antigen profiling is performed in triplicate.
12. Incubate the cells on ice for 1 h. Resuspend the cells every 15 min by gentle tapping on the tube.
13. Centrifuge the cell suspension for 5 min at 800 × *g*, 4°C.
14. Repeat step 7.
15. Resuspend cell pellet in 0.5 ml PBS with 2% FBS and run the FACS analysis on a FACSCalibur using the CELLQuest software.
16. Utilize cells stained with mouse IgG1-FITC or mouse IgG1-PE as nonspecific fluorescence control.

3.6. Differentiation Assay

1. In our experience, hESC-derived MSCs can be induced to differentiate into adipocytes, osteocytes, and chondrocytes using commercially available MSC differentiation assay kit. We have tried assay kits from both Invitrogen and Lonza and by following the manufacturer's instructions, our hESC-derived MSCs can differentiate into adipocytes, chondrocytes, and osteocytes (see Note 14).
2. Grow hES-MSCs to 80–100% confluency in hES-MSC medium.
3. Trypsinize cells as described above.

4. Count the number of cells using a hemocytometer.
5. Seed the cells according to the manufacturer's instruction for each differentiation assay kit.
6. Process cells as per manufacturer's instructions.

4. Notes

1. The knockout serum replacement (Invitrogen), FGF and EGF in hES-MSC medium can be eliminated and replaced with ES cell qualified fetal bovine serum (Invitrogen). The other supplements in this serum-containing medium include nonessential amino acids (Invitrogen), PSG (Invitrogen), and β-mercaptoethanol (Invitrogen). Store at 4°C for up to 4 weeks.
2. The efficiency in deriving MSCs from hESCs using the method described below depends on the hESC lines. In our hands, HuES9 (11) is most amenable to generating MSCs using this method and we have also derived MSCs from H1 ESC line (12).
3. Dissociation of hESC to derive MSCs by trypsin. Use of Trypsin is key for differentiation. Use of TrypLE Express or collagenase does not induce differentiation of human ESC.
4. Dissociation of hESC into single cells is critical in the derivation.
5. The 6-cm gelatinized plate is prepared by adding 1–2 ml of gelatin to completely cover a 6-cm plate and leaving the plate to stand for at least 15 min or longer at room temperature. Just before use, remove the gelatin.
6. It is expected that there will be extensive cell death and differentiation of hESCs after 48 h culture in hES-MSC medium.
7. Human ESCs can be propagated on MSCs. During the early stages of MSC derivation, poorly dissociated hESC colonies will continue to propagate with the newly derived MSCs.
8. Repeated trypsinization of hESCs more than 5 times as described above generally generate a fairly homogenous MSC culture.
9. During the process of trypsinization of hESCs to generate MSCs and then the subsequent expansion to 10^7 cells, the growth rate could be quite variable and could take about 4–8 weeks before stabilizing to a population doubling time of about 2–3 days.
10. The trypsinized cells can be sorted for CD105+ and CD24– hES-MSCs as early as 1 week after hESCs have been trypsinized.

11. Sorting of cells by FACS significantly reduce the viability of cells and an alternative method to enhance homogeneity of the MSC culture is to use the more laborious method of limiting dilution.
12. It is recommended to always maintain hESC-derived MSCs between 25 and 80% confluency or ~15–50,000 cells per cm^2. Upon reaching 80% confluence, it is recommended to passage hES-MSCs at a 1:3 or 1:4 split ratio.
13. Longer incubation time of MSCs on the bacterial culture dish will cause cell clumping and affect the subsequent FACS analysis.
14. The cells tend to differentiate poorly at higher passage number or after prolonged propagation in serum-containing medium.

Acknowledgments

This work was supported by funding from A*STAR. We thank members of our laboratories for their contributions to this work.

References

1. Brooke, G., Cook, M., Blair, C., Han, R., Heazlewood, C., Jones, B., et al. (2007) Therapeutic applications of mesenchymal stromal cells. *Semin Cell Dev Biol.* **18**, 846–858.
2. Wagner, W., Ho, A. D. (2007) Mesenchymal stem cell preparations – comparing apples and oranges. *Stem Cell Rev.* **3**, 239–248.
3. Dominici, M., Le Blanc, K., Mueller, I., Slaper-Cortenbach, I., Marini, F., Krause, D., et al. (2006) Minimal criteria for defining multipotent mesenchymal stromal cells. The international society for cellular therapy position statement. *Cytotherapy.* **8**, 315–317.
4. Barberi, T., Willis, L. M., Socci, N. D., Studer, L. (2005) Derivation of multipotent mesenchymal precursors from human embryonic stem cells. *PLoS Med.* **2**, e161.
5. Lian, Q., Lye, E., Suan Yeo, K., Khia Way Tan, E., Salto-Tellez, M., Liu, T. M., et al. (2007) Derivation of clinically compliant MSCS from cd105+, cd24- differentiated human ESCS. *Stem Cells.* **25**, 425–436.
6. Olivier, E. N., Rybicki, A. C., Bouhassira, E. E. (2006) Differentiation of human embryonic stem cells into bipotent mesenchymal stem cells. *Stem Cells.* **24**, 1914–1922.
7. Caplan, A. I., Dennis, J. E. (2006) Mesenchymal stem cells as trophic mediators. *J Cell Biochem.* **98**, 1076–1084.
8. Sze, S. K., de Kleijn, D. P., Lai, R. C., Khia Way Tan, E., Zhao, H., Yeo, K. S., et al. (2007) Elucidating the secretion proteome of human embryonic stem cell-derived mesenchymal stem cells. *Mol Cell Proteomics.* **6**, 1680–1689.
9. Timmers, L., Lim, S.-K., Arslan, F., Armstrong, J. S., Hoefler, I. E., Doevendans, P. A., et al. (2007) Reduction of myocardial infarct size by human mesenchymal stem cell conditioned medium. *Stem Cell Res.* **1**, 129–137.
10. Choo, A., Ngo, A. S., Ding, V., Oh, S., Kiang, L. S. (2008) Autogeneic feeders for the culture of undifferentiated human embryonic stem cells in feeder and feeder-free conditions. *Methods Cell Biol.* **86**, 15–28.
11. Cowan, C. A., Klimanskaya, I., McMahon, J., Atienza, J., Witmyer, J., Zucker, J. P., et al. (2004) Derivation of embryonic stem-cell lines from human blastocysts. *N Engl J Med.* **350**, 1353–1356.
12. Thomson, J. A., Itskovitz-Eldor, J., Shapiro, S. S., Waknitz, M. A., Swiergiel, J. J., Marshall, V. S., et al. (1998) Embryonic stem cell lines derived from human blastocysts. *Science.* **282**, 1145–1147.

Chapter 12

Isolation and Culture of Rodent Bone Marrow-Derived Multipotent Mesenchymal Stromal Cells

Nance Beyer Nardi and Melissa Camassola

Abstract

Initial attempts to cultivate mesenchymal stem cells (MSCs) were more successful from human than murine tissues. Methods for the in vitro expansion of murine MSCs were described more recently, but are now well established. Despite limitations such as a poor equivalence to be expected between cultured stem cells and their in vivo counterparts, MSC culture allows the expansion of a cell population capable of providing important information on the biology of stem cells and their therapeutic application.

Murine MSCs may be obtained from the bone marrow and virtually from any other organ or tissue. This chapter describes the most widely used method, which involves the preparation of single-cell suspension followed by incubation for 1–3 days and removal of nonadherent cells. The adherent fraction is then expanded by continuous culture and may be maintained for prolonged periods of time.

Key words: Cell culture, Mesenchymal stem cell, Mouse, Rat, Stromal cells

1. Introduction

Mesenchymal stem cells (MSCs) are one of the most attractive types of adult stem cells, for characteristics such as ease of collection and expansion, differentiation potential, and immunoregulatory activity. Dramatic benefits reported in preclinical and early phase clinical trials (reviewed in (1)) have added to this interest.

As with all other types of adult stem cells, MSCs are operationally defined by their capacity to undergo long-term self-renewal and to differentiate into specialized cell types. Protocols for their isolation involve the selection of plastic-adherent cells, and only more recently have methods for the culture of murine MSCs been established (2). Although methods for the prospective identification of MSCs from collected tissues have been proposed, they are still recognized by their behavior after in vitro

Mohan C. Vemuri et al. (eds.), *Mesenchymal Stem Cell Assays and Applications*, Methods in Molecular Biology, vol. 698, DOI 10.1007/978-1-60761-999-4_12,

cultivation (3), leading to the concept that "MSC biology, at the present time, remains biology out of context" (4). Furthermore, culture conditions are not able to replace the physiological niche of stem cells, which will probably not behave in vitro in the same way they would in their natural habitat. Therefore, cultured stem cells should not be considered as completely equivalent to their in vivo counterparts (5).

Having recognized these limitations of the ex vivo study of adult stem cells in general, it is important to stress that, nevertheless, the in vitro cultivation of MSCs allows the expansion of a cell population capable of providing important information on the biology of stem cells and their therapeutic application (reviewed in (1)). Although conventionally isolated from the bone marrow, MSCs may be obtained from virtually any organ or tissue (6). The most widely used method, which will be described in this chapter, involves the preparation of a single-cell suspension, using collagenase in case of solid tissue, followed by incubation for 1–3 days and removal of nonadherent cells. The adherent fraction is then expanded by continuous culture. For reasons not yet clear, established cultures may present different morphologies: spindle-shaped, fibroblast-like cells or large, flat cells (Fig. 1). Although some differences in growth and plasticity of these two cell types have been reported (reviewed in (7)), they seem to represent the same basic stem cell population.

Finally, the relationship between MSCs and fibroblasts has attracted some attention more recently, and the existence of common characteristics has led to the suggestion, supported by our own experience, that they also represent different elements of the same type of stem cell pool (8).

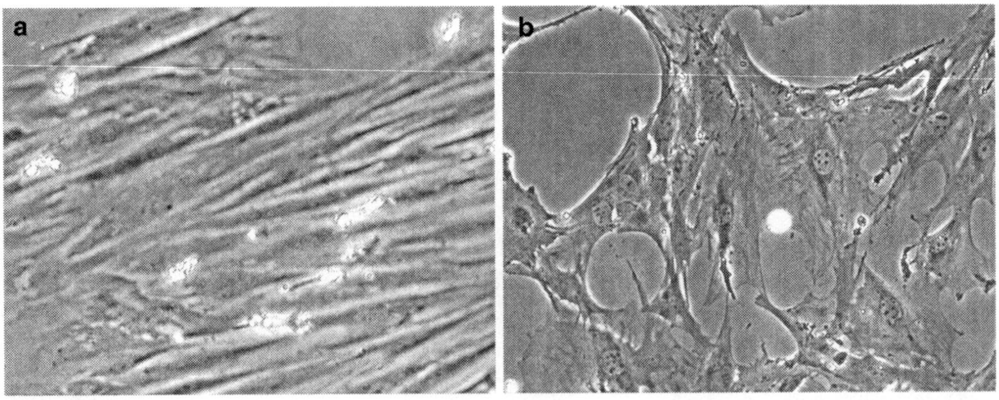

Fig. 1. Mesenchymal stem cells (MSCs) in culture may present two different morphologies. (**a**) Spindle-shaped cells are more similar to conventional fibroblasts in culture. (**b**) Large, flat cells are most usually seen with the methodology of isolation and culture described in this chapter. Original magnification 100×.

2. Materials

The methods described later are designed for a well-established cell culture laboratory. Equipment and basic items such as tissue culture supplies will not be specified. In our experience, reagents provided by different suppliers have similar efficacy, and therefore their source is not specified. One piece of equipment is of particular importance in the study of MSCs: a phase-contrast microscope of good quality. Cultures may look very different if analyzed by bright-field or phase-contrast microscopy (Fig. 2).

2.1. Mouse Perfusion

1. Ketamine hydrochloride injection, USP, 100 mg/ml.
2. Xylazine hydrochloride injection, USP, 20 mg/ml.
3. Heparin (10 U/ml).
4. Ca^{2+} Mg^{2+}-free Hank's balanced salt solution (HBBS).
5. Syringes (5 and 20–30 ml) equipped with 27 G needles.
6. Sterilized surgical instruments including forceps, surgical scissors, and scalpel blades.

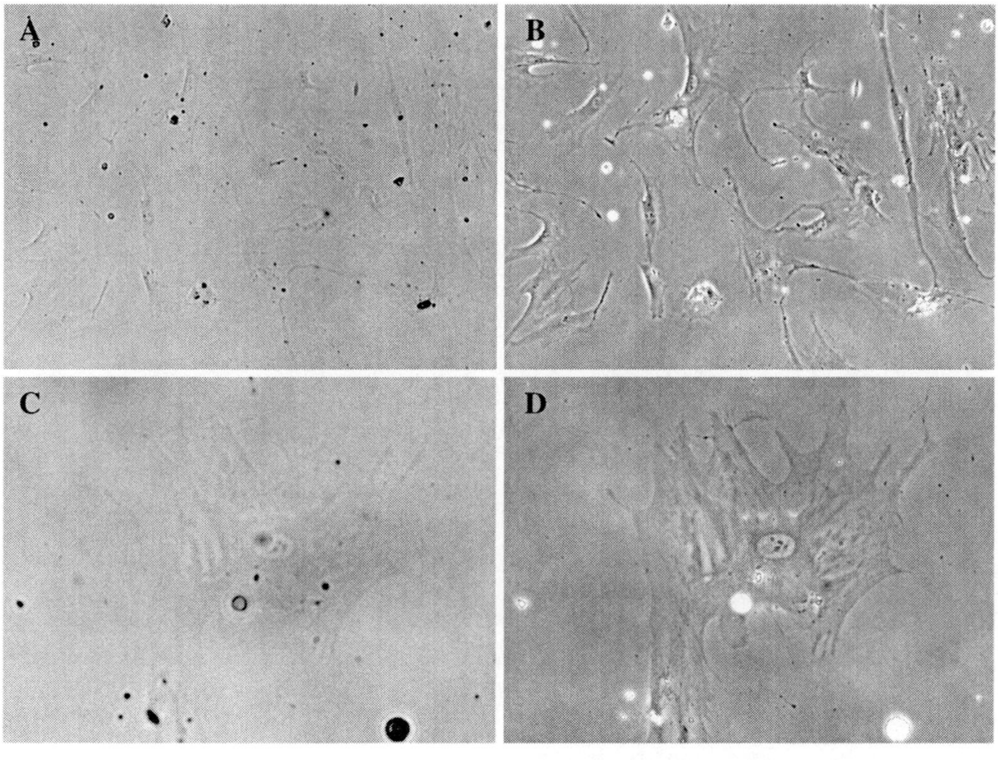

Fig. 2. Initial cultures of MSCs, still with other contaminant cell types. MSCs are more easily identified by phase contrast microscopy (**b**, **d**) than by bright-field microscopy (**a**, **c**) Original magnifications: 100× (**a**, **b**) and 200× (**c**, **d**).

2.2. Collection of Bone Marrow and Other Organs and Tissues

1. 70% ethanol.
2. Ca^{2+} Mg^{2+}-free Hank's balanced salt solution (HBBS).
3. Collagenase type I (0.5 mg/ml in DMEM/10 mM HEPES).
4. Cork board covered with foil paper.
5. Sterilized surgical instruments including forceps, surgical scissors, and scalpel blades.
6. Sterile gauze swabs
7. Petri dish
8. 10-ml syringe equipped with 27 G needle

2.3. Establishment and Expansion of MSC Culture

1. Culture medium: low-glucose Dulbecco's modified Eagle's medium (DMEM) containing 10% fetal bovine serum (see Note 1) and 5.96 g/L HEPES buffer. Antibiotics may be used.
2. Solution of trypsin (0.25%) with 1 mM ethylenediamine tetraacetic acid (EDTA).

3. Methods

The methodology described here has been successfully used for several isogenic mouse strains, including C57BL/6 and BALB/c. Some inter-strain differences in the establishment of MSC culture have been described, but in our experience the methods have similar results for different strains. We have also established MSC culture from organs and tissues of eGFP mice (green mouse FM131, kindly provided by M. Okabe, Osaka University, Japan), but the level of fluorescence tends to decrease after a few weeks of culture. This has already been reported by other groups (9).

3.1. Isolation and Culture of Mouse Bone Marrow Mesenchymal Stem Cells

3.1.1. Bone Marrow Collection

1. Euthanize mouse (see Note 2) by exposure to carbon dioxide gas or by cervical dislocation.
2. Submerge the mouse in 70% ethanol or a germicidal detergent and take the container to a sterile hood.
3. After letting some of the disinfectant drain into the container, place the mouse ventral side up on a cork board covered with foil paper, pinning each limb to the board.
4. Using sterile forceps and scissors or scalpel, make an incision around the perimeter of the hind limbs. Grab the skin and pull toward the foot, pinning it to the board so that it stays out of the way.
5. Collect femurs (see Note 3) by excising muscle and connective tissue. Remove as much muscle and other tissues as possible,

using forceps and sterile gauze swabs, and place the bones in a Petri dish with 5 ml cold HBSS.

6. Hold the femur with the forceps and, using a pair of sharp scissors or a bone cutter, carefully avoid splintering the bone, cut it at the epiphysis level.
7. Fill a 10-ml syringe with cold HBSS, attach a 27 G needle to it and insert the needle into the distal end of the femur. Slowly inject HBSS until the marrow plug is flushed out the cut end of the bone, collecting it in the Petri dish.

3.1.2. Establishment of MSC Culture

1. Prepare a single-cell suspension by drawing the marrow plug with HBSS into the syringe and expelling it out for 3 or 4 times. Wait a few seconds, tilt the Petri dish at a 45° angle, and collect most of the cell suspension, while leaving a small part in the bottom to avoid collecting larger clumps of tissue or bone fragments (see Note 4).
2. Place the collected cell suspension into a 15-ml centrifuge tube. Complete the volume to 10 ml with HBSS, centrifuge at $400\times g$ for 10 min, at room temperature (RT), remove the supernatant and add 10 ml fresh HBSS. Collect a small sample for determination of yield and viability by Trypan blue exclusion and counting in a Neubauer chamber (see Note 5). Centrifuge the cell suspension again, in the same conditions.
3. Discard the supernatant and resuspend the cells in RT or prewarmed (37°C) culture medium, to a final concentration of 5×10^{6} cells/ml. Plate in six-well tissue culture dishes, at 3.5 ml/well (or 1.94×10^{6} cells/cm^2).
4. Incubate plate at 37°C in a humidified 5% CO_2 incubator. After 3 days, remove nonadherent cells by changing the culture medium.

3.2. Isolation and Culture of Mesenchymal Stem Cells from Other Mouse Organs and Tissues

The only tissue from which MSCs seem difficult to isolate is blood (6). When adequately treated, MSCs may be isolated from any other tissue or organ. Perfusion of animals is essential for achieving good morphology of the brain, kidneys, heart, and many other organs.

3.2.1. Mouse Perfusion

1. Anesthetize the mouse with an intraperitoneal injection of ketamine and xilazine (1.16 and 2.3 g per kg body weight, respectively).
2. Open the abdominal cavity, rupture the diaphragm and inject 100 U of heparin in 200 μl HBSS into the beating heart.
3. Insert a 27-G intravenous catheter through left ventricle into ascending aorta.
4. Cut the caudal vena cava for drainage, and pump in (slowly but constantly) around 50 ml of perfusion medium. The procedure

may also be performed with a pump. If the perfusion is working well, organs such as the liver and kidney will turn grayish-white.

5. For lung perfusion, catheterize the pulmonary artery instead.

3.2.2. Collection and Processing of Organs and Tissues

1. Prepare the mouse as described in Subheading 3.1.1, steps 1–3.
2. Using sterile forceps and a scissors or scalpel, make a short (3–4 mm) ventral mideline incision through the skin. Hold the skin on either side of the incision and pull it toward the head and tail until the abdomen is completely exposed. Reflect the skin flaps out of the way and pin them to the board.
3. Collect each organ or tissue as more appropriate in each case, keeping sterility and avoiding contamination with surrounding structures (see Notes 6 and 7). The amount of tissue to be collected depends on the volume of MSC culture to be established. In our experience, the following quantities are appropriate for final plating into one well of a six-well culture plate, and should be adapted for other tissues:
   - Kidneys, spleen, and lungs: half of the organ; remove as much air as possible from lungs.
   - Pancreas: one third of the organ.
   - Liver and muscle: a small fragment, of around 3–5 mm^3.
   - Aorta and vena cava: whole structure.
   - Brain: half of one hemisphere.

The following steps are designed for these amounts of tissues, and volumes should be proportionately increased for larger-scale cultures:

4. Rinse organs and tissues in cold HBSS, transfer to a Petri dish, and cut into small pieces of around 0.2–0.8 cm^3.
5. Wash with cold HBSS and cut into smaller fragments, as possible in each case.
6. Transfer fragments to 15-ml centrifuge tubes, remove all HBSS, and add 3 ml collagenase type I.
7. Incubate for 30 min to 3 h at 37°C (see Note 8), in a shaking water bath.
8. Pipette vigorously to improve cell dissociation, add 10 ml cold HBSS and centrifuge at 400 × *g* for 10 min at RT (see Note 9).
9. Remove supernatant and resuspend in 4 ml of RT or pre-warmed (37°C) culture medium.
10. Wait for 1–3 min to allow gross remnants to settle and transfer 3.5 ml of the supernatant to a well of a six-well culture plate.

3.2.3. Establishment of MSC Culture

1. Incubate plate at 37°C in a humidified atmosphere containing 5% CO_2.
2. Change culture medium, with removal of nonadherent cells, after 24 h.
3. Usually, growing spots of adherent cells become visible in around 2 or 3 days.

3.3. Isolation and Culture of Rat Bone Marrow Mesenchymal Stem Cells

The protocol used for establishing MSC cultures from mouse bone marrow is also adequate for rat bone marrow. In our experience, rates of expansion of mouse or rat cultures are very similar, and cultured cells also show similar morphology.

3.4. Expansion of MSC Cultures

Cultures are more easily established and expanded from some tissues than from others, in a curve that seems to be related to the degree of vascularization of the tissue or organ. MSC cultures are more easily obtained from lungs or kidneys, for instance, than from the bone marrow (6). Differences may be seen on the time taken for confluence to be reached after the isolated cells are plated. Although primary cultures generally become confluent in around 6–7 days, in some cases it may take much longer than that (see Note 10). Once established, although expansion rates may vary, maintenance of MSC cultures follows the same rules.

The aspect of cells may also present differences. During this first stage, cultures will be contaminated with other cell types, which tend to disappear after the first or second passage. The adherent layer may be more fibroblast-like or have a more flattened morphology (Fig. 1). These differences do not seem to affect proliferation or differentiation of the cultures in the long term.

The following protocol may be used for rat or mouse-derived MSC culture:

1. After the primary culture becomes confluent, begin subculture. For that, wash plate once with 3 ml HBSS, add 1 ml pre-warmed (37°C) 0.25% trypsin/EDTA, and incubate at 37°C for 10 min (see Note 11).
2. Add 1 ml culture medium to inactivate the trypsin and collect the cells by thorough pipetting. Wash the plate with fresh medium to remove residual cells.
3. Resuspend cells in prewarmed (37°C) culture medium to a final volume of 7.0 or 10.5 ml and split into two or three new wells, respectively, depending on the degree of confluence of the initial culture.
4. Subsequent passages are performed when cultures reach around 90% confluence, at ratios empirically determined for two subcultures a week at most (see Note 12). In our experience, split ratios of 1:6 are adequate at passage 5 or 6, 1:9 at

around passage 11 and, if long-term cultures are established, a maximal ratio of 1:24 is used from passage 28 onwards (see Note 13).

5. Mouse and rat MSC cultures may be maintained for extended periods of time. Different from human cells, murine MSCs tend to become polyploidy, but that does not seem to affect their capacity to proliferate or differentiate. During long-term culture, care must be taken with the effect of these genetic differences, and resulting cells are generally seen as unfit for therapeutic use.

4. Notes

1. In our experience, media and fetal bovine serum produced by different manufacturers may be used with the same efficiency. Some of us have repeatedly used these methods in different laboratories, in different countries, and with reagents from different sources, with similar results.
2. For general studies, use healthy, young adults (4–8 weeks of age). Until larger experience with the method of bone marrow collection is acquired, it is better to sacrifice one mouse at a time, collect the cells, and then move to the next one. Later on, the procedure can include several mice.
3. One femur will yield approximately $1.5–2 \times 10^7$ cells. Cells can also be collected, although less easily, from the tibia.
4. To get cleaner preparations, the cell suspension may be filtered through a 70-μm strainer or a nylon screen. Other groups recommend that murine bone marrow cells are centrifuged on Ficoll-Paque. In our experience, these steps are not necessary for the establishment of MSC cultures.
5. Always resuspend cells by pipetting up and down against the side of the centrifuge tube, to give a uniform cell suspension. Collect the sample immediately after resuspension.
6. To separate the adipose layer surrounding the aorta, incubate with collagenase for around 30 min and agitate vigorously. Wash the remnant of the vessel in 20 ml HBSS and transfer to a new tube for a second digestion with collagenase. The fractions may be mixed for the establishment of one single culture.
7. Adipose-derived MSC-like stem cells (ADSCs) are generally similar, though not identical, to MSCs, and have been increasingly used in the clinical setting as an alternative stem cell. Similar to human ADSCs, murine cells have also been shown to present improved proliferative capacity and plasticity when compared with bone marrow MSCs. Subcutaneous and

visceral adipose tissue may be collected from mice. Adipose tissue may also be obtained from epididymal fat pads or lymph nodes.

8. Some types of tissue are digested more easily than others, so that the time of incubation must be empirically determined for each one.
9. Cells may be further cleared from debris by centrifugation on Ficoll-Paque, followed by an additional washing step.
10. For most tissues, a well-succeeded isolation will yield cultures nearly confluent and demand passaging in around 1 week. However, MSC cultures may also be established from preparations, which take longer to show initial expansion. If the cells seem healthy but are not actively proliferating, change the medium after 1 week. However, if in another week they do not show expansion, it is better to establish a new culture.
11. The trypsin solution should be of good quality and well stored. If the culture does not detach from the plate surface after adequate incubation with trypsin, or scraping is necessary for collecting the cells, try using a fresh bath of the enzyme.
12. Avoid leaving the culture to expand after reaching confluence. This may lead to several unwanted situations, including poor cell dissociation due to excess deposition of extracellular matrix or spontaneous differentiation.
13. The rate of expansion of cultures established from different organs may also show variation. MSCs isolated from lungs, pancreas, and kidney proliferate more, whereas those from liver and brain usually expand more slowly.

Acknowledgments

The authors would like to thank Drs. Lindolfo Meirelles and Luisa Maria Gomes de Macedo Braga, whose work was important for the establishment of many of the techniques described here. This work was supported by Conselho Nacional de Desenvolvimento Científico e Tecnológico (CNPq) and Fundação de Amparo à Pesquisa do Estado do Rio Grande do Sul (FAPERGS).

References

1. Meirelles, L. da S. and Nardi, N. B. (2009) Methodology, biology and clinical applications of mesenchymal stem cells. *Front. Biosci.* **14**, 4281–4298.
2. Meirelles, L. da S. and Nardi, N. B. (2003) Murine marrow-derived mesenchymal stem cell: isolation, in vitro expansion, and characterization. *Br. J. Haematol.* **123**, 702–711.
3. Dominici, M., Le Blanc, K., Mueller, I., Slaper-Cortenbach, I., Marini, F., Krause, D., et al. (2006) Minimal criteria for defining multipotent mesenchymal stromal cells. The

International Society for Cellular Therapy position statement. *Cytotherapy* **8**, 315–317.

4. Javazon, E. H., Beggs, K. J. and Flake, A. W. (2004) Mesenchymal stem cells: Paradoxes of passaging. *Exp. Hematol.* **32**, 414–425.
5. da Silva Meirelles, L., Caplan, A. I. and Nardi, N. B. (2008) In search of the in vivo identity of mesenchymal stem cells. *Stem Cells (Dayton, Ohio)* **26**, 2287–2299.
6. da Silva Meirelles, L., Chagastelles, P. C. and Nardi, N. B. (2006) Mesenchymal stem cells reside in virtually all post-natal organs and tissues. *J. Cell Sci.* **119**, 2204–2213.
7. Bobis, S., Jarocha, D. and Majka, M. (2006) Mesenchymal stem cells: characteristics and clinical applications. *Folia Histochem. Cytobiol.* **44**, 215–230.
8. Haniffa, M. A., Collin, M. P., Buckley, C. D. and Dazzi, F. (2009) Mesenchymal stem cells: the fibroblasts' new clothes? *Haematologica* **94**, 258–263.
9. Harting, M. T., Jimenez, F. and Cox, C. S. (2009) Isolation of mesenchymal stem cells (MSC) from green fluorescent protein positive (GFP+) transgenic rodents: The grass is not always green(er). *Stem Cells Dev.* **18**, 127–136.

Chapter 13

Cryopreservation and Revival of Mesenchymal Stromal Cells

Mandana Haack-Sørensen and Jens Kastrup

Abstract

Over the past few years, the pace of preclinical stem cell research is astonishing and adult stem cells have become the subject of intense research. Due to the presence of promising supporting preclinical data, human clinical trials for stem cell regenerative treatment of various diseases have been initiated. As there has been a precedent for the use of bone marrow stem cells in the treatment of hematological malignancies and ischemic heart diseases through randomized clinical safety and efficacy trials, the development of new therapies based on culture-expanded human mesenchymal stromal cells (MSCs) opens up new possibilities for cell therapy.

To facilitate these applications, cryopreservation and long-term storage of MSCs becomes an absolute necessity. As a result, optimization of this cryopreservation protocol is absolutely critical. The major challenge during cellular cryopreservation is the lethality associated with the cooling and thawing processes. The major objective is to minimize damage to cells during low temperature freezing and storage and the use of a suitable cryoprotectant. The detrimental effects of cellular cryopreservation can be minimized by controlling the cooling rate, using better cryoprotective agents, maintaining appropriate storage temperatures, and controlling the cell thawing rate. As is described in this chapter, human MSCs can either be frozen in cryovials or in freezing bags together with cryopreserve solutions containing dimethyl sulfoxide (DMSO).

Key words: Mesenchymal stromal cells, Cryopreservation, Cryoprotectant, DMSO, Viability measurement

1. Introduction

Regenerative medicine has a wide definition covering therapies that aim to repair, replace, restore, and regenerate damaged cells, tissues, and organs. Innovative therapies derived from stem cells fall within this definition and may, in the coming years, provide options for the treatment of a variety of diseases such as autoimmune, cardiovascular, CNS, and hematological malignancies.

Mohan C. Vemuri et al. (eds.), *Mesenchymal Stem Cell Assays and Applications*, Methods in Molecular Biology, vol. 698, DOI 10.1007/978-1-60761-999-4_13,

Stem cell transplants have been used since the 1960s to treat a variety of hematological diseases. In 1989, cord blood stem cells were used for the first time in hematopoietic stem cell transplantation (1). An effective and consistent cryopreservation technique is crucial for any stem cell laboratory with future perspectives for clinical application. Due to technical advancements, the cryopreservation of cells and tissue has been improved (2, 3). As a result, hematopoietic stem cells have been extensively cryopreserved and successfully used for transplantation (4, 5).

While the use of cryopreserved cells has become commonplace, the quality and utility of cryopreserved cells continue to have significant challenges. As the use of in vitro cell systems continues to rise, more attention has been directed towards developing new approaches to assure effective cryopreservation of cellular systems (maintaining a high degree of cell viability and function). Optimal cell recovery depends on the type, concentration, exposure time, and temperature of the cryoprotectant, the addition and removal procedures of the cryoprotectant, and the freezing and warming protocols (6, 7). Most cell cryopreservation techniques use a mixture of cell culture media, animal sera, and dimethylsulfoxide (DMSO) as a freezing solution. This solution, combined with slow cooling rates, is used to prevent ice crystal formation within cells and affects other physical aspects of the freezing process. DMSO has been extensively used as a cryoprotectant because of its high membrane permeability (8). However, despite the protection this cryoprotectant offers, DMSO can be damaging to cells when used in high concentration. Exposing cells to high concentrations of permeating solutes can cause extensive initial dehydration followed by rehydration (3). The resultant cellular shrinking and swelling can cause damage and even cell death (9).

Water is the major component of all living cells and must be present for chemical reactions to occur. During cryopreservation, water changes to ice and cellular metabolism ceases. In addition, dehydration occurs, changing the concentration of salts and other metabolites, and creates an osmotic imbalance that can be detrimental to cell recovery. Intracellular ice formation (IIF) occurs during rapid cooling and can be lethal if the cell is unable to respond by exosmosis to the formation of extracellular ice and the concomitant concentration of extracellular solutes. This process results in the intracellular retention of supercooled water and an increased probability of intracellular ice nucleation (10). For this reason, there is enormous clinical interest in avoiding IIF during cryopreservation of cells used for transplantation and cryoprotectant solutions have been used to avoid cell and subcellular structural disruption during freezing and thawing.

MSCs have a capacity for proliferation and are able to differentiate into several distinct cell lineages under appropriate conditions (11). This differentiation potential makes MSCs a candidate for cell-based therapy for a variety of degenerative diseases (12). Despite the tremendous potential applications for these cells, cryopreservation studies are limited. Results have shown that 10% DMSO was effective in preserving functional stromal cells capable of supporting subsequent in vitro culture (13, 14). Recently, it has also been shown that MSCs frozen in reduced (5%) DMSO and slow freezing methods are still capable of expansion and differentiation and expressed all surface antigens detected before storage, which can be easily used in large-scale cryopreservation of MSCs and is ideally suited for good manufacturing practice (GMP) laboratories (15).

The aim of this chapter is to describe different methods, including those applicable to clinical application, to cryopreserve MSCs collected from human bone marrow and stored at low temperature in liquid nitrogen through an optimal cryopreservation solution.

2. Materials

2.1. Cell Culture and Harvest

1. Complete medium: Dulbecco's Modified Eagle's Medium (DMEM, low Glucose with HEPES and L-Glutamine) (PAA Laboratories, Pasching) supplemented with 10% Fetal Bovin Serum (FBS, Pharma Grade Gamma Irradiated AUS Origin) (PAA Laboratories), and 1% penicillin/streptomycin (GIBCO, Invitrogen GmbH, Pasching). The complete medium can be stored at 4°C for 14 days.
2. PBS minus Ca^{2+} and Mg^{2+} (GIBCO, Invitrogen GmbH, Pasching).
3. Glass pasteur pipettes (VWR).
4. Membrane-vacuum pump (KNF lab Laboport, VWR).
5. TrypLE Select (animal origin free, GIBCO, Invitrogen, Taastrup).
6. 10 ml centrifuge tube.
7. Cell counting chamber, Bürker, 0.100 mm Tiefe Depth (Marienfeld).
8. Bench centrifuge.
9. Inverted light microscope.

2.2. Cryopreservation of Human MSCs for Research Applications

1. DMSO (Wak-Chemie Medical GmbH, Steinbach).
2. FBS, Pharma Grade Gamma Irradiated AUS Origin (PAA Laboratories).
3. 1.5 ml cryovials (Nunc, Roskilde).

4. 1 ml sterile syringe.
5. 14G needles.
6. Foam ice box.
7. Isoproponol freezing container.
8. Liquid nitrogen tank, Air Liquide.

2.3. Cryopreservation of Human MSCs for Clinical Applications

1. DMSO (pyogen free, 10 ml ampule) (WAK-Chemie medical GmbH).
2. Hemofreeze DF 700 freezing bags (100 ml) (Fresenius Kabi, Hamburg).
3. Transfer bags (400 ml) (Baxter Healthcare Corporation, Deerfield).
4. Transfer set with spike coupler and luer adaptor (Baxter Healthcare Corporation).
5. 14G and 18G needles.
6. 1, 10, 25, and 50 ml sterile syringes.
7. Foam ice box.
8. Sterile sheet.
9. Heating sealer (Vingmed WO3).
10. Aluminum freezing canisters (especially made for hospital Blood bank).
11. Programmable controlled-rate freezer (Kryo 560-16) (Planer).
12. Liquid nitrogen tank, Air Liquide.

2.4. Recovery of Cryopreserved MSCs

1. Small liquid nitrogen tank (Voyageur plus), Air Liquide.
2. Foam ice box.
3. Membrane tube connectors, one for each bag.
4. Liquid nitrogen safety groves.
5. 70% Ethanol.
6. Heated water bath (preheated to 37–40°C).
7. Ziploc bags.
8. Esprit wraps.
9. 50 ml centrifuge tubes.

2.5. Determining Viability of Thawed MSCs

1. NucleoCounter (Chemometec, Allerød).
2. NucleoCassettes (Chemometec).
3. Reagent A100 lysis buffer (Chemometec).
4. Reagent B stabilizing buffer (Chemometec).

5. Phosphate-buffered saline (PBS) minus Ca^{2+} and Mg^{2+} (GIBCO, Invitrogen GmbH).
6. Trypan blue (TB) solution (0.4%) prepared in 0.81% sodium chloride, 0.06% potassium phosphate (Sigma-Aldrich).

3. Methods

The following protocols describe cryopreservation using DMSO in FBS as a cryoprotectant. DMSO is used as the most common cryopreservant for cells, which prevents the formation of intracellular ice, but it is known to be toxic at room temperature. Therefore, the cryopreservation solution should be cold before adding to cell suspension and should be removed or diluted quickly after thawing. The optimal protocol should include a low DMSO concentration and minimizing the amount of time between introduction of DMSO and initiation of the freezing protocol (2). Following these guidelines will result in reduced IIF, better MSC membrane vitality, and will help maintain MSC stemness and differentiation characteristics (see Note 1).

3.1. Human MSC Culture and Harvest

1. MSCs are cultured in T75 culture flasks with 15 ml complete medium and incubated at 37°C in humidified air with 5% CO_2. MSCs are passaged when approaching appropriate confluence (80%) (see Note 2).
2. Remove growth medium with a sterile Pasteur pipette, wash with PBS, and aspirate.
3. Harvest MSCs by incubating at 37°C for 10 min with 3 ml TrypLE Select. To confirm that all cells are detached, check the flask under a microscope. If the cells have not begun to detach or is not fully detached, incubate for an additional 5 min at 37°C.
4. Label a sterile 10 ml centrifuge tube with cell line and passage number.
5. Add 7 ml complete medium to the flask to neutralize the TrypLE Select. Transfer cell suspension to the 10 ml tube.
6. Centrifuge at $300 \times g$ for 5 min.
7. Aspirate supernatant. Gently reconstitute the cell pellet in a known volume of complete medium.
8. The number of cells can be manually determined per unit volume of a suspension in a Bürker counting chamber. Suspensions should be diluted to a level where the cells do not overlap each other on the grid and should be uniformly distributed. To perform the count, determine the magnification

needed to recognize the desired cell count. Count the cells in the selected squares so that the total count is over 100 cells (number of cells needed for a statistically significant count).

3.2. Cryopreservation of MSCs for Research Applications

1. Centrifuge previously counted MSCs for 5 min at $300\times g$ to remove the culture medium.
2. Aspirate the supernatant and place the cell pellet on ice (see Note 3).
3. Make an appropriate volume of cryopreservation solution by combining 5% DMSO with 95% cold FBS and store on ice until needed. Make enough medium to allow reconstitution of cell pellet to $0.5–1\times10^6$ cells/ml.
4. Label cryovials and place vials on ice until needed.
5. Gently reconstitute the cell pellet in cryopreservation solution ($0.5–1\times10^6$cells/ml).
6. Transfer 1 ml of cell solution to chilled cryovials and place the vials in an isopropanol freezing container.
7. Subject the vials to slow cooling by placing them in a –80°C freezer. Using this protocol, samples are cooled at –1°C/min down to –80°C.
8. The following day, transfer cryovials into the liquid nitrogen container (see Note 4).

3.3. Cryopreservation of MSCs for Clinical Applications

To support the cryopreservation of expanded MSC for clinical applications, cells can be frozen in cryopreservation bags for long-term storage. The freezing bag is designed to ensure a homogeneous, controlled-rate freeze with maximum cell viability. The bags are made of Kapton/Teflon, are biocompatible, and are highly resistant to mechanical strain at low temperatures. To support cryopreservation of MSCs in freezing bags, a controlled-rate freezer is used. To monitor temperature during cryopreservation, this type of freezer will utilize two temperature probes, one for measuring the temperature within the freezer chamber and a second that is inserted into a control freezing bag to monitor the temperature of a simulated product throughout the freezing run.

1. For each donor MSC population, two freezing bags are used, one for cell suspension and a second control bag. Bags should be labeled with the date and donor name.
2. Insert a transfer set with spike coupler and a luer adaptor into the membrane port of the freezing bags (Fig. 1a).
3. After harvesting and counting MSCs (see Subheading 3.1), cells are resuspended in 25 ml of cold FBS and transferred to a freezing bag through a spike coupler connected to the membrane port with syringe (Fig. 2a). For the simulated control, 25 ml of cold FBS is transferred to a freezing bag.

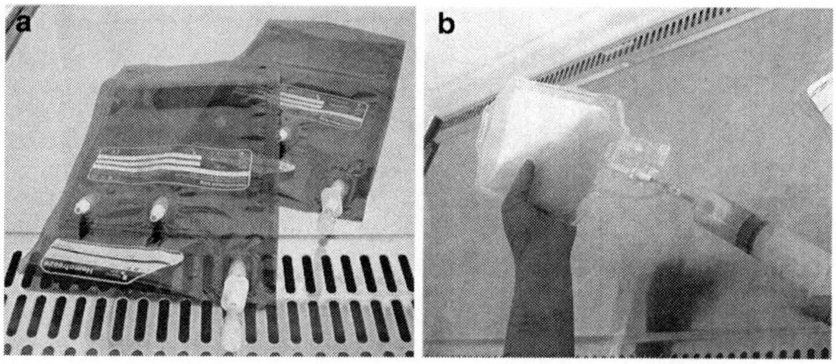

Fig. 1. (**a**) Hemofreezing bags with spike coupler. (**b**) Cryopreservation solution is made in a transfer bag by injecting the cold FBS and DMSO into the bag with syringe.

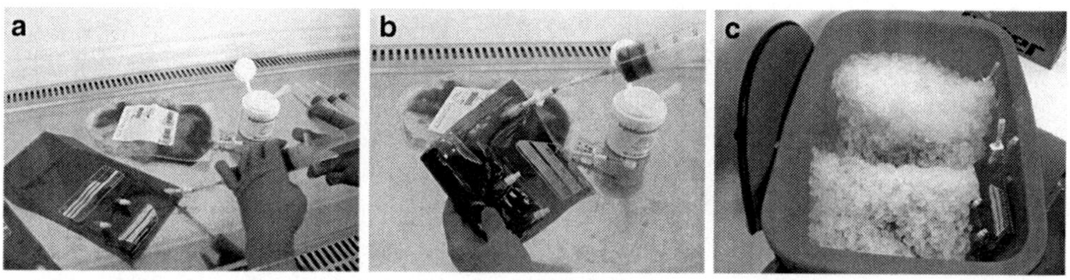

Fig. 2. (**a**) Cell suspension is injected into the freezing bag with syringe. (**b**) The excess air in the bag is squeezed out by rolling the bags from the bottom. (**c**) Freezing bags are stored on ice.

The excess air in the bags should be squeezed out by rolling the bags from the bottom (Fig. 2b). Store the freezing bags on ice until needed (Fig. 2c).

4. MSCs are frozen in a total volume of 50 ml, comprised of 25 ml cell suspension and 25 ml cryopreservation solution. Cryopreservation solution is made directly in a transfer bag (Fig. 1b). To prepare cryopreservation solution, add 45 ml of cold FBS to a transfer bag using 50 ml syringe. Next, draw up a 5 ml volume of DMSO into a 10 ml syringe with an 18G needle. After drawing up the DMSO, draw up some additional air in the syringe. Inject 5 ml DMSO gradually into the transfer bag. Use the additional air in the syringe to empty the coupler tubing. The transfer bag is stored on ice until needed (see Note 5).
5. Switch on the controlled-rate freezer (Fig. 6a). Follow the manufacturer's guidelines for machine operation.
6. An ice box is prepared by using lid from a foam box (Fig. 3a). Foam ice box is covered with a sterile sheet (Fig. 3b). The cryopreservation solution is drawn up into 2×25 ml syringes. The syringes and transfer bag are placed on ice (Fig. 3c). The syringes should not be in direct contact with melting ice.

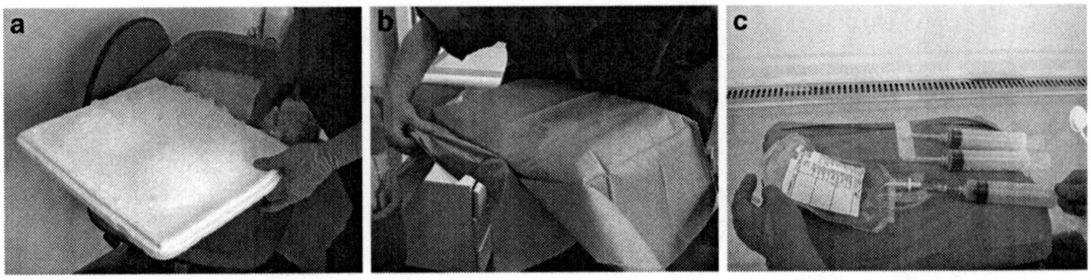

Fig. 3. (**a**) Foam ice box. (**b**) Ice box is covered with a sterile sheet. (**c**) Cryopreservation solution is drawn up into syringes.

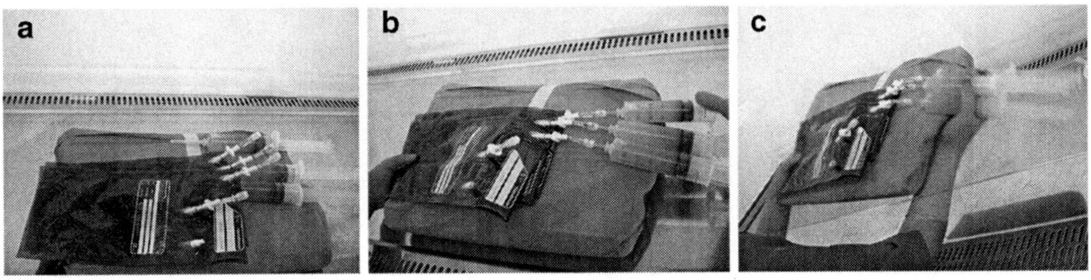

Fig. 4. (**a**) The syringes are connected to freezing bags. (**b**) The cryopreservation solution is injected slowly into the freezing bags. (**c**) Rock the freezing bags on ice between cryopreservation solution injections.

7. Place the freezing bags on foam ice box (Fig. 4a) and connect the syringes to the freezing bags (Fig. 4b). Inject the cryopreservation solution slowly (5 ml at the time) into the freezing bags containing cell solution. The freezing bags connected to syringes are rocked between cryopreservation solution injections to make sure that cell suspension is mixed very well with cryopreservation solution (Fig. 4c). It is very important to maintain the cells at approximately 4–8°C during the addition of the cryopreservation solution.
8. Inject 25 ml of cryopreservation solution into the control freezing bag.
9. Heat seal the entry tubing on the freezing bags (Fig. 5a). Place freezing bags and aluminum freezing canisters on ice (Fig. 5b).
10. Before freezing, place freezing bags into a cold aluminum freezing canister, so the content of the bag is spread into a thin layer (Fig. 5c). It is very important that the freezing bags and the aluminum canisters are kept cold under the whole process.
11. Attach a temperature probe to the control freezing bag (Fig. 6b). Place the freezing canisters into the controlled-rate freezer and follow the manufacturer's guidelines (Fig. 6c).

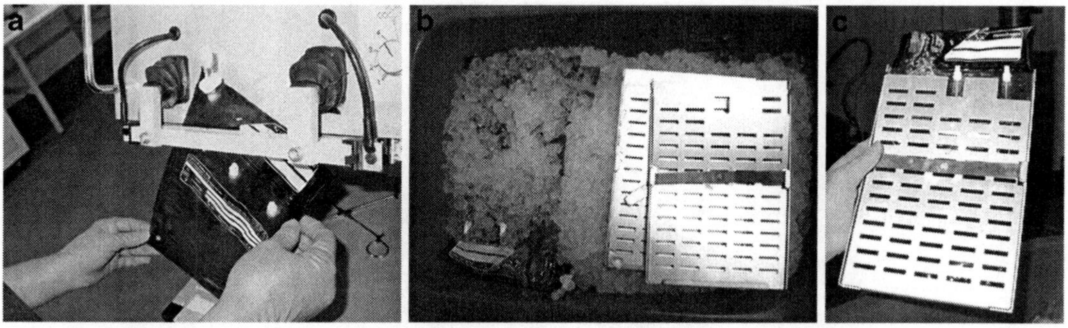

Fig. 5. (**a**) Heat seal the entry tubing on the freezing bags. (**b**) The freezing bags and aluminum canister must be kept cold. (**c**) Freezing bag is placed into a aluminum canister.

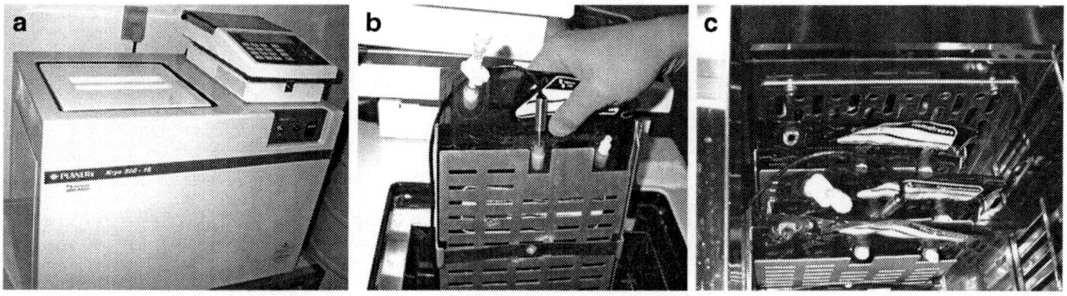

Fig. 6. (**a**) Programmable controlled-rate freezer. (**b**) Temperature probe is attached to control freezing bag. (**c**) The freezing canisters are placed into the freezer.

Begin the precool phase program to maintain refrigerated temperature in the chamber before placing the canister into the freezer (see Note 6).

12. When the program completes and the desired temperature has been achieved, the aluminum canister is removed and the freezing bag is transferred to a liquid nitrogen tank.
13. Switch off the controlled-rate freezer.

3.4. Recovery of Cryopreserved MSCs in Cryovials

1. Remove the cryovials from the liquid nitrogen container and place the vials on ice until thawing is to begin.
2. Thaw the frozen MSCs quickly in the water bath until a small amount of ice remains (see Note 7).
3. Transfer the thawed cell suspension into a 10 ml centrifuge tube and dilute the cell suspension slowly (drop-wise) in cold complete medium.
4. Centrifuge the cell suspension at $300\times g$ for 5 min (see Note 8) and reconstitute in 1 ml complete medium. Analyze cell viability manually with TB or by utilizing an automated cell counter such as NucleoCounter (see Subheading 3.6).

Fig. 7. Small liquid nitrogen tank used for transferring the freezing bag from the big liquid nitrogen over to water bath.

3.5. Recovery of Cryopreserved MSCs in Freeing Bags

To avoid contamination by virus or other microorganisms, it is important to hold the bag in the vapor phase of nitrogen in a smaller liquid nitrogen tank (Fig. 7) before immersing into the warm water bath. Covering the bag with a second, sterile zippered bag is an effective way to save valuable cells if the freezing bag has been damaged.

1. Remove the freezing bag from the liquid nitrogen tank and place in a small liquid nitrogen tank containing liquid nitrogen. Confirm correct donor information on the freezing bag before thawing.
2. The freezing bag with the cells is transferred into a sterile ziploc bag and placed into the water bath. Be sure to hold the top of the ziploc bag out of the water to keep the inside of the bag dry.
3. Visually inspect the integrity of the cells as the cell suspension begins to thaw. Continue thawing until a small amount of ice crystals remain.

4. Remove the ziploc bag from the water bath and take out the freezing bag.
5. Clean around the entry port on the freezing bag with esprit wraps and then connect the membrane spike coupler to the bag.
6. Slowly add 50 ml of cold complete medium to the freezing bag with the cells and mix.
7. Draw gently the cell suspension through the tubing connected to the membrane port with 2 × 50 ml syringes.
8. Transfer the cell suspension into 50 ml centrifuge tubes and centrifuge for 5 min at 300 × *g*.
9. Remove the supernatant and reconstitute the cell pellet in a known volume of complete medium. Pool the cell suspension from both centrifuge tubes.
10. Measure the cell viability (see Subheading 3.6) and plate cells in culture flasks in complete medium.

3.6. Determining Viability of Thawed MSCs

Viability of cells is a general term that refers to functional assessment techniques performed and is not a precise indicator of the overall health status of a cell (16). Different assessment techniques to measure cell viability include an evaluation of the structure or function of different biological elements and may give different measures of cell health and viability. TB and other fluorescence-based viability assays are the most commonly used assays for the assessment of membrane integrity of nucleated cells (17) (see Note 9).

3.6.1. Trypan Blue Staining

1. Mix cell suspension with an equal volume of 0.4% (w/v) TB solution.
2. Count live vs. dead cells using a light microscope in a Bürker cell counting chamber. Blue stained cells are nonviable and unstained cells are viable.

3.6.2. NucleoCounter Analysis

The NucleoCounter offers an easy and reliable means to determine the total viable cell concentration in a sample. No calibration of cell size or volume is needed as the volume of sample measured in every analysis is known. In summary, a cell sample is loaded into a NucleoCassette which contains the fluorescent dye propidium iodide (PI) immobilized inside the flow channels of the cassette. The PI is mixed with the cell sample and stains the DNA within the cell nuclei. After placement in the NucleoCounter, the stained mixture is automatically transferred to the measurement chamber where the fluorescent image is recorded. As PI stains the cell nuclei, it needs to be permeable to the dye. The dye cannot penetrate a viable cell, thus it is necessary to lyse the cell membrane prior to staining when counting the total number of cells.

As nonviable cells are permeable, they can be stained directly with the PI and counted by the NucleoCounter. Combining these two results makes it possible to determine the cell viability of a sample. In addition, NucleoView software can be connected to the NucleoCounter and can offer a variety of advantages for the user.

1. In order to count the nonviable cells, the cell suspension must be loaded directly into the NucleoCassette without any pretreatment and only the nonviable cells with impaired plasma membrane are stained with PI and counted.
2. To count the total number of cells in a suspension, mix 100 µl of cell suspension with 100 µl reagent A100 lysis buffer and mix (see Note 10).
3. Mix the lysed cells (200 µl) with 100 µl reagent B stabilizing buffer (see Note 10).
4. Draw approximately 50 µl of the stabilized lysate into the NucleoCassette and place in the NucleoCounter.
5. The percent viability of MSCs can be calculated from total number of MSCs and the number of nonviable cells.

4. Notes

1. The method described here uses 5% (v/v) DMSO as the cryoprotectant for research use and for clinical application. Other methods have been described using DMSO in combination with Hydroxyethyl starch (18, 19); polyvinylpyrrolidone (PVP), a nonpermeating cryoprotective agent (20); and trehalose (21).
2. Cultivation of MSCs has recently been optimized for clinical use (22).
3. Cooling is known to have the most significant influence on cell survival. A cooling rate too high or too low can result in low cell recovery due to dehydration or ice crystal formation. Controlled-rate freezing before long-term storage minimizes these variables to ensure maximum viability for a wide variety of cells.
4. The temperature at which frozen cells are stored affects their viability after recovery. The lower the temperature, the longer the viable storage period. Cryopreservation solution containing 10% DMSO will not fully solidify at –80°C and the rate of cell loss will increase with time in storage (23). At –196°C (–320°F) in liquid nitrogen, thermally-driven chemical reactions

are not known to occur. Therefore, it is believed that the majority of the population of stored cells will remain viable and unchanged for an indefinite period of time.

5. It is very important that the cryopreservation solution is cold (4°C) before it is mixed with the cell suspension. The cryopreservation can be made before starting freezing process and stored at 4°C.
6. The controlled-rate freezer starts at 4°C, after the freezing program is started, then it freezes with 1°C/min until the temperature is reached 0°C. Thereafter, it freezes with 2°C/min until the temperature is reached −45°C. At last it freezes with 5°C/min until −100°C.
7. A cell that has survived freezing ultimately will not survive if it is not thawed correctly. The warming rate can influence survival equally as compared to the cooling rate (3).
8. Washing cells after thawing is essential to remove cryoprotectant (DMSO). Washing cells by centrifugation will result in significant cell loss. Studies have observed a loss of 30% of cells from postthaw washing (24). Another option would be that the thawing cells are diluted by directly transferring to the culture flask containing complete medium, and the flask is stored in incubator until next day, where medium is changed. With this method, non cells are loosed under centrifugation and the DMSO is diluted enough, so it is not toxic for the cells.
9. TB is one of the most commonly used viability assays. TB can distinguish between viable and nonviable nucleated cells. Cells with damaged plasma membranes absorb TB and appear blue, but TB can be toxic to human cells. Alternatively, 0.2% Nigrosin solution (Science Lab, Houston, TX) can be used for measurement of viability of MSCs. Cells with damaged plasma membranes absorb nigrosin and appear blue just like TB (25).
10. The mixing of cell suspension with reagent A100 lysis buffer causes permeabilization of the plasma membrane and allows the nuclei to be stained with PI. The addition of reagent B stabilizing buffer raises the pH of the mixture and allows PI to efficiently stain the nuclei.

Acknowledgments

The Lundbeck Foundation and the Research Foundation at Rigshospitalet supported the study.

References

1. Gluckman E, Broxmeyer HA, Auerbach AD et al. (1989) Hematopoietic reconstitution in a patient with Fanconi's anemia by means of umbilical-cord blood from an HLA-identical sibling N. Engl. J. Med. 321 (17), 1174–8.
2. Hubel A. (1997) Parameters of cell freezing: implications for the cryopreservation of stem cells Transfus. Med. Rev. 11 (3), 224–33.
3. Woods EJ, Benson JD, Agca Y, Critser JK. (2004) Fundamental cryobiology of reproductive cells and tissues Cryobiology 48 (2), 146–56.
4. Adami V, Malangone W, Falasca E et al. (2005) A closed system for the clinical banking of umbilical cord blood Blood Cells Mol. Dis. 35 (3), 389–97.
5. Broxmeyer HE, Srour EF, Hangoc G et al. (2003) High-efficiency recovery of functional hematopoietic progenitor and stem cells from human cord blood cryopreserved for 15 years Proc. Natl. Acad. Sci. USA 100 (2), 645–50.
6. Pegg DE. (2001) The current status of tissue cryopreservation Cryoletters 22 (2), 105–14.
7. Yang PF, Hua TC, Wu J et al. (2006) Cryopreservation of human embryonic stem cells: a protocol by programmed cooling Cryoletters 27 (6), 361–8.
8. Wang X, Hua TC, Sun DW et al. (2007) Cryopreservation of tissue-engineered dermal replacement in Me2SO: Toxicity study and effects of concentration and cooling rates on cell viability Cryobiology 55 (1), 60–5.
9. Mazur P, Schneider U. (1986) Osmotic responses of preimplantation mouse and bovine embryos and their cryobiological implications Cell Biophys. 8 (4), 259–85.
10. Acker JP, Croteau IM. (2004) Pre- and post-thaw assessment of intracellular ice formation J. Microsc. 215 (Pt 2), 131–8.
11. Pittenger MF, Mackay AM, Beck SC et al. (1999) Multilineage potential of adult human mesenchymal stem cells Science 284 (5411), 143–7.
12. Sotiropoulou PA, Perez SA, Salagianni M, Baxevanis CN, Papamichail M. (2006) Cell culture medium composition and translational adult bone marrow-derived stem cell research Stem Cells 24 (5), 1409–10.
13. Kotobuki N, Hirose M, Takakura Y, Ohgushi H. (2004) Cultured autologous human cells for hard tissue regeneration: preparation and characterization of mesenchymal stem cells from bone marrow Artif. Organs 28 (1), 33–9.
14. Nicol A, Nieda M, Donaldson C et al. (1996) Cryopreserved human bone marrow stroma is fully functional in vitro Br. J. Haematol. 94 (2), 258–65.
15. Haack-Sorensen M, Bindslev L, Mortensen S, Friis T, Kastrup J. (2007) The influence of freezing and storage on the characteristics and functions of human mesenchymal stromal cells isolated for clinical use Cytotherapy. 9 (4), 328–37.
16. Bank HL, Schmehl MK. (1989) Parameters for evaluation of viability assays: accuracy, precision, specificity, sensitivity, and standardization Cryobiology 26 (3), 203–11.
17. Yang H, Acker JP, Cabuhat M, McGann LE. (2003) Effects of incubation temperature and time after thawing on viability assessment of peripheral hematopoietic progenitor cells cryopreserved for transplantation Bone Marrow Transplant. 32 (10), 1021–6.
18. Rowley SD, Feng Z, Chen L et al. (2003) A randomized phase III clinical trial of autologous blood stem cell transplantation comparing cryopreservation using dimethylsulfoxide vs dimethylsulfoxide with hydroxyethylstarch Bone Marrow Transplant. 31 (11), 1043–51.
19. Takaue Y, Abe T, Kawano Y et al. (1994) Comparative analysis of engraftment after cryopreservation of peripheral blood stem cell autografts by controlled- versus uncontrolled-rate methods Bone Marrow Transplant. 13 (6), 801–4.
20. Li Y, Lu RH, Luo GF, Pang WJ, Yang GS. (2006) Effects of different cryoprotectants on the viability and biological characteristics of porcine preadipocyte Cryobiology 53 (2), 240–7.
21. Buchanan SS, Gross SA, Acker JP et al. (2004) Cryopreservation of stem cells using trehalose: evaluation of the method using a human hematopoietic cell line Stem Cells Dev. 13 (3), 295–305.
22. Haack-Sorensen M, Friis T, Bindslev L et al. (2008) Comparison of different culture conditions for human mesenchymal stromal cells for clinical stem cell therapy Scand. J. Clin. Lab. Invest. 68 (3), 192–203.
23. Hubel A. (2001) Cryopreservation of HPCs for clinical use Transfusion 41 (5), 579–80.
24. Perotti CG, Del FC, Viarengo G et al. (2004) A new automated cell washer device for thawed cord blood units Transfusion 44 (6), 900–6.
25. Hirata AA. (1963) Cytolytic antibody assay by tryptic digestion of injured cells and electronic counting J. Immunol. 91, 625–32.

Chapter 14

Dynamic Expansion Culture for Mesenchymal Stem Cells

Hicham Majd, Thomas M. Quinn, Pierre-Jean Wipff, and Boris Hinz

Abstract

To be applied in sufficient numbers for regenerative medicine, primary mesenchymal stem cells (MSCs) need to be amplified in culture. Standard cell culture involves regular passing because MSC proliferation in size-limited culture vessels stagnates due to contact inhibition of growth. The use of harmful enzymes for passaging and the mechanical properties of standard culture vessels change the MSC phenotype. Initially, fast growing multipotent and regenerative MSCs will turn into slowly growing cells with reduced multipotency and fibrotic character. We here describe an innovative culture system that maintains overall constant cell densities which are near-optimal for proliferation, while preventing contact-inhibition of cell growth. This is achieved by dynamically enlarging a novel highly elastic culture dish using a motorized mechanical device and adapting the culture surface to the increasing cell numbers. Dynamic MSC culture expansion reduces the number of enzymatic passages by a factor of 3 and delivers higher MSC yields than conventional culture. On the expanded culture surface, MSCs maintain stem cell characteristics and high growth rates over months and are still inducible to follow different lineages thereafter.

Key words: Highly elastic silicone rubber, MSC culture, Passaging, Surface functionalization, Fibrosis

1. Introduction

Mesenchymal stem cells (MSCs) are promising candidates to repair and to regenerate damaged tissue either by injection into the lesion (e.g., for cardiac repair) (1) or by delivery with a scaffold (e.g., for cartilage and bone repair) (2). To avoid immune reactions, the use of autologous MSCs is desired but the number of cells obtained from biopsies is generally too low for direct implantation. From bone marrow, one of the most important sources of MSC, rarely more than 5×10^4 cells are purified from every 20 ml aspirate, because only about 0.01% of nucleated bone marrow cells are MSCs (3). Harvesting efficiency declines with donor age and is very low for donors over 50 years of age (4).

Mohan C. Vemuri et al. (eds.), *Mesenchymal Stem Cell Assays and Applications*, Methods in Molecular Biology, vol. 698,
DOI 10.1007/978-1-60761-999-4_14, © Springer Science+Business Media, LLC 2011

This poor yield undermines an estimated need of 1–10 million MSCs/kg bodyweight per injection, adding up to more than one billion MSCs per patient during repeated injection therapy (5). The only way to satisfy this high demand for MSCs is cell multiplication in culture.

Contact-inhibition of growth gradually reduces the MSC proliferation rate in conventional plastic dish cell culture, requiring weekly transfer to new dishes (passaging). Ideally, the dilution factor for MSC passaging is kept low (~1:2) to retain efficient proliferation rates and to preserve cell stemness. MSCs are adherent, and harvesting these cultured cells for population expansion requires enzymes like trypsin that degrade extracellular matrix and cell surface proteins. Too frequent enzymatic passaging leads to the loss of MSC multipotency and proliferation capacity (6, 7). It has been reported that early passage (≤5) MSCs preserve multipotency whereas late passage (≥15) MSCs are only able to differentiate into adipocytes (8), a phenotype that rarely finds application in therapy.

Loss of MSC multipotency is not the only problem arising from standard culture. A number of studies have demonstrated spontaneous phenotypic switch of MSCs into myofibroblasts (9), key cells in the development and progression of tissue contracture and fibrosis (10). The percentage of MSCs that spontaneously express myofibroblast markers gradually increases with long-term culture (11). MSCs that have developed a myofibroblast phenotype in culture possibly contribute to the formation of fibrotic scar tissue instead of regenerating organ function once engrafted (10). This danger is increased by the fact that many of the potential therapeutic applications for MSCs imply their engraftment into fibrotic tissue, such as infarcted heart (12), fibrotic lung (13, 14), kidney (15), and liver (16).

Cell phenotype alteration and reduced growth rates in standard culture are not new observations and different techniques have evolved to solve these problems, including a number of bioreactor systems (17, 18). Bioreactors do not usually allow for direct observation of cells by conventional microscopy, which means that continuous control over the cell phenotype is difficult. We here describe a new cell culture method that provides for direct observation of cells using a transparent culture surface that is dynamically enlarged during cell proliferation. This approach maintains constantly high cell density while preventing contact-inhibition of growth. Compared to standard culture, our dynamic expansion culture produces ~10-fold more MSCs over 9 weeks of growth and reduces the numbers of enzymatic passages by a factor of 3. The multipotent and lineage-inducible character of MSCs is preserved over long-term culture, and development of undesired fibrogenic myofibroblasts is suppressed (19).

2. Materials

2.1. Preparation of Culture Substrate Extension Device

1. Device for expansion of high-extension silicone rubber (HESR) surfaces (e.g., "The Cellerator," CR-010 and associated motor CR-011, Cytomec GmbH, Spiez, CH) (Fig. 1a).
2. 70% Ethanol.
3. Deionized water.

2.2. Preparation of the High-Extension Silicone Rubber (HESR) Culture Surface

1. HESR culture surfaces (e.g., MM-CR-010, Cytomec GmbH) (Fig. 1b).
2. Deionized water.
3. 30% H_2SO_4 in deionized water.
4. 1% APTES ((3-aminopropyl)triethoxysilane) (Sigma, Buchs, CH) in deionized water.
5. 6% Glutaraldehyde in deionized water, sterile-filtered.
6. 2 μg/ml monomeric collagen type I (Sigma) in phosphate-buffered saline (PBS): 150 mM NaCl, 8.06 mM Na_2HPO_4, 1.74 mM $NaH_2PO_4 \cdot 2H_2O$, pH 7.4.

2.3. Culture with Expansion Device and HESR Surface

1. Human MSCs, isolated from bone marrow or other sources.
2. MSC proliferation medium: α-minimum essential medium (α-MEM), supplemented with 10% fetal calf serum (FCS), 20 mM L-glutamine, 1,000 U/ml penicillin/streptomycin, 2.5 μg/ml amphotericin B, and 10 ng/ml basic fibroblast growth factor (R&D Systems, Minneapolis, MN, USA).
3. Cell culture PBS (PBS, supplemented with 2 mM Ca^{2+} and Mg^{2+}, sterilized).

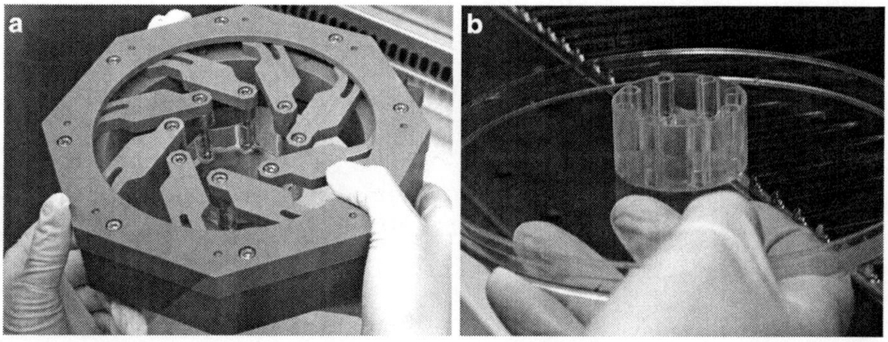

Fig. 1. Extension hardware and HESR culture surfaces. (**a**) The Cellerator extension apparatus functions mechanically like a camera iris, allowing highly uniform equiaxial stretch of the culture surface mounted at the center. (**b**) HESR culture surfaces are shaped like petri dishes with eight slots for mounting in extension hardware molded into the walls. Relaxed (unstretched) surface area is 8 cm^2.

4. 0.25% Trypsin solution in sterile PBS.
5. Computer interfaced with HESR expansion device.

2.4. MSC Lineage Induction

1. Basic lineage medium: Dulbecco's minimal essential medium (DMEM), supplemented with 10% FCS, 20 mM L-glutamine, 1,000 U/ml penicillin/streptomycin.
2. 0.25% Trypsin solution in PBS.

2.4.1. Adipogenesis

1. 5 mg/ml insulin (Sigma) stock solution in distilled sterile water, pH 7.2, aliquots at 4°C.
2. 1 μM dexamethasone (Sigma) stock solution in DMSO, aliquots at –20°C.
3. 100 mM indomethacin (Sigma) stock solution in DMSO, aliquots at –20°C.
4. 100 mM 3-isobutyl-1-methyl-xanthine (IBMX) (Sigma) stock solution in 0.35 M KOH, aliquots at –20°C.

2.4.2. Chondrogenesis

1. 5 μg/ml TGFβ3 (R&D Systems) stock solution prepared in 10 mM citric acid at 0.1 mg/ml and subsequently diluted 1:20 in 2 mg/ml bovine serum albumin (BSA) in PBS (see Note 1), aliquots at –20°C.
2. 5 mg/ml insulin stock solution in distilled sterile water, pH 7.2, aliquots at 4°C.
3. 1 mM dexamethasone stock solution in DMSO, aliquots at –20°C.
4. L-ascorbic acid 2-phosphate sodium salt (ascorbate 2-phosphate) (Sigma) to be directly added to the medium.

2.4.3. Osteogenesis

1. 1 M β-glycerophosphate (Sigma) stock solution in DMSO, aliquots at –20°C.
2. 1 mM dexamethasone stock solution in DMSO, aliquots at –20°C.
3. Ascorbate 2-phosphate to be directly added to the medium.

2.4.4. Myogenesis

1. Horse serum.
2. 66 mg/ml hydrocortisone (Sigma) stock solution in distilled water, aliquots at –20°C.
3. 1 mM dexamethasone stock solution in DMSO, aliquots at –20°C.

2.5. MSC Phenotyping

2.5.1. Naïve MSC

1. Cell culture PBS.
2. 3% Paraformaldehyde (PFA) in PBS.
3. 0.5% BSA in PBS.

4. Primary antibody anti-CD34, phycoerythrin (PE)-conjugated (e.g., mouse IgG3k, Alexis Corporation, Lausen, CH).
5. Primary antibody anti-CD44, Pacific Blue-conjugated (e.g., rat IgG2b, BioLegend, CA, USA).
6. Primary antibody anti-CD73, PE-conjugated (e.g., mouse IgG1, BD Biosciences, NJ, USA).
7. Primary antibody anti-CD90 (e.g., mouse IgG2a, Abcam, Cambridge, UK).
8. Primary antibody anti-CD105 (e.g., mouse IgG1, Abcam).
9. Primary antibody anti-CD166 (e.g., mouse IgG2a, Abcam).
10. Secondary antibody goat antimouse IgG1-TRITC (e.g., Southern Biotechnology Associates, Inc., Birmingham, AL).
11. Secondary antibody goat antimouse IgG2a-FITC (e.g., Southern Biotechnology Associates).
12. Flow cytometer and FACS software (e.g., Summit V4.3 software (CyanADP, DAKO)).

2.5.2. Identification of Fibrotic Cells

1. 3% PFA in PBS.
2. 0.2% Triton X-100 in PBS.
3. Primary antibody anti-α-SMA (e.g., mouse IgG2a, DAKO, Glostrup, DK).
4. Fluorescently labeled Phalloidin (e.g., Alexa 488-conjugated Phalloidin, Molecular Probes, Invitrogen, Basel, CH).
5. Secondary antibody goat antimouse IgG2a-FITC (e.g., Southern Biotechnology Associates).
6. DAPI.

2.5.3. MSC Lineages

1. 3 mg/ml stock solution of Oil Red O (Sigma) in 99% isopropanol, room temperature. Final working concentration is 60% of stock solution in deionized water.
2. Alkaline phosphatase kit (Sigma).
3. 0.3% w/v Alcian blue (Sigma) stock solution in 3% acetic acid, pH 2.5, room temperature.
4. 3% PFA in PBS.
5. 0.2% Triton X-100 in PBS.
6. Primary antibody anti-smooth muscle-myosin heavy chain (e.g., rabbit, BTI, Stoughton, MA).
7. Primary antibody anti-SM22 (e.g., rabbit, Abcam).
8. Primary antibody anti-α-SMA (e.g., mouse IgG2a, DAKO, Glostrup, DK).
9. Secondary antibody antirabbit-TRITC (e.g., Southern Biotechnology Associates).

3. Methods

3.1. Preparation of Culture Substrate Extension Device

Techniques for cell culture on HESR surfaces are new, and commercially available hardware is still under development. The following protocol describes a sterilization procedure for the Cellerator device (Cytomec GmbH) (Fig. 1a), which was necessary for nonautoclavable prototypes. Though cumbersome, these methods were used successfully to demonstrate the feasibility and efficacy of HESR culture methods for MSC (19). Autoclavable prototypes are the logical next step for this hardware, and would streamline these methods significantly.

1. Disassemble the Cellerator expansion device and immerse all of its parts in a 70% ethanol bath for 2 h.
2. Drain the 70% ethanol and allow the parts to dry within a sterile cell culture hood. Reassemble the sterilized device.

3.2. HESR Culture Surface Preparation

Biocompatible polydimethylsiloxane (PDMS) silicone rubber is the basic material of the HESR (Fig. 1b). PDMS is convenient for cell culture applications due to its transparency, moldability, and tunable mechanical properties. However, cell adhesion and proliferation on bare PDMS are poor, and surface modification to improve these properties to the levels obtained on cell culture polystyrene is usually necessary. The following protocol represents one possibility for achieving this and is adapted from previous methods for functionalization of glass surfaces (20) and Sylgard 184 PDMS (21).

1. Wash the HESR culture surface thoroughly with deionized water and leave to dry sitting on a petri dish in a chemical fume hood.
2. Mount the HESR culture dish in the Cellerator expansion device and expand the surface to the maximal area.
3. Pipette 30% H_2SO_4 solution (see Note 2) onto the HESR culture surface to a depth of 0.5 cm and leave at room temperature for 15 min. Remove H_2SO_4 solution, wash thoroughly with deionized water, and leave to dry.
4. Pipette 1% APTES solution onto HESR culture surface to a depth of 0.5 cm. Place expanded HESR with APTES solution onto a heating plate for 60 min at 70°C. Make sure that the heat-sensitive frame of the device is not in contact with the heating plate (see Note 3).
5. Remove APTES solution, wash thoroughly with deionized water, and leave to dry.
6. Pipette 6% glutaraldehyde solution onto HESR culture surface to a depth of 0.5 cm and leave at room temperature for 15 min. Remove glutaraldehyde solution, wash thoroughly with deionized water, and leave to dry.

7. Pipette 2 μg/ml collagen type I solution (see Note 4) onto the expanded HESR culture surface to a depth of 0.5 cm and leave overnight in a standard cell culture incubator at 37°C and 5% CO_2.
8. Reduce area of the HESR culture surface to the initial area.
9. Remove collagen solution immediately prior to cell seeding.

3.3. Culture with Expansion Device and HESR Surface

Multiplication of MSC populations on HESR was recently achieved using stepwise, monotonic increases of the culture surface over a 16-day culture period. Enlargement of the culture surface by 5% every 6 h was imposed from days 5 to 16 (Fig. 2a), providing an eightfold increase in culture surface that matched three 1:2 passages of control cultures over the same period (19). These particular parameters were chosen because previous work

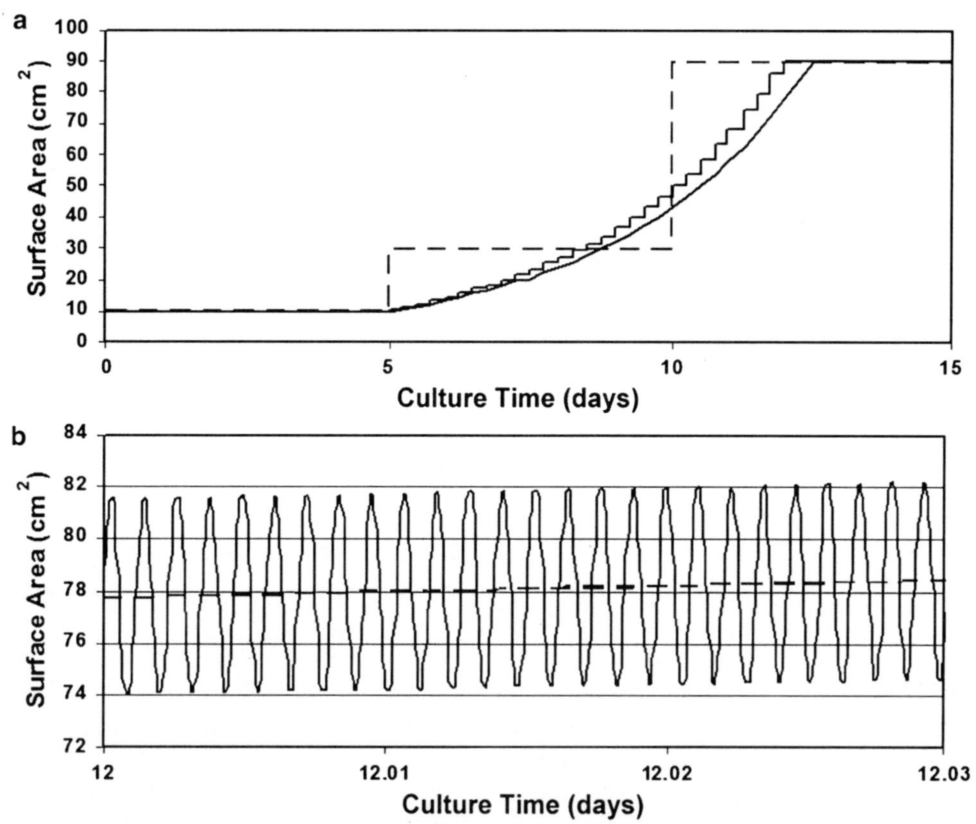

Fig. 2. Representative surface extension protocols for MSC proliferation and differentiation. (**a**) Standard culture methods for two 1:3 passages of MSC from 10 to 90 cm² culture surface area would involve exposure to trypsin (or other degradative enzymes) on days 5 and 10 of a 15-day culture period (*dotted line*). HESR culture can avoid this by a series of small step increases in culture surface area (e.g., 8% every 6 h; stepping curve) or by continuous exponential surface area augmentation (e.g., surface doubling time of 2.4 days; smooth curve). (**b**) In addition to continuous expansion during MSC proliferation (*dotted line*), HESR culture can be used to superimpose oscillatory equiaxial stretch (e.g., 5% amplitude at 0.01 Hz; *solid line*) for the promotion of specific differentiation pathways.

had indicated that (1) 5% expansion would not induce MSC injury and (2) 6 h would provide enough time for MSC to remodel their extracellular matrix adhesion sites following each stepwise perturbation. This protocol enhanced MSC yields by 2-times over standard culture. Other (faster) surface expansion protocols could likely have been used to achieve similar effects, and it is presently unclear whether an "optimum" surface expansion protocol exists in this context. At least three parameters are adjustable to increase the HESR culture surface including the amount of stretch applied per expansion step, the duration of each step, and the delay between each stretch event. Stretches of up to 10% per step are tolerated by cells with sufficient delay time between the steps. We previously modulated the delay time from 6 h down to 2 h (19). It is reasonable to expect that relatively low-amplitude and high frequency oscillatory stretching of the culture surface might be usefully superimposed on the large-amplitude expansion (Fig. 2b) for further control of MSC differentiation pathways (if desired). Such stimuli have been shown to promote MSC differentiation toward bone (22), cartilage (23), and vascular smooth muscle (24) cell lineages.

1. Initiate primary cell culture obtained from donor bone marrow aspirates with 1×10^5 cells/cm^2 for one passage until reaching confluence.
2. Harvest all cells, sort for MSCs and control for the absence of hematopoietic cells according to established procedures (25).
3. Seed 5×10^3 MSCs/cm^2 onto the prepared HESR surface, mounted on the Cellerator device with initial surface of 8 cm^2.
4. Connect device motor with computer controller and place device in the incubator at 37°C and 5% CO_2.
5. Change medium every 3 days.

Expand HESR dish culture surface by applying single stretches of 5% over 60 s in regular intervals of (2, 4 or) 6 h until reaching final surface expansion.

3.4. MSC Lineage Induction

To verify that human MSCs obtained from expanded HESR preserve their multilineage potential, they are subsequently differentiated to adipogenic, chondrogenic, osteogenic and myogenic lineages using specific induction media and culture conditions. Protocols are adapted from ref. (26). More detailed and alternative protocols can be found elsewhere within this text.

1. Rinse cells twice with serum-free basic lineage medium.
2. Harvest MSCs from the expanded HESR dish by rinsing once with cell culture PBS, followed by detachment with 0.25% trypsin solution at 37°C for 5 min.

3. Centrifuge at $800 \times g$ for 5 min.
4. For MSC lineage induction, resuspend pellet in MSC proliferation medium, divide into four fractions, and process each subfraction with one of the protocols described under Subheadings 3.4.1–3.4.4.
5. For naïve MSC characterization, resuspend pellet in cell culture PBS and continue with Subheading 3.5.1.
6. For fibrotic cell characterization, resuspend pellet in cell culture PBS and continue with Subheading 3.5.2.

3.4.1. Adipogenesis

1. Seed 2×10^4 MSCs/cm^2 in standard culture dishes and grow for 1 week until 100% confluence in MSC proliferation medium in the incubator at 37°C and 5% CO_2.
2. Change medium for a period of 3 days to basic lineage medium, supplemented with 10 μg/ml insulin (1:500 stock dilution), 1 μM dexamethasone (1:1,000 stock dilution), 100 μM indomethacin (1:1,000 stock dilution), and 100 μM IBMX (1:1,000 stock dilution). This medium is referred to as "adipogenic medium."
3. Change adipogenic medium for a period of 3 days to basic lineage medium, supplemented with 10 μg/ml insulin (stock dilution of 1:500). This medium is referred to as "adipogenic maintenance medium."
4. Repeat steps 2 and 3 three times over a culture period of 18 days and end with another 4 days culture in adipogenic maintenance medium (see step 3).

3.4.2. Chondrogenesis

1. Transfer 3×10^5 MSCs to a 15-ml polypropylene culture tube and centrifuge at $1,600 \times g$ at room temperature for 5 min.
2. Culture the MSC pellet for 1 day in 1 ml basic lineage medium in the incubator at 37°C and 5% CO_2.
3. Change medium to 0.5 ml basic lineage medium, supplemented with 10 ng/ml TGFβ3 (1:500 stock dilution), 50 μg/ml ascorbate 2-phosphate, and 10 μg/ml insulin (stock dilution of 1:500). This medium is referred to as "chondrogenic medium" (see Note 5).
4. Culture MSC pellets in 0.5 ml chondrogenic medium for 20 days with medium changes every 3 days.

3.4.3. Osteogenesis

1. Seed 5×10^3 MSCs/cm^2 in standard culture dishes and grow until reaching 100% confluence in proliferation medium in the incubator at 37°C and 5% CO_2.
2. Change medium to basic lineage medium, supplemented with 5 mM β-glycerophosphate (1:200 stock dilution), 50 μg/ml ascorbate 2-phosphate (~0.2 mM), and 0.1 μM

dexamethasone (1:10,000 stock dilution). This medium is referred to as "osteogenic medium."

3. Culture MSCs in osteogenic medium for 3 weeks without passaging and with medium changes every 3 days.

3.4.4. Myogenesis

1. Seed 5×10^3 MSCs/cm^2 in standard culture dishes and grow in proliferation medium until reaching 100% confluence.
2. Change medium to basic lineage medium, supplemented with 5% horse serum, 55 μM hydrocortisone (1:333 stock dilution), and 0.1 μM dexamethasone (1:10,000 stock dilution). This medium is referred to as "myogenic medium."
3. Culture MSCs in myogenic medium for 3 weeks without passaging and with medium changes every 3 days.

3.5. MSC Phenotyping

Multiple molecular markers and protocols exist to identify naïve MSCs and may vary between research groups. Other protocols than the one described here for detection of surface markers with fluorescence-activated cell sorting (FACS) are available elsewhere in this text.

3.5.1. Naïve MSCs

1. Trypsinize MSCs from HESR culture surfaces and wash with cell culture PBS.
2. Fix in 3% PFA in PBS for 10 min at RT and wash with 2 mM Ca^{2+}/Mg^{2+} in PBS. Divide suspension into 3 fractions in 1.5 ml reaction tubes.
3. Spin cells down at $800 \times g$ and incubate fractions separately with the following mixtures of primary antibodies: (1) anti-CD105 (1:30 dilution) and PE-conjugated anti-CD34 (1:50 dilution), (2) PE-conjugated anti-CD73 (1:5 dilution) and anti-CD90 (1:30 dilution), (3) anti-CD166 (1:50 dilution) and Pacific Blue-conjugated anti-CD44 (1:200 dilution) at room temperature for 60 min.
4. Wash with PBS at room temperature for 10 min, 3 times.
5. Incubate the three fractions respectively with secondary antibodies: (1) antimouse-IgG1-APC (1:300 dilution), (2) IgG2a-Alexa647 (1:300 dilution), (3) IgG2a-Alexa647 (1:300 dilution), antimouse-IgG1-APC, and antimouse IgG2a-Alexa647.
6. Wash with PBS at room temperature for 15 min, 3 times.
7. Quantify fluorescence with a multi laser-lines flow cytometer and analyze using appropriate FACS software.

3.5.2. Identification of Fibrotic Cells

In addition to preserving the multipotent character of the MSCs, culture on the expanded HESR culture surface with monotonic expansion suppresses development of fibrotic cells (Fig. 3). Fibrogenic cells can be identified on the basis of stress fiber

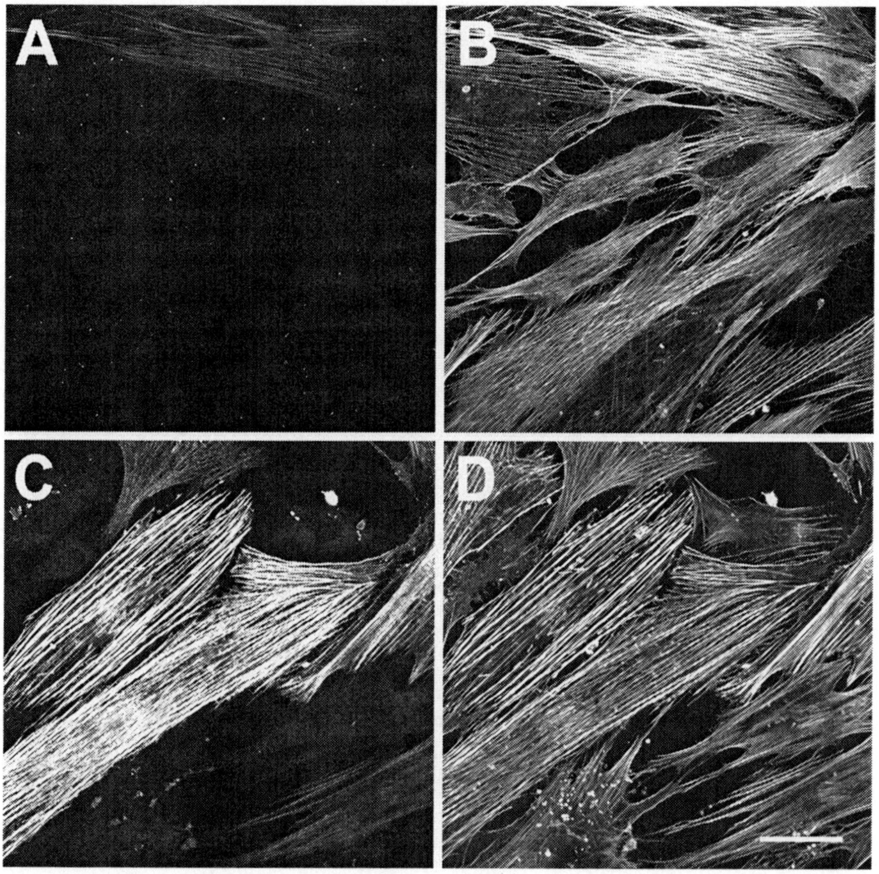

Fig. 3. HESR dish expansion culture suppresses development of the fibrotic MSCs phenotype. MSCs were cultured in normal culture medium for one full expansion of the HESR surface (**a**, **b**). For comparison, MSCs were cultured on standard culture dishes for the same period of 3 weeks (**c**, **d**). Spontaneous development of differentiated myofibroblasts is assessed by probing expression of α-SMA (**a**, **c**) and stress fibers (**b**, **d**, Phalloidin). Scale bar: 25 μm.

formation and α-SMA expression. α-SMA is the most prominent cytoskeletal marker and key functional protein of the differentiated myofibroblast (10). If MSCs are directly immunostained on the expanded membrane, the final staining must be observed mounted to the expansion device. Removal of the stretched membrane with stained cells will lead to significant destruction of the cell morphology (see Note 6).

1. Wash MSCs once with serum-free medium.
2. Fix cells in 3% PFA in PBS at room temperature for 10 min (see Note 7).
3. Rinse once with PBS.
4. Permeabilize cells in 0.2% Triton X-100 in PBS at room temperature for 5 min.

5. Rinse once with PBS.
6. Incubate with primary antibody against α-SMA at room temperature for 60 min.
7. Wash 3 times with 0.02% Triton X-100 in PBS at room temperature for 10 min.
8. Incubate with a mix of secondary antibody goat antimouse-Alexa568, DAPI (nuclear stain), and Alexa488-conjugated Phalloidin (stress fiber stain).
9. Wash with 0.02% Triton X-100 in PBS at room temperature for 15 min, 3 times.
10. Rinse once with distilled water.
11. Mount samples with suitable mounting medium and observe with an epifluorescence microscope.

3.5.3. MSC Lineages, Detailed Protocols Can Be Found Elsewhere Within this Text

1. To determine adipogenic lineage differentiation, stain lipid vesicles with Oil Red O.
2. Identify osteogenic cells by means of alkaline phosphatase detection.
3. To identify chondrogenic cells, fix and paraffin-embed one micro-mass of MSC pellet culture, cut 10 μm sections with a microtome, and stain for cartilage glycosaminoglycans using Alcian blue.
4. Identify myogenic cells by immunostaining for α-SMA, smooth muscle-myosin heavy chain, and SM22 (for more details on immunostaining, see Subheading 3.5.2 and see Note 7).

4. Notes

1. TGFβ3 is delivered lyophilized in its latent form, associated with the latent TGFβ1 binding peptide (LAP). Dilution in citric acid will dissociate active TGFβ from the LAP due to the low pH. Active TGFβ3 is stable in 2 mg/ml BSA/PBS at −20°C.
2. Alternatively, to using 30% H_2SO_4, it is possible to use the more efficient Piranha solution (30% H_2SO_4 in H_2O_2). Piranha solution, however, is a very strong and aggressive oxidant and must be handled with extreme care. Always wear eye and hand protection as well as a lab coat when working with this solution. Considerable heat develops when diluting H_2SO_4 in H_2O_2.
3. The prototype of the commercialized device is made of a material that deforms at ~60°C; subsequent models will be of

heat resistant and autoclavable material. Alternative to the functionalization protocol described in Subheading 3.2, the HESR dish can be treated in the relaxed state. For this, move step 2 (expansion) simply after step 5. The nonmounted HESR dish alone resists temperatures above 150°C and can be baked in an oven instead of using a heating plate.

4. The HESR functionalization protocol is compatible with any other cell-adhesive protein, including other types of collagen, laminins, vitronectin, and fibronectin. Specific protein coating can be used to induce desired phenotypes.
5. Chondrogenic medium must be prepared fresh and used within 12 h after addition of TGFβ3.
6. At full expansion, multiple regions can be stained on the large HESR surface. Alternatively, to staining MSCs directly on the expanded HESR dish, cells can be trypsinized, seeded onto coverslips, and stained after a short growth period. Although this will simplify the staining and mounting procedure, the passaging step and exposure to hard plastic likely induce morphological changes.
7. Detailed immunofluorescence protocols for cytoskeletal proteins and for fibrogenic cells are described in ref. (27).

Acknowledgments

This work was financed by the GEBERT RÜF STIFTUNG and the Collaborative Health Research Programme CIHR/NSERC (CHRP), grant #1004005 (to BH and TMQ) and the Canadian Institutes of Health Research (#488342) (to BH).

References

1. Segers, V.F. and Lee, R.T. (2008) Stem-cell therapy for cardiac disease. *Nature* **451**, 937–942.
2. Caplan, A.I. (2005) Review: mesenchymal stem cells: cell-based reconstructive therapy in orthopedics. *Tissue Eng* **11**, 1198–1211.
3. Braccini, A., Wendt, D., Jaquiery, C., Jakob, M., Heberer, M., Kenins, L., Wodnar-Filipowicz, A., Quarto, R., and Martin, I. (2005) Three-dimensional perfusion culture of human bone marrow cells and generation of osteoinductive grafts. *Stem Cells* **23**, 1066–1072.
4. Caplan, A.I. (2007) Adult mesenchymal stem cells for tissue engineering versus regenerative medicine. *J Cell Physiol* **213**, 341–347.
5. Giordano, A., Galderisi, U., and Marino, I.R. (2007) From the laboratory bench to the patient's bedside: an update on clinical trials with mesenchymal stem cells. *J Cell Physiol* **211**, 27–35.
6. Javazon, E.H., Beggs, K.J., and Flake, A.W. (2004) Mesenchymal stem cells: paradoxes of passaging. *Exp Hematol* **32**, 414–425.
7. Bruder, S.P., Jaiswal, N., and Haynesworth, S.E. (1997) Growth kinetics, self-renewal, and the osteogenic potential of purified human mesenchymal stem cells during extensive subcultivation and following cryopreservation. *J Cell Biochem* **64**, 278–294.
8. Vacanti, V., Kong, E., Suzuki, G., Sato, K., Canty, J.M., and Lee, T. (2005) Phenotypic

changes of adult porcine mesenchymal stem cells induced by prolonged passaging in culture. *J Cell Physiol* **205**, 194–201.

9. Kinner, B., Zaleskas, J.M., and Spector, M. (2002) Regulation of smooth muscle actin expression and contraction in adult human mesenchymal stem cells. *Exp Cell Res* **278**, 72–83.
10. Hinz, B. (2010), The myofibroblast – friend or foe in regenerative medicine ? In *Regenerative medicine and biomaterials for the repair of connective tissues*, C.A.J. Ralphs, Editor. Woodhead: Cambridge.
11. Galmiche, M.C., Koteliansky, V.E., Briere, J., Herve, P., and Charbord, P. (1993) Stromal cells from human long-term marrow cultures are mesenchymal cells that differentiate following a vascular smooth muscle differentiation pathway. *Blood* **82**, 66–76.
12. Breitbach, M., Bostani, T., Roell, W., Xia, Y., Dewald, O., Nygren, J.M., Fries, J.W., Tiemann, K., Bohlen, H., Hescheler, J., Welz, A., Bloch, W., Jacobsen, S.E., and Fleischmann, B.K. (2007) Potential risks of bone marrow cell transplantation into infarcted hearts. *Blood* **110**, 1362–1369.
13. Ortiz, L.A., Gambelli, F., McBride, C., Gaupp, D., Baddoo, M., Kaminski, N., and Phinney, D.G. (2003) Mesenchymal stem cell engraftment in lung is enhanced in response to bleomycin exposure and ameliorates its fibrotic effects. *Proc Natl Acad Sci U S A* **100**, 8407–8411.
14. Yan, X., Liu, Y., Han, Q., Jia, M., Liao, L., Qi, M., and Zhao, R.C. (2007) Injured microenvironment directly guides the differentiation of engrafted Flk-1(+) mesenchymal stem cell in lung. *Exp Hematol* **35**, 1466–1475.
15. Ninichuk, V., Gross, O., Segerer, S., Hoffmann, R., Radomska, E., Buchstaller, A., Huss, R., Akis, N., Schlondorff, D., and Anders, H.J. (2006) Multipotent mesenchymal stem cells reduce interstitial fibrosis but do not delay progression of chronic kidney disease in collagen4A3-deficient mice. *Kidney Int* **70**, 121–129.
16. di Bonzo, L.V., Ferrero, I., Cravanzola, C., Mareschi, K., Rustichell, D., Novo, E., Sanavio, F., Cannito, S., Zamara, E., Bertero, M., Davit, A., Francica, S., Novelli, F., Colombatto, S., Fagioli, F., and Parola, M. (2008) Human mesenchymal stem cells as a two-edged sword in hepatic regenerative medicine: engraftment and hepatocyte differentiation versus profibrogenic potential. *Gut* **57**, 223–231.
17. Eisenstein, M. (2006) Thinking outside the dish. *Nat Methods* **3**, 1035–1043.
18. Martin, I., Wendt, D., and Heberer, M. (2004) The role of bioreactors in tissue engineering. *Trends Biotechnol* **22**, 80–86.
19. Majd, H., Wipff, P.J., Buscemi, L., Bueno, M., Vonwil, D., Quinn, T.M., and Hinz, B. (2009) A novel method of dynamic culture surface expansion improves mesenchymal stem cell proliferation and phenotype. *Stem Cells* **27**, 200–209.
20. Tan, W. and Desai, T.A. (2003) Microfluidic patterning of cells in extracellular matrix biopolymers: effects of channel size, cell type, and matrix composition on pattern integrity. *Tissue Eng* **9**, 255–267.
21. Wipff, P.J., Majd, H., Acharya, C., Buscemi, L., Meister, J.J., and Hinz, B. (2009) The covalent attachment of adhesion molecules to silicone membranes for cell stretching applications. *Biomaterials* **30**, 1781–1789.
22. Simmons, C.A., Matlis, S., Thornton, A.J., Chen, S., Wang, C.Y., and Mooney, D.J. (2003) Cyclic strain enhances matrix mineralization by adult human mesenchymal stem cells via the extracellular signal-regulated kinase (ERK1/2) signaling pathway. *J Biomech* **36**, 1087–1096.
23. Friedl, G., Schmidt, H., Rehak, I., Kostner, G., Schauenstein, K., and Windhager, R. (2007) Undifferentiated human mesenchymal stem cells (hMSCs) are highly sensitive to mechanical strain: transcriptionally controlled early osteo-chondrogenic response in vitro. *Osteoarthritis Cartilage* **15**, 1293–1300.
24. Kurpinski, K., Park, J., Thakar, R.G., and Li, S. (2006) Regulation of vascular smooth muscle cells and mesenchymal stem cells by mechanical strain. *Mol Cell Biomech* **3**, 21–34.
25. Frank, O., Heim, M., Jakob, M., Barbero, A., Schafer, D., Bendik, I., Dick, W., Heberer, M., and Martin, I. (2002) Real-time quantitative RT-PCR analysis of human bone marrow stromal cells during osteogenic differentiation in vitro. *J Cell Biochem* **85**, 737–746.
26. Pittenger, M.F., Mackay, A.M., Beck, S.C., Jaiswal, R.K., Douglas, R., Mosca, J.D., Moorman, M.A., Simonetti, D.W., Craig, S., and Marshak, D.R. (1999) Multilineage potential of adult human mesenchymal stem cells. *Science* **284**, 143–147.
27. Smith-Clerc, J. and Hinz, B. (2010) Immunofluorescence detection of the cytoskeleton and extracellular matrix in tissue and cultured cells. *Methods Mol Biol* **611**, 43–57.

Chapter 15

Ex Vivo Expansion of Human Mesenchymal Stem Cells on Microcarriers

Francisco dos Santos, Pedro Z. Andrade, Gemma Eibes, Cláudia Lobato da Silva, and Joaquim M. S. Cabral

Abstract

Due to the very low titers of human mesenchymal stem cells (MSC) in their niches, namely the bone marrow, an effective approach to isolate and expand those cells ex vivo is required to meet the needs of the increasing MSC clinical applications (e.g., therapy-resistant severe acute graft-versus-host disease). Herein we describe a microcarrier-based stirred culture system protocol for the efficient ex vivo expansion of human bone marrow-derived MSC. This protocol is potentially adaptable to different culture conditions, namely focusing the use of serum-free medium formulations, other sources of MSC, or different types of microcarriers.

Key words: Mesenchymal stem cells, Expansion, Spinner flask, Microcarriers

1. Introduction

The low titers of mesenchymal stem cells (MSC) in the bone marrow (BM) demand a fast ex vivo expansion process to meet highly demanding and clinically relevant cell dosages. Indeed, the frequency of human MSC is considered to be as low as 0.01% of BM mononuclear cells in a newborn, declining with age to 0.001–0.0005% (1). On the other hand, though minimal and maximal doses for therapeutic application have not yet been determined, several million ($1–5 \times 10^6$) BM MSC per kg of patient body weight have been infused (2). Therefore, an efficient and Good Manufacturing Practices (GMP)-compliant ex vivo expansion process is required to achieve clinically relevant MSC numbers.

Since the first clinical application of expanded MSC as adjuvant in hematopoietic transplantation in 2000 (3), the procedures

Mohan C. Vemuri et al. (eds.), *Mesenchymal Stem Cell Assays and Applications*, Methods in Molecular Biology, vol. 698, DOI 10.1007/978-1-60761-999-4_15, © Springer Science+Business Media, LLC 2011

for the isolation and expansion of these cells have remained static, without further improvements. MSC cultures have typically been performed under static conditions in traditional culture flasks, which are limited in terms of cell number generated, their nonhomogeneous nature, difficulty in monitoring, and an extensive handling requirement for feeding/harvesting procedures (4, 5). In this context, bioreactor systems have been developed as alternatives to standard flasks for in vitro culture (5, 6). At the laboratory scale, spinner flasks offer attractive advantages over standard culture configurations, such as a ready scalability and higher homogeneity and minimizing concentration gradients (pH, dissolved oxygen, metabolites). In fact, a microcarrier-based stirred culture system has been successfully established by our group to expand mouse embryonic stem cells (7, 8). Recently, different approaches have been studied to expand MSCs in a bioreactor culture system. Human MSC have been cultured as aggregates in a rotary bioreactor under a microgravity environment (9) or attached to microcarriers (10–12). In addition, porcine MSC (13), goat MSC (14), or rat MSC (15) have also been expanded in microcarrier-based culture systems.

The supplementation of culture media with fetal bovine serum (FBS), commonly used for MSC isolation and expansion, rises concerns mainly focused on cell product safety as it may be a source of xenogeneic antigens and a vehicle for the transmission of mycoplasma, virus, and prions. Hence, the search for feasible FBS alternatives to supplement MSC media is required. These include human serum, platelet-derived products, and growth factors (16, 17). An effective and consistent MSC growth-supplement will represent a major breakthrough accelerating MSC use for multiple cellular therapies.

Herein, we describe an efficient protocol to expand human MSC in a microcarrier-based stirred culture system. This protocol has the advantage of being easily adapted to serum-free culture conditions, as well as to MSC derived from different sources (e.g., adipose tissue derived) or different types of microcarriers.

2. Materials

2.1. Thawing and Expansion of MSC Under Static Conditions

1. Expansion medium: Supplement Dulbecco's Modified Eagle Medium (DMEM, Invitrogen, Carlsbad, CA) with 10% MSC-Qualified Fetal Bovine Serum (FBS, Invitrogen), 1% of L-glutamine (stock solution of 200 mM (100×), Invitrogen), 1% of Penicillin-Streptomycin (10,000 units/ml Penicillin + 10,000 μg/ml Streptomycin, Invitrogen), and 0.1% of Fungizone (250 μg/ml (1,000×), Invitrogen) (hereafter

referred as DMEM-10%FBS). *Alternative*: MesenPRO RS™ complete medium (Invitrogen), 1% of L-glutamine (stock solution of 200 mM (100×), Invitrogen) (hereafter referred to as MesenPRO RS™).

2. Thawing medium: Supplement Iscove's Modified Dulbecco's Medium (IMDM, Invitrogen) with 20% FBS (Invitrogen), 1% mM of L-glutamine (stock solution of 200 mM (100×), Invitrogen), 1% of Penicillin-Streptomycin (10,000 units/ml Penicillin + 10,000 μg/ml Streptomycin, Invitrogen), and 0.1% of Fungizone (250 μg/ml (1,000×), Invitrogen).
3. Phosphate buffered saline (PBS) solution. Make solution (1×) by dissolving PBS powder (Invitrogen) in 1 L of water. Filter the solution using a 0.22-μm filter and store at room temperature.
4. Accutase® solution (Sigma, St. Louis, MO). Store at 4°C after thawing.
5. 0.4% Trypan blue dye solution (Invitrogen). Store at room temperature.
6. Laminar flow hood (BIOAIR, EuroClone, Siziano PV, Italy).
7. Temperature adjustable water bath set to 37°C (Memmert, Schwabach Germany).
8. Polypropylene conical tubes (15/50 ml, BD Biosciences, San Jose, CA).
9. Cell culture centrifuge (Z400K, Hermle, Wehingen, Germany).
10. Cell culture incubator with control of CO_2, temperature and humidity (Inco 2, Memmert).
11. Tissue culture treated plastic ware (T-25/T-75, BD Biosciences).
12. Hemocytometer.
13. Optical microscope equipped with ultraviolet (UV) light (DMI3000 B, Leica Microsystems, Bannockburn, IL).

2.2. Expansion of MSC Under Stirred Conditions

1. Spinner flask *Stem Span* (StemCell Technologies, Vancouver, BC) of 50 ml working volume equipped with an impeller with 90° normal paddles and a magnetic stir bar.
2. Sigmacote® (Sigma).
3. Stirring plate (Variomag Biosystem Direct, Thermo-Fisher, Waltham, MA).
4. Microcarriers Cultispher®-S (Sigma). Store at room temperature.
5. Expansion medium (DMEM-10%FBS or MesenPRO RS™).

2.3. Monitoring of Cell Culture in the Spinner Flask

2.3.1. Cell Count and Viability

1. PBS (1×) solution (see above).
2. 2.5% Trypsin solution (Invitrogen). Prepare 37.5 ml of 1% trypsin in PBS with 15 ml of 2.5% trypsin solution, 1.5 ml of heat-inactivated chicken serum (Invitrogen), and 1.5 ml of 0.1 M EDTA in PBS. Filter-sterilize (0.22 μm) the solution and store in 10 ml aliquots at −20°C. After thawing, the trypsin aliquots can be maintained at 4°C. Prepare a 0.25% trypsin solution in PBS.
3. 0.4% Trypan blue dye solution (Invitrogen). Store at room temperature. Prepare a 0.1% trypan blue solution from the 0.4% trypan blue dye by diluting in PBS (i.e., 1:3 in PBS). Store at room temperature.
4. Hemocytometer.
5. Thermomixer comfort (Eppendorf, Hamburg, Germany).

2.3.2. Cell Distribution on Microcarriers and Metabolic Activity

1. PBS (1×) solution (see above).
2. 2% Paraformaldehyde (PFA) solution: Dissolve 2 g of PFA (Sigma) in 100 ml of PBS (see Note 1). Filter (0.22 μm) before use and maintain at 4°C.
3. 1.5 μg/ml 4′,6-diamidino-2-phenylindole, dihydrochloride (DAPI) solution (Sigma) in PBS, store at 4°C. Prepare from 1 mg/ml stock solution in deionized (DI) water, store at −20°C.
4. 5 mg/ml MTT (3-(4,5-dimethylthiazol-2-yil)-2,5-diphenyltetrazolium bromide) solution in DI water.

2.4. Determination of MSC Multipotency After Expansion

1. PBS (1×) solution (see above).
2. Expansion medium (either DMEM-10%FBS or MesenPRO RS™).
3. StemPro® Adipogenesis Differentiation Kit (Invitrogen).
4. StemPro® Osteogenesis Differentiation Kit (Invitrogen).
5. StemPro® Chondrogenesis Differentiation Kit (Invitrogen).
6. 0.3% Oil Red O (Sigma) solution in isopropanol (Sigma).
7. Alkaline Phosphatase (ALP) Staining: Solution of 10% cold neutral-buffered formalin (Sigma), store at room temperature. Fast Violet B Salt (Sigma), store at 4°C. Dissolve one capsule in 48 ml of Milli-Q water. Aliquot and store at −20°C. 0.25% Naphthol AS-MX Phosphate Alkaline Solution (Sigma), store at 4°C. Prepare Reagent X: Add 4% (v/v) Naphthol AS-MX Phosphate Alkaline Solution to a prethawed aliquot of Fast Violet Solution. Protect from light and use immediately.
8. 2.5% Silver nitrate (Merck, Whitehouse Station, NJ) solution in DI water.
9. 1% Alcian Blue (Sigma) solution in DI water.

3. Methods

3.1. Thawing and Expansion of MSC Under Static Conditions

1. Retrieve a cryogenic vial of MSC (approximately 1 ml) from the liquid nitrogen tank and quickly thaw in a 37°C water bath.
2. Dilute the content of the cryogenic vial 1:4 in thawing medium.
3. Centrifuge at 200 × *g* for 7 min, discard the supernatant and resuspend the pellet in 5 ml of expansion medium.
4. Plate thawed cells in T-25 (5 ml of medium) or T-75 (10 ml of medium) flasks within a cell density range of $3–6 \times 10^3$ cells/cm^2.
5. Incubate cells at 37°C, 5% CO_2 in a humidified atmosphere.
6. Refresh culture medium every 3 days.
7. Passage cells at 70–80% cell confluence. Remove the exhausted culture medium from the flasks and add PBS (same volume as culture medium) to wash the cell layer. Remove PBS and add Accutase® (2 and 4 ml for T-25 and T-75 flasks, respectively). Incubate at 37°C for 7 min.
8. After complete cell detachment, recover the cell suspension to a polypropylene tube. Wash flasks once with expansion medium. Centrifuge at 200 × *g* for 7 min.
9. Discard the supernatant and resuspend the pellet in culture medium. Determine cell number and viability using the trypan blue dye exclusion test. Mix cell suspension (1:1) with a 0.4% trypan blue stain solution, and then viable (unstained cells) and dead cells (blue-stained cells) are identified and counted using a hemocytometer under a microscope.
10. Replate MSC in T-25 (5 ml of medium) or T-75 (10 ml of medium) flasks within a cell density range of $3–6 \times 10^3$ cells/cm^2.

3.2. Expansion of MSC Under Stirred Conditions

1. In order to prevent the microcarriers from sticking to the glass, the spinner flask must be siliconized using Sigmacote® prior to use. Approximately 25 ml of Sigmacote is pipetted onto the spinner flask in order to wet the inner glass surfaces of the spinner flask. The excess of Sigmacote® is then removed by pipetting, and the spinner is allowed to air dry inside the laminar flow hood. Then DI water (30 ml) is added/removed to rinse the inner surfaces of the spinner, and the procedure is repeated 3 times. The spinner is then allowed to air dry inside the laminar flow hood and then proceed to autoclaving before use (121°C, 20 min).

2. Prior to cell inoculation, hydrate the Cultispher®-S microcarriers overnight with PBS and sterilize by autoclaving (121°C, 15 min) (see Note 2). Afterward, wash the microcarriers twice with fresh PBS and incubate in FBS overnight at 37°C, 5% CO_2 (see Note 3). Remove FBS and equilibrate the microcarriers in expansion medium for at least 4 h at 37°C, 5% CO_2.
3. The initial MSC concentration for the expansion under stirred conditions should be within $4–5 \times 10^4$ cells/ml, using 1 g/L of Cultispher®-S microcarriers.
4. Mix the required cell number with Cultispher®-S microcarriers in one sixth of the final volume of expansion medium (either DMEM-10%FBS or MesenPRO RS™) in a polypropylene tube and incubate at 37°C, 5% CO_2 for 30 min. Gently agitate every 5 min.
5. Add prewarmed expansion medium to the cell-microcarriers suspension to reach 50% of the final volume and transfer it carefully to the spinner flask (see Note 4).
6. For 24 h agitate intermittently at 30 rpm (15 min ON, 60 min OFF). Then add the remaining half volume of prewarmed expansion medium and agitate continuously at 30 rpm (see Note 5).
7. After day 3, replace 25% of the medium everyday with prewarmed fresh expansion medium (see Note 6).

3.3. Monitoring the Cell Culture in the Spinner Flask

3.3.1. Cell Count and Viability

1. Take daily 0.5 ml duplicate samples of an evenly mixed culture from the spinner flask (see Note 7).
2. Let the microcarriers settle down, remove 0.3 ml of supernatant and wash with 2 ml of prewarmed PBS. Remove PBS and add 0.8 ml of 0.25% trypsin solution to the washed sample and incubate at 37°C and 650 rpm using the thermomixer until the microcarriers dissolve completely.
3. After complete dissolution of the microcarriers, add 0.25 ml of 0.4% trypan blue solution and 0.75 ml of PBS (see Note 8).
4. Mix thoroughly and determine cell number and viability using a hemocytometer.

3.3.2. Cell Distribution on Microcarriers and Metabolic Activity

1. Take a 0.4-ml sample of an evenly mixed culture from the spinner flask to a 24-well plate.
2. Let the microcarriers settle down, remove 0.3 ml of supernatant and wash twice with PBS.
3. For cellular distribution analysis:
   (a) Fix cells with 0.5 ml of 1% PFA solution for 20 min at room temperature.

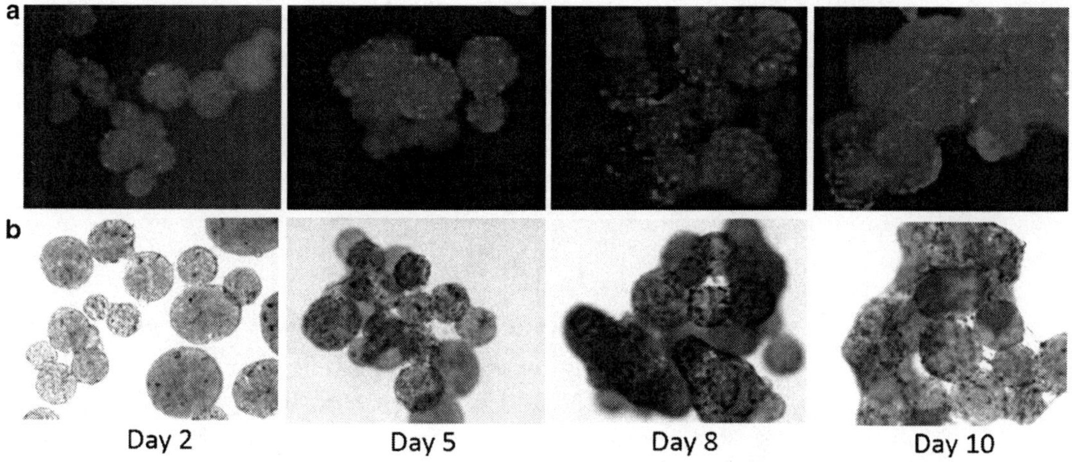

Fig. 1. Monitoring of MSC culture in spinner flask throughout time in culture. (**a**) DAPI staining; (**b**) MTT staining. Images attest a homogenous proliferation of MSC on the microcarriers, and cells remain metabolically active after 10 days under stirred conditions.

(b) Wash twice with PBS. Add 0.5 ml of 1.5 μg/ml DAPI solution and incubate in the dark at room temperature for 5 min (see Note 9).

(c) Wash three times with PBS and keep it protected from the light at 4°C. Observe using a microscope under UV light (Fig. 1a).

4. For cellular metabolic activity analysis:

   (a) Wash twice with PBS. Add 5 mg/ml MTT solution diluted 1:10 in PBS and incubate for 2–4 h at 37°C, 5% CO_2.

   (b) Observe under a microscope (Fig. 1b).

3.4. Determination of MSC Multipotency After Expansion

3.4.1. Osteogenic/ Adipogenic Differentiation

1. Retrieve the microcarriers from the spinner flask. Wash with PBS and add 5 ml of 0.25% trypsin solution. Incubate at 37°C and 650 rpm using the thermomixer until the microcarriers dissolve completely.
2. Determine cell number and viability using the trypan blue exclusion method (see Subheading 3.1).
3. Plate MSC in a 12-well plate at a cell density range of $3–6 \times 10^3$ cells/cm^2.
4. Incubate cells at 37°C, 5% CO_2 in a humidified atmosphere.
5. Upon reaching 80% cell confluence, replace the expansion medium with the respective differentiation medium (osteogenesis or adipogenesis differentiation medium).
6. Refresh differentiation media every 3–4 days.

3.4.2. ALP/von Kossa Staining (Osteogenesis)

1. After 14 days of osteogenic differentiation, remove the differentiation medium and wash cells with PBS.
2. Fix cells with 10% cold neutral-buffered formalin for 20 min at room temperature.
3. Wash once with PBS and keep cells for 15 min in DI water at room temperature.
4. Incubate the cells with Reagent X for 45 min at room temperature in the dark.
5. Wash three times in DI water and observe under a microscope.
6. Remove DI water and incubate cells with 2.5% silver nitrate solution for 30 min at room temperature.
7. Wash three times in DI water and observe under a microscope.

3.4.3. Oil Red O Staining (Adipogenesis)

1. After 14 days of adipogenic differentiation, remove the differentiation medium and wash cells with PBS.
2. Fix cells with 2% PFA for 20 min at room temperature.
3. Wash twice with PBS and add 0.3% oil red O solution for 60 min at room temperature.
4. Wash three times with PBS, add DI water and observe under a microscope.

3.4.4. Chondrogenic Differentiation

1. Retrieve the microcarriers from the spinner flask. Wash with PBS and add 5 ml of 0.25% trypsin solution. Incubate at 37°C and 650 rpm using the thermomixer until the microcarriers dissolve completely.
2. Determine cell number and viability using the trypan blue exclusion method (see Subheading 3.1).
3. Centrifuge cells at $200\times g$ for 7 min and generate a cell suspension of 1.6×10^7 cells/ml in expansion medium.
4. Seed 5 μl droplets of cell suspension in a 24-well plate. Incubate for 2 h at 37°C, 5% CO_2 in a humidified atmosphere. Then add chondrogenesis differentiation medium and incubate at 37°C, 5% CO_2.
5. Refresh differentiation medium every 2–3 days.

3.4.5. Alcian Blue Staining (Chondrogenesis)

1. After 14 days of chondrogenic differentiation, remove the differentiation medium and wash cells with PBS.
2. Fix cells with 2% PFA for 20 min at room temperature.
3. Wash twice with PBS and add 1% Alcian Blue solution for 30 min at room temperature.
4. Wash three times with PBS, add DI water and observe under a microscope.

4. Notes

1. PFA powder should be initially dissolved in a low volume of water at a high temperature (lower than 70°C) in order to facilitate dissolution. The pH should be set to 7.3 and the final volume completed with 10× PBS.
2. This protocol can also be adapted to other microcarrier types (for instance, Cytodex-3 microcarriers (GE Healthcare) (7) and animal-free microcarriers (e.g., plastic) have also been tested with success for human – both bone marrow- and adipose tissue-derived MSC in our laboratory). Depending on microcarrier's characteristics, some steps of the protocol may have to be changed.
3. In order to improve the initial cell adhesion, microcarriers are incubated with FBS, which is known to have high content of adhesion factors. Envisaging a serum-free medium culture, coating microcarriers with human plasma is a feasible alternative to FBS.
4. Some of the microcarriers may attach to the polypropylene tube in this step. It is recommended to collect the microcarriers using a smaller volume of medium and then wash the plastic tube (two or three times) using the remaining volume to minimize the loss of microcarriers. Polypropylene tubes pretreated with Sigmacote® can also be used to reduce the adhesion of microcarriers to the plastic.
5. The intermittent agitation regimen favors the initial cell attachment to the microcarriers. Different intermittent time periods can also be used (e.g., 5 min ON, 90 min OFF) with culture conditions more adverse to cellular adhesion (for instance, when using serum-free expansion media). Throughout the expansion, microcarriers tend to form large aggregates (particularly gelatin microcarriers, such as Cultispher®-S). The increase of the agitation speed (e.g., for 40 or 50 rpm) may contribute to reduce microcarrier aggregation.
6. The medium replenishment must be done without removing microcarriers. After the spinner flask has been placed inside the laminar flow hood, allow the microcarriers to settle for 10–15 min. Then, proceed to the medium renewal.
7. A homogeneous sampling is essential for accurate cell number. Before taking 0.5 ml samples, it is important to assure an evenly mixed culture inside the spinner flask. For that purpose, a stirring plate may be used inside the laminar flow chamber.
8. After the trypsin treatment under constant agitation, a cell pellet may be formed. If necessary, the cell suspension must

be homogenized before determining the cell number and viability.

9. The well-plate should be protected from light using aluminum foil. It is also advised to turn off the laminar flow chamber light while preparing the sample.

References

1. Caplan, A. I. (2007) Adult mesenchymal stem cells for tissue engineering versus regenerative medicine, *J Cell Physiol 213*, 341–347.
2. Subbanna, P. K. (2007) Mesenchymal stem cells for treating GVHD: in-vivo fate and optimal dose, *Med Hypotheses 69*, 469–470.
3. Koc, O. N., Gerson, S. L., Cooper, B. W., Dyhouse, S. M., Haynesworth, S. E., Caplan, A. I., and Lazarus, H. M. (2000) Rapid hematopoietic recovery after coinfusion of autologous-blood stem cells and culture-expanded marrow mesenchymal stem cells in advanced breast cancer patients receiving high-dose chemotherapy, *J Clin Oncol 18*, 307–316.
4. Sotiropoulou, P. A., Perez, S. A., Salagianni, M., Baxevanis, C. N., and Papamichail, M. (2006) Characterization of the optimal culture conditions for clinical scale production of human mesenchymal stem cells, *Stem Cells 24*, 462–471.
5. Cabrita, G. J. M., Ferreira, B. S., da Silva, C. L., Goncalves, R., Almeida-Porada, G., and Cabral, J. M. S. (2003) Hematopoietic stem cells: from the bone to the bioreactor, *Trends Biotechnol 21*, 233–240.
6. King, J. A., and Miller, W. M. (2007) Bioreactor development for stem cell expansion and controlled differentiation, *Curr Opin Chem Biol 11*, 394–398.
7. Abranches, E., Bekman, E., Henrique, D., and Cabral, J. M. (2007) Expansion of mouse embryonic stem cells on microcarriers, *Biotechnol Bioeng 96*, 1211–1221.
8. Fernandes, A. M., Fernandes, T. G., Diogo, M. M., da Silva, C. L., Henrique, D., and Cabral, J. M. (2007) Mouse embryonic stem cell expansion in a microcarrier-based stirred culture system, *J Biotechnol 132*, 227–236.
9. Chen, X., Xu, H., Wan, C., McCaigue, M., and Li, G. (2006) Bioreactor expansion of human adult bone marrow-derived mesenchymal stem cells, *Stem Cells 24*, 2052–2059.
10. Eibes, G., dos Santos, F., Andrade, P. Z., Boura, J. S., Abecasis, M. M., da Silva, C. L., and Cabral, J. M. S. (2010) Maximizing the ex vivo expansion of human mesenchymal stem cells using a microcarrier-based stirred culture system, *J Biotechnol 146*, 194–197.
11. Schop, D., van Dijkhuizen-Radersma, R., Borgart, E., Janssen, F. W., Rozemuller, H., Prins, H. J., and de Bruijn, J. D. (2009) Expansion of human mesenchymal stromal cells on microcarriers: growth and metabolism, *J Tissue Eng Regen Med 4*, 131–140.
12. Wu, Q. F., Wu, C. T., Dong, B., and Wang, L. S. (2003) [Cultivation of human mesenchymal stem cells on macroporous CultiSpher G microcarriers], *Zhongguo Shi Yan Xue Ye Xue Za Zhi 11*, 15–21.
13. Frauenschuh, S., Reichmann, E., Ibold, Y., Goetz, P. M., Sittinger, M., and Ringe, J. (2007) A microcarrier-based cultivation system for expansion of primary mesenchymal stem cells, *Biotechnol Prog 23*, 187–193.
14. Schop, D., Janssen, F. W., Borgart, E., de Bruijn, J. D., and van Dijkhuizen-Radersma, R. (2008) Expansion of mesenchymal stem cells using a microcarrier-based cultivation system: growth and metabolism, *J Tissue Eng Regen Med 2*, 126–135.
15. Sart, S., Schneider, Y.-J., and Agathos, S. N. (2009) Ear mesenchymal stem cells: An efficient adult multipotent cell population fit for rapid and scalable expansion, *J Biotechnol 139*, 291–299.
16. Bernardo, M. E., Avanzini, M. A., Perotti, C., Cometa, A. M., Moretta, A., Lenta, E., Del Fante, C., Novara, F., de Silvestri, A., Amendola, G., Zuffardi, O., Maccario, R., and Locatelli, F. (2007) Optimization of in vitro expansion of human multipotent mesenchymal stromal cells for cell-therapy approaches: further insights in the search for a fetal calf serum substitute, *J Cell Physiol 211*, 121–130.
17. Ng, F., Boucher, S., Koh, S., Sastry, K. S., Chase, L., Lakshmipathy, U., Choong, C., Yang, Z., Vemuri, M. C., Rao, M. S., and Tanavde, V. (2008) PDGF, TGF-beta, and FGF signaling is important for differentiation and growth of mesenchymal stem cells (MSCs): transcriptional profiling can identify markers and signaling pathways important in differentiation of MSCs into adipogenic, chondrogenic, and osteogenic lineages, *Blood 112*, 295–307.

Part III

MSC Lineage Differentiation and Analysis

Chapter 16

Osteogenic Differentiation of Human Multipotent Mesenchymal Stromal Cells

Deepak M. Gupta, Nicholas J. Panetta, and Michael T. Longaker

Abstract

A comprehensive knowledge of the molecular biology underlying osteogenic differentiation in a controlled, laboratory setting may promise optimization of future cell-based tissue engineering strategies for clinical problems. The scope of this review encompasses a discussion of the methodology utilized to perform such studies. Our laboratory routinely performs both in vitro and in vivo assays underlying osteogenic differentiation, and the widespread use of singular methodology across multiple investigators and institutions promises great advancements for the skeletal tissue engineering community.

Key words: Osteogenic differentiation, Skeletal tissue engineering, Regeneration, Mesenchymal stromal cells, Bone marrow, Adipose derived stromal cells, Alizarin red

1. Introduction

Contributions from multidisciplinary investigations have focused attention on the potential of tissue engineering to yield novel therapeutics. Congenital and acquired skeletal defects represent excellent targets for the implementation of tissue engineering applications secondary to the technically challenging nature and inherent inadequacies of current reconstructive interventions. Apropos to the search for answers to these clinical conundrums, studies have focused on elucidating the molecular signals driving the biological activity of the aforementioned maladies. Strategies devised over the past century to treat congenital or acquired skeletal defects with grafts, bone matrices and pastes, semi-synthetic scaffolds, and alloplastics have all yielded suboptimal results. The need to repair bone loss in a functional and lasting manner has therefore driven a shift in treatment paradigm from one of traditional tissue replacement to tissue regeneration.

Mohan C. Vemuri et al. (eds.), *Mesenchymal Stem Cell Assays and Applications*, Methods in Molecular Biology, vol. 698,
DOI 10.1007/978-1-60761-999-4_16, © Springer Science+Business Media, LLC 2011

Accumulating evidence suggests that progenitor cells derived from adult tissues such as bone marrow and adipose have the capacity to differentiate into a multitude of cell types and participate in tissue regeneration directly or by helping to recruit additional cell types. This regenerative potential suggests that adult progenitor cells may be used to fulfill the mounting needs of patients with genetic disorders, degenerative diseases, and traumatic or post-surgical tissue deficits of the adult skeleton.

Furthermore, as stem cell biology, materials science, developmental biology, bioengineering, and gene therapy become increasingly interconnected in contemporary research institutions, the need for clear, conveyable, scientifically sound methodology will underscore efficient interdisciplinary collaboration. Further understanding of the mechanisms by which mesenchymal stem cells undergo osteogenic differentiation will be essential in optimizing clinically applicable strategies in skeletal regeneration. In addition, comprehending the unique characteristics of the osteoprogenitor cells, such as their cell surface markers and gene expression profile, will identify pure populations of skeletal progenitor cells and further enhance the potential of cell-based skeletal tissue engineering.

As research evolves in the field of tissue engineering, alternative strategies for osteogenic cellular sources are emerging. For example, induced pluripotent stem (iPS) cells have been generated from adult human fibroblasts and characterized. Using these methods, direct reprogramming of differentiated somatic cells may potentially be used to generate patient-specific stem cells for regenerative medicine applications. However, at present there are concerns for tumorigenesis and the use of retroviral vectors.

Thus, many obstacles still need to be overcome to translate laboratory findings of cell-based therapies for skeletal engineering to the clinical setting. The field of skeletal tissue engineering is rapidly progressing, and cell-based therapies for skeletal defects may potentially lead to clinical applications in the near future.

2. Materials

2.1. Isolation of Human and Mouse Adipose-Derived Stromal Cells

1. Phosphate-buffered saline (PBS) (Invitrogen, Carlsbad, CA).
2. Hanks' balanced salt solution (HBSS) (Mediatech, Manassas, VA).
3. Collagenase Type II (Sigma Aldrich, St. Louis, MO).
4. Growth Medium: Dulbecco's Modified Eagle Medium (DMEM) (Invitrogen) supplemented with 10% (v/v) fetal bovine serum (FBS) (Omega Scientific, Tarzana, CA) and 1% (v/v) Penicillin/Streptomycin Solution (Invitrogen).

5. Povidone-iodine solution.
6. 100-μm Cell strainer (BD Biosciences, San Jose, CA).

2.2. Isolation of Mouse Bone Marrow Stromal Cells

1. PBS (Invitrogen).
2. Growth medium: DMEM (Invitrogen) supplemented with 10% (v/v) FBS (Omega Scientific) and 1% (v/v) Penicillin/Streptomycin Solution (Invitrogen).
3. 25–27 Gauge needle on 10 ml syringe.

2.3. MSC Expansion and Propagation

1. PBS (Invitrogen).
2. Growth medium: DMEM (Invitrogen) supplemented with 10% (v/v) FBS (Omega Scientific) and 1% (v/v) Penicillin/Streptomycin Solution (Invitrogen).
3. TrypLE Express Stable Trypsin Replacement Enzyme (Invitrogen).
4. 100-μm Filter baskets.

2.4. Cell Counting

1. 0.4% (w/v) Trypan Blue solution (Sigma Aldrich).
2. Hemacytometer.
3. PBS (Invitrogen).

2.5. Cell Freezing

1. PBS (Invitrogen).
2. Freezing medium: 90% (v/v) FBS (Omega Scientific), 10% (v/v) dimethyl sulfoxide (DMSO).
3. TrypLE Express Stable Trypsin Replacement Enzyme (Invitrogen).
4. Cryogenic vials (Nunc, Rochester, NY).
5. Cryo 1°C freezing container (Nalgene, Rochester, NY).

2.6. Osteogenic Differentiation

1. Osteogenic differentiation medium (ODM): DMEM (Invitrogen) with 10% (v/v) FBS (Omega Scientific), 1% (v/v) penicillin/streptomycin solution (Invitrogen), 10 mM β-glycerophosphate (Sigma Aldrich) and 250 μM ascorbic acid (Fisher Scientific, Pittsburgh, PA).

2.7. rhBMP-2 Preparation and Use

1. Resuspend lyophilized rhBMP-2 (R&D Systems, Minneapolis, MN) in 4 mM HCl (per manufacturer instructions).
2. Supplement ODM with 10–200 μg/ml rhBMP-2. The specific concentration should be determined by your laboratory for your specific experimental design.

2.8. Alkaline Phosphatase Staining

1. Citrate working solution (2 ml of citrate concentrate + 98 ml distilled water) (Sigma Aldrich).
2. Fixative solution: 40% (v/v) citrate working solution + 60% (v/v) acetone.

3. Alkaline phosphatase staining solution: 12 mg of Fast Violet B salt (Sigma Aldrich) to 48 ml water, add 2 ml of Naphthol AS-MX (Sigma Aldrich) for every 48 ml of salt solution.
   (a) Add naphthol immediately before use as it reacts quickly.
   (b) Shield from light.
4. PBS (Invitrogen).

2.9. Von Kossa Staining

1. Formalin or ethanol.
2. Ultraviolet lamp.
3. Safranin-O (Sigma Aldrich).
4. Silver nitrate (Sigma Aldrich).
5. Sodium thiosulfate (Sigma Aldrich).

2.10. Alizarin Red Staining

1. Alizarin Red staining solution: 0.2% (w/v) Alizarin Red S (Sigma Aldrich) powder is dissolved in distilled water.
   (a) Adjust pH to 6.36–6.40 with ammonia.
   (b) Keep stain solution in refrigerator for up to 1 month.
2. 100% Ethanol.
3. Optional: To quantify stain, prepare a leaching solution of 20% (v/v) methanol and 10% (v/v) acetic acid in distilled water.

3. Methods

3.1. Isolation of Human Adipose-Derived Stromal Cells

1. Collect fresh lipoaspirate in a sterile container and immediately place on ice until it can be transported to the laboratory for processing (see Notes 1–3).
2. In a negative pressure cell culture hood, allow fresh lipoaspirate to "settle" so that the aqueous and lipid-laden layers separate. Using a sterile pipette, aspirate the aqueous layer from the bottom of the container (see Note 4).
3. With only the lipid-laden layer remaining, add 300 ml of fresh lipoaspirate into a sterile 1 l jar by pipetting using aseptic technique (see Note 5).
4. Wash #1: Add 300 ml of sterile PBS containing 6 ml povidone/iodine (2% v/v solution) to the container and gently stir with a sterile pipette.
5. Again, allow fresh lipoaspirate to "separate" into aqueous and lipid-laden layers. Using a sterile pipette, aspirate the aqueous layer from the bottom of the container.

6. Wash #2: Add 300 ml of sterile PBS to the container and gently stir with a sterile pipette.
7. Allow fresh lipoaspirate to "separate" into aqueous and lipid-laden layers. Using a sterile pipette, aspirate the aqueous layer from the bottom of the container.
8. Wash #3: Add 300 ml of sterile PBS for a second time to the container and gently stir with a sterile pipette.
9. Allow fresh lipoaspirate to "separate" into aqueous and lipid-laden layers. Using a sterile pipette, aspirate the aqueous layer from the bottom of the container.
10. Repeat steps 12 and 13 until fresh lipoaspirate is free of gross blood contamination.
11. Enzymatic digestion: Add 300 ml of sterile-filtered 0.075% (w/v) Collagenase type II solution (in HBSS) to the freshly washed lipoaspirate. Gently stir with a sterile pipette and then place the lid on the bottle (see Note 6).
12. Place the sealed jar in a shaking (120 rpm) 37°C water bath for 1 h. During this hour, vent the jar every 10–15 min by copiously spraying around the lid with 70% (v/v) ethanol and then briefly loosening and retightening the lid.
13. Neutralization: In a negative pressure cell culture hood, add 300 ml of growth media to the lipoaspirate. Stir gently. Total volume in the jar should now be 900 ml.
14. Pipette 45 ml of aliquots into 50 ml centrifuge tubes using aseptic technique. Centrifuge each tube for 5 min at 4°C in a swinging bucket rotor at $200 \times g$.
15. Wash #4: Aspirate the lipid and aqueous layers into a waste container without disturbing the cell pellet at the bottom of the tube.
16. Resuspend the cell pellet in 10 ml of fresh growth media. Combine media from 4 to 5 resuspended cell pellets into a single 50 ml tube.
17. Using aseptic technique, filter the resuspended cells through a 100-μm filter basket. Collect the filtrate in a new 50-ml tube.
18. Centrifuge each 50 ml tube for 5 min at 4°C in a swinging bucket rotor at $200 \times g$.
19. Aspirate the lipid and aqueous layers into a waste container without disturbing the cell pellet at the bottom of the tube.
20. Plating: Resuspend each cell pellet in 20 ml of fresh growth media. Plate cells in 15 cm (diameter) or T-225 (or similar) culture vessels and place in a 37°C, humidified air incubator.
21. Maintenance: Growth media should be changed every 2–3 days and passaged when approximately 70% confluent (see Note 7).

3.2. Isolation of Mouse Adipose-Derived Stromal Cells

1. Euthanize mouse. Dissect inguinal fat pads from five mice that are 30-days old. Place fat pads in ice cold PBS until further processing (see Notes 3 and 8).
2. In a negative pressure cell culture hood, mince fat pads with sterile scissors until the consistency is free of large pieces of tissue. This should be done with aseptic technique.
3. Place minced fat pads in 50 ml centrifuge tube using aseptic technique (e.g., pipette).
4. Wash #1: Add 25 ml of sterile PBS containing 0.5 ml povidone/iodine (2% v/v solution) to the container and gently mix.
5. Allow minced fat pads to "separate" from the PBS/povidone/iodine into aqueous and lipid-laden layers. Using a sterile pipette, aspirate the aqueous layer from the bottom of the container and discard (see Note 4).
6. Wash #2: Add 25 ml of sterile PBS to the container and gently mix.
7. Again, allow minced fat pads to "separate" from PBS into aqueous and lipid-laden layers. Using a sterile pipette, aspirate the aqueous layer from the bottom of the container and discard.
8. Wash #3: Add 25 ml of sterile PBS for a second time to the container and gently stir with a sterile pipette.
9. Allow minced fat pads to "separate" into aqueous and lipid-laden layers. Using a sterile pipette, aspirate the aqueous layer from the bottom of the container and discard.
10. Enzymatic digestion: Add 15 ml of sterile-filtered 0.075% (w/v) Collagenase type II solution (in HBSS) to the freshly minced fat pads. Gently mix (see Note 6).
11. Place the digestion tube in a shaking (120 rpm) 37°C water bath for 1 h. During this hour, vent the tube every 10–15 min by copiously spraying around the cap with 70% (v/v) ethanol and then briefly loosening and retightening the cap.
12. Neutralization: In a negative pressure cell culture hood, add 15 ml of growth media to the lipoaspirate. Mix thoroughly. Total volume in the jar should now be 30 ml + the volume of minced fat pads.
13. Centrifuge the tube for 5 min at 4°C in a swinging bucket rotor at $200 \times g$.
14. Wash #4: Aspirate the lipid and aqueous layers into a waste container without disturbing the cell pellet at the bottom of the tube.
15. Resuspend the cell pellet in 25 ml of fresh growth media.

16. Using aseptic technique, filter the resuspended cells through a 100-μm filter basket. Collect the filtrate in a new 50-ml tube.
17. Centrifuge the tube for 5 min at 4°C in a swinging bucket rotor at 200 × *g*.
18. Aspirate the lipid and aqueous layers into a waste container without disturbing the cell pellet at the bottom of the tube.
19. Plating: Resuspend the cell pellet in 10 ml of fresh growth media. Plate cells in a 10 cm (diameter) or T-75 (or similar) culture vessel and place in a 37°C, humidified air incubator.
20. Maintenance: Growth media should be changed every 2–3 days and passaged when approximately 70% confluent.

3.3. Isolation of Mouse Bone Marrow Stromal Cells

1. Euthanize mouse. Dissect femurs and tibiae using sterile instruments and aseptic technique. Remove attached muscle, skin, fur, etc. Place bones in ice cold PBS until further processing (see Notes 3 and 8).
2. In a negative pressure cell culture hood, remove the epiphyseal plate from each end of the bone with a scalpel. Flush the bone marrow plug with fresh growth media using a 25–27 gauge needle on a syringe (10 ml). Collect multiple bone marrow plugs into a centrifuge tube.
3. Pipette vigorously up and down until most of the plugs are washed clean of cells. Plugs will turn white and media will become cloudy with cells.
4. Filter the cell suspension through a 100 μm cell strainer.
5. Centrifuge the tube for 5 min at 4°C in a swinging bucket rotor at 200 × *g*.
6. Plating: Resuspend the cell pellet from five bone marrow plugs in 10 ml of fresh growth media. Plate cells in a 10 cm (diameter) or T-75 (or similar) culture vessel and place in a 37°C, humidified air incubator.
7. Maintenance: Growth media should be changed every 2–3 days and passaged when approximately 70% confluent.

3.4. Splitting and Passaging Cells (see Note 9)

1. Using a sterile Pasteur pipette, aspirate the old media from the cell culture vessel.
2. Wash cells briefly with sterile PBS (without calcium or magnesium). Aspirate the PBS.
3. Add TrypLE to the cells so that a thin film covers the plate.
4. Place cells in a 37°C, humidified air incubator for 5 min.
5. Mechanically disrupt cell attachments from the plate by tapping on the side of the dish or using a cell scraper. Using a microscope, ensure that the cells have detached from the cell

culture vessel surface. The cells should appear uniformly round ("balled up") and not "ragged" (this appearance usually indicates unhealthy or dead cells).

6. If the cells remain aggregated, disrupt by vigorous pipetting or another 5 min in the incubator (still in TrypLE). You may also elect to pass the cells through a 100-μm filter basket at this time as an alternative solution; however, this may cause a decrease in cell number (see Note 10).
7. Neutralize the TrypLE with a 2:1 ratio volume of growth medium. Collect cells, TrypLE and growth media into a centrifuge tube.
8. Centrifuge the tube for 5 min at 4°C in a swinging bucket rotor at 200 × *g*.
9. Aspirate the TrypLE and media from the tube without disrupting the cell pellet.
10. Resuspend the cells in an appropriate volume of growth media for the cell culture vessel, taking into account the ratio of the split (i.e., 2:1, 4:1, 10:1, etc.). Pipette the fresh media over the cell pellet vigorously to ensure homogeneity in cell density and singlet microscopic appearance (not doublet, aggregate, etc.).

3.5. Cell Counting

1. Follow the protocol in Subheading 3.4 to collect cells from a cell culture vessel.
2. Mix thoroughly and draw up 10 μl of cells in media using a micropipette. Mix with 15 μl of PBS and 25 μl of 0.4% (w/v) Trypan Blue solution (1:5 dilution of cells).
3. Place 10 μl of Trypan Blue-stained cells under a glass cover slip on a hemacytometer.
4. Count the total cells in each set of 16 boxes in each corner (one quadrant) of the hemacytometer as shown below. Briefly, the 16 boxes labeled "1" should be totaled together. Similarly, count the cells for boxes labeled "2," "3," and "4." Also keep a running tally of the number of Trypan Blue-positive cells (these cells will appear blue). See Fig. 1.
5. Add the four numbers together. Aim for a cell density of 100–200 total cells during this count. If you are outside this range, adjust the amount of media in which you have resuspended the cell pellet in the last step of Subheading 3.4 and repeat this section.
6. Subtract the number of Trypan Blue-positive cells.
7. Calculate the average number of cells per quadrant that are Trypan Blue-negative. Multiply this number by 5×10^4. This number represents the number of cells present *per milliliter* at the end of Subheading 3.4.

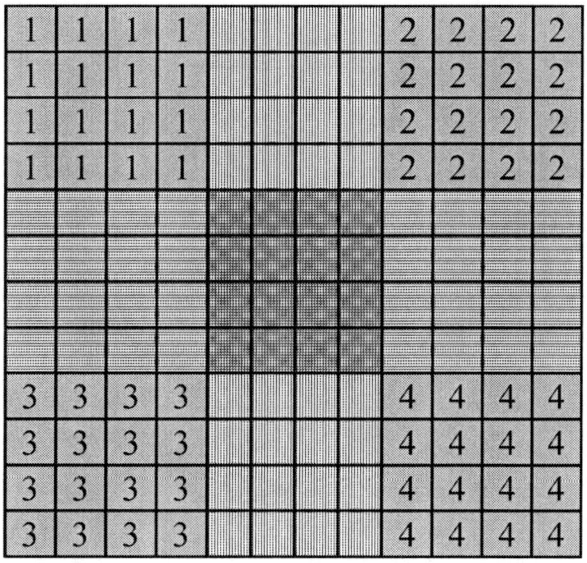

Fig. 1. Hemocytometer counting pattern.

3.6. Cell Freezing

1. Follow the protocol in Subheading 3.4 to collect cells from a cell culture vessel. However, do not resuspend cells in fresh growth media.
2. Resuspend cells in sterile-filtered freezing medium. Mix thoroughly to ensure homogeneity in cell density and singlet microscopic appearance (not doublet, aggregate, etc.).
3. Work quickly from this point to minimize the time cells are exposed to freezing medium at room temperature as DMSO is toxic. Aliquot cells into cryogenic vials.
4. Place cryogenic vials in an isopropanol-loaded freezing apparatus in a –20°C freezer overnight.
5. The next day, work quickly to transfer the cryogenic vials to a chamber containing liquid nitrogen. Use appropriate safety precautions when working with liquid nitrogen. Do not allow tubes to thaw during transfer.

3.7. Seeding and Maintenance of Osteogenic Differentiation Assays

1. Follow the protocol in Subheading 3.5 to collect cells from a cell culture vessel and calculate the number of cells (per milliliter) suspended in fresh growth media.
2. Plate cells in cell culture vessel for differentiation assay in fresh growth medium (see Note 11).
   - For a 10 cm plate, use 200,000 cells.
   - For one well of a 6-well plate, use 100,000 cells.
   - For one well of a 12-well plate, use 50,000 cells.
   - For one well of a 24-well plate, use 20,000 cells.
   - For one well of a 96-well plate, use 3,500 cells.

3. After 24 h, the cells should be adherent and subconfluent (70–80%). Aspirate growth media and replace with ODM (see Note 12).
4. Using a sterile Pasteur pipette, aspirate the old media from the cell culture vessel.
5. Replace with fresh osteogenic differentiation media every 2–3 days for a total of 10–14 days (depending on experimental design).

3.8. Alkaline Phosphatase Staining (see Note 13)

1. Aspirate media and wash cells with PBS twice.
2. Fix cells with fixative solution for 30 s.
3. Rinse cells briefly with distilled water. Do not allow cells to dry out.
4. While shielded from light, stain cells with alkaline phosphatase staining solution for 30 min at room temperature.
5. Aspirate staining solution and rinse briefly with water.
6. Image cells by microscopy or gross scanning.

3.9. Von Kossa Staining (see Note 13)

1. Aspirate media and wash cell layer in culture vessel with PBS 3 times.
2. Fix with formalin or 100% ethanol for 30 min.
3. Quickly wash with distilled water. Allow wells to air dry.
4. Stain with fresh 1% (w/v) aqueous silver nitrate (in distilled water) for 15 min under direct ultraviolet light.
5. Wash twice with distilled water.
6. Stain with 5% (w/v) fresh sodium thiosulfate (in distilled water) for 2 min (see Note 14).
7. Wash twice with distilled water.
8. Optional: Counterstain with 1% (v/v) Safranin-O for 10 min (see Note 15).
9. Wash twice with distilled water and acquire photographs.

3.10. Alizarin Red Staining

1. Aspirate media and wash cell layer in culture vessel with PBS 3 times.
2. Wash cell layer in culture vessel with distilled water once quickly.
3. Fix cells with 100% ethanol for 15 min.
4. Stain cells with Alizarin Red staining solution for 30–60 min at room temperature.
5. Wash stained cells with distilled water 3 times.
6. Take pictures promptly after staining.

3.11. Alizarin Red Quantification

1. Follow the steps in Subheading 3.10.
2. Wash the cells with PBS twice.
3. Add leaching solution to each well and gently agitate on an orbital rocker for 15 min.
4. Measure the absorbance of the leached stain on a spectrophotometer set at 450 nm (wavelength). The spectrophotometer should be "zeroed" to clean leaching solution.
5. All absorbance values should be normalized to protein concentration of the well.

3.12. Seeding Apatite-Coated PLGA Scaffolds

1. The fabrication process has been previously described (1).
2. Sterilize scaffolds with 70% ethanol, ultraviolet light, and/or ethylene oxide.
3. Rinse scaffolds twice with PBS.
4. Slowly pipette 150,000 cells suspended in 5 μl of growth media onto the scaffold in a 96-well plate.
5. Allow the cells to adhere for 30 min in a 37°C humidified warm air incubator.
6. For the next 6 h, slowly add 5 μl of growth media directly onto the cell-loaded scaffold every 10–15 min.
7. Submerge the cell-loaded scaffold in growth media and incubate overnight in a 37°C humidified warm air incubator.
8. The next day confirm that the cells have attached to the scaffold by microscopy.
9. From this point, cell-seeded scaffolds may be treated as described above depending on experimental design. For example, they may be subjected to proliferation assays by adding growth medium, allowing for expansion and propagation, and subsequently trypsinization of these cells from the scaffold for counting. In another example, the seeded cells may be subjected to osteogenic differentiation by adding ODM. After a brief period of differentiation, the cell-seeded scaffold may be stained with Alizarin Red, which may then also be quantified as described earlier. Furthermore, in our experience, the cell-seeded scaffold may be subjected to a variety of other standard laboratory protocols such as nucleic acid extraction, viral-mediated transduction, and protein extraction, without modification as noted by members of our laboratory. The cell-seeded scaffold may also be surgically placed in vivo as a single construct to assay for proliferation, differentiation, gene expression changes, tissue regeneration, etc.

4. Notes

1. All human specimens should be handled according to universal safety precautions. Body fluids should be assumed to contain dangerous pathogens, i.e., human immunodeficiency virus (HIV), hepatitis B virus (HBV), hepatitis C virus (HCV), etc. Appropriate training and approvals should be established before undertaking this procedure.
2. Tissue procurement: Informed consent should be signed prior to surgery by any patient whose tissue will be made available for research purposes. This should be approved under the auspices of the local governing Institutional Review Board (IRB) or other congruent entity.
3. All tissue procurement, whether from animals or humans, should be performed with sterile surgical instruments and reagents.
4. All waste should be placed into a waste container containing a final volume of 10% bleach for at least 10 min. Proper disposal of biohazardous material should be followed.
5. The volumes in the following procedures may be scaled up or down depending on specific experimental and/or laboratory needs.
6. Collagenase solution used to establish primary cultures derived from adipose tissue should be made fresh before use. It is important that the solution is shaken vigorously or vortexed before use and then sterile filtered.
7. The presence of numerous red blood cells in primary cultures is expected. Initially, they may not be easily washed off; however, after the first split/passage, they will not adhere and will get washed away with media changes.
8. Use of animals should be approved by Institutional Animal Care and Use Committee (IACUC) or other congruent local governing entity.
9. Mouse and human adipose derived stromal cells should be limited to fourth passage or earlier. Beyond this stage, our unpublished results suggest that these cells have limited capacity to proliferate and differentiate toward the osteogenic lineage. Literature also suggests that there are molecular changes that may not necessarily reflect upon the experimental design.
10. Related to the above point, splitting, passaging, and seeding cells may also be complicated by cell clumping. All cell suspensions should be pipetted vigorously. If cells are allowed to

sit undisturbed for more than a few minutes, again they should be pipetted vigorously or passed through a cell strainer to remove clumps. Prolonged trypsinization may also be helpful in removing the extracellular matrix that may bind clumps.

11. Our unpublished results have suggested that cell monolayer confluence at the time of initiation of osteogenic differentiation may affect the differentiation process. We have observed consistent results when differentiating cells between 60 and 70% confluent. Individual conditions should be optimized in the laboratory.
12. Use of rhBMP-2 should follow specific experimental design. The use of rhBMP-2 is not routine in osteogenic differentiation media for human or mouse cells. When using rhBMP-2 resuspended in 4 mM HCl, experimental design should include use of control sample supplemented with 4 mM HCl (without rhBMP-2).
13. All staining solutions should be made fresh immediately before use and shielded from light with aluminum foil. One exception to this rule is Alizarin Red, which may be stored at 4°C for up to 6 months.
14. During the Von Kossa staining procedure, staining with 5% (w/v) fresh sodium thiosulfate (in distilled water) may be omitted. It can potentially bleach out black staining. One should be sure to photograph the stained cells before this step.
15. During the Von Kossa staining procedure, counterstaining with 1% (v/v) Safranin-O is optional. This step should turn nuclei, cytoplasm, and osteoid red.
16. Although not explicitly covered in this chapter, our laboratory uses standard fluorescence activated cell sorting (FACS) to identify osteoprogenitor cells among adipose derived stromal cells and bone marrow derived stromal cells. However, we have noticed significant difficulty and persistent cell "clumping" during these experiments. To address this challenge, we use 1–5 mM ethylenediaminetetraacetic acid (EDTA) in our FACS buffer (optimal concentration should be determined) and perform all steps at 4°C or on ice. DNAse may also help.
17. Although no methods are explicitly presented for protein extraction, western blot, RNA extraction, DNA extraction, reverse transcription, PCR, in vitro transfection, and viral-mediated transduction, our laboratory uses standard protocols for these experiments without difficulty.

Acknowledgments

This work was supported by NIH R01 DE014526, NIH R21 DE019274, and grants from the Oak and Hagey Foundations to M.T.L., NIH F32 AR055871, Stanford University Dean's Award and Lucile Packard Children's Health Research Program Award to D.M.G., and an American College of Surgeons Resident Research Award and Lucile Packard Children's Health Research Program Award to N.J.P.

Reference

1. Cowan, C.M., Shi, Y.Y., Aalami, O.O. et al. Adipose-derived adult stromal cells heal critical-size mouse calvarial defects. *Nat Biotechnol* 22, 560–7 (2004).

Chapter 17

Assays of Osteogenic Differentiation by Cultured Human Mesenchymal Stem Cells

Ulf Krause, Anja Seckinger, and Carl A. Gregory

Abstract

One of the most noteworthy characteristics of mesenchymal stem cells (MSCs) is their ability to differentiate into osteoblasts in vitro and in vivo. In vitro, this is easily achieved by culturing in the appropriate induction medium. It is because of the reliability and ease of this process that osteogenic differentiation has become a popular assay for the demonstration of MSC plasticity. Although the conditions required for inducing osteogenic differentiation by MSCs typically do not vary particularly between investigators, many methods are employed to measure the extent of differentiation. These methods include, but are not limited to, reverse transcriptase PCR (RT-PCR) for detection of osteogenic transcripts, enzyme linked immunosorbent assay (ELISA) for secreted protein markers, colorimetric assays for osteogenic enzymes, and direct staining of matrix components. This chapter reviews the protocols most commonly utilized for the evaluation of osteogenic differentiation for cultured MSCs.

Key words: Osteogenic differentiation, Alkaline phosphatase, Osteoprotegerin

1. Introduction

The osteogenic potential of human mesenchymal stem cells (hMSCs) is one of the most noteworthy and robust characteristics of these cells. Friedenstein and coworkers initially described MSCs as fibroblast precursors with osteogenic potential and provided the first evidence demonstrating that MSCs could form osteoblasts and bone matrix in vivo (1, 2). In these studies, pre-expanded MSCs were encapsulated in porous chambers and implanted into the intraperitoneal cavity of recipient animals. After 30–90 days, the chambers were excised and histologically examined, demonstrating that the cells had produced bone and

Mohan C. Vemuri et al. (eds.), *Mesenchymal Stem Cell Assays and Applications*, Methods in Molecular Biology, vol. 698, DOI 10.1007/978-1-60761-999-4_17, © Springer Science+Business Media, LLC 2011

cartilage (2, 3). In a series of in vitro experiments, it was also shown that osteogenesis could be induced through the addition of certain nutrients and hydrocortisone (2), resulting in the extensive expression of alkaline phosphatase (ALP), extracellular matrix (ECM), and the extracellular deposition of calcium phosphate. Since this discovery, numerous techniques have been employed to measure the degree and rate of osteogenic differentiation by MSCs. Because the initiation of osteogenic differentiation can be easily induced in monolayer culture by the addition of ascorbic acid, β-glycerophosphate, and in some cases, hydrocortisone or dexamethasone, the osteogenic assay has been employed as a common tool for the appraisal of MSC plasticity. Rapid differentiation assays for MSCs are of the utmost importance since cultures frequently vary due to donor variation, irreversible cell contact-induced effects, and prolonged expansion (4–6). Although additional molecules are occasionally added to cultures to accelerate or influence the osteogenic process, the culture conditions have remained relatively unchanged for over 30 years. In contrast, methods for the measurement of osteogenic potential by MSCs have evolved extensively.

Three decades ago, the most common protocols for examining in vitro osteogenesis were qualitative assays, based on the binding of colored stains to insoluble components of the secreted ECM such as calcium, phosphate, and collagen. Colorimetric processing of reporter molecules by ALP was also employed. These assays had a major limitation in that they relied heavily on microscopic visualization of the staining and objectivity of the investigator. Because these techniques required high levels of detectable substrate, they also had limited sensitivity and were generally restricted to the detection of the very latest stages of osteoblast maturation. Nowadays, quantitative RT-PCR and microarray analysis are more favored technologies for measuring osteogenesis, since these methods are sensitive and more quantifiable. Nevertheless, measurement of transcriptional activity of osteogenic genes may not necessarily correlate with functionality, especially as the activity of secreted proteins can be regulated at numerous stages by posttranslational modifications. In consideration of the limitations of mRNA analyses, there has been a reappraisal of classical techniques for measuring in vitro osteogenesis, resulting in improved sensitivity and the capacity for quantitation. With an emphasis on the generation of quantitative and semi-quantitative data at the level of protein activity and biomineralization, this chapter reviews a range of such protocols for the measurement of osteogenic differentiation in cultures of MSCs.

2. Materials

2.1. Osteogenic Culture of hMSCs

1. 70% (v/v) reagent ethanol (Thermo Fisher Scientific, Waltham, MA).
2. Sterile polypropylene centrifuge tubes (15 and 50 mL) (Corning, Corning, NY).
3. Sterile plastic serological pipettes (5, 10, 25, and 50 mL) (Corning).
4. Six-well polypropylene tissue culture plates with 9.6 cm^2 wells (Costar brand, Corning).
5. 154 cm^2 Polypropylene tissue culture dishes (Costar brand, Corning).
6. Sterile phosphate buffered saline (PBS) without magnesium and calcium (Invitrogen, Carlsbad CA).
7. Trypan blue solution in 0.85% saline (Invitrogen).
8. 500 and 1,000 mL vacuum filter units with 0.22-μm diameter filtration membrane (Millipore, Billerica, MA).
9. Porcine trypsin (0.25% w/v) and EDTA (1 mM) solution (Invitrogen).
10. Complete culture medium (CCM): Minimal essential alpha medium (αMEM) with 2 mM glutamine but without ribonucleosides or deoxy-ribonucleosides (Invitrogen) containing 20% hybridoma qualified and non-heat inactivated Fetal bovine serum (Atlanta Biologicals, Norcross, GA), 1× antibiotic solution (100 units/mL penicillin G sodium, 100 μg/mL streptomycin sulfate, Invitrogen), and additional 2 mM L-glutamine (Invitrogen). Filter sterilize the medium.
11. Osteo-base medium (OBM): CCM containing 5 mM β-glycerophosphate (Sigma, St Louis, MO) and 50 μg/mL ascorbate-2-phosphate (Sigma). Filter sterilize the medium.
12. Osteo-mineralization medium (OMM): OBM containing 10 nM dexamethasone (Sigma, account for the β-methyl cyclodextrin carrier in the dexamethasone when preparing a 1,000× stock solution in water. Amount varies by batch). Filter sterilize the medium.

2.2. Semi-Quantification of Mineralization by Alizarin Red S Staining and Recovery

1. Sterile PBS without magnesium and calcium (Invitrogen).
2. 10% Buffered formalin (Sigma).
3. Acetic acid (Sigma).
4. Rigid cell scrapers (Fisher).
5. Screw top, 2 mL polypropylene tubes with gaskets (Fisher).

6. Alizarin red S solution (ARS): 40 mM Alizarin red S (Sigma) in water carefully adjusted to pH 4.1–4.3 with 0.5 N ammonium hydroxide (Sigma).
7. Colorless flat-bottomed 96-well polypropylene plates (Fisher).
8. Visible range spectrophotometer, preferably with 96-well capability.

2.3. Quantification of Mineralization by Acid Mediated Recovery and Arsenazo III Assay

1. Sterile PBS without magnesium and calcium (Invitrogen).
2. Analytical grade 6 M hydrochloric acid (Sigma).
3. Analytical grade calcium chloride (Sigma).
4. Analytical grade Trizma base (Sigma).
5. Analytical grade arsenazo III (Sigma).
6. Milli-Q grade water ($MQdH_2O$).
7. Rigid cell scrapers (Fisher).
8. Analytical balance with 1 mg resolution.
9. Silica resin for desiccation (Fisher).
10. Round bottom flasks (5–10 mL capacity) with quick fit adapters (Pyrex brand, Fisher).
11. Liebig condensers compatible with flasks.
12. Conical flasks (5–10 mL capacity) (Pyrex brand, Fisher) with Teflon stirrers (Fisher).
13. Accurate pH meter with microprobe designed for volumes as low as 5 mL.
14. Colorless flat-bottomed 96-well polypropylene microtiter plates (Fisher).
15. Visible range spectrophotometer, preferably with 96-well capability.

2.4. Alkaline Phosphatase (ALP) Assays on Intact Cultures

1. Materials for osteogenic cultures of hMSCs (see Subheading 2.1).
2. Analytical grade 6 M hydrochloric acid (Sigma).
3. Automated microplate reader with kinetic capability or spectrophotometer with 100 μL cuvette for manual readings.
4. ALP reaction buffer (ALPB): 100 mM Tris–HCl pH 9.0 (Sigma) containing 1 mM $MgCl_2$ (Sigma) and 100 mM NaCl (Sigma). All reagents should be of analytical grade.
5. One step *p*-nitrophenol phosphate (PNPP) solution (Pierce, Rockford, IL) or a freshly prepared 1 mg/mL solution of PNPP in 1 M diethanolamine (pH 9.8).
6. *For manual only*: 96-well microtiter plate or 0.5 mL microcentrifuge tubes.
7. *For manual only*: 1 M NaOH solution (Sigma).

2.5. Osteoprotegerin (OPG) ELISA on hMSCs

1. Materials for osteogenic cultures of hMSCs (see Subheading 2.1).
2. Human osteoprotegerin (OPG) ELISA duo set consisting of goat-anti-human-OPG and biotinylated goat-anti-human OPG antibodies, streptavidin-peroxidase conjugate and OPG reference material (R&D systems).
3. Analytical grade bovine serum albumin.
4. Block buffer: PBS containing 0.1% (v/v) Tween 20 (Sigma) and 4% (w/v) bovine serum albumin.
5. Wash buffer: PBS containing 0.1% (v/v) Tween 20 (Sigma).
6. Media dilution buffer: PBS containing 0.1% (v/v) Tween 20 (Sigma) and 1% (w/v) bovine serum albumin.
7. Colorless, flat bottomed ELISA 96-well microtiter plates (Immunosorp, Nunc, Nalgene, Rochester, NY).
8. 3, 3′, 5, 5′ tetramethyl benzidine substrate (TMB) (Pierce).
9. 2 M H_2SO_4 (Fisher).
10. Microplate reader with 450 nm absorbance capability.

2.6. Cell Counting Assays on Osteogenic Monolayers of hMSCs Using Fluorescent DNA Binding Dyes

1. Osteogenic monolayer of hMSCs from one of the protocols.
2. Materials for osteogenic cultures of hMSCs (see Subheading 2.1).
3. PBS (Invitrogen).
4. Triton X-100 (Sigma).
5. Analytical grade NaCl (Sigma).
6. Analytical grade $MgCl_2$ (Sigma).
7. Proteinase K (Sigma).
8. *Eco*RI and *Hind*III restriction enzymes (40–50 units/μL, Roche Applied Science, Indianapolis, IN).
9. RNAse A (Invitrogen).
10. CyQuant cell GR DNA dye (Invitrogen) or equivalent.
11. Frozen (–80°C) pellets of human cells (120,000 per pellet) for reference material.
12. Black, opaque, flat bottomed microtiter plates (Nunc, Nalgene, Rochester, NY).
13. Microplate reader with 480/520 nm (fluorescein equivalent) fluorescence capability.
14. Extraction and lysis buffer: PBS containing 0.1% (v/v) Triton X100, 100 mM NaCl, 1 mM $MgCl_2$, 5 μg/mL RNAse A and for extraction of osteogenic or highly confluent plates, include 10 U/mL *Eco*RI and *Hind*III.

3. Methods

3.1. Osteogenic Culture of hMSCs

This protocol assumes that hMSCs are already isolated and expanded as described in a separate chapter within this text and described previously (6). In brief, mononuclear cells from heparinized whole bone marrow aspirates are separated by Ficoll gradient centrifugation, plated and cultured in CCM. During media changes, contaminating hematopoietic cells are washed away. After 2–3 days, colonies of plastic-adherent fibroblast-like MSCs become apparent. Upon reaching 60–70% confluence, the cells are lifted with trypsin/EDTA and are subcultured and expanded at 100 cells per cm^2 in CCM (see Note 1) or cryo-preserved in αMEM containing 30% (v/v) FCS and 5% (v/v) DMSO. Osteogenic differentiation is a two-step process: in the first phase, MSCs turn into/differentiate into mature osteoblasts. In the second phase, the confluent layers of osteoblasts synthesize mineralized matrix. Therefore, a two-step protocol with two different media is preferred. Once the cultures reach confluence, the medium is changed to osteogenic basal medium (OBM). After 14 days of maturation into osteoblasts, mineralization is induced by the addition of dexamethasone to the OBM, OMM. Levels of osteogenic differentiation can be detected and quantified in several ways. For detection of calcium levels in biomineralized monolayers, the second, dexamethasone mediated, stage of the differentiation process is required but assays of OPG and ALP may be performed on hMSCs treated with OBM only.

1. For osteogenic differentiation, both freshly isolated or cryopreserved cells can be used. The use of cryopreserved cells is considered superior because hMSCs from different donors can be compared synchronously in parallel, whereas primary cell cultures often differ in growth rates, leading to different passage numbers and thereby population doublings, which can alter the differentiation capability of the cells.
2. Prepare two 15 cm dishes with 20 mL prewarmed CCM for each vial (containing approximately 1 million cryopreserved cells).
3. Retrieve vial(s) with 1 mL frozen cells from liquid nitrogen, thaw in a 37°C water bath until only a small piece of ice is visible, and transfer 500 μL each into the 15 cm plates.
4. Let cells recover overnight, then lift with Trypsin/EDTA and reseed at 100 cells per cm^2 in 15 cm plates for expansion or at 1,000–10,000 cells per cm^2 in 6-well plates for differentiation.
5. Continue to grow the cells in the 6 wells to 70–80% confluence in 2 mL of CCM. Change media every 2–3 days.

6. After the cells reach 80% confluence (see Note 2), osteogenic differentiation is induced by changing to OBM. Wells with negative control cells are continued in CCM. All tests should be done in triplicate wells.
7. Change respective media every 2 days for 14 days. After 7 days, OBM treated wells should appear milk-glass like.
8. After 14 days, matrix deposition of osteoblasts is enhanced by addition of 10 nM dexamethasone to the OBM, now termed OMM.
9. Change respective media every 2 days for another 7–21 days.

3.2. Semi-Quantification of Mineralization by Alizarin Red S Staining and Recovery

Biomineralization of hMSC monolayers in response to β-glycerophosphate, ascorbate, and dexamethasone has been employed for decades as an indicator of plasticity. The resultant calcium phosphate-rich monolayers can be visualized by a number of techniques, but the calcium binding dye, alizarin red S (ARS), has gained most popularity because of its ease of use, wide availability, and low cost (7). In addition, ARS can be easily extracted and quantified spectrophotometrically. This protocol describes a straightforward, robust, and sensitive method for semi-quantitative assay of ARS staining in hMSC monolayers based on acetic acid-mediated extraction and spectrophotometry (8).

1. After hMSC monolayers have been induced for osteogenic differentiation for a total of 21 days, aspirate the media and wash wells carefully with 2 mL of PBS (see Note 3).
2. Fix with 1 mL of 10% (v/v) formaldehyde at room temperature for 15 min.
3. Wash the monolayers twice with excess dH_2O
4. Add 1 mL of 40 mM ARS (pH 4.1) per well.
5. Incubate the plates at room temperature for 20 min with gentle shaking.
6. Aspirate off the unincorporated dye.
7. Wash wells four times with 4 mL dH_2O while shaking for 5 min (see Note 4).
8. Visualize the degree of differentiation of stained monolayers by microscopy using a dissecting scope or phase-contrast inverted microscope (Fig. 1, see Note 5).
9. For quantification of staining, add 800 μL 10% (v/v) warm (50°C) acetic acid to each well and incubate the plate at room temperature for 30 min with shaking.
10. Transfer the supernatant from each well to a 2 mL microcentrifuge tube and repeat the extraction step with an additional 800 μL 10% (v/v) warm (50°C) acetic acid.

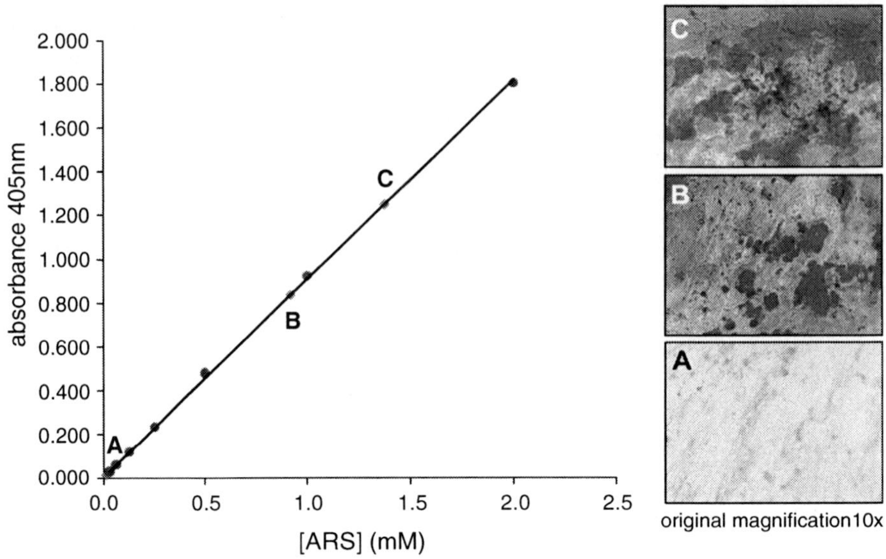

Fig. 1. Typical semi-quantitative result of an ARS staining. (**a**) Absorbance (405 nm) of relevant concentrations of ARS was measured and plotted to demonstrate linearity of the assay. (**b**) ARS was extracted from a monolayer that received no osteogenic supplements (**A**) and two other moderately osteogenic monolayers (**B**, **C**). Absorbance measurements from the extractions are within the semi-quantitative range of the assay (points **A–C**).

11. Scrape the now loosely attached monolayer from the plate with a cell scraper and transfer it with the 10% (v/v) acetic acid to the same 2 mL microcentrifuge tube with a wide-mouth pipette. Vortex for 30 s (see Note 6).
12. Centrifuge the slurry at 20,000 × *g* for 15 min.
13. Transfer 500 μL of the supernatant to a new 1.5 mL microcentrifuge tube (see Note 7).
14. Read 150 μL aliquots of the supernatant in a 96-well plate reader at 405 nm in triplicates using transparent-bottomed plates (Fig. 1, see Note 8).

3.3. Quantification of Mineralization by Acid Mediated Recovery and Arsenazo III Assay

This protocol facilitates the direct measurement of calcium ions released from partially acid-hydrolyzed preparations of mineralizing hMSCs. The assay is based on the color change that occurs when Arsenazo III dye chelates calcium ions (9, 10). Although time consuming when compared with the other protocols described here, the method provides a very sensitive, direct quantification of insoluble calcium in the monolayer without the necessity for specialized instrumentation. Furthermore, this method is especially convenient for osteogenic assays performed in three-dimensional matrices or on monolayers of a thickness that precludes semi-quantitative staining with ARS. Our laboratory has even employed this method for the analysis of calcium levels in rodent bones.

1. If monolayers or three-dimensional constructs are to be assayed, aspirate spent medium and wash the cultures thrice in excess PBS.
2. If necessary scrape the material thoroughly from the tissue culture plastic. Visualize using a microscope to ensure complete recovery. For three-dimensional matrices, chop the material into particles no more than 2 mm cubed. With a wide-mouthed pipette, transfer material to two 15 mL polypropylene tube and add at least 10 volumes of $MQdH_2O$.
3. Centrifuge material for 5 min at maximum speed ($3,000 \times g$), carefully remove supernatant and wash twice with $MQdH_2O$ (see Note 9).
4. At this stage, large samples can be freeze-dried or desiccated in the presence of silica gel at 80°C to remove excess water that may influence normalization to mass. Weigh samples as accurately as possible using an analytical balance.
5. Transfer the material to a round-bottomed Pyrex flask containing 5–10 mL of 6 M HCl. Install the condenser and reflux the mixture for 2–15 h at 80°C until the sample has completely dissolved.
6. At this stage, the sample can be preserved at 4°C if necessary.
7. Allow the solution to cool to room temperature then carefully neutralize to pH 5.0 by slow addition of 2.5 M Trizma base. Perform a mock titration with 5 mL 6 M HCl to establish approximate volumes. The resultant solution should be retained for dilution of standards.
8. Derive initial volumes for all of the hydrolysate solutions by mass (estimate 1 g of solution equals 1 mL).
9. Prepare stocks of Arsenazo III (100 μM) in $MQdH_2O$ and $CaCl_2$ standards (10–100 μM) in the retained solution of neutralized Tris–HCl.
10. In a flat-bottomed 96-well colorless microtiter plate add 10 μL of sample or standard solution followed by 90 μL of the arsenazo III solution.
11. After 5 min at room temperature, a color change should be apparent, with transition from pink/purple to blue color as the concentration of Ca^{2+} increases.
12. Measure absorbance at 595 nm and plot standard curves. Derive concentrations of unknown samples, and thus the absolute amount of calcium in the sample from the initial volume of the hydrolysate. The values can then be expressed as moles of calcium extracted per gram of starting material.

3.4. Quantification of ALP Activity by Kinetics of PNPP to Nitrophenolate Conversion

ALP is expressed on the membranes of many tissues including osteoblasts. The function of ALP in osteoblasts is to generate phosphate for incorporation into hydroxylapatite, the inorganic component of bone (11, 12) and to degrade inhibitory pyrophosphatases (13). The activity of ALP can be rapidly determined by following the kinetics of *p*-nitrophenylphosphate (PNPP) conversion to *p*-nitrophenolate (14), which results in a color change from colorless to deep yellow, measured by absorbance at 405 nm. In intact osteogenic hMSCs, ALP can be measured by direct addition of PNPP to the monolayer in the presence of an appropriate buffer. The rate of *p*-nitrophenolate accumulation can be determined in an automated plate reader or manually. Because osteogenic monolayers vary substantially in protein content and ease of extraction, it is advisable to normalize results based on cell number rather than protein content.

1. Perform osteogenic assays in 6, 12, or 24-well plates. If you intend to measure rate manually, use 6-well plates only.
2. Wash monolayers twice with PBS, then once with room temperature ALPB. For 6-well plates, use 2 mL, 12-well plates, use 1 mL and 24-well plates, use 0.5 mL wash buffer.
3. For automated kinetic measurements, set an automated plate reader (e.g., BMG Labtech Omega, Durham, NC) to detect absorbance at 405 nm for 15 min, with measurements every 10–30 s. Set the path length to 0.25 cm with orbital averaging if possible. For manual kinetic measurements, add 50 μL of 1 M NaOH to 30 wells of a microtiter plate or 0.5 mL microcentrifuge tubes.
4. For automated measurements, add 1, 0.5, or 0.25 mL room temperature ALPB to wells of a 6, 12, or 24-well plate, respectively. Add an equal volume of PNPP equilibrated to 4°C and immediately initiate readings.
5. For manual measurements, add 1 mL room temperature ALPB to wells of a 6-well plate. Immediately remove 50 μL and mix with an aliquot of NaOH. Repeat every 30 s until 15 min have passed.
6. Plot the rate of accumulation of *p*-nitrophenolate and derive the initial rate by calculating the slope at the linear phase of the reaction (Fig. 2, see Note 10).
7. After the reaction, wash the monolayers in PBS and store at –20°C. The monolayers can be stored indefinitely prior to cell number measurements.

3.5. Quantification of OPG Secretion by Enzyme Linked Immunosorbent Assay (ELISA)

OPG, also known as osteoclast inhibitory factor or tumor necrosis factor receptor superfamily, member 11b is a secreted glycoprotein with affinity for the osteoclast activating factor, receptor activator of NF kappaB ligand (RANKL). OPG is a reversible inhibitor of RANKL, acting as a soluble decoy receptor (15).

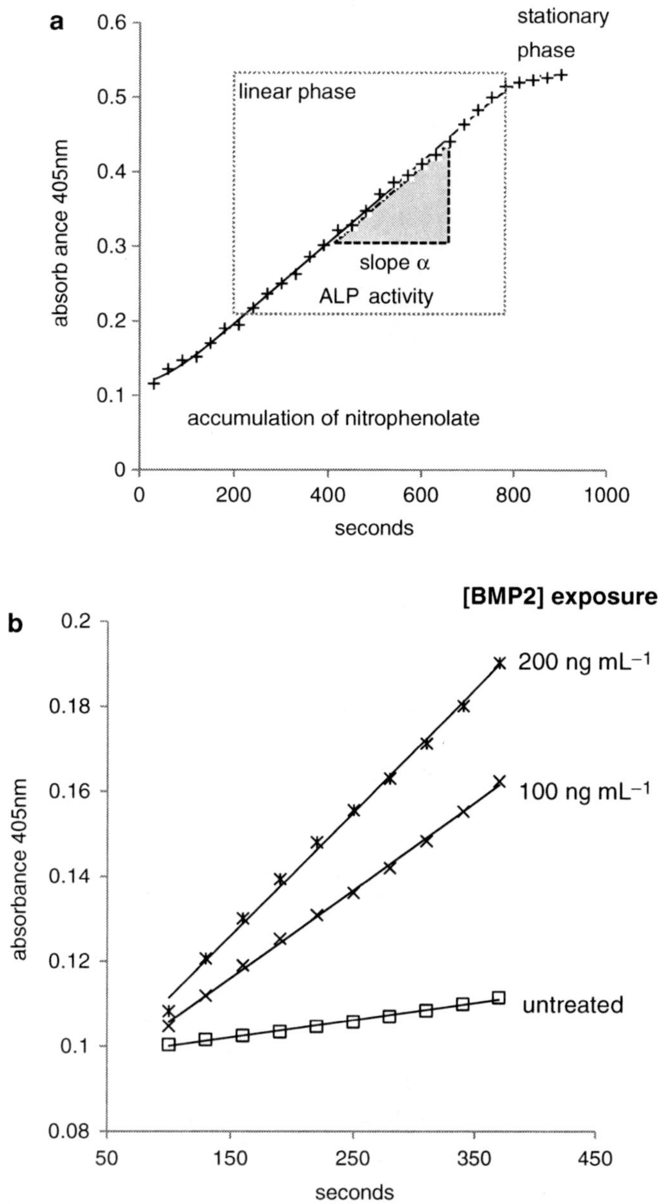

Fig. 2. (**a**) Representative kinetic profile of an ALP assay demonstrating accumulation of the yellow *p*-nitrophenolate product over time. Note that the reactions can exhibit a slight lag phase in the initial seconds of the assay and a plateau as substrate becomes depleted. The slope of the linear section of the curve (indicated by a *dotted rectangle*) should be calculated for ALP measurements. (**b**) Example of ALP measurements on hMSC monolayers treated with osteogenic supplements and two concentrations of bone morphogenic protein 2 (BMP2).

In MSCs, OPG is an early indicator of osteogenic differentiation, but its detection in the medium is attenuated as differentiation progresses (Fig. 3). OPG is therefore a minimally invasive and rapid indicator of osteogenic differentiation.

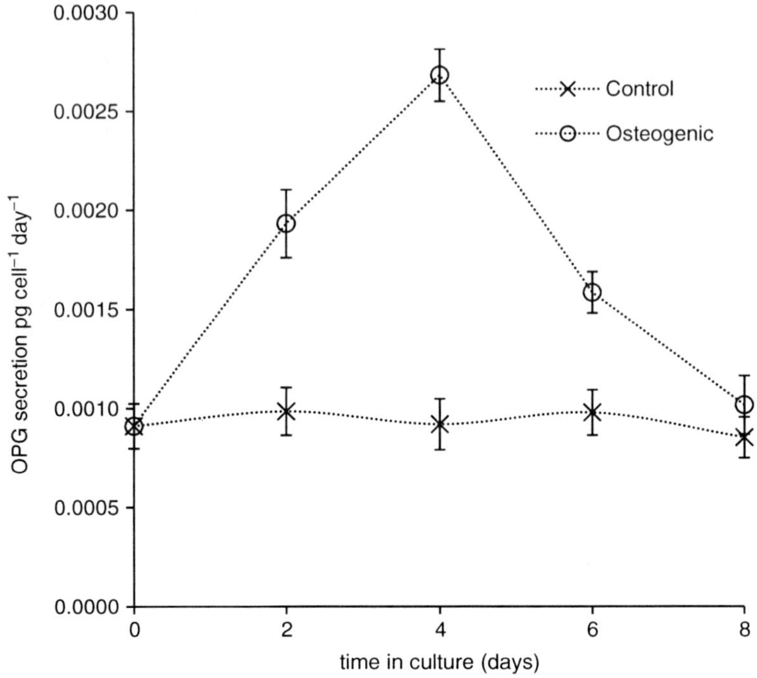

Fig. 3. Timecourse of OPG secretion by hMSCs after treatment with osteogenic supplements. Note that OPG levels rise dramatically over the initial 4–7 days of exposure, then decline as ALP activity increases. In some cases, OPG secretion returns to control levels.

1. Prepare osteogenic cultures in 6–24 well format in accordance with protocol 3.1.
2. After 2–8 days of osteogenic stimulus, recover at least 100 μL of 2-day old conditioned media and store at −20°C until required for the assay.
3. Prepare antibody solutions and OPG standards as directed by the manufacturer.
4. Coat 96-well microtiter plates with 100 μL of coating antibody per well at the recommended concentration for 15 h. Use PBS only as the diluent.
5. Block for 2 h with block solution per well.
6. Wash the wells twice with 100 μL wash buffer.
7. While blocking the plate, prepare 600 μL of a standard solution of OPG at 4,000 pg/mL in media dilution buffer.
8. Dilute media samples 5–20-fold in media dilution buffer to produce 300 μL of sample.
9. Prepare a triplicate 7-point standard curve (4,000–62.5 pg/mL) by doubling dilution. For this purpose, load wells A1–A3 with 200 μL of the 4,000 pg/mL standard and wells B1–3 to

H1–3 with 100 μL of media dilution buffer. With a multi-channel pipette, transfer 100 μL of the high standard to wells B1–3, 100 μL from B1–3 to C1–3 and repeat with subsequent wells. Allow wells H1–3 to remain as the blank. Load 100 μL of sample in remaining wells in triplicate. Incubate samples at 4°C for 4–15 h.

10. Remove samples and wash the wells three times for 2 min with 100 μL wash buffer.
11. To each well add 100 μL of the recommended concentration of biotinylated detection antibody and incubate on the bench for 2 h.
12. Wash the wells three times for 2 min with 100 μL wash buffer.
13. To each well add 100 μL of the recommended concentration of peroxidase conjugated streptavidin and incubate on the bench for 30 min.
14. Wash wells for three times for 2 min with 100 μL wash buffer.
15. Prepare two separate multichannel loading reservoirs with at least 20 mL per plate TMB solution and at least 10 mL per plate 2 M H_2SO_4.
16. To each well add 100 μL of TMB and incubate on the bench for 2–10 min or until the blue color develops.
17. To each well add 50 μL of 2 M H_2SO_4 to stop the reaction and develop the yellow color.
18. Read at 450 nm on a microtiter plate reader. Plot standards and calculate unknowns.

3.6. Cell Counting Assays on Osteogenic Monolayers of hMSCs Using Fluorescent DNA Binding Dyes

In culture-based assays of osteogenesis, slight variations in the number of MSCs present in the culture can occur. Variation can arise experimentally through effects of treatment conditions, cell preparation, or simply due to inherent variation in the assay. Therefore, normalization is frequently necessary. Because osteogenic cultures are usually rich in ECM components, single cell suspensions are generally impossible to generate, effectively dismissing the possibility for hemacytometer or flow-based methods. Furthermore, the dense nature of the monolayers usually precludes image-analysis for cell number evaluation. Normalization at the level of whole protein has been used in the past, but the variations on the level of protein output by cells under various conditions may introduce bias. In most cases, evaluation of the DNA content of the culture is reliable and relatively easy to perform. Our laboratory has enjoyed great success with dyes that have altered fluorescent properties while intercalated between the base pairs of DNA. However, release of a homogeneous solution of DNA from dense osteogenic cultures can be problematic as the ECM components are resistant to even high concentrations of proteinase K, and very long fragments of chromatin become associated with protease resistant aggregates. In the method described here, we take the alternative approach of releasing fragments of chromosomal DNA into solution by incubation of the cells in the presence of a buffered solution of detergent, salts, and restriction endonucleases. After lysis and digestion, soluble fragments of DNA are generated in a soluble and entirely homogeneous solution. Irrespective of the nature of the ECM of the monolayer, the concentration of extracted DNA is linearly proportional to the number of cells in the original culture.

1. Prepare osteogenic cultures in 6–24 well format in accordance with protocol 3.1.
2. Wash monolayers twice with PBS using 2, 1, or 0.5 mL per well of a 6, 12, or 24-well plate, respectively. Remove PBS and store plates at −20°C until completely frozen. Storage at this step permits preservation of the monolayers until all samples or time points have been processed and is necessary for efficient lysis.
3. To the frozen monolayers, add the appropriate volume of extraction buffer (see step 2 for volumes) and incubate for at least 4 h in a humidified incubator at 37°C. Process a pellet of undifferentiated reference cells in parallel with the samples by resuspending an aliquot in 200 μL of extraction buffer. Monitor lysis of the cells by microscopy.
4. Add proteinase K to the wells to a final concentration of 10 μg/mL. Dilute the proteinase K stock in extraction buffer to 5 mg/mL and add 40, 20, or 10 μL to 6, 12, or 24 well cultures, respectively (to a final concentration of 100 μg/mL). Add 4 μL of proteinase K solution to the reference cells. Incubate the samples with intermittent agitation until the monolayers have detached from the surface of the wells (30 min to 3 h).
5. Transfer each digestion to a microcentrifuge tube and centrifuge at 15–20,000 × *g* for 15 min. Transfer each supernatant to a fresh tube.
6. Prepare the appropriate volume of 1.5× CyQuant GR working solution in accordance with the manufacturer's instructions.
7. For preparation of the reference and samples, add 200 μL of the extract to 400 μL of the CyQuant GR solution in a fresh microcentrifuge tube and incubate at room temperature in the dark for 10 min.
8. Prepare a standard curve by transferring 200 μL of the standard cell extract into wells A1–3 of an opaque black 96-well microtiter plate. Prepare a 1:2 solution of extraction buffer and

CyQuant GR dye solution and transfer 100 μL to wells B1–3 to H1–3. Perform doubling dilutions from wells A1–3 to wells G1–3 to generate a 7 point standard curve from 20,000 to 625 cell equivalents per well. Maintain wells H1-3 as blanks.

9. Load 100 μL of sample in remaining wells in triplicate.
10. Read top fluorescence at 480/520 nm on a microtiter plate reader. Plot standards and calculate unknowns.
11. Confluent cultures of hMSCs can contain thousands of cells. If necessary, dilute samples in additional extraction buffer to generate values that are within the linear range of the assay.

4. Notes

1. MSCs are very sensitive to culture conditions. Suboptimal growth conditions (e.g., confluence levels of more than 70–80% for more than 24 h, stress induced by serum or other nutrient starvation) will lead to a higher percentage of bigger, less proliferative cells in the culture. The best way of avoiding this is to initially plate the cells at 100 cells per cm^2, change media every other day, and passage the cells at 60–70% confluence.
2. To avoid detachment of the cell layer, it is often helpful to switch to osteogenic media at subconfluent levels (80–90% confluence). Cells will continue to proliferate for 2 or 3 days in OBM.
3. While washing, care should be taken not to scrape and dislodge the sometimes brittle calcium phosphate matrix which could swim off, making semi-quantification impossible.
4. After the last wash, leave plates at an angle for 2 min to facilitate removal of excess water.
5. Plates can be stored at −20°C at this point prior to dye extraction. Freeze/thawing will adversely affect the quality of microscopic pictures.
6. To improve the sensitivity of the assay, the recovery of the ARS from the matrix can be maximized by overlaying the slurry with 200 μL mineral oil, heat to exactly 85°C for 10 min, and transfer to ice for 5 min. Take great care at this point to avoid opening of the tubes until fully cooled. Continue with centrifugation Subheading 3.2, step 12.
7. To improve the sensitivity of the assay, 200 μL of 10% (v/v) ammonium hydroxide can be added to neutralize the acid.
8. Although it might be tempting to plot the OD_{405} values of the samples against a "standard curve" of known Alizarin Red S concentrations, the latter should only serve as a linearity

control, since complete extraction of the dye cannot be achieved. This method therefore should be regarded as semi-quantitative. For a quantitative approach, refer to Subheading 3.3.

9. This calcium assay is extremely sensitive (as low as 10 μM Ca^{2+}). It is important to use Milli-Q grade ($MQdH_2O$) water and analytical grade materials. Work as cleanly as possible and wash glassware thoroughly with $MQdH_2O$.

10. Because the units of this rate are based on arbitrary values, some investigators prefer to run solutions containing known amounts of commercially available ALP. However, be aware that in some cases, the specific activity of commercially acquired ALP reference material may not be comparable to the specific activity of the membrane bound ALP on the MSCs.

References

1. Friedenstein, A. J., Chailakhyan, R. K., Gerasimov, U. V. (1987) Bone marrow osteogenic stem cells: in vitro cultivation and transplantation in diffusion chambers. *Cell Tissue Kinet.* **20**, 263–272.
2. Owen, M. and Friedenstein, A. J. (1988) Stromal stem cells: marrow-derived osteogenic precursors. *Ciba Found. Symp.* **136**, 42–60.
3. Friedenstein, A. J., Chailakhjan, R. K., Lalykina, K. S. (1970) The development of fibroblast colonies in monolayer cultures of guinea-pig bone marrow and spleen cells. *Cell Tissue Kinet.* **3**, 393–403.
4. Phinney, D. G., Kopen, G., Righter, W., Webster, S., Tremain, N., Prockop, D. J. (1999) Donor variation in the growth properties and osteogenic potential of human marrow stromal cells. *J. Cell Biochem.* **75**, 424–436.
5. Gregory, C. A., Ylostalo, J., Prockop, D. J. (2005) Adult bone marrow stem/progenitor cells (MSCs) are pre-conditioned by microenvironmental "niches" in culture: a two-stage hypothesis for regulation of MSC fate. *Sci. STKE.* **294**, pe37.
6. Sekiya, I., Larson, B. L., Smith, J. R., Pochampally, R., Cui, J. G., Prockop, D. J. (2002) Expansion of human adult stem cells from bone marrow stroma: conditions that maximize the yields of early progenitors and evaluate their quality. *Stem Cells.* **20**, 530–541.
7. Puchtler, H., Meloan, S., Terry, M. S. (1969) On the history and mechanism of alizarin and alizarin red S stains for calcium. *J. Hist. Cytochem.* **17**, 110–124.
8. Gregory, C. A., Gunn, W. G., Peister, A., Prockop, D. J. (2004) An Alizarin red based method for the assay of mineralization by adherent cells in culture. Comparison with cetylpyridinium chloride extraction. *Anal. Biochem.* **329**, 77–84.
9. Bauer, J. (1981) Affinity and stoichiometry of calcium binding by Arsenazo III. *Anal. Biochem.* **110**, 61–72.
10. Leary, N. O., Pembroke, A., Duggan, P. F. (1992) Single stable reagent (Arsenazo III) for optically robust measurement of calcium in serum and plasma. *Clin. Chem.* **38**, 904–908.
11. Poole, A. R., Matsui, Y., Hinek, A., Lee, E. R. (1989) Cartilage macromolecules and the calcification of cartilage matrix. *Anat. Rec.* **224**, 167–179.
12. Anderson, H. C. (2003) Matrix vesicles and calcification. *Curr. Rheumatol. Rep.* **5**, 222–226.
13. Balcerzak, M., Hamade, E., Zhang, L., Pikula, S., Azzar, G., Radisson, J., Bandorowicz-Pikula, J., Buchet, R. (2003) The roles of annexins and alkaline phosphatase in mineralization process. *Acta Biochim. Pol.* **50**, 1019–1038.
14. Bessey, O. A., Lowry, O. H., Brock, M. J. (1946) A method for the rapid determination of alkaline phosphatase with five cubic millimeters of serum. *J. Biol. Chem.* **164**, 321–329.
15. Khosla, S. (2001) The OPG/RANKL/RANK system. *Endocrinology.* **142**, 5050–5055.

Chapter 18

Bioreactor Cultivation of Functional Bone Grafts

Warren L. Grayson, Sarindr Bhumiratana, Christopher Cannizzaro, and Gordana Vunjak-Novakovic

Abstract

The clinical demand for functional tissue-engineered bone grafts to regenerate bone defects resulting from trauma and surgical resection of congenital anomalies remains very high. One approach involves the use of human mesenchymal stem cells (hMSCs) that are seeded into biomaterial scaffolds and are induced to generate new bone tissue by osteo-inductive cues. The size of tissue constructs that can be cultured under conventional static conditions is seriously limited by diffusional constraints of nutrient supply resulting from high metabolic activity of bone cells. To cultivate bone constructs of clinically-relevant sizes, it is necessary to utilize perfusion bioreactors, which provides convective transfer of nutrients, and most critically oxygen, to the cells throughout the construct volume. This chapter describes a method for engineering 4-mm thick cylindrical bone grafts using hMSCs (isolated from bone marrow aspirates), biomaterial scaffolds (made of fully decellularized bovine trabecular bone), and a perfusion bioreactor (designed for simultaneous cultivation of six constructs for up to 5 weeks). This approach results in the formation of completely viable, biological bone grafts of clinically relevant sizes.

Key words: Human mesenchymal stem cells, Bone grafts, Scaffold, Bioreactor, Perfusion, Tissue engineering

1. Introduction

Adult bone marrow-derived human mesenchymal stem cells (hMSCs) can be readily isolated from bone marrow aspirates, and differentiated into multiple lineages including bone, cartilage, adipose tissue, muscle, and fibroblasts for ligament and tendon. This differentiation capacity, combined with the extensive in vitro proliferation characteristics of hMSCs, makes these cells a prime cell source for use in a number of clinical scenarios including bone regeneration.

Mohan C. Vemuri et al. (eds.), *Mesenchymal Stem Cell Assays and Applications*, Methods in Molecular Biology, vol. 698, DOI 10.1007/978-1-60761-999-4_18,

Human MSCs can be directed into osteogenic lineages by providing soluble osteogenic factors including ascorbic acid, β-glycerophosphate, and dexamethasone. To guide the in vitro development of functional bone grafts, hMSCs need to be seeded into a scaffold providing a structural template for the new bone growth. Under static culture conditions, diffusion is the primary mechanism for transfer of nutrients and metabolites between the bulk medium and the cells within the construct. The availability of oxygen becomes limited extremely close to the periphery of the constructs, with oxygen penetration depths being far less than 1 mm, depending on local cell density (1–3). This creates a limit in the size of a viable construct that can be successfully cultured with diffusional control of mass transport. In native bone, mass transport is enhanced by interstitial flow generated by mechanical loading. The same mechanism for the enhancement of mass transport has been utilized in bioreactors with medium perfusion providing interstitial flow through the cultured construct.

Perfusion bioreactors enable cultivation of viable bone constructs several millimeters thick, by virtue of local environmental control. Several designs have been utilized in studies of bone tissue engineering (4–7). Here we describe a custom-designed bioreactor capable of simultaneous cultivation of six cylindrical constructs with tightly controlled interstitial flow. Using this system we have cultivated fully viable, clinically sized human bone constructs for up to 10 weeks in vitro (Figs. 1 and 2) (8). We describe here the methods for engineering human bone by an integrated use of hMSCs, decellularized bone scaffolds, and perfusion bioreactors.

2. Materials

2.1. General Supplies

1. Kim Wipes (sterilized in autoclave).
2. Forceps (sterilized in autoclave).
3. Scalpels (sterilized in autoclave).
4. Sterilization pouches.
5. 12-Well plates (nontissue Culture treated).
6. Petri dishes (6 cm) (sterile).

2.2. Cell Culture Supplies

1. Cell expansion medium: Low-glucose Dulbecco's Modified Eagle Medium (DMEM) supplemented with 10% fetal bovine serum (FBS) and 1% antibiotics (penicillin/streptomycin) (all from Invitrogen, Carlsbad, CA).
2. Trypsin (0.25%) with ethylenediame tetraacetic acid (EDTA) (1 mM) (Invitrogen).

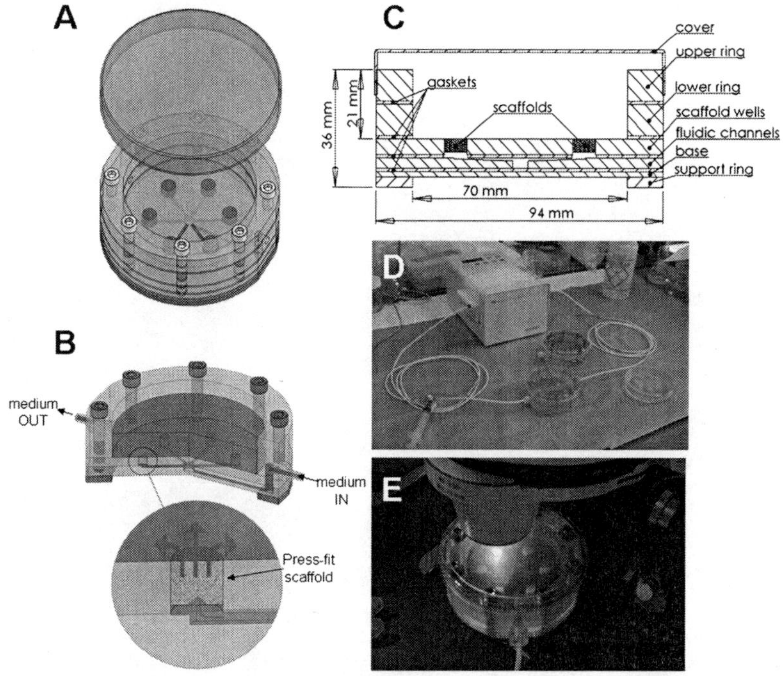

Fig. 1. Perfusion bioreactor design. (**a**) Schematic presentation of the bioreactor system used to provide perfusion through six cell-seeded scaffolds that were press-fit into the culture wells. Scaffolds up to 8 mm diameter × 4 mm high can be accomodated. (**b**) Cut out view of bioreactor showing medium flow patterns through inlet and distributed through individual channels. Blow-up illustrates interstitial flow of medium through the scaffolds into the reservoir. (**c**) Cross-sectional schematic of bioreactor demonstrating layer-by-layer design. (**d**) Complete experimental set-up showing bioreactor, perfusion loop and peristaltic pump. (**e**) Bioreactor on a microscope stage during imaging of tissue constructs in situ. Reproduced from Grayson et al. (7), with permission.

3. Dexamethasone (Sigma, St. Louis, MO). Dissolve in 100% ethanol at 1 mM (10,000×). Filter sterilize, wrap in aluminum foil to prevent light exposure, and store temporarily at 4°C. Aliquot and store at −20°C for up to 6 months.
4. Sodium-β-glycerolphosphate (Sigma). Dissolve in DMEM at 200 mM (20×). Filter sterilize and store temporarily at 4°C. Aliquot and store at −20°C for up to 6 months.
5. Ascorbic acid 2-phosphate (AA-2-P) (Sigma). Dissolve in sterile water at 50 mg/ml (1,000×). Filter sterilize and temporarily store at 4°C. Aliquot and store at −20°C for up to 6 months.
6. Stock osteogenic medium: Prepare in 500 ml aliquots by supplementation of control medium with 10 nM dexamethasone and 10 mM sodium-β-glycerophosphate.
7. Complete osteogenic medium: Prepare on the day of medium addition or change by adding AA-2-P to Stock Osteogenic Medium at a concentration of 0.05 mg/ml (see Note 1).

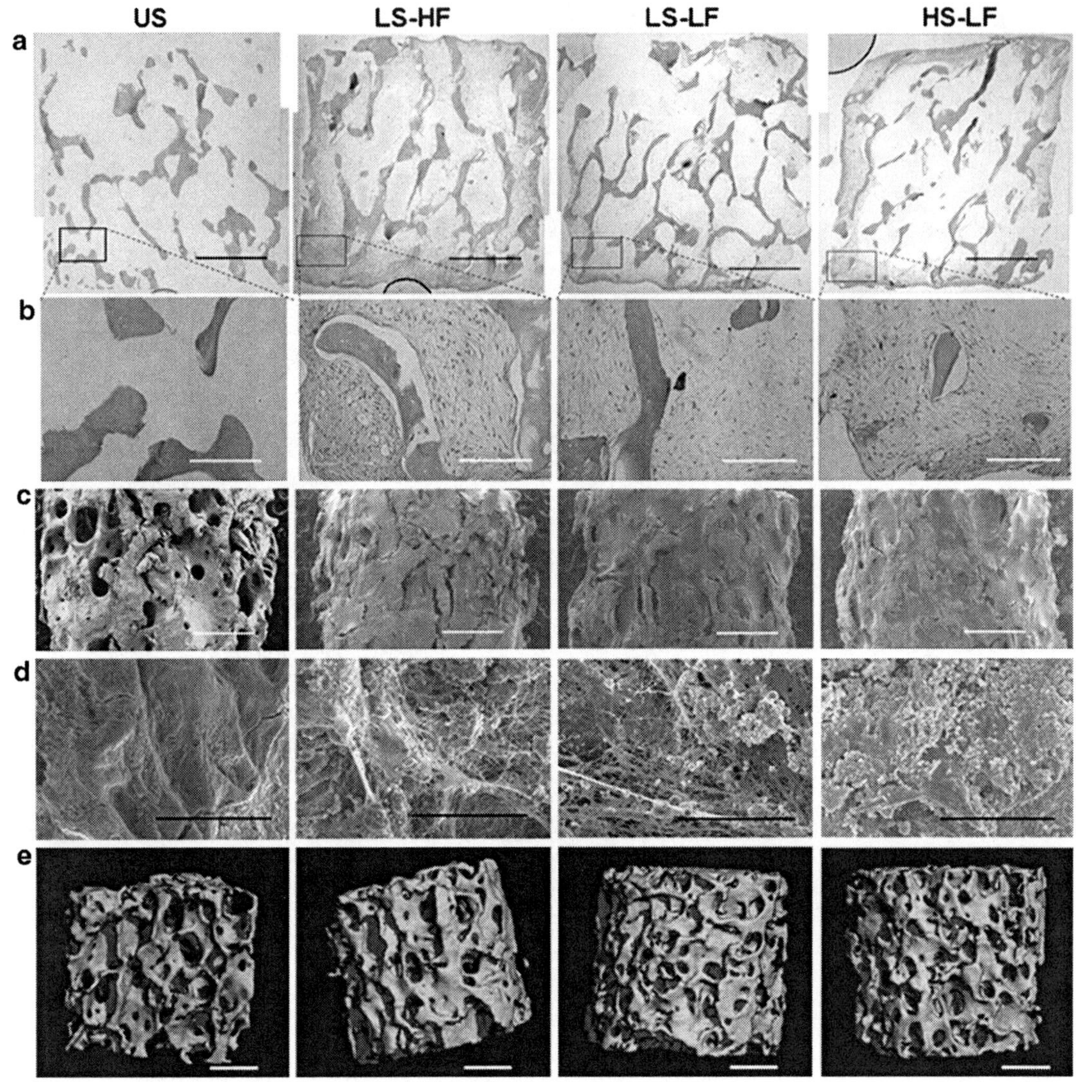

Fig. 2. Tissue engineered bone (**a**) H&E staining of seeded scaffolds after 5 weeks of culture. Cells grow along the periphery of scaffolds in all groups. Uniformity is greatest in LS-HF group where cells grow ≥800 μm into scaffold from upper and lower edges, and least in HS-LF group with growth up to ~350 μm (scale bar = 1 mm). (**b**) Cell density in specific regions differs significantly among groups. LS-HF group and HS-LF group have high localized cell densities, while LS-LF group has low cell density (scale bar = 200 μm). (**c**) Low magnification SEM images (30×) of external surfaces of tissue constructs showing cell growth and ECM deposits in seeded scaffolds (scale bar = 1 mm). (**d**) High magnification SEM images (1,000×) of inner regions showing cell interaction with the mineralized walls and formation of 3D networks in the seeded groups (scale bar = 50 μm). (**e**) μCT images of tissue constructs from all groups after 5 weeks of culture (scale bar = 1 mm) (see Note 9). Reproduced from Grayson et al. (7), with permission.

2.3. Decellularized Bone Scaffold

1. Carpometacarpal joints of 4–6 months old cows (Green Village Packing Company, Green Village, NJ) are used for harvesting the trabecular regions at the ends of the long bones.
2. 4 mm Diamond-tipped mounted core drills (Starlite Industries, Rosemont, PA).

3. Vice-grip.
4. Hand-held power drill.
5. Decellularization reagents:
   (a) 0.1% EDTA (w/v) in PBS (Sigma), store at room temperature.
   (b) 0.1% EDTA (w/v) in 10 mM Tris (pH 8.0) (Sigma), store at room temperature.
   (c) 0.5% SDS (w/v) in 10 mM Tris (pH 8.0), store at 4°C.
   (d) Deoxyribonuclease I (Sigma) at 50 U/ml and ribonuclease (RNAse, Sigma) at 1 U/ml in 10 mM Tris–HCl (pH 7.5), make fresh prior to use.

2.4. Perfusion Bioreactor

1. The bioreactor chamber consists of a layered design as shown in Fig. 1. The main components include the following:
   - Stainless steel ring.
   - Polycarbonate base of 1.5 mm thickness.
   - PDMS gasket of 1 mm thickness.
   - Polycarbonate plate (3-mm thick) with 2 mm wide channels, which allow medium inlet and distribution to the six scaffolds.
   - PDMS gasket (1 mm thick) (with holes for scaffolds).
   - Polycarbonate plate (4.5 mm thick) with six wells (8 mm in diameter) to allow placement of scaffolds.
   - PDMS ring insert adapters (to facilitate smaller diameter constructs).
   - Polycarbonate ring with connectors for medium inlet and outlet.
   - PDMS gasket ring (1 mm-thick).
   - Polycarbonate ring (reservoir wall).
   - 8 Hex screws + hex screw-driver (sterilized).
2. Glass 10-cm Petri dish cover.
3. High-temperature soft silicone rubber tubing size 1/16" ID, 1/8" OD, and 1/32" wall thickness (McMaster-Carr, Atlanta, GA). The silicone tubing can be steam-autoclaved (dry cycle, 30 min at 121°C), and it provides visibility (to detect bubbles) and permeability for transport of oxygen and carbon dioxide.
4. Tubing connectors (McMaster-Carr):
   - Polypropylene barbed tube fitting coupling for 1/16" tube ID.
   - Plastic quick-turn (Luer lock) coupling polypropylene, female X barb, for 1/16" tube ID.

- Plastic quick-turn (Luer lock) coupling polypropylene, male X barb, for 1/16" tube ID.

5. Sterile Hi-Flo™ three-way stopcock with swivel male Luer lock (Smiths Medical ASD Inc, Dublin, OH).
6. MasterFlex® three-stop pump tubing 1.3 mm ID (Cole-Parmer, Vernon Hills, IL).
7. Multi-channel peristaltic pump (Cole-Parmer).

3. Methods

3.1. hMSC Cell Preparation

1. The hMSCs are expanded in control medium up to the second (P2) or third (P3) passage so that the cultured cells do not exceed passage four (P4) (see Note 2).
2. Verify the osteogenic potential of expanded hMSCs using pellet cultures. Trypsinize and count cells and prepare aliquots of cell suspension containing 2.5×10^5 in sterile screw-capped centrifuge vials.
3. Centrifuged vials at $300 \times g$ for 5 min to form a pellet.
4. Mechanically dislodge pellets from the base of the tube and culture in 1 ml osteogenic medium (using expansion medium as a negative control) for 4 weeks with full medium changes 3 times per week.
5. To create bone grafts, trypsinize hMSCs and resuspended as a single-cell suspension in expansion medium at a predetermined density, which will be used for seeding scaffolds. A reasonable density used in most studies is 3×10^7 cells/ml.

3.2. Scaffold Preparation

1. Use scalpels to skin the bovine limb and cut open the joint capsule to expose the condyle of the carpometacarpal joints (see Note 3). Clamp the bones vertically in a vice grip so that the cartilage regions face upward.
2. Use a diamond-tipped drill bit to core out trabecular bone cylindrical pieces. Cool the coring device with running water to prevent the bone cores from being burned at the edges (see Note 3).
3. Remove the drill from inside the bone by reversing the direction of drilling while retracting the bit. Remove the bit from the drill base and push the bone cores out of the coring bit using a long, thin metal rod with a flat face that will not damage the bone. Approximately 8–10 plugs can be obtained from a single joint.
4. Wash bone cores using a high velocity water jet to remove marrow from the pore spaces (see Note 3). This is done until the cores appear completely white.

5. Subject the cores to sequential washes in various decellularizing solutions using 50 ml of solution per gram of trabecular bone sample as follows:
   - PBS/0.1% EDTA for 3 h at room temperature.
   - 10 mM Tris/0.1% EDTA for 6 h at room temperature.
   - 10 mM Tris/0.5% SDS at 4°C for 18 h.
   - DNAse-RNAse at 37°C for 3–6 h

   For each step, place bone samples and solution into a 50-ml conical centrifuge tube, securing the tubes to the surface of a rotating plate and setting the rotating speed to 200 rpm.
6. After the last wash, rinse bone plugs in PBS repeatedly until no bubbles appear. The bubbles indicate the presence of SDS, which is toxic to cells.
7. Briefly wash the completely rinsed scaffolds in water and lyophyllize overnight.
8. Cut the dried, decellularized plugs into approximately 5 mm length pieces. Grind up each piece using a Dremel rotary tool to precisely 4.0 ± 0.1 mm.
9. Measure the weight and dimensions of each scaffold to determine the density (=mass of dried scaffold/volume). Since the pore size and void volume vary within samples, the density is used to roughly estimate scaffold properties for each experiment. A typical density of dried bone scaffolds is 300–400 mg/ml.
10. Sterilize the scaffolds in 70% Ethanol overnight. Rinse in sterile PBS and then incubated in expansion medium at 37°C for several hours before seeding the cells.

3.3. Bioreactor Assembly

1. Assemble bioreactors using a "top–down" procedure by starting with the top layer and building downward. Place the screws into the top layer and invert the assembly so that each layer can be added sequentially. It is very important that the holes enabling inlet of medium into the channels are aligned in all of the layers. Do not fully tighten the screws until after the sterilization (autoclaving) step.
2. Assemble the tubing by connecting approximately 50 cm long portions of the silicon tubing to either side of the three-stop tubing via the barbed couplings. On the inlet side, place Luer lock connectors approximately 10 cm from the connection to the bioreactor unit (see Note 4).
3. Attach the tubing to the assembled bioreactors and place each bioreactor into a separate sterilization pouch.
4. Autoclave the bioreactors for 20 min at 121°C on the "liquid cycle." The heating and cooling rates are slower on the "liquid cycle" of most autoclaves. This regime decreases temperature

differences between different materials during the heating and cooling phases of autoclaving and mitigates any warping that might occur due to adjacent surfaces expanding and contracting at different rates.

5. It is recommended that two persons work together to assemble the sterile bioreactor. One person should use sterile gloves and maintain sterility throughout the process so that he/she can easily manipulate the unit during assembly, while the other person acts as an "assistant" who presents all necessary sterile tools when needed.
6. Use a sterile hex key to tighten the screws of the bioreactor. Connect a three-way stop-cock and 10 ml syringe to the Luer lock connecters near the inlet.
7. Add 20 ml of sterile PBS into the bioreactor reservoir. PBS has a dual purpose: it verifies that there are no leaks in the assembled units and is used to remove air bubbles from the channels and tubing. This is accomplished by sucking the PBS into the channels using a 25-ml syringe and then ejecting it through the tubing until it reenters the reservoir (see Note 5).
8. Keep the assembled bioreactor in the biological safety cabinet until needed.

3.4. Cultivation of Bone Grafts

1. Blot-dry the scaffolds on sterile Kim Wipes and place in a well of nontissue culture treated six-well plate. Scaffolds should be blotted and seeded as a single operation and not allowed to dry for an extended period of time.
2. Add a 50-μl aliquot of the cell suspension to the top of the scaffold and ensure that the aliquot is absorbed into the scaffold.
3. Repeat steps 1 and 2 with the other scaffolds.
4. Allow scaffolds to incubate with cell suspension for 15 min.
5. Use forceps to flip the scaffolds and add 5 μl of medium. This facilitates uniform distribution of cells (which settle due to gravity) and prevents the scaffolds from drying during cell seeding. Repeat this process every 15 min for up to 2 h to facilitate uniform cell distribution.
6. After seeding, add 5 ml of medium to each well. Keep scaffolds in static culture for 3–4 days to enable cell attachment and matrix deposition within the scaffold.
7. Place the seeded scaffolds inside the bioreactor wells (containing PBS). For 4 mm diameter scaffolds, PDMS adapters are used in each well. Replace the 10 ml syringe with a syringe containing culture medium (see Note 6).
8. The medium flow patterns can be tested by slowly injecting culture medium through the inlet channels and observing the flow through all scaffolds into the reservoir (see Note 7).

Remove the PBS in the reservoir with a pipet and replace it with 40 ml of osteogenic medium (the residual PBS in the tubing is negligible).

9. Replace the syringe with an empty syringe. This syringe may be used throughout culture to remove air bubbles that occur in the channels (see Note 8).
10. Cover each bioreactor with a glass Petri dish cover.
11. Transfer the bioreactor units to the incubator and connect tubing to peristaltic pump.
12. For medium changes, the entire bioreactor unit is disconnected from the pump (by simply taking the tubing loops out of pump cartridges and not breaking any of the connections) and transfer to the biological safety cabinet for medium changes.
13. Medium changes are typically conducted twice per week, 3–4 days apart. Traditional cell-culture techniques are used for medium changes. For each bioreactor, remove 20 ml of medium (1 ml aliquots may be saved for future analysis) using a 25-ml pipette and add 20 ml of fresh osteogenic medium.
14. Under the culture hood, the operator must observe if there are any bubbles trapped inside the channels. If necessary, the 25 ml syringe that was currently connected to the tubing can be used to eject the bubbles out of the channel. The operator must keep track of medium being ejected into the syringe.
15. Once all bubbles are removed, take out the medium in the reservoir such that a total of 20 ml is being removed.
16. Add 20 ml of fresh warm medium into the chamber.
17. Place the cover glass back on the bioreactor and wrap with Parafilm.
18. Place the bioreactor inside the incubator and connect the tubing to the pump (see Note 4).
19. Operate the pump in the reverse direction for 10 min and then switch to the normal forward direction for the remainder of the culture.

4. Notes

1. The addition of AA-2-P is made immediately before use because AA-2-P at the working concentration decomposes in the culture medium within 1–2 days, thereby losing efficacy.
2. Lower passages may be used, but it is not advised to use higher passages, due to the decrease in osteogenic capacity of the cells with passage number (9, 10).

3. For sterile harvesting of trabecular bone, while maintaining viability of the cores, the process can be modified in several ways:
   - Care should be taken in cutting open the joint. A "popping" sound should be heard when the joint capsule is cut. If this is not heard, it probably indicates that the integrity of the capsule has been compromised and the joint should not be used. After opening, the joint can be washed and sprayed with 70% ethanol.
   - The coring process can be done in a biological safety cabinet using tools that have been presterilized with 70% ethanol.
   - Sterile PBS can be used to cool the drill bit during the coring process.
   - A dental pick may be used with sterile PBS to wash out cellular material in the pore spaces.
4. The tubing length is dictated by the distance between the pump and the bioreactor units. The pumps may be left outside of the incubator so that tubing passes through the front door or a ventilation hole to the rear of the incubator. If the pump system can be placed in the incubator, much shorter tubing will be needed. Using these multi-channel pumps, several bioreactor units may be run simultaneously. Care must be taken so that tubing does not become entangled making it difficult to handle bioreactor units individually.
5. Six scaffolds are being perfused simultaneously. It is critical to ensure equally distributed flow through all channels. Bubbles inside any of the channels may cause unequal distribution of fluid flows among six scaffolds. It is extremely important to remove all the bubbles using the connected syringe. However, it is possible that the bubbles are trapped in the channel near or under the scaffold and would be difficult to remove by the syringe. Operators may be required to sterilely remove the scaffold and replace after removing the bubbles.
6. Occasionally bioreactor leakage occurs slowly and may not be detected during assembly. If leakage occurs during cultivation, first try to tighten the screws and inlet/outlet connectors to stop the leak. If the leakage does not stop, it may be necessary to transfer the scaffolds to a new sterile bioreactor. Always have one or two sterile bioreactors ready in case of any accidents.
7. It may be necessary to toggle scaffolds so that they fit snuggly into the wells and to ensure that medium does not flow preferentially through any of the channels.
8. Except for periods where bubbles are being removed, ensure that the valve on the three-way stopcock is set so that it does not block the recirculating flow.

9. Representative results of bone tissue engineering using this method are presented in Fig. 2. The study whereby this tissue was generated investigated the effects of low (LS) vs. high (HS) seeding density and low (LF) vs. high (HF) flow-rate. The new bone growth resulting from three groups (LS-LF, LS-HF, and HS-LF) were compared with those of unseeded (US) scaffolds. Several evaluation methods are demonstrated including histology, scanning electron microscopy, and micro-computed tomography.

References

1. Obradovic, B., Carrier, R. L., Vunjak-Novakovic, G., and Freed, L. E. (1999) Gas exchange is essential for bioreactor cultivation of tissue engineered cartilage, *Biotechnol. Bioeng. 63*, 197–205.
2. Radisic, M., Malda, J., Epping, E., Geng, W., Langer, R., and Vunjak-Novakovic, G. (2006) Oxygen gradients correlate with cell density and cell viability in engineered cardiac tissue, *Biotechnol. Bioeng. 93*, 332–343.
3. Martin, I., Wendt, D., and Heberer, M. (2004) The role of bioreactors in tissue engineering, *Trends Biotechnol. 22*, 80–86.
4. Bancroft, G. N., Sikavitsas, V. I., and Mikos, A. G. (2003) Design of a flow perfusion bioreactor system for bone tissue-engineering applications, *Tissue Eng. 9*, 549–554.
5. Botchwey, E. A., Pollack, S. R., Levine, E. M., and Laurencin, C. T. (2001) Bone tissue engineering in a rotating bioreactor using a microcarrier matrix system, *J. Biomed. Mater. Res. 55*, 242–253.
6. Glowacki, J., Mizuno, S., and Greenberger, J. S. (1998) Perfusion enhances functions of bone marrow stromal cells in three-dimensional culture, *Cell Transplant. 7*, 319–326.
7. Grayson, W. L., Bhumiratana, S., Cannizzaro, C., Chao, G. P., Lennon, D., Caplan, A. I., and Vunjak-Novakovic, G. (2008) Effects of initial seeding density and fluid perfusion rate on formation of tissue-engineered bone, *Tissue Eng. A 14*, 1809–1820.
8. Bhumiratana, S., Cannizzaro, C., Wan, L. Q., Grayson, W. L., Kaplan, D. L., Vunjak-Novakovic, G. (2009) Enhancement of mechanical support and biocompatibility of mineralized silk scaffolds with high-throughput fabrication, in *Orthopaedic Research Society*, Las Vegas.
9. Meinel, L., Karageorgiou, V., Fajardo, R., Snyder, B., Shinde-Patil, V., Zichner, L., Kaplan, D., Langer, R., and Vunjak-Novakovic, G. (2004) Bone tissue engineering using human mesenchymal stem cells: Effects of scaffold material and medium flow, *Ann. Biomed. Eng. 32*, 112–122.
10. Meinel, L., Karageorgiou, V., Hofmann, S., Fajardo, R., Snyder, B., Li, C. M., Zichner, L., Langer, R., Vunjak-Novakovic, G., and Kaplan, D. L. (2004) Engineering bone-like tissue in vitro using human bone marrow stem cells and silk scaffolds, *J. Biomed. Mater. Res. A 71A*, 25–34.

Chapter 19

Adipogenic Differentiation of Human Mesenchymal Stem Cells

Trine Fink and Vladimir Zachar

Abstract

Mesenchymal stem cells have the capability to differentiate into a number of cell types including adipocytes. The adipocytic phenotype is characterized by intracellular accumulation of lipid droplets as well as transcription of adipocyte-specific genes. This paper details a basic protocol for adipogenic induction of bone marrow and adipose tissue-derived stem cells, as well as protocols for staining lipid accumulation and the transcriptional analysis of PPAR-γ and aP2 by real-time RT-PCR.

Key words: Mesenchymal stem cells, Adipogenesis, Differentiation, Lipid staining, PPAR-γ, aP2

1. Introduction

Adipogenesis is a process whereby mesenchymal stem cells (MSCs) or preadipocytes differentiate to acquire phenotypic characteristics of mature adipocytes. The molecular mechanisms underlying this process in vitro typically occur in a well-defined sequence, which has been thoroughly described in a number of reviews (1–3). In summary, the adipogenic process begins with a commitment to adipocytic differentiation characterized by activation of the transcription factors CCAAT/enhancer binding proteins (C/EBP)-β and -δ, and sterol-regulatory element binding protein1/adipocyte determination and differentiation factor 1 (SREBP1/ADD1). Adipogenic commitment in turn leads to transcription of the pivotal adipocyte-specific transcription factors (C/EBPα) and peroxisome proliferative-activated receptor-γ (PPAR-γ). This stage is then, for some cell types, followed by clonal expansion and growth arrest before terminal differentiation. The later stages

Mohan C. Vemuri et al. (eds.), *Mesenchymal Stem Cell Assays and Applications*, Methods in Molecular Biology, vol. 698, DOI 10.1007/978-1-60761-999-4_19,

of adipogenic differentiation are characterized by transcription of late markers such as fatty acid binding protein aP2 and adiponectin as well as intracellular accumulation of lipid vesicles.

The protocols for adipogenic induction of MSCs vary in terms of both composition of induction cocktail and length of induction. However, most protocols call for insulin, dexamethasone, isobutylmethylxantine (IBMX), indomethacine, and a PPAR-γ ligand (4–8). In the following chapter, we describe a protocol for the induction of adipogenesis that, in our experience, works well with MSCs of both human and animal (rat, rabbit, brown bear, sheep) origin, as well as for bone marrow and adipose tissue-derived MSCs.

2. Materials

2.1. Adipogenic Induction of MSCs

1. Phosphate buffered saline (PBS) (GIBCO/Invitrogen, Leiden, The Netherlands).
2. Trypsin/EDTA blend: Trypsin (Invitrogen) is obtained as a solution of 2.5% in PBS. Dilute 10 times with PBS and sterile filter. Divide into 10 mL aliquots and store at –20°C. Prepare EDTA (anhydrous, crystalline, cell culture tested) (Sigma-Aldrich, Brøndby, DK) by dissolving in PBS without $Ca^{2+}$and Mg^{2+} at a concentration of 0.02%. Sterile filter, divide into 10 mL aliquots and store at 4°C. Prior to use, mix the trypsin and EDTA solutions 1:1 to produce a ready-to-use 0.125% trypsin/0.01% EDTA blend. The Trypsin/EDTA blend is stable for up to 1 week at 4°C.
3. Trypan blue solution, 0.4%, sterile filtered, cell culture tested (Sigma-Aldrich).
4. Growth medium: Consists of Minimum essential Medium alpha (A-MEM) supplemented with 10% fetal bovine serum (FBS), penicillin (100 U/mL), streptomycin (100 μg/mL) and gentamicin (50 μg/mL) (all from Invitrogen). (For the use of other media formulations, please see Note 1.)
5. Dexamethasone (Sigma-Aldrich). Dissolve 1 mg of dexamethasone in 1 mL 96% EtOH. While mixing gently, 25 mL sterile A-MEM is added to obtain a final stock concentration of 0.1 mM (1,000× working solution). Divide into 100 μL aliquots and store at –20°C.
6. 3-Isobutyl-1-methylxanthine (IBMX) (Sigma-Aldrich). To achieve a stock concentration of 45 mM (100× working solution), dissolve 250 mg of IBMX in 25 mL 96% EtOH. Divide into 1 mL aliquots and store at –20°C.

7. Insulin, recombinant human (Sigma-Aldrich). The concentration of 10 mg/mL corresponds to 1.7 mmol/L (10,000× working solution). Store at 4°C.
8. Indomethacin (Sigma-Aldrich). Dissolve 1 g of indomethacin in 56 mL 96% EtOH to yield a stock solution of 50 mM (250× working solution). Divide into 250 μL aliquots and store at −80°C, protected from light.
9. Rosiglitazone/BRL49653 (Cayman Chemical Company, Ann Arbor, MI). Prepare a 2-mM stock solution (2,000× working solution) by dissolving 5 mg rosiglitazone in 7 mL DMSO. Divide the stock solution into aliquots of 10 and 20 μL in thin-walled PCR tubes and store at −80°C.

   Adipogenic induction medium (AIM): To prepare 100 mL AIM, further supplement 100 mL growth medium with 100 μL of 0.1 mM dexamethasone (0.1 μM final), 1 mL of 45 mM IBMX (0.45 mM final), 10 μL of 10 mg/mL insulin (1 μg/mL final) (see Note 2) and 250 μL of 50 mM indomethacin (0.2 mM final). Sterile filter and store at 4°C for up to 3 weeks. If desired, rosiglitazone can be added to AIM just prior to use with 0.5 μL of 2 mM rosiglitazone/mL AIM, to yield a final concentration of 1 μM rosiglitazone. After rosiglitazone is added to the medium, it is only stable for 1–2 days when stored at 4°C (see Note 3).

2.2. Histochemical Analysis With Oil Red O Stain

1. Oil red O stock solution: Dissolve 0.5 mg oil red O (Sigma-Aldrich) in 100 mL isopropanol. Incubate stock solution at room temperature for 1 h, filter through a 0.2-μm filter and store at room temperature for up to 1 year. To prepare a working solution of oil red O, mix 6 mL of stock solution with 4 mL distilled water. Incubate at room temperature for 1 h, filter through a 0.2-μm filter. The working solution is stable for approximately 3 h.
2. Distilled water.
3. Four percent neutral buffered formaldehyde (Merck, Darmstadt, Germany).

2.3. Histochemical Analysis with BODIPY and Hoechst Stain

1. Hoechst 33342 (Molecular Probes/Invitrogen).
2. BODIPY 493/503 (D-3922) (Molecular Probes/Invitrogen).
3. Growth medium.
4. PBS without Ca^{2+} and Mg^{2+}.
5. Four percent neutral buffered formaldehyde.
6. Lab-Tek II eight-well chamber slides (Nalge Nunc, Rochester, New York).
7. Fluorescent mounting medium (DAKO, Copenhagen, Denmark).

8. Nail polish.
9. Paper towels.

2.4. Real-Time Quantatitve RT-PCR Aurum Total RNA Mini Kit, Bio-Rad, Hercules, CA (see Note 4)

1. iScript cDNA synthesis kit (Bio-Rad).
2. PCR primers for early adipogenesis marker PPAR-γ, late marker aP2, reference genes Tyrosine 3-monooxygenase/tryptophan 5-monooxygenase activation protein, zeta polypeptide (YWHAZ), and Cyclophilin A (PPIA):
   (a) PPAR-G forward primer: 5′-TCAGGTTTGGGCGGATGC-3′
   (b) PPAR-G reverse primer: 5′-TCAGCGGGAAGGACTTTATGTATG-3′
   (c) aP2 forward primer: 5′-ATGGGATGGAAAATCAACCA-3′
   (d) aP2 reverse primer: 5′-GTGGAAGTGACGCCTTTCAT-3′
   (e) PPIA forward primer: 5′-TCCTGGCATCTTGTCCATG-3′
   (f) PPIA reverse primer: 5′-CCATCCAACCACTCAGTCTTG-3′
   (g) YWHAZ forward primer: 5′-ACTTTTGGTACATTGCTTCAA-3′
   (h) YWHAZ reverse primer: 5′-CCGCCAGGACAAACCAGTAT-3′
3. SYBR Green Supermix (Bio-Rad).

3. Methods

3.1. Adipogenic Induction of MSCs

1. Isolate and propagate adipose-derived MSCs as described elsewhere in this text, see Chapter "Isolation and Growth of Adipose Tissue-Derived Stem Cells."
2. Remove medium and wash cells twice with PBS.
3. Add enough Trypsin/EDTA to cover the cells when gently rocking the flask back and forth.
4. Incubated cells at 37°C for 2 min. Monitor the progress of detachment visually under the microscope. If cells have not detached, return the flask to the incubator for an additional 2 min.
5. As soon as cells have detached, stop the reaction by adding 5–10 mL growth medium and homogenize the cell suspension by repetitive pipetting with a 5 or 10-mL serological pipette.

6. Determine the concentration of cells by removing a small aliquot, mix the aliquot 1:1 with trypan blue and count viable cells in a hemocytometer. The number of viable cells should be greater than 98%.
7. Adjust the concentration of cells to 30,000 cells/mL through the addition of growth medium.
8. Seed cells in multi-well plates or chamber slides (for BODIPY and Hoechst staining) at a concentration of approximately 18,000 cells/cm^2 in growth medium (see Note 5).
9. Place reseeded cultures back in the incubator and allow cells to expand until they reach 80–90% confluence (see Note 6).
10. Initiate adipogenic induction by replacing the growth medium with AIM. For 48-well plates, a volume of 0.4 mL/well is appropriate. For other formats, adjust the volume accordingly based upon surface area.
11. Refeed plates with fresh AIM twice a week.
12. Follow adipogenic differentiation visually under the microscope through the formation of intracellular lipid droplets. The adipogenic differentiation of MSCs can be assessed by accumulation of lipids and/or transcription of adipocyte-specific genes. The adipogenic conversion of the cells is usually apparent after 2–3 weeks postadipogenic induction (see Note 7).

3.2. Histochemical Analysis with Oil Red O Stain

1. Remove AIM and gently wash cells twice in PBS.
2. Fix cells by adding 4% neutral buffered formaldehyde and incubate at 4°C for 1 h.
3. Aspirate fixative and gently wash twice with PBS.
4. Add enough Oil red O working solution to cover the surface of the cells and incubate for 15 min at room temperature.
5. Carefully rinse cells with distilled water. Continue until the removed wash solution is clear (usually achieved after 6–8 washes).
6. After the final wash, add a sufficient volume of distilled water to cover the surface of the well.
7. Observe cells under an inverted microscope at ×100 – ×400 magnification. The lipid vesicles are visible as bright red spheres that tend to cluster together (see Fig. 1a) (see Note 8).

3.3. Histochemical Analysis with BODIPY and Hoechst Stain

While oil red O staining is widely used, it is not suitable for high magnification. The below described staining protocol utilizing fluorescent dyes allows for a more detailed, high resolution imaging of intracellular lipid accumulation.

1. Prepare staining solution by supplementing complete growth medium with 2 μg/mL Hoechst 33342 and 10 μg/mL BODIPY.

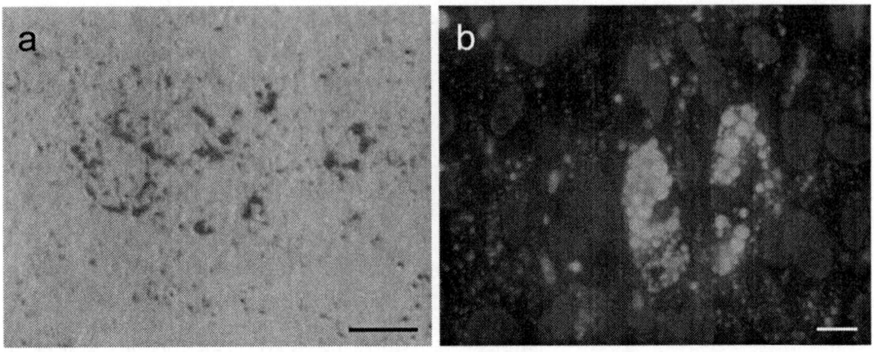

Fig. 1. Human adipose tissue-derived mesenchymal cells were induced into adipogenesis by dexamethasone, isobutyl methylxanthine, insulin, indomethacin, and rosiglitazone. (**a**) After 7 days, the lipid droplets were stained with oil red O, and bright-field microscopy was used to capture the image. The scale bar indicates 200 μm. (**b**) After 14 days, the lipid inclusions and nuclei were stained with the fluorescent dyes BODIPY and Hoechst 33342, respectively. The scale bar indicates 20 μm.

2. After adipogenic induction in a chamber slide (see Subheading 3.1), remove AIM and replace with Hoechst 33342 and BODIPY staining solution.
3. Incubate at 37°C for 1 h.
4. After incubation, remove medium and wash 3 times with PBS.
5. Fix cells by adding 4% neutral buffered formaldehyde to each well and incubate at 4°C for 15 min.
6. Wash twice with PBS.
7. Carefully separate the chamber slide and remove any remaining PBS on the slide with a paper towel. Take care not to touch any areas containing cells with the towel.
8. Place a drop of fluorescent mounting medium onto each area of the slide with cells.
9. Carefully place a coverslip over the mounting medium and gently push away any air bubbles (see Note 9).
10. Using a towel, remove residual mounting medium from the edges of the coverslip.
11. Allow slide to air dry for 5 min.
12. Seal the edges of the coverslip with nail polish.
13. Observe cells under a fluorescence microscope (see Fig. 1b).

3.4. Real-Time Quantitative RT-PCR

In addition to the above staining protocols, we have found it very important to include confirmation of adipogenic conversion through the transcriptional analysis of adipocyte-specific genes as some cells have been shown to accumulate lipid droplets without transcribing adipocytic genes (9). We perform such an analysis using a real-time PCR analysis of PPAR-γ and aP2. Also, for the normalization of these genes to cDNA input, we have previously shown that PPIA and YWHAZ are suitable reference genes during adipogenic differentiation of MSCs (10).

1. Harvested RNA as per the manufacturer's instructions and synthesize cDNA using random decamer primers (see Note 10).
2. Real-time PCR reactions are performed using primers for the adipocyte-specific genes PPAR-γ and aP2 and the reference genes PPIA and YWHAZ. The concentration of primers is 5 pmol pr 25 μL reaction. Perform amplification using two-temperature cycling consisting of a denaturation step at 95°C for 15 s, followed by an annealing/extension step at 68°C (PPAR-γ and aP2) or 60°C (for PPIA and YWHAZ) for 30 s.
3. For each sample, the levels of PPAR-γ and aP2 are normalized to the geometric mean of the levels of PPIA and YWHAZ (see Note 11).

4. Notes

1. We recommend the use of A-MEM basal medium based on a comparative study of different types of media; however, DMEM/F12 and F12 basal media also appear to be well suited for adipogenic differentiation (11).
2. The concentration of insulin is tenfold lower than that in many other protocols. In a study from 2001, Janderova et al. (4) compared the concentration of insulin in induction media to the rate of adipogenic conversion of stem cells and found that this low concentration of insulin is sufficient. We found that this also holds true in our hands.
3. The use of rosiglitazone is not a prerequisite for adipogenesis when following this protocol. However, the adipogenic differentiation appears slightly faster (2–3 days faster) when rosiglitazone is included.
4. Most Silica-based RNA isolation kits will yield RNA of good quality. In addition to the Aurum Total RNA mini kit, we have good experience with GenElute Mammalian Total RNA purification, Sigma; and RNeasy Mini Kit, Qiagen, Valencia, CA.
5. We routinely culture cells in 48-well plates for Oil Red O staining or for harvest of RNA for subsequent RT-PCR analysis. For BODIPY staining, we culture cells in eight-well chamber slides.

6. Some cell lines, such as mouse embryo 3T3-L1, require clonal expansion to undergo adipogenic differentiation. In our experience, clonal expansion is not necessary for adipogenesis to occur in MSCs.
7. The rate of adipogenic conversion varies between different cell types. As the cells acquire a more mature phenotype, there is an increased tendency for the cells to detach from the culture plate. We recommend observing the cells microscopically to identify when lipid droplets form and thereafter stain cells or harvest RNA before cells detach.
8. The nuclei may be counterstained with Mayer's haematoxylin for 5 min at room temperature after oil red O staining.
9. When placing the coverslip, one edge of the coverslip is placed onto the slide and the other edge is very slowly lowered until the coverslip is flush, thus minimizing formation or air bubbles.
10. Before synthesis of cDNA is performed, isolated RNA must be DNase treated.
11. The benefit of using multiple housekeeping genes for optimal normalization has been extensively described by Vandesompele et al. (12). The use of different genes requires that the normalization values are calculated as a geometric mean, which is for two numbers is calculated as the square root of their product. This calculation adjusts for the fact that some housekeeping genes are expressed at a lower level than others. An example of using geometric mean vs. average values is seen in the table below, where it is shown than reducing the PPIA values by one-third from 6 to 4, has almost no influence on calculation of the mean and changes the geometric mean considerably.

	YWHAZ value	PPIA value	Average	Geometric mean
Condition 1	400	6	203	49
Condition 2	400	4	202	40

References

1. Gregoire, F. M. (2001) Adipocyte differentiation: from fibroblast to endocrine cell. *Exp Biol Med (Maywood)* **226**, 997–1002.
2. MacDougald, O. A., Mandrup S. (2002) Adipogenesis: forces that tip the scales. *Trends Endocrinol Metab* **13**, 5–11.
3. Otto, T. C., Lane M. D. (2005) Adipose development: from stem cell to adipocyte. *Crit Rev Biochem Mol Biol* **40**, 229–42.
4. Janderova, L., McNeil M., Murrell A. N., Mynatt R. L., Smith S. R. (2003) Human mesenchymal stem cells as an in vitro model for human adipogenesis. *Obes Res* **11**, 65–74.
5. Pittenger, M. F., Mackay A. M., Beck S. C., et al. (1999) Multilineage potential of adult human mesenchymal stem cells. *Science* **284**, 143–7.
6. Zuk, P. A., Zhu M., Mizuno H., et al. (2001) Multilineage cells from human adipose tissue:

implications for cell-based therapies. *Tissue Eng* **7**, 211–28.

7. Reger, R. L., Tucker A. H., Wolfe M. R. (2008) Differentiation and characterization of human MSCs. *Methods Mol Biol* **449**, 93–107.
8. Bunnell, B. A., Estes B. T., Guilak F., Gimble J. M. (2008) Differentiation of adipose stem cells. *Methods Mol Biol* **456**, 155–71.
9. Fink, T., Abildtrup L., Fogd K., et al. (2004) Induction of adipocyte-like phenotype in human mesenchymal stem cells by hypoxia. *Stem Cells* **22**, 1346–55.
10. Fink, T., Lund P., Pilgaard L., Rasmussen J. G., Duroux M., Zachar V. (2008) Instability of standard PCR reference genes in adipose-derived stem cells during propagation, differentiation and hypoxic exposure. *BMC Mol Biol* **9**, 98.
11. Lund, P., Pilgaard L., Duroux M., Fink T., Zachar V. (2009) Effect of growth media and serum replacements on the proliferation and differentiation of adipose-derived stem cells. *Cytotherapy* **11**, 189–97.
12. Vandesompele, J., De Preter K., Pattyn F., et al. (2002) Accurate normalization of real-time quantitative RT-PCR data by geometric averaging of multiple internal control genes. *Genome Biol* **3**, RESEARCH0034.

Chapter 20

Chondrogenic Differentiation of Bone Marrow-Derived Mesenchymal Stem Cells: Tips and Tricks

Luis A. Solchaga, Kitsie J. Penick, and Jean F. Welter

Abstract

It is well known that adult cartilage lacks the ability to repair itself; this makes articular cartilage a very attractive target for tissue engineering. The majority of articular cartilage repair models attempt to deliver or recruit reparative cells to the site of injury. A number of efforts are directed to the characterization of progenitor cells and the understanding of the mechanisms involved in their chondrogenic differentiation. Our laboratory has focused on cartilage repair using mesenchymal stem cells and studied their differentiation into cartilage. Mesenchymal stem cells are attractive candidates for cartilage repair due to their osteogenic and chondrogenic potential, ease of harvest, and ease of expansion in culture. However, the need for chondrogenic differentiation is superposed on other technical issues associated with cartilage repair; this adds a level of complexity over using mature chondrocytes. This chapter will focus on the methods involved in the isolation and expansion of human mesenchymal stem cells, their differentiation along the chondrogenic lineage, and the qualitative and quantitative assessment of chondrogenic differentiation.

Key words: Mesenchymal stem cell, Chondrogenesis, Tissue culture, Fibroblast growth factor-2, Transforming growth factor-beta, Aggregate culture, Cartilage

1. Introduction

Bone marrow is the most common source for the isolation of adult human mesenchymal stem cells (hMSCs) (1) as first described by Friedenstein and colleagues (2). In the mid-to-late 1980s, Owen proposed that marrow stromal cells give rise to cells of the mesenchymal lineages including fibroblasts, reticular cells, adipocytes, osteoblasts, and others (3, 4). This model was later termed mesengenesis (5). For some time, it has been known that rat and human bone marrow form bone and cartilage when loaded into ceramic carriers and implanted subcutaneously (6–8).

Mohan C. Vemuri et al. (eds.), *Mesenchymal Stem Cell Assays and Applications*, Methods in Molecular Biology, vol. 698, DOI 10.1007/978-1-60761-999-4_20, © Springer Science+Business Media, LLC 2011

Goshima et al. achieved similar results with cells cultured from rat bone marrow (9, 10). It was also shown by Haynesworth et al. (11) that cultured human bone marrow-derived cells assay could generate bone when placed in the same assay. These cultured cells were isolated from bone marrow, recovered from the top of a density gradient, and seeded into tissue culture dishes in medium containing selected lots of fetal bovine serum (FBS) (12). After subcultivation, these cultured cells referred to as hMSCs exhibit a strong, yet not unlimited, proliferative potential (13–15). As Human MSCs are primary cultures, the preparation-to-preparation variability in respect to proliferation, differentiation potential, and eventual senescence (14) poses a significant challenge.

A number of in vitro assays can be used to assess the multipotentiality of these cell preparations (16). Osteogenic differentiation of hMSCs can be triggered by exposure to specific culture supplements including dexamethasone, ascorbate-2-phosphate, and beta glycerolphosphate (13, 17). Adipogenic differentiation can also be induced through the use of induction medium containing dexamethasone, indomethacin, isobutylmethylxanthine, and insulin, but these cultures require a higher seeding density than that used for osteogenic differentiation (16). Finally, hMSCs placed in aggregate or pellet cultures in a defined medium containing dexamethasone, ascorbate-2-phosphate, insulin, selenious acid, transferrin, sodium pyruvate, and transforming growth factor-beta will undergo chondrogenic differentiation (16, 18–21). The ability of MSCs to differentiate along these lineages is strongly suggestive of their multipotency and stem cell nature. However, hMSCs do not maintain these characteristics indefinitely and hMSCs senescence with extensive subcultivation in vitro whereby they lose their proliferation and differentiation potential (1, 22, 23).

Tissue engineering combines the fields of cell biology, engineering, material sciences, and surgery to provide new functional tissues using living cells, biomatrices, and signaling molecules (24–26). In recent years, this discipline has greatly expanded, with numerous research groups focusing on the development of strategies for the repair and regeneration of a variety of tissues (27, 28). Many of these tissue-engineered approaches have targeted the musculoskeletal system in general, with special emphasis on articular cartilage (29–37). Articular cartilage is a especially attractive target for tissue engineering strategies because it has been well documented that injuries of articular cartilage, an avascular tissue without direct access to a significant source of reparative cells, do not spontaneously heal (38–44). The vast majority of approaches to repair or regenerate articular cartilage are cell-based, aiming to provide a population of reparative cells to the injured site. Cells used to develop these strategies can be either differentiated chondrocytes isolated from unaffected areas of the joint surface (35, 45–56) or progenitor cells capable of differentiating into chondrocytes and can be isolated from a variety of tissues (57–69). As harvesting a tissue biopsy from valuable healthy articular cartilage will result in an additional injury, which ultimately cannot repair itself, this cell source does not seem to be a good choice. Therefore, a number of research efforts are directed to the isolation of progenitor cells and the understanding of the mechanisms involved in their chondrogenic differentiation. Protocols for the isolation and mitotic expansion of adult human bone marrow-derived mesenchymal stem cells (2, 11, 70, 71) and the specific culture conditions developed for their chondrogenic differentiation (18, 19, 72, 73) will be further discussed in this chapter.

2. Materials

2.1. MSC Isolation and Culture

1. 15 and 50-ml centrifuge tubes (BD Biosciences, Franklin Lakes, NJ).
2. 50-ml polycarbonate capped centrifuge tubes (Thermo Fisher Scientific, Waltham, MA).
3. Pipettes and tissue culture dishes (BD Biosciences).
4. Tyrode's salt solution (Sigma, St. Louis, MO).
5. Complete culture medium: Dulbecco's Modified Eagle's Medium, Low (1.5 g/l) Glucose (DMEM-LG) (Invitrogen, Carlsbad, CA) supplemented with 10% FBS (best available). Store at 4°C for up to 8 weeks.
6. Percoll gradient (see Note 1): Mix 22.05 ml Percoll (Sigma), 2.45 ml of 1.5 M NaCl (Fisher), and 10.5 ml Tyrode's salt solution. Add 35 ml of Percoll solution into individual 50-ml polycarbonate capped sterile centrifuge tubes. Spin tubes at 13,000 rpm (21,200 × *g*) for 15 min in a SS 34 rotor in a Sorvall centrifuge (Thermo Fisher Scientific, Waltham, MA). The resulting density gradient is 1.03–1.12 g/l. Percoll should only be stored for up to 4 weeks at 4°C.
7. 0.25% trypsin 53 mM EDTA (Gibco).
8. Bovine calf serum (BCS) (Thermo Fisher Scientific, Waltham, MA).
9. 4% acetic acid (Fisher) in H_2O.
10. 70% ethanol in H_2O.
11. 15% sodium hypochlorite (bleach) in H_2O.
12. Fibroblast Growth Factor-2 (FGF-2): Prepare a 10-μg/ml stock solution (1,000×) of FGF-2 (Peprotech) in complete culture medium. Store at –80°C.
13. DMSO (Sigma).

14. Recovery™ Freezing medium (Invitrogen).
15. Cryogenic vials (Thermo Fisher Scientific).
16. "Mr. Frosty™" isopropanol freezing container (Thermo Fisher Scientific).
17. Liquid nitrogen storage.
18. Liquid nitrogen.

2.2. MSC Chondrogenic Differentiation

1. Chondrogenic differentiation medium: Dulbecco's Modified Eagle's Medium, High (4.5 g/l) Glucose (DMEM-HG, Invitrogen) supplemented with 10% ITS+ Premix Tissue Culture Supplement (Becton Dickinson), 10^{-7} M dexamethasone (Sigma), 1 μM ascorbate-2-phosphate (Wako, Richmond, VA), 1% sodium pyruvate (Invitrogen), and 10 ng/ml transforming growth factor-beta 1 (TGF-β1, Peprotech, Rocky Hill, NJ).
2. Dexamethasone stock solution (100×): dexamethasone (Sigma) is dissolved at 10^{-3} M in absolute ethanol and diluted 1:100 with DMEM-HG to obtain the 10^{-5} M stock solution. Store at −20°C.
3. Ascorbate-2-phosphate stock solution (100×): Prepare 100 μM ascorbate-2-phosphate (Wako) in H_2O. Store at −20°C.
4. Transforming Growth Factor-beta 1 (TGF-β1): Prepare a 1-μg/ml stock solution (100×) of TGF-β1 (Peprotech, Rocky Hill, NJ) in 4 mM hydrochloric acid (HCl) and 1% bovine serum albumin. Store at −80°C.
5. Polypropylene 96-well plates (Phenix, Candler, NC).
6. Electronic repeater pipette (Brand, Essex, CT).
7. Repeater pipette tips (Brand; Eppendorf, Westbury, NY).
8. Large orifice 200-μl pipette tip (Fisher).

2.3. Histologic and Immunohistochemical Analyses

1. 10% neutral-buffered formalin (Fisher).
2. Toluidine Blue (Fisher).
3. Anticollagen type I antibody (clone col-1, Sigma) used diluted 1:500.
4. Anticollagen type II antibody (Developmental Studies Hybridoma bank, University of Iowa, Iowa City, IA). Used diluted 1:50.
5. Anticollagen type X antibody (Dr. Gary J. Gibson, Ph.D., Breech Research Laboratory, Bone & Joint Center, Henry ford Hospital & Medical Centers, Detroit, MI). Used diluted 1:100.
6. Mouse preimmune IgG (Vector Laboratories, Burlingame, CA).

7. Fluorescein isothiocyanate (FITC)-conjugated goat antimouse Immunoglobulin secondary antibody (Cappel 55499, detects IgG, IgA, and IgM isotypes. MP Biomedicals, Irvine, CA, USA), used diluted 1:500.
8. Ethanol (Aaper, Shelbyville, KY).
9. Xylene (Fisher).
10. Pronase E (Sigma) prepared as 1 mg/ml solution in PBS.
11. 5% BSA (Sigma) in PBS for 30 min.
12. *N*-propyl gallate. Prepared as a 5% solution in glycerol (both Sigma). This can take up to a day to dissolve.

2.4. DNA and GAG Assays

1. Papain buffer: 25 μg/ml papain (Sigma); 2 mM cysteine (Sigma); 50 mM sodium phosphate (Fisher); 2 mM EDTA (Sigma) in 150 ml nuclease-free water. Adjust pH to 6.5 and filter sterilize with a 0.2-μm Nalgene filter.
2. Neutralizing solution: 4 M NaCl (Fisher), 100 mM Na_2PO_4 (pH 7.2) (Fisher); 0.1 N HCl.
3. Hoechst Dye: Prepare a 1-mg/ml Hoechst #33258 (bisBenzimide, Sigma B-2883) stock solution in H_2O. Dilute stock solution 1:1,500 (0.7 μg/ml) in H_2O to prepare a working solution.
4. Certified Calf Thymus DNA standard (Sigma). Prepare DNA standards according to Table 1.
5. Visible/UV spectrophotometer (Tecan, Maennedorf, CH).
6. Dot-blot apparatus (Bio-Rad, Hercules, CA).
7. 0.45 μm pore size Nitrocellulose membrane (Bio-Rad).
8. Safranin O reagent: 0.02% safranin O in 50 mM (pH 4.8) sodium acetate (Fisher). Filter solution through a 0.45-μm filter. Store at room temperature for up to 6 months.

Table 1
Matrix for the preparation of calf thymus standards for DNA assay

DNA concentration[a] (ng/ml)	Papain buffer (μl)	Neutralizing solution (μl)	0.1 N NaOH (μl)	1 mg/ml calf thymus DNA (μl)
0	200	400	400	–
500	200	400	400	1
1,000	200	400	400	2
1,500	200	400	400	3
2,000	200	400	400	4

[a] This will be the final concentration in the well once the dye solution is added

Table 2
Matrix for the preparation of chondroitin sulfate standards for GAG assay

CS-C[a] (μg)	H_2O (μl)	0.2 mg/ml CS-C in H_2O (μl)	1 mg/ml CS-C in H_2O (μl)
0	75	–	–
1	60	15	–
2	45	30	–
3	30	45	–
4	15	60	–
5	60	–	15
6	57	–	18

[a] This is the total amount of Chondroitin sulfate-C (CS-C) that is loaded in the wells; it is not a concentration

9. CPC: 10% cetylpyrinidium chloride (Sigma) in H_2O. Warm solution to dissolve. Filter solution through a 0.45-μm filter. Store at room temperature for up to 6 months.
10. Chondroitin sulfate C (CS-C): Prepare 1 mg/ml chondroitin sulfate c (shark cartilage) (Seikagaku America, East Falmouth, MA) stock solution in water. Dilute stock solution 1:5 (0.2 mg/ml) to prepare a working solution. Prepare (CS-C) standards according to Table 2.

2.5. RNA Extraction

1. TRIzol reagent (Invitrogen)
2. 4-ml cryovial (Nunc)
3. Omni TH handheld homogenizer (OMNI international, Marietta, GA)
4. Hard tissue plastic tips (OMNI international)
5. Isopropanol (Fisher)
6. Ethanol (Fisher)
7. Nuclease-free water (Invitrogen)

3. Methods

3.1. Isolation and Seeding of Human Mesenchymal Stem Cells

MSCs are isolated by their ability to adhere to tissue culture plastic and proliferate under specific culture conditions (12). The original bone marrow cell suspension seeded on the culture vessels contains a heterogeneous mixture of cells that includes red blood cells, nucleated cells of the hematopoietic lineage, monocytes, macrophages, and fibroblast-like cells from the bone marrow stroma. For the most part, erythrocytes and leukocytes do not attach to the tissue culture surface and are eventually removed during the medium changes and subsequent subculturing (see Notes 2–4).

1. Prepare the biological safety cabinet for aseptic work.
2. Bring the syringe with the bone marrow into the biological safety cabinet (see Fig. 1).
3. Label and open a sterile 50-ml disposable centrifuge tube.
4. Add 25 ml of complete medium to the 50-ml tube.
5. Eject the contents of the syringe into the 50-ml disposable centrifuge tube.
6. Thoroughly mix the medium and bone marrow sample by pipetting up and down with a 25-ml pipette.
7. Transfer a small aliquot (0.2 ml) to a 1.5-ml microcentrifuge tube for a preliminary cell count.
8. Centrifuge the entire bone marrow sample at 500×*g* for 5 min in a benchtop centrifuge.
9. While the sample is being centrifuged, determine the number of cells in the preliminary sample as follows:
   (a) Transfer 50 μl of the sample from step 2 to a 0.5 or 1.5-ml microcentrifuge tube.
   (b) Add 50 μl of serum-containing medium.
   (c) Add 100 μl 4% acetic acid and mix to lyse the erythrocytes.

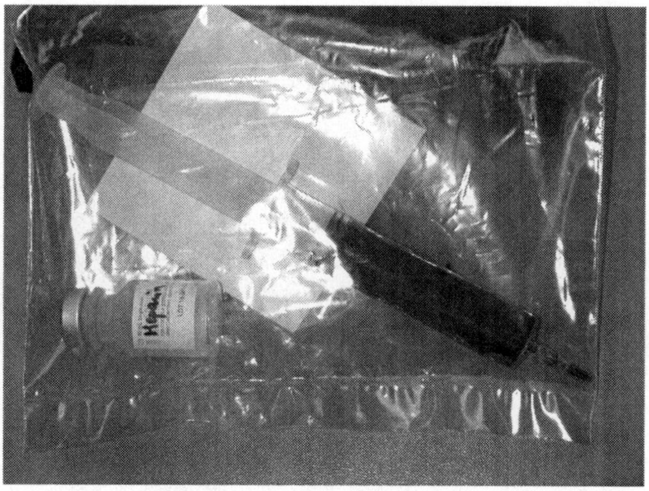

Fig. 1. Bone marrow aspirate in the syringe immediately after aspiration from a healthy volunteer donor.

(d) Count the nucleated cells with a hemacytometer and determine the total number of nucleated cells in the 50-ml tube.

10. Determine the number of tubes of preformed Percoll (density 1.03–1.12 g/ml) required to fractionate the nucleated cells based on the preliminary cell count (11). Load no more than 2.0×10^8 cells per tube.
11. After the sample has been centrifuged, carefully remove the fat layer and aspirate the supernatant being careful not to disturb the pellet (see Fig. 2).
12. Adjust the volume of the pellet to allow 5 ml of cell suspension per tube of Percoll. If the volume of the pellet will exceed 5 ml, divide the volume between two tubes of Percoll.
13. Load the proper volume of cell suspension carefully onto the top of a Percoll gradient. Transfer the cell suspension slowly so that cells will remain at the top of the gradient (see Fig. 3).
14. Carefully transfer the tubes to a centrifuge and spin in SS 34 rotor in a Sorvall centrifuge at $400 \times g$ for 15 min with the brake off.
15. Return the sample (see Fig. 4) to the biological safety cabinet.
16. Carefully pipette off the top layer or layers of cells (about 10–14 ml) and transfer them to another labeled sterile 50-ml centrifuge tube.
17. Add serum-containing medium to the tube for a final volume of 50 ml.

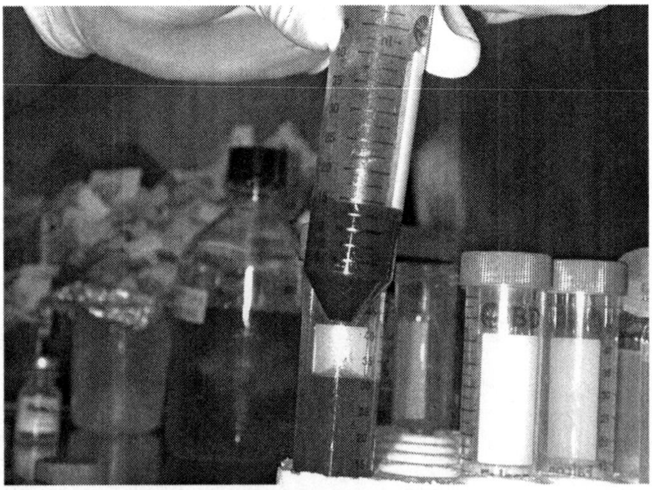

Fig. 2. Loose bone marrow cell pellet after initial centrifugation.

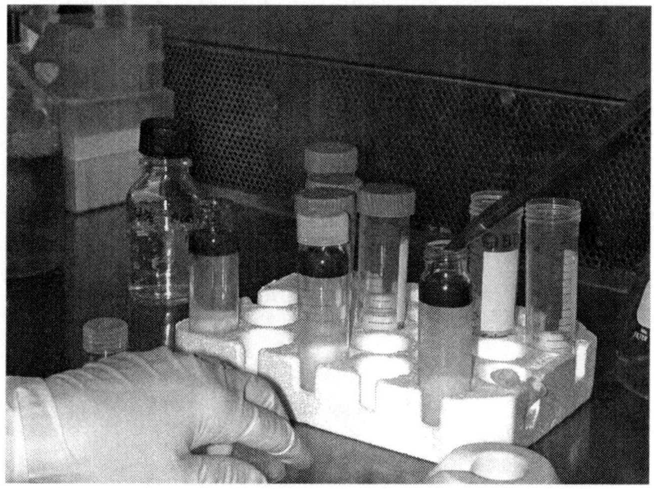

Fig. 3. Bone marrow cells being layered over the preformed Percoll gradient.

Fig. 4. Sample after gradient fractionation; note the band of nucleated cells at the top of the gradient.

18. Spin tubes in a benchtop centrifuge at $500 \times g$ for 5 min.
19. Remove and discard supernatant.
20. Resuspend the cell pellet (see Fig. 5) in 10 ml of complete medium.
21. Use 1-ml pipette to take a small sample (about 100 μl) for determination of the final cell number as described in step 9.
22. Plate cells at 1.7×10^6 per cm^2 in tissue culture dishes or flasks in complete medium.

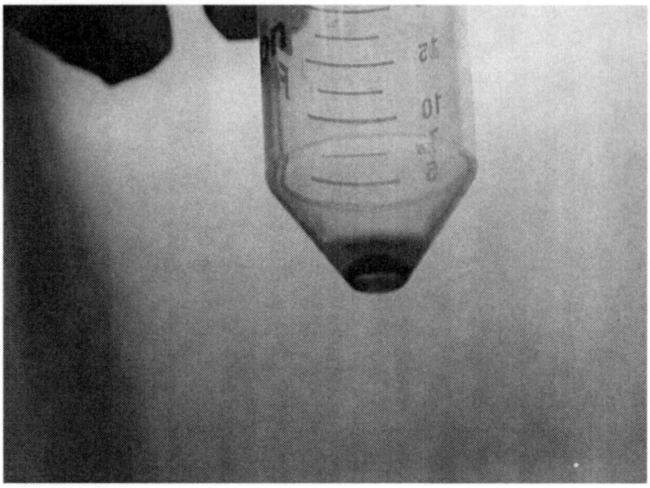

Fig. 5. Pellet of mononuclear cells after the final wash and centrifugation.

3.2. Mesenchymal Stem Cell Culture

Mesenchymal Stem Cells are cultured at 37°C in a humidified atmosphere of 95% air and 5% CO_2. Only a small percentage of the cells in the original cell suspension will attach to the tissue culture surface. These fibroblast-like cells proliferate and form loose colonies of spindle-shaped cells that are usually visible under the microscope between days 4 and 6 of culture. These colonies will increase in size and cell number over the next 7–10 days and should be subcultured before the cells become confluent (it is important to note that proliferation of these cells is not contact-inhibited). No specific efforts are made to remove nonadherent cells, medium is pipetted or aspirated from the dish without vigorous swirling or rinsing. It is hypothesized that these "contaminating" cells provide cytokines and other factors needed for the growth of the attached cells during the primary culture.

Ourselves and others have demonstrated that supplementation of the culture medium with additional FGF-2 enhances proliferation and, more importantly, the chondrogenic potential of hMSCs (74–77). In the first (and subsequent) medium changes, the cultures receive FGF-supplemented culture medium (see Note 5 and Fig. 6). Human MSCs must be subcultured before they reach confluence. Typically, they are passaged when they reach 80–90% confluence. In primary cultures, the density within the colonies rather than the level of confluence is the primary consideration to determine when the cells should be subcultured (see Fig. 7). Primary cultures are usually subcultured around day 14 of culture (±3 days).

1. During primary MSC culture, make complete medium changes twice per week. Aspirate spent medium from plates

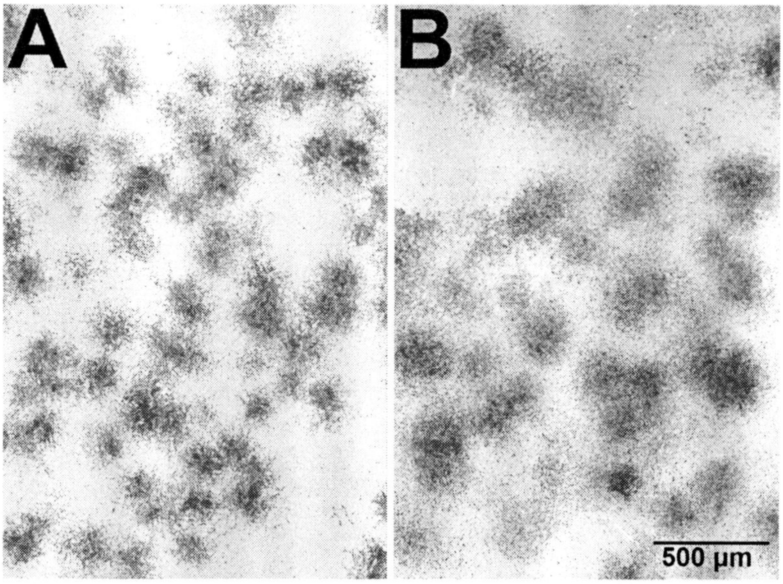

Fig. 6. Appearance of hMSC colonies in primary culture. (**a**) Control culture. (**b**) FGF-treated culture; note that the colonies are looser and there is a higher number of cells between the colonies in the FGF-treated cultures. Crystal black staining.

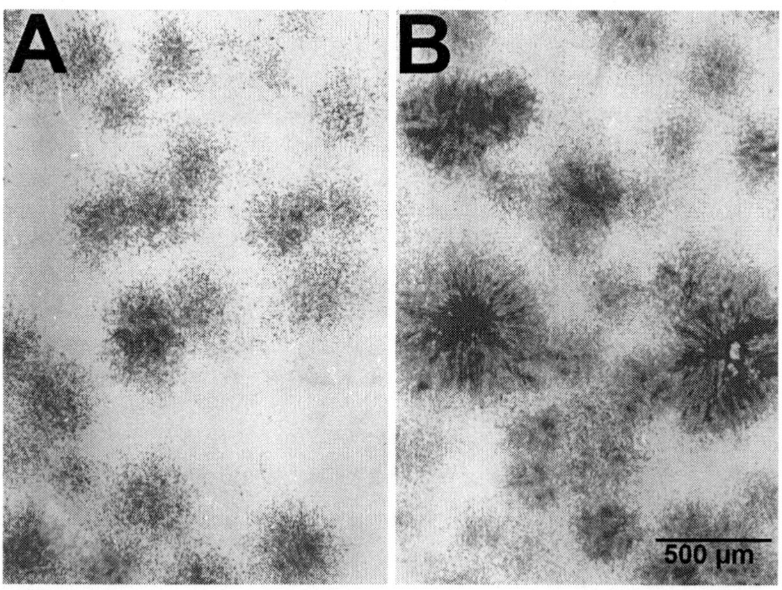

Fig. 7. Appearance of hMSC colonies in primary culture. (**a**) Colonies are not yet ready to be subcultured. (**b**) Colonies must be subcultured; note the higher cell density in the center of the colonies. Crystal black staining.

and add fresh FGF-2-supplemented complete medium. Repeat this step until cultures reach 80–90% confluence.

2. Rinse the cell layer with an appropriate volume (5 ml for a 56 cm^2 tissue culture dish) of Tyrode's salt solution to wash cells.

3. Repeat the rinse.
4. Add the appropriate volume of trypsin-EDTA (4 ml for a 56-cm^2 tissue culture dish).
5. Return cultures to the incubator for 5–7 min. Keep the time of exposure as brief as possible (see Note 6).
6. When the majority of the cells have become well-rounded or have detached from the tissue culture surface, stop the reaction by adding BCS equal to half the volume of trypsin-EDTA.
7. Draw up the cell suspension with a pipette and, with the same pipette, use the suspension to gently wash the remaining cells from the dish.
8. Transfer the cell suspension from all of the cultures to an appropriate size centrifuge tube(s).
9. Spin tubes in a benchtop centrifuge at $500 \times g$ for 5 min.
10. Resuspend the cell pellet in 5–10 ml of FGF-2-supplemented complete medium.
11. Determine cell number with a hemacytometer.
12. Plate cells at $3.5–4.0 \times 10^3$ cells/cm^2.

Further subculturing is conducted according to the above protocol (see Note 7). MSCs have a high proliferative capacity and may be subcultured multiple times. During the process of cell expansion, MSCs maintained in complete medium remain in an undifferentiated state (17). Subcultured MSCs remain spindle-shaped fibroblastic cells, distributed evenly over the culture dish.

3.3. Human MSC Cryopreservation and Recovery

At this point, expanded MSCs can either be subcultured or cryopreserved for later use. This decision depends on the investigation underway, and given donor-to-donor variability of these preparations, it may be advisable to bank an aliquot of the cells as a matter of course.

3.3.1. Cryopreservation

1. Follow steps 1–11 of Subheading 3.2.
2. Label a sufficient quantity of appropriately-sized freezing vials using a LN_2 compatible pen.
3. Spin cells in a benchtop centrifuge at $500 \times g$ for 5 min.
4. Resuspend the cells at 1.0×10^6 cells/ml in freezing medium and immediately place on ice (see Note 8).
5. Aliquot cells into cryogenic vials.
6. Cap vials and place in an isopropanol freezing container.
7. Transfer the freezing container to –80°C freezer overnight.
8. Transfer frozen vials to a LN_2 storage container for long-term storage.

3.3.2. Recovery of Cryopreserved Human MSCs

1. Remove vial from LN_2 storage container.
2. Thaw vial rapidly in a 37°C water bath.
3. Transfer the cell suspension to a 15-ml conical tube containing approximately 10 ml of complete medium prewarmed at 37°C.
4. Centrifuge for 5 min at 500×*g*.
5. Remove most of the freezing medium (so as not to aspirate any of the sedimented cells) and resuspend the cell pellet in prewarmed (37°C) complete medium.
6. Plate cells at 1.0×10^6 cells per T175 flask. Do not handle the flask for at least 24 h.

3.4. Chondrogenic Induction

Bone marrow-derived mesenchymal stem cells can be induced to differentiate into chondrocytes under specific culture conditions. These conditions include three-dimensional conformation of the cells in aggregates where high cell density and cell–cell interaction play an important role in the mechanism of chondrogenesis. Together with these physical culture conditions, a defined culture medium containing TGF-β1 is required to achieve chondrogenic differentiation (18, 19, 78). The culture conditions described below were initially developed for a small-scale chondrogenic differentiation assay, although we have also found that they work well for larger-scale bioreactor-based tissue engineering. Traditionally, these assays were performed in 15-ml polypropylene centrifuge tubes, but we have developed an improved method for preparing cell aggregates for in vitro chondrogenesis studies (79). This method replaces the original 15-ml polypropylene tubes with 96-well plates (see Fig. 8). These modifications allow

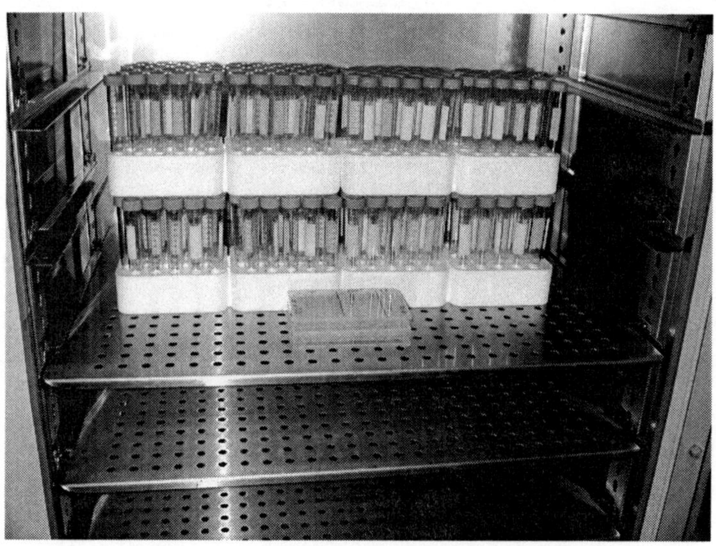

Fig. 8. Comparison of plasticware and incubator space needed for 200 aggregates in 15-ml centrifuge tubes or in 96-well polypropylene plates.

a high-throughput approach to chondrogenic cultures, which reduces both the cost and time with no detrimental effects on the histological and histochemical qualities of the aggregates.

1. Centrifuge harvested cells (see Subheading 3.2 steps 1–11). in a benchtop centrifuge at 500 × *g* for 5 min.
2. Resuspend cells at a density of 1.25×10^6 cells/ml in chondrogenic differentiation medium.
3. Using a repeater pipette, dispense 0.2-ml aliquots of the cell suspension (2.5×10^5 cells/well) into polypropylene 96-well plates.
4. Spin aliquots in a benchtop centrifuge at 500 × *g* for 5 min.
5. Place the multiwell plates in the incubator at 37°C in a humidified atmosphere of 95% air and 5% CO_2 for up to 3 weeks.
6. Twenty four hours after centrifugation, ensure that the aggregates can float freely by releasing them from the bottom of the wells by aspirating 100 μl of media and gently releasing it back into the wells using an eight-channel pipette.
7. Change chondrogenic medium every other day. Carefully aspirate expired medium using a sterile 200-μl pipette tip affixed to a vacuum system.
8. Aliquot 0.2 ml of fresh chondrogenic medium to each well.

3.5. Analysis of Chondrogenic Differentiation

Under the described culture conditions, human MSCs undergo chondrogenic differentiation within 2–3 weeks, producing abundant extracellular matrix composed primarily of cartilage-specific molecules such as type II collagen and aggrecan. The expression of these cartilage markers can be used as evidence of the chondrogenic differentiation of MSCs.

3.5.1. Qualitative Assessment of Chondrogenic Differentiation

The initial cell aggregates contain type I collagen and no cartilage-specific molecules. Type II collagen is typically detected in the matrix by day 5. By day 14, type II and type X collagen are detected throughout the cell aggregates, except for an outer layer of flattened cells that remains rich in type I collagen. Aggrecan and link protein can also be detected in extracts of the aggregates (18, 19).

Chondrogenic differentiation can be assessed by toluidine blue staining of pellet sections, whereby cartilaginous extracellular matrix will stain purple (metachromasia) while undifferentiated or fibrous tissue will stain blue (see Fig. 9), or through the immunohistochemical staining for various types of Collagen (see Fig. 10). Typically, chondrogenic cultures are harvested at 1-week intervals and, in most cases, the experiments are terminated after 3 weeks. Frequency and timing of aggregate harvest are otherwise determined by the experimental objectives.

Fig. 9. MSC aggregate after 3 weeks in chondrogenic conditions. Toluidine blue staining.

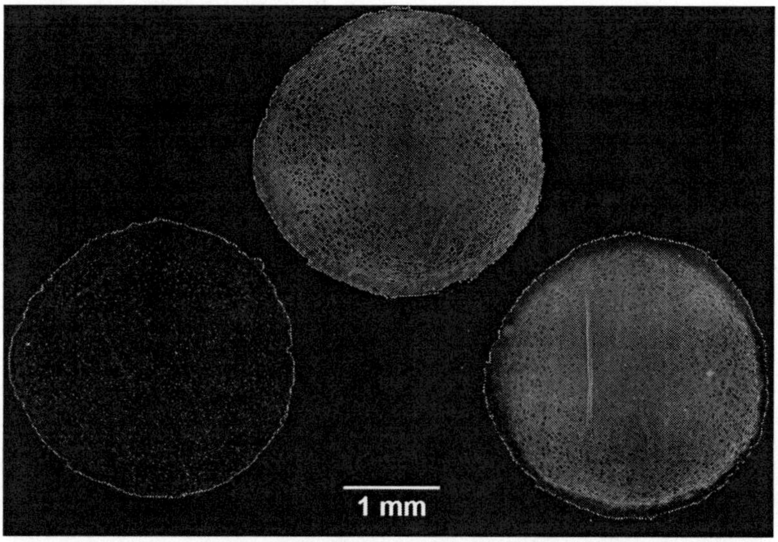

Fig. 10. Immunohistochemical staining of serial sections of a 3-week aggregate. Type I collagen (*left*), type II collagen (*middle*), and type X collagen (*right*). Note the absence of type I collagen staining throughout the aggregate and the lack of type X staining in the periphery of the aggregate. The contours of the aggregates have been outlined to help identify areas of the aggregates devoid of specific types of collagen.

1. Harvest aggregates at predetermined time points.
2. Use a large orifice 200-µl pipette (Fisher) tip affixed to a micropipette to harvest the aggregates.
3. Transfer the aggregates to histology cassettes lined with biopsy sponges (three to four aggregates per cassette).
4. Fix the aggregates in 10% neutral-buffered formalin for 24 h (see Note 9).
5. Transfer the aggregates to 70% Ethanol in H_2O.
6. Dehydrate in graded ethanol (Fisher) series (25, 50, 75, 90, 95 and 100%, 3 min each) followed by three clarification steps in xylene (Fisher) and paraffin-embed (Surgipath, Richmond, IL) following routine histological procedures.
7. Obtain adjacent 7 µm sections using a microtome, e.g., a Leica model 2250 (Deerfield, IL) or equivalent, and mount on StatLab HistoBond slides (McKinney, TX).
8. Deparaffinize the sections in three changes of xylene.
9. Rehydrate the sections using decreasing alcohol series (100, 100, 95, 95, 70, 50, and 25%, 3 min each), followed by a final rinse with deionized water for 5 min.
10. Stain with toluidine blue reagent for 1–2 min.
11. Rinse with tap water.
12. Dehydrate in graded ethanol series (25, 50, 75, 90, 95 and 100%, 3 min each) followed by three clarification steps in xylene and mount the slide in Permount® (Fisher).
13. Visualize using a microscope.
14. Additional chondrogenic differentiation sections can be further assessed by immunohistochemistry (see Fig. 10).
15. For many antibodies, epitope unmasking should be performed by digesting the rehydrated sections with pronase for 15 min.
16. Wash with PBS, 3 times for 5 min each.
17. Block with 10% normal serum (ideally from the species in which the secondary antibody was raised, usually goat) in PBS for 30 min.
18. Incubate with primary antibodies or control IgG in 1% normal serum in PBS for 1 h.
19. Wash with PBS, 3 times for 5 min each.
20. Incubate with secondary antibody diluted 1:50 (depending on brand, follow manufacturer's instructions) in 1% normal serum in PBS for 45–60 min.
21. Wash with PBS, 3 times for 5 min each.

22. If desired, counterstain with 4′, 6-diamidino-2-phenylindole (DAPI, Sigma). Prepare fresh from 5 mg/ml stock by diluting 1:10,000 in water.
23. Wash with PBS, 3 times for 5 min each.
24. Apply a coverslip to the slide with a small drop of 5% *N*-propyl gallate (Sigma) in glycerol (Sigma).
25. Photograph the wet-mounts immediately (see Note 9).

3.5.2. Quantitative Assessment of Chondrogenic Differentiation

As described above, chondrogenic differentiation of hMSCs can be assessed by histology and immunohistochemistry. This type of assessment is, in most cases, very informative, but it is not quantitative. We have adapted previously published methodologies to quantitatively assess the success of the chondrogenic differentiation. To this end, we can determine cell numbers within the aggregates by measuring DNA content (80) and the extent of extracellular matrix accumulation by measuring the glycosaminoglycan (GAG) content (81) of the aggregates. Our modification (74) of these published assays allows us to measure both DNA and GAG content simultaneously in a single aggregate.

1. Harvest aggregates at predetermined time points.
2. Use a large orifice 200-μl pipette tip affixed to a micropipette to harvest the aggregates.
3. Transfer the aggregates to 1.5-ml microcentrifuge tubes (one aggregate per tube).
4. Carefully aspirate excess medium that may have been transferred with the aggregate.
5. Store harvested aggregates frozen at –80°C until ready for analysis.
6. Add 200-μl of papain buffer to each microcentrifuge tube.
7. Place the tubes in a 65°C water bath.
8. Check the aggregates every 30 min until the aggregates are no longer macroscopically visible.
9. Vortex briefly.
10. Add 400 μl of 0.1 N NaOH to each tube.
11. Incubate at room temperature for 20 min.
12. Neutralize with 400 μl of Neutralizing Solution.
13. Centrifuge for 5 min in a microcentrifuge.

Transfer supernatant to a clean tube. This lysate is used for both DNA and GAG measurements.

For DNA measurements, a fluorometric dye-binding assay such as that using Hoechst 33258, a bisbenzimide DNA intercalator, is convenient:

1. Prepare a standard curve with the certified Calf Thymus DNA standard (see Note 10).
   (a) Prepare a linear series of standards (*e.g.*, 0, 500, 1000, 1,500, and 2,000 ng/ml).
   (b) The standards should be prepared fresh at the time of the assay (see Table 1).
2. Transfer 100 μl of each standard to four replicate wells of a black, flat-bottom 96-well plate.
3. Transfer 100 μl of test lysate from each tube to four replicate wells of the 96-well plate.
4. Add 100 μl of Hoechst dye working solution (0.7 μg/ml Hoechst 33258 in water) to each well, mix gently.
5. Read the plate at an excitation wavelength of 340 nm and an emission wavelength of 465 nm.

For GAG measurements, we use a modified Safranin-O dye-binding assay:

1. Prepare a standard curve with chondroitin sulfate C (shark cartilage) standards (see Note 12).
   (a) Again, a linear series of dilutions is recommended (i.e., 0, 1, 2, 3, 4, 5, and 6 μg in water).
   (b) The standards should be prepared fresh (see Table 2).
2. Cut a piece of 0.45 μm nitrocellulose large enough to cover the necessary number of wells.
3. Moisten the nitrocellulose in distilled H_2O.
4. Assemble the dot-blot apparatus following the manufacturer's instructions and connect to a vacuum source (see Fig. 11).

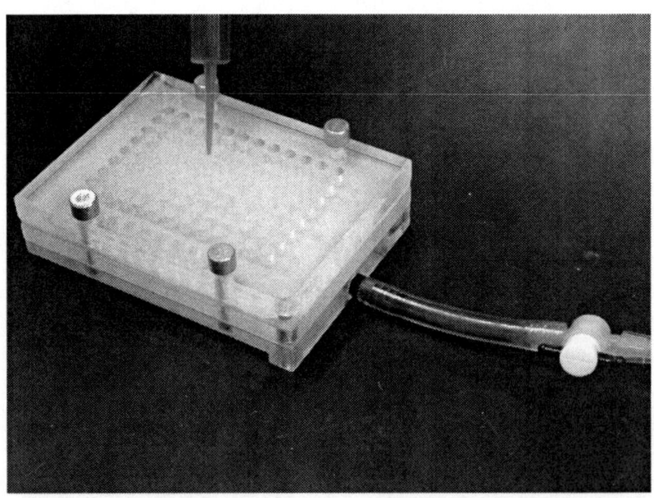

Fig. 11. Assembled dot-blot apparatus.

5. Add 250 μl of Safranin O reagent to the wells of the dot-blot apparatus (see Note 11).
6. Add samples (25 μl) to Safranin O reagent already in the wells.
7. Let stand about 1 min to allow precipitation.
8. Cover unused wells.
9. Turn on vacuum to collect precipitates.
10. Rinse wells 2–3 times by filling with H_2O and allowing it to filter through the membrane by applying vacuum.
11. Turn vacuum off.
12. Disassemble the dot-blot apparatus and remove the nitrocellulose membrane.
13. Air-dry the membrane briefly, do not over-dry as static charge makes the dots difficult to handle.
14. Punch out dots from nitrocellulose with a hole punch (see Fig. 12).
15. Transfer the dots to 1.5-ml microcentrifuge tubes.
16. Add 1 ml of 10% cetylpyridinium chloride (CPC) in H_2O to elute the dye from the nitrocellulose dots.
17. Incubate at 37°C for 20 min; vortex after 10 min.
18. Read absorbance at 536 nm (see Note 13).

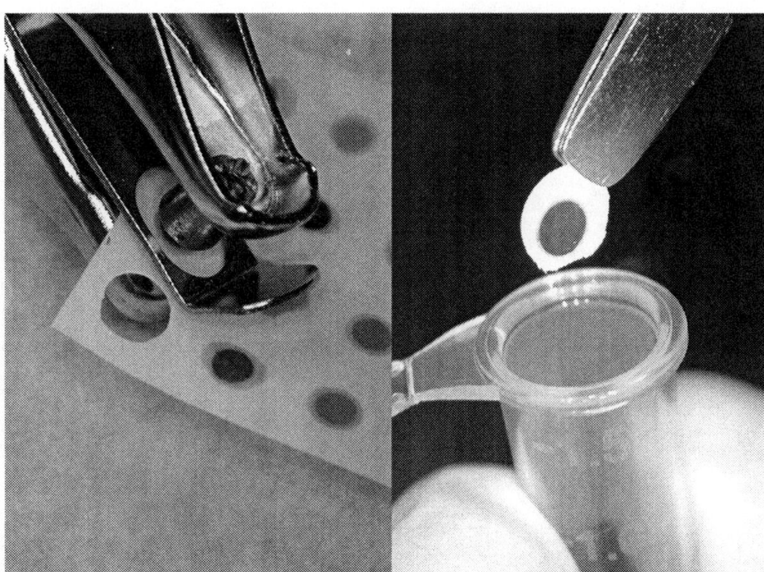

Fig. 12. Dots being punched from the membrane and transferred to the microcentrifuge tubes.

3.5.3. RNA Isolation

If desired, the chondrogenic process can be assessed at the gene expression level by harvesting total RNA from the aggregates. We found that the abundant negatively charged extracellular matrix in the pellets makes it almost impossible to obtain good RNA yields using affinity column-based kits and have obtained better results using TRIzol.

1. Harvest aggregates as described above.
2. Pool 12 aggregates into a 15-ml polypropylene conical tube.
3. Add 1.5 ml of TRIzol reagent.
4. Quick-freeze in liquid nitrogen or in a dry ice/ethanol bath.
5. Store at –80°C (at least overnight, can keep longer).
6. Thaw samples on ice.
7. Homogenize the aggregates (see Note 14 and Fig. 13).
8. Incubate for 5 min at room temperature.
9. Store at –80°C (at least overnight, can keep longer).
10. Thaw samples in ice.
11. Add 500 μl of chloroform.
12. Incubate for 3–5 min at room temperature.
13. Centrifuge for 15 min at 4°C (12,000 × *g*).
14. Aspirate the aqueous (top) layer and transfer it to a clean tube.
15. Add 1,250 μl of isopropanol.

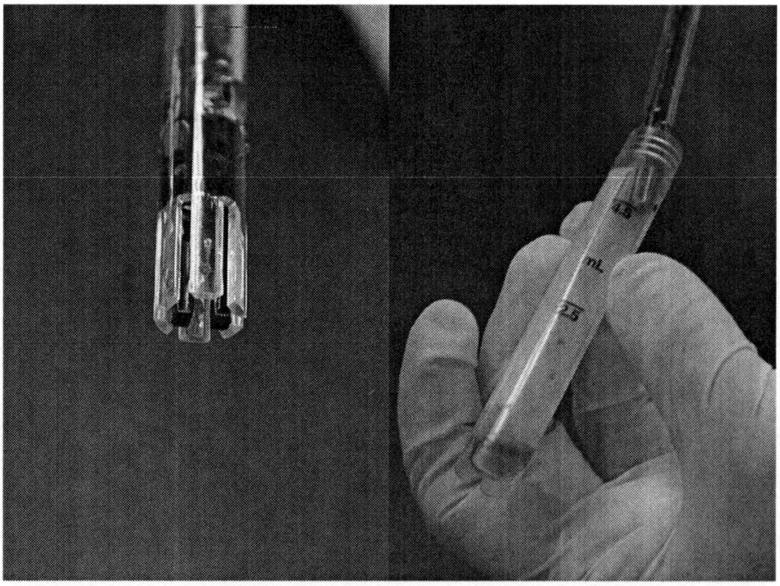

Fig. 13. Homogenizer tip and 4-ml cryovial used for the RNA isolation.

16. Mix well and incubate at room temperature for 10 min.
17. Freeze at –80°C (at least overnight, can keep longer).
18. Thaw samples.
19. Centrifuge for 10 min at 4°C (12,000 × *g*).
20. Carefully discard supernatant.
21. Add 1.5 ml of 70% ethanol.
22. Vortex.
23. Centrifuge for 5 min at 4°C (7,500 × *g*).
24. Add 1.5 ml of 70% ethanol.
25. Vortex.
26. Centrifuge for 5 min at 4°C (7,500 × *g*).
27. Air-dry the pellet.
28. Resuspend in nuclease-free (or DEPC-treated) water.

4. Notes

1. The Percoll solution can be prepared in any convenient multiples of these volumes and should be mixed thoroughly.
2. All solutions and plasticware which will come in contact with cells should be sterile. Waste should be treated as biohazardous, and universal precautions should be applied.
3. The bone marrow sample, and all cells derived from it, is treated with universal biohazard precautions. Appropriate waste containers for all sharp and nonsharp disposable supplies that come into contact with human tissue, cells, or medium that has contacted these cells must be used.
4. The marrow sample is processed in a Class II biological safety cabinet. Personnel processing these samples must wear proper personal protective equipment, including a lab coat, goggles or a face shield, gloves, and a surgical mask.
5. FGF-2 acts as a scattering factor on hMSCs. Primary cultures exposed to FGF-2 present colonies less tightly packed that control cultures with many more cells dispersed in between the colonies (see Fig. 6).
6. When subculturing hMSCs, the following considerations must be taken into account: Subcultured hMSCs are evenly distributed on the tissue culture vessel surface and are not in colonies as for primary cultures. Therefore, the degree of confluence determines when the cells should be trypsinized. As a general rule, hMSCs should be trypsinized before they become confluent. Passaged hMSCs are more easily trypsinized

than primary cultures, so exposure of these cells to trypsin is usually limited to 5 min. It is not necessary (nor desirable) to remove all of the adherent cells from the dish as most of the nonfibroblastoid cells (which are not likely to be MSCs) in these cultures are more trypsin-resistant than the spindle-shaped fibroblast-like cells. Thus, trypsinization represents, after attachment of fibroblastic cells to plastic, the second component of the process of selection of MSCs from the total marrow cell population. Passaged hMSCs are slightly larger than primary cells, a feature that becomes more pronounced with additional subcultivation.

7. Gibco freezing medium works well or 15% HPLC-grade DMSO with 85% FBS. DMSO dilution is exothermic, so if making your own freezing medium, take this into account and allow freshly made freezing medium to chill before adding it to the cells.
8. Complete fixation of the aggregates is important. Allow at least 24 h. Wet-mount immunofluorescent sections should be documented photographically as soon as possible, preferably the same day. Stained sections can be retained for a few days at 4°C. Stability can be improved by ringing the coverslips with collodion or nail polish.
9. Fluorescently labeled antibodies fade overtime and diffuse away, therefore it is important to document the sections as soon as possible (ideally within 24 h).
10. A new batch of calf thymus DNA should be checked against the old one prior to use as there appears to be lot-to-lot and vendor-to-vendor variations. The calculation of DNA content in the lysates of the aggregates can be easily performed in Excel (Microsoft). Absorbance data are converted to DNA concentration using the following spreadsheet formula: =((FORECAST(C1,$A$1:$A$4,$B$1:$B$4))*2)/1,000; where C1 is the relative reference to the cell containing the absorbance value for a given replicate of a given extract, $A$1:$A$4 is the absolute reference to the range containing the DNA concentration of the standards, and $B$1:$B$4 is the absolute reference to the range containing the mean absorbance for the replicates of each standards. The "2" multiplier is to account for the dilution of the extract with the Dye solution and the "1,000" denominator is to convert ng/ml to μg/ml. Because the final volume of the lysate is 1 ml, the concentrations can be directly converted to total DNA amounts.
11. Because the reagent slowly flows through the nitrocellulose, wells should be filled only five at a time to avoid having too much reagent flow out of the wells by the time the sample is added.

12. The calculations of GAG content in the lysates of the aggregates can be easily performed in excel (Microsoft). Absorbance data are converted to GAG amounts using the following spreadsheet formula: =FORECAST(C1,$A$1:$A$5,$B$1:$B$5)* 40; where C1 is the relative reference to the cell containing the absorbance value for a given replicate of a given extract, $A$1:$A$5 is the absolute reference to the range containing the GAG content of the standards, and $B$1:$B$5 is the absolute reference to the range containing the mean absorbance for the replicates of each standards. The "40" multiplier is to account for the total volume of the extract; remember that only 25 μl of the 1 ml of extract are loaded into the dot-blot apparatus.
13. If using a 96-well plate reader, transfer the eluates to replicate wells of a 96-well plate for analysis; otherwise use disposable plastic cuvettes.
14. Aggregates, particularly those at later time points, do not dissolve in TRIzol by themselves. We use a handheld Omni TH homogenizer (OMNI international, Marietta, GA) with hard tissue-homogenizer tips. The homogenization is performed in the 4-ml tubes to avoid splashing and loss of sample, and it is performed until remnants of the aggregates cannot be visually identified (see Fig. 11).

Acknowledgments

This work was supported by grants from the Arthritis Foundation (PIs Jean F. Welter and Luis A. Solchaga), the Ohio Department of Development (PI Luis A. Solchaga), NIH; R01 AR050208 (PI Jean F. Welter) and P01 AR053622 (PIs Jean F. Welter and Luis A. Solchaga); and the Hematopoietic Stem Cell Core Facility of the Case Comprehensive Cancer Center (P30 CA43703; PI Stanton L. Gerson). The authors also want to thank Ms. Harris who processes the majority of the bone marrow specimens used in the laboratory.

References

1. Minguell JJ, Erices A, Conget P. Mesenchymal stem cells. Exp Biol Med (Maywood). 2001;226(6):507–20.
2. Friedenstein AJ. Precursor cells of mechanocytes. Int Rev Cytol. 1976;47:327–59.
3. Owen M. Marrow stromal stem cells. J Cell Sci Suppl. 1988;10:63–76.
4. Owen M, Friedenstein AJ. Stromal stem cells: marrow-derived osteogenic precursors. Ciba Found Symp. 1988;136:42–60.
5. Caplan AI. The mesengenic process. Clin Plast Surg. 1994;21(3):429–35.
6. Ohgushi H, Goldberg VM, Caplan AI. Heterotopic osteogenesis in porous ceramics

induced by marrow cells. J Orthop Res. 1989;7(4):568–78.

7. Ohgushi H, Okumura M. Osteogenic capacity of rat and human marrow cells in porous ceramics. Experiments in athymic (nude) mice. Acta Orthop Scand. 1990;61(5):431–4.
8. Ohgushi H, Okumura M, Tamai S, Shors EC, Caplan AI. Marrow cell induced osteogenesis in porous hydroxyapatite and tricalcium phosphate: a comparative histomorphometric study of ectopic bone formation. J Biomed Mater Res. 1990;24(12):1563–70.
9. Goshima J, Goldberg VM, Caplan AI. The osteogenic potential of culture-expanded rat marrow mesenchymal cells assayed in vivo in calcium phosphate ceramic blocks. Clin Orthop Relat Res. 1991(262):298–311.
10. Goshima J, Goldberg VM, Caplan AI. Osteogenic potential of culture-expanded rat marrow cells as assayed in vivo with porous calcium phosphate ceramic. Biomaterials. 1991;12(2):253–8.
11. Haynesworth SE, Goshima J, Goldberg VM, Caplan AI. Characterization of cells with osteogenic potential from human bone marrow. Bone. 1992;13:81–8.
12. Lennon DP, Haynesworth SE, Bruder SP, Jaiswal N, Caplan AI. Human and animal mesenchymal progenitor cells from bone marrow: Identification of serum for optimal selection and proliferation. In Vitro Cell Dev Biol. 1996;32(10):602–11.
13. Bruder SP, Jaiswal N, Haynesworth SE. Growth kinetics, self-renewal, and the osteogenic potential of purified human mesenchymal stem cells during extensive subcultivation and following cryopreservation. J Cell Biochem. 1997;64(2):278–94.
14. Digirolamo CM, Stokes D, Colter D, Phinney DG, Class R, Prockop DJ. Propagation and senescence of human marrow stromal cells in culture: a simple colony-forming assay identifies samples with the greatest potential to propagate and differentiate. Br J Haematol. 1999;107(2):275–81.
15. Phinney DG, Kopen G, Righter W, Webster S, Tremain N, Prockop DJ. Donor variation in the growth properties and osteogenic potential of human marrow stromal cells. J Cell Biochem. 1999;75(3):424–36.
16. Pittenger MF, Mackay AM, Beck SC, Jaiswal RK, Douglas R, Mosca JD, et al. Multilineage potential of adult human mesenchymal stem cells. Science. 1999;284(5411):143–7.
17. Jaiswal N, Haynesworth SE, Caplan AI, Bruder SP. Osteogenic differentiation of purified culture-expanded human mesenchymal stem cells in vitro. J Cell Biochem. 1997;64:295–312.
18. Johnstone B, Hering TM, Caplan AI, Goldberg VM, Yoo JU. In vitro chondrogenesis of bone marrow-derived mesenchymal progenitor cells. Exp Cell Res. 1998;238(1):265–72.
19. Yoo JU, Barthel TS, Nishimura K, Solchaga L, Caplan AI, Goldberg VM, et al. The chondrogenic potential of human bone-marrow-derived mesenchymal progenitor cells. J Bone Joint Surg Am. 1998;80(12):1745–57.
20. Lee SH, Lumelsky N, Studer L, Auerbach JM, McKay RD. Efficient generation of midbrain and hindbrain neurons from mouse embryonic stem cells. Nat Biotechnol. 2000;18(6):675–9.
21. Mackay AM, Beck SC, Murphy JM, Barry FP, Chichester CO, Pittenger MF. Chondrogenic differentiation of cultured human mesenchymal stem cells from marrow. Tissue Eng. 1998;4(4):415–28.
22. Pittenger MF, Mbalaviele G, Black M, Mosca JD, Marshak DR. Mesenchymal stem cells. In: Koller MR, Palsson BO, Masters JRW, editors. Primary mesenchymal cells. Dordrech: Kluwer Academic Publishers; 2001. p. 189–207.
23. Banfi A, Bianchi G, Notaro R, Luzzatto L, Cancedda R, Quarto R. Replicative aging and gene expression in long-term cultures of human bone marrow stromal cells. Tissue Eng. 2002;8(6):901–10.
24. Langer R, Vacanti JP. Tissue engineering. Science. 1993;260(5110):920–6.
25. Vacanti CA, Vacanti JP. The science of tissue engineering. Orthop Clin North Am. 2000;31(3):351–6.
26. Vacanti J, Langer R. Tissue engineering: the design and fabrication of living replacement devices for surgical re-construction and transplantation. Lancet. 1999;354(Suppl. 1):SI32–SI4.
27. Bonassar LJ, Vacanti CA. Tissue engineering: the first decade and beyond. J Cell Biochem Suppl. 1998;30–31:297–303.
28. Pearson RG, Bhandari R, Quirk RA, Shakesheff KM. Recent advances in tissue engineering: an invited review. J Long Term Eff Med Implants. 2002;12(1):1–33.
29. Temenoff JS, Mikos AG. Review: tissue engineering for regeneration of articular cartilage. Biomaterials. 2000;21(5):431–40.
30. O'Driscoll SW. Preclinical cartilage repair: current status and future perspectives. Clin Orthop. 2001(391 Suppl):S397–401.

31. Luyten FP, Dell'Accio F, De Bari C. Skeletal tissue engineering: opportunities and challenges. Best Pract Res Clin Rheumatol. 2001;15(5):759–69.
32. Musgrave DS, Fu FH, Huard J. Gene therapy and tissue engineering in orthopaedic surgery. J Am Acad Orthop Surg. 2002;10(1):6–15.
33. Risbud M. Tissue engineering: implications in the treatment of organ and tissue defects. Biogerontology. 2001;2(2):117–25.
34. Guilak F, Butler DL, Goldstein SA. Functional tissue engineering: the role of biomechanics in articular cartilage repair. Clin Orthop. 2001(391 Suppl):S295–305.
35. Risbud MV, Sittinger M. Tissue engineering: advances in in vitro cartilage generation. Trends Biotechnol. 2002;20(8):351–6.
36. Caplan AI, Goldberg VM. Principles of tissue engineered regeneration of skeletal tissues. Clin Orthop Relat Res. 1999(367 Suppl): S12–6.
37. Caplan AI, Elyaderani M, Mochizuki Y, Wakitani S, Goldberg VM. Principles of cartilage repair and regeneration. Clin Orthop Relat Res. 1997;342:254–69.
38. Hunter W. Of the structure and diseases of articulating cartilages. Philos Trans. 1743;42: 514–21.
39. Buckwalter J. Articular cartilage: injuries and potential for healing. J Orthop Sports Phys Ther. 1998;28:192–202.
40. Buckwalter JA, Mankin HJ. Articular cartilage repair and transplantation. Arthritis Rheum. 1998;41(8):1331–42.
41. Buckwalter JA, Mankin HJ. Articular cartilage: degeneration and osteoarthritis, repair, regeneration, and transplantation. Instr Course Lect. 1998;47:487–504.
42. Mankin HJ. The reaction of articular cartilage to injury and osteoarthritis (first of two parts). N Engl J Med. 1974;291(24):1285–92.
43. Mankin HJ. The reaction of articular cartilage to injury and osteoarthritis (second of two parts). N Engl J Med. 1974;291(25): 1335–40.
44. Mankin HJ, Buckwalter JA. Restoration of the osteoarthrotic joint [editorial]. J Bone Joint Surg Am. 1996;78(1):1–2.
45. Chesterman PJ, Smith AU. Homotransplantation of articular cartilage and isolated chondrocytes. An experimental study in rabbits. J Bone Joint Surg Br. 1968;50(1): 184–97.
46. Bentley G, Greer RB 3rd. Homotransplantation of isolated epiphyseal and articular cartilage chondrocytes into joint surfaces of rabbits. Nature. 1971;230(5293):385–8.
47. Wakitani S, Kimura T, Hirooka A, Ochi T, Yoneda M, Yasui N, et al. Repair of rabbit articular surfaces with allograft chondrocytes embedded in collagen gel. J Bone Joint Surg Br. 1989;71(1):74–80.
48. Vacanti CA, Kim W, Schloo B, Upton J, Vacanti JP. Joint resurfacing with cartilage grown in situ from cell-polymer structures. Am J Sports Med. 1994;22(4):485–8.
49. Freed LE, Grande DA, Lingbin Z, Emmanual J, Marquis JC, Langer R. Joint resurfacing using allograft chondrocytes and synthetic biodegradable polymer scaffolds. J Biomed Mater Res. 1994;28(8):891–9.
50. Frenkel SR, Toolan B, Menche D, Pitman MI, Pachence JM. Chondrocyte transplantation using a collagen bilayer matrix for cartilage repair. J Bone Joint Surg Br. 1997; 79(5):831–6.
51. Nehrer S, Breinan HA, Ramappa A, Shortkroff S, Young G, Minas T, et al. Canine chondrocytes seeded in type I and type II collagen implants investigated in vitro. J Biomed Mater Res. 1997;38(2):95–104.
52. Breinan HA, Minas T, Hsu HP, Nehrer S, Sledge CB, Spector M. Effect of cultured autologous chondrocytes on repair of chondral defects in a canine model. J Bone Joint Surg Am. 1997;79(10):1439–51.
53. Minas T. Chondrocyte implantation in the repair of chondral lesions of the knee: economics and quality of life. Am J Orthop. 1998;27(11):739–44.
54. Giannini S, Buda R, Grigolo B, Vannini F. Autologous chondrocyte transplantation in osteochondral lesions of the ankle joint. Foot Ankle Int. 2001;22(6):513–7.
55. Breinan HA, Minas T, Hsu HP, Nehrer S, Shortkroff S, Spector M. Autologous chondrocyte implantation in a canine model: change in composition of reparative tissue with time. J Orthop Res. 2001;19(3): 482–92.
56. Brittberg M, Lindahl A, Nilsson A, Ohlsson C, Isaksson O, Peterson L. Treatment of deep cartilage defects in the knee with autologous chondrocyte transplantation [see comments]. N Engl J Med. 1994;331(14):889–95.
57. Amiel D, Coutts RD, Abel M, Stewart W, Harwood F, Akeson WH. Rib perichondrial grafts for the repair of full-thickness articular-cartilage defects. A morphological and biochemical study in rabbits. J Bone Joint Surg Am. 1985;67(6):911–20.
58. O'Driscoll SW, Keeley FW, Salter RB. The chondrogenic potential of free autogenous periosteal grafts for biological resurfacing of major full-thickness defects in joint surfaces

under the influence of continuous passive motion. An experimental investigation in the rabbit. J Bone Joint Surg Am. 1986;68(7): 1017–35.

59. Amiel D, Coutts RD, Harwood FL, Ishizue KK, Kleiner JB. The chondrogenesis of rib perichondrial grafts for repair of full thickness articular cartilage defects in a rabbit model: a one year postoperative assessment. Connect Tissue Res. 1988;18(1):27–39.
60. Shahgaldi BF, Amis AA, Heatley FW, McDowell J, Bentley G. Repair of cartilage lesions using biological implants. A comparative histological and biomechanical study in goats. Journal of Bone & Joint Surgery - British Volume. 1991;73(1):57-64.
61. Wakitani S, Goto T, Pineda SJ, Young RG, Mansour JM, Caplan AI, et al. Mesenchymal cell-based repair of large, full-thickness defects of articular cartilage. J Bone Joint Surg Am. 1994;76(4):579–92.
62. Chu CR, Coutts RD, Yoshioka M, Harwood FL, Monosov AZ, Amiel D. Articular cartilage repair using allogeneic perichondrocyte-seeded biodegradable porous polylactic acid (PLA): a tissue-engineering study. J Biomed Mater Res. 1995;29(9):1147–54.
63. Butnariu-Ephrat M, Robinson D, Mendes DG, Halperin N, Nevo Z. Resurfacing of goat articular cartilage by chondrocytes derived from bone marrow. Clin Orthop Relat Res. 1996;330:234–43.
64. Hunziker EB, Rosenberg LC. Repair of partial-thickness defects in articular cartilage: cell recruitment from the synovial membrane. J Bone Joint Surg Am. 1996;78(5):721–33.
65. Chu CR, Dounchis JS, Yoshioka M, Sah RL, Coutts RD, Amiel D. Osteochondral repair using perichondrial cells. A 1-year study in rabbits. Clin Orthop Relat Res. 1997(340): 220–9.
66. Goldberg VM, Caplan AI. Biologic restoration of articular surfaces. Inst Course Lect. 1999;48:623–7.
67. Johnstone B, Yoo JU. Autologous mesenchymal progenitor cells in articular cartilage repair. Clin Orthop Relat Res. 1999(367 Suppl):S156–62.
68. Dounchis JS, Bae WC, Chen AC, Sah RL, Coutts RD, Amiel D. Cartilage repair with autogenic perichondrium cell and polylactic acid grafts. Clin Orthop Relat Res. 2000(377):248–64.
69. Solchaga LA, Gao J, Dennis JE, Awadallah A, Lundberg M, Caplan AI, et al. Treatment of osteochondral defects with autologous bone marrow in a hyaluronan-based delivery vehicle. Tissue Eng. 2002;8(2):333–47.
70. Caplan AI. Mesenchymal stem cells. J Orthop Res. 1991;9(5):641–50.
71. Ashton BA, Allen TD, Howlett CR, Eaglesom CC, Hattori A, Owen M. Formation of bone and cartilage by marrow stromal cells in diffusion chambers in vivo. Clin Orthop. 1980(151):294–307.
72. Angele P, Smith C, Mansour J, Jepsen K, Johnstone B, Yoo J. Effects of cyclic hydrostatic pressure on the chondrogenic differentiation of mesenchymal progenitor cells. Trans Orthop Res Soc. 2000;25(2):645.
73. Hanada K, Solchaga LA, Caplan AI, Hering TM, Goldberg VM, Yoo JU, et al. BMP-2 induction and TGF-beta 1 modulation of rat periosteal cell chondrogenesis. J Cell Biochem. 2001;81(2): 284–94.
74. Solchaga LA, Penick K, Porter JD, Goldberg VM, Caplan AI, Welter JF. FGF-2 enhances the mitotic and chondrogenic potentials of human adult bone marrow-derived mesenchymal stem cells. J Cell Physiol. 2005; 203(2):398–409.
75. Bianchi G, Banfi A, Mastrogiacomo M, Notaro R, Luzzatto L, Cancedda R, et al. Ex vivo enrichment of mesenchymal cell progenitors by fibroblast growth factor 2. Exp Cell Res. 2003;287(1):98–105.
76. Martin I, Muraglia A, Campanile G, Cancedda R, Quarto R. Fibroblast growth factor-2 supports ex vivo expansion and maintenance of osteogenic precursors from human bone marrow. Endocrinology. 1997;138(10):4456–62.
77. Tsutsumi S, Shimazu A, Miyazaki K, Pan H, Koike C, Yoshida E, et al. Retention of multilineage differentiation potential of mesenchymal cells during proliferation in response to FGF. Biochem Biophys Res Commun. 2001;288(2):413–9.
78. Ballock RT, Reddi AH. Thyroxine is the serum factor that regulates morphogenesis of columnar cartilage from isolated chondrocytes in chemically defined medium. J Cell Biol. 1994;126(5):1311–8.
79. Penick KJ, Solchaga LA, Welter JF. High-throughput aggregate culture system to assess the chondrogenic potential of mesenchymal stem cells. Biotechniques. 2005;39(5): 687–91.
80. Ponticiello MS, Schinagl RM, Kadiyala S, Barry FP. Gelatin-based resorbable sponge as a carrier matrix for human mesenchymal stem cells in cartilage regeneration therapy. J Biomed Mater Res. 2000;52(2):246–55.
81. Carrino DA, Arias JL, Caplan AI. A spectrophotometric modification of a sensitive densitometricSafraninOassayforglycosaminoglycans. Biochem Int. 1991;24(3):485–95.

Chapter 21

Use of Human Mesenchymal Stem Cells as Alternative Source of Smooth Muscle Cells in Vessel Engineering

Zhaodi Gong and Laura E. Niklason

Abstract

Adult stem cell-derived smooth muscle cells (SMC) may be a promising source of cells for applications in regenerative medicine, including cardiovascular tissue engineering. Primary SMC from native vessels may have limited proliferative capacity and reduced collagen production when sourced from elderly donors, who are the patients in need of vascular grafts due to coronary disease or peripheral arterial disease. Our recent work showed that the ability of human bone marrow-derived mesenchymal stem cells (hMSCs) to differentiate into SMC was modulated by various growth factors, matrix proteins, and mechanical forces. In addition, the components of the culture medium play a very important role in SMC differentiation from hMSCs. In this chapter, we will summarize our experience with the impact of various factors on SMC differentiation from hMSCs. Based upon our findings regarding growth factors, cyclic strain and matrix proteins, a two-phase vessel regeneration culture protocol including a 4-week proliferation phase and a 4-week differentiation phase was developed to optimize proliferation and SMC differentiation of hMSCs consecutively.

Key words: Mesenchymal stem cells, Smooth muscle cells, Small-diameter vessel, Differentiation

1. Introduction

Bone marrow-derived mesenchymal stem cells (hMSCs) have become an attractive cell source in tissue engineering because they are relatively easy to obtain, autologous in nature, and have the potential to differentiate into diverse cell types, such as adipogenic, osteogenic, chondrogenic, and myogenic lineages. Beyond these studies of differentiation potential of MSCs, the application of bone marrow-derived stem cells in vitro and in vivo for vascular engineering is just emerging (1–8).

We hypothesized that the local environment (9) and the resident cellular population in the intact or regenerating vascular wall

Mohan C. Vemuri et al. (eds.), *Mesenchymal Stem Cell Assays and Applications*, Methods in Molecular Biology, vol. 698, DOI 10.1007/978-1-60761-999-4_21, © Springer Science+Business Media, LLC 2011

should be involved in directing the differentiation of hMSCs toward a smooth muscle cell (SMC) phenotype. Indeed, bone marrow-derived MSCs have been shown to differentiate into SMCs in response to transforming growth factor β (TGF-β) (10), mechanical stress (11), direct contact with vascular endothelial cells (12), and interaction with endothelial cell matrix (13). We have found that growth factors that are elaborated by platelets and vascular cells after vessel injury (PDGF, TGF-β1, and bFGF), extracellular proteins found in native vessel wall, and cyclic mechanical strain all impact SMC differentiation from hMSCs (14). In addition, a novel two-phase vessel engineering protocol was developed and is described here. The novel protocol is based upon our original vascular tissue engineering protocol (15, 16), but is modified in order to promote hMSC proliferation and subsequent SMC differentiation (14) in the regenerating blood vessel.

2. Materials

2.1. Isolation and Cell Culture of hMSCs

1. 25 ml fresh unprocessed human bone marrow (Lonza, Basel, Switzerland).
2. Ficoll-Paque Plus density media (StemCell Technologies, Vancouver, BC, Canada) stored at 4°C.
3. Dulbecco's phosphate buffered saline (D-PBS; Invitrogen, Carlsbad, CA, USA).
4. Basal hMSC culture medium: Dulbecco's Modified Eagle's Medium (DMEM), low-glucose (Invitrogen, stored at 4°C, avoid direct light) containing 10% selected lot of fetal bovine serum (selected lot of FBS; Hyclone, South Logan, UT, USA) and 1% penicillin-streptomycin-glutamate (Invitrogen). The screening of the FBS lot was based on the support of cell proliferation of hMSC.

2.2. Flow Cytometry

1. Fluorescein isothiocyanate (FITC)-conjugated mouse anti-human IgGs: CD14 (Abcam Inc., Cambridge, MA, USA); CD45 (Santa Cruz Biotechnology, Santa Cruz, CA, USA); and CD34 (Miltenyi Biotec Inc., Auburn, CA, USA).
2. SH2 and SH3 hybridoma (American Type Culture Collection, ATCC, Manassas, VA, USA) cultured in ATCC Hybri-Care Medium plus 20% FBS (ATCC). Supernatant was obtained with a series dilution where the optimal dilution provided the brightest fluorescence and highest dilution of the supernatant.
3. Isotype control for each primary antibody: For CD14- and CD45-IgG1-FITC antibody, isotype control was mouse IgG1-FITC (Santa Cruz Biotechnology); For CD34-IgG2a-FITC

antibody, isotype control was purchased from Miltenyi Biotech. For SH2 and SH3, the isotype controls were goat-anti-mouse IgG1 and IgG2b-FITC, respectively (Santa Cruz Biotechnology).

4. Fixative: 4% paraformaldehyde (Boston BioProducts, Boston, MA, USA) stored at 4°C.
5. FACStar flow cytometer (BD Biosciences, Franklin Lakes, NJ, USA).

2.3. Immunofluorescence Staining of SMC Markers

1. Primary antibodies: 1:100 dilution for mouse anti-human monoclonal SMA and calponin, 1:50 dilution for smooth muscle myosin heavy chain (SM-MHC) antibodies (Dako, Copenhagen, Denmark) in dilution buffer.
2. Secondary antibody: 1:2,000 dilution for FITC-conjugated goat anti-mouse IgG (Santa Cruz Biotechnology) in dilution buffer.
3. Vectashield Mounting Medium for Fluorescence with 4′6-diamidine-2-phenylindole (DAPI) kit (Vector Laboratories, Inc., Burlingame, CA, USA).

2.4. Western Blot for SMC Markers

1. Tris-buffered saline (TBS) buffer: Dilute in deionized water from 10× TBS (Boston BioProducts) containing 0.5 M Tris and 1.5 M NaCl (pH 7.4).
2. TBS/Tween-20 buffer (TBST): Dilute from 10× TBST buffer (Boston BioProducts) containing TBS buffer with 0.5% Tween-20.
3. Blocking buffer: 5% nonfat dry milk (Bio-Rad, Hercules, CA, USA) in TBST buffer stored at 4°C.
4. Antibody dilution buffer: Dilute blocking buffer to 1% with TBST buffer
5. Primary antibodies for SMC markers: 1:100 dilution for mouse anti-human SMA and calponin, 1:50 dilution for mouse anti-human smooth muscle myosin heavy chain (SM-MHC) antibodies (Dako, Copenhagen, Denmark) in dilution buffer freshly prepared before Western blot (see Note 1).
6. Mouse anti-human β-actin (Sigma) 1:5,000 dilution before use.
7. Secondary antibody, horseradish peroxidase (HRP)-conjugated goat anti-mouse IgG (Santa Cruz Biotechnology) 1:2,000 diluted in dilution buffer before use.
8. Polyvinylidene difluoride membranes (Millipore, Bedford, MA).
9. SuperSignal West Pico Chemiluminescent detection system (Pierce, Rockford, IL).
10. Restore™ Western Blot Stripping Buffer (Pierce).

11. RIPA buffer with Triton (Boston BioProducts, Worcester, MA) consisting of 50 mM Tris–HCl (pH 7.4), 150 mM NaCl, 1% Tritonx100, 0.5% sodium deoxycholate, and 0.1% SDS.

2.5. Effect of Various Soluble Factors on hMSCs

1. Human coronary artery SMCs (CASMC, Lonza).
2. Smooth muscle growth medium (SmBM) supplemented with SmGM-2 SingleQuots (Lonza).
3. Preparation of growth factors (see Note 2): stock solution for TGFβ1 (R&D Systems, Inc., Minneapolis, MN, USA) was prepared by dissolving 2 μg of TGFβ1 in 400 μl dilution buffer (1% BSA in PBS, store at 4°C), resulting in a final concentration of 5 μg/ml. Aliquot 100 μl/tube and store at −70°C. PDGF-BB stock solution (10 μg/ml) was made by dissolving 50 μg PDGF-BB (R&D Systems, Inc.) in 5 ml dilution buffer. Aliquot 500 μl/tube and store at −70°C. PDGF-CC stock solution (50 μg/ml) was made by dissolving 25 μg PDGF-CC (R&D Systems, Inc.) in 500 μl dilution buffer. Aliquot 100 μl/tube and store at −70°C. bFGF stock solution (50 μg/ml) was made by dissolving 25 μg bFGF (R&D Systems, Inc.) in 500 μl dilution buffer. Aliquot 100 μl/tube and store at −70°C.

2.6. Effect of Various Matrix Proteins on hMSCs

1. 6-well plates untreated or coated with collagen type I, IV, elastin, fibronectin, and laminin (Flexcell International, Hillsborough, NC, USA).

2.7. Effect of Cyclic Strain on hMSCs

1. Flexcell 4000T unit (Flexcell International): The strain unit is a computer-regulated device that applies cyclic tensile strain to the cell culture through regulated vacuum pressure on the bottom culture plates with a flexible membrane that is untreated or pretreated with fibronectin or type I collagen. The strain causes the flexible plate to stretch across a cylindrical loading post to provide equibiaxial strain to the cells.

2.8. Engineered Vessel Culture Bioreactor

1. The flow system consists of two tubing sets – large and small (Ark-Plas Products, Flippin, AR, USA) connected by a series of different connectors (Cole-Parmer, Vernon Hills, IL, USA). Please see Fig. 1 for details of measurement and connection between the tubes with connectors. The fat tubing in the large tubing set is L/S 18. The rest of the tubes are L/S 16. There is a Y connector for the small tubing set. For the large tubing, there are one Y-connector, one pair of locked connectors (male and female), and two thick connectors.
2. Bioreactor hand-blown by an expert glass-blower. Dimension and measurement of the bioreactor is illustrated in Fig. 2.
3. Stir bar (Fisher Scientific, Pittsburgh, PA, USA).

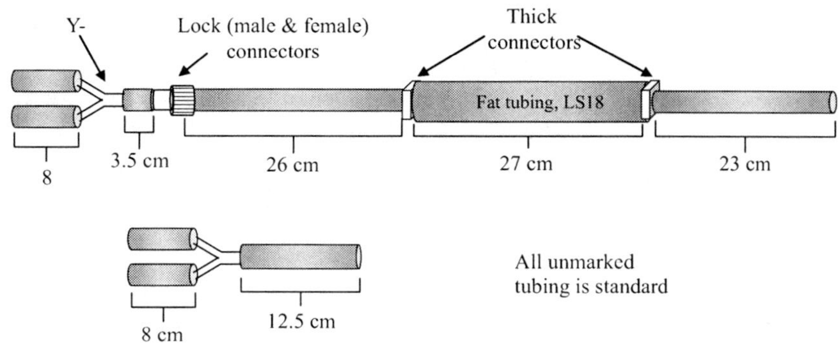

Fig. 1. Measurement of the large and small tubing for the flow system.

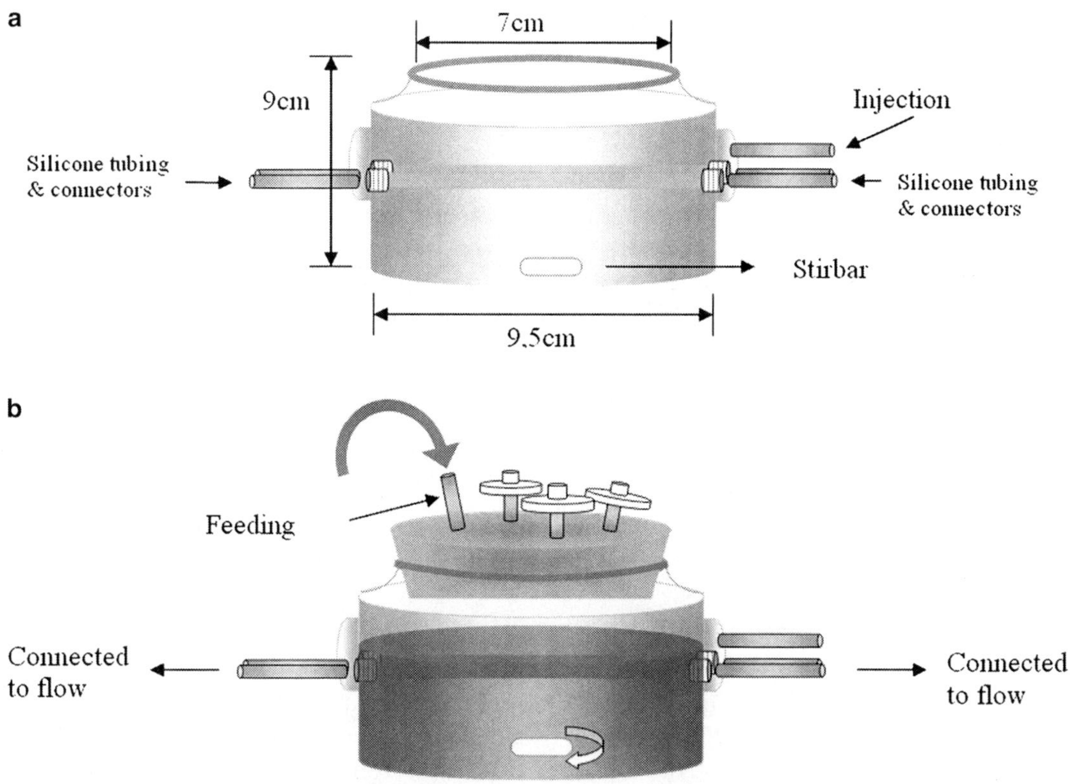

Fig. 2. General schematic of the bioreactor/mesh setup before (**a**) and after (**b**) the assembly on the launch day.

4. Bioreactor lids were made by slicing a silicone stopper (Cole-Parmer) into half horizontally and four holes were created on the lid with a cork borer (use borer #2, 6 mm diameter, Cole-Parmer) for gas exchange via PTFE filters (Cole-Parmer).
5. PGA mesh purchased from Concordia Medical, Coventry Rhode Island. Keep dry in a sealed bag.

6. Silicone tubing (Saint-Gobain Performance Plastics) cleaned with dH_2O (Sigma) and let dry.
7. Dexon 6.0 suture: 18 in. in length, Dexon S uncoated braided synthetic absorbable polyglycolic acid suture from United States Surgical Corporation (US Surgical, Norwalk, CT, USA).
8. Prepare 1 M NaOH solution by dissolving 40 g NaOH (Sigma) in 1 liter dH_2O.
9. Three 1 liter beakers (Fisher Scientific) filled with dH_2O.
10. Kimwipes (Kimberly-Clark, Neenah, WI, USA).
11. Dacron sleeves: "Cooley Veri-Soft Woven Vascular Graft" (Boston Scientific, Natick, MA, USA).
12. Prolene 4.0 suture from US Surgical Corporation.
13. Petri dishes: Large (100 mm diameter) and small (60 mm diameter) (BD Biosciences).
14. Sterile tissue culture water (Sigma).
15. 5 or 10 ml serological pipettes (BD Biosciences).
16. Flow equipment consisting of a PBS bag (Miltenyi Biotech), a pressure transducer and a 3-way stopcock (Edwards Lifesciences, Irvine, CA, USA).
17. Two gallons of fresh 190-proof ethanol (Fisher Scientific).
18. Tools for attaching the PGA mesh: Two sets of forceps, scissors, and hemostat forceps (VWR Scientific, West Chester, PA, USA).
19. Alcohol wipes (Fisher Scientific).
20. Walrus tubing (Arrow International, Inc., Reading, PA, USA).
21. PBS w/fungizone: Use 5 ml fungizone (Sigma) per 500 ml PBS.
22. Fibronectin: Prepared fresh on the launch day by dissolving 5 mg fibronectin (BD Biosciences) in 10 ml sterile Ca^{2+} and Mg^{2+}-free PBS to make a concentration of 0.5 mg/ml. 100 μl of fibronectin is needed to coat PGA mesh (approximate 10 cm^2 surface area) with fibronectin at 5 $\mu g/cm^2$.
23. 10, 20, and 60 ml syringes (BD Biosciences).
24. Injection port with luer lock (Baxter Healthcare, Deerfield, IL, USA).
25. Parafilm (Pechiney Plastic Packaging, Menasha, WI, USA).
26. Tube clamp (Small Parts, Inc., Miramar, FL, USA).
27. 100× pen/strep solution (Invitrogen), aliquot 5 ml/tube, store at 4°C.
28. Blue Nalgene 0.20 μm filter (Pall Corporation, Ann Arbor, MI, USA).

29. Needles: Pink (18G1/2) and green (21G1/2) (BD Biosciences).
30. Easy-load Masterflex pump, Model 7518-10 (Cole-Parmer).
31. Pressure monitor for intravenous or invasive blood pressure (Abbott Labs, Abbott Park, IL, USA).
32. Proliferation medium (14): Each 500 ml of medium consists of DMEM (low-glucose) (Invitrogen) plus 50 ml (10%) selected lot of FBS (Hyclone), 5 ml 1% penicillin-streptomycin-glutamate (Invitrogen), vitamin C (25 mg/5 ml PBS or DMEM), 1.5 μg $CuSO_4$, 25 mg Proline and Glycine, 10 mg Alanine (5 ml standard solution), 5 ml 1 M HEPES solution, and 10 ng/ml PDGF-BB (R&D Systems).
33. Differentiation medium (14): Prepare proliferation medium with the substitution of PDGF-BB with 1 ng/ml TGFβ1 (R&D Systems).
34. 3% acetic acid with methylene blue (StemCell Technologies, Inc., Vancouver, BC, Canada).
35. Quick Start Bradford Dye Reagents (Life Science, Hercules, CA).
36. Image J software (National Institutes of Health, Bethesda, MD).

3. Methods

3.1. Isolation and Cell Culture of hMSCs

1. Slowly load fresh unprocessed human bone marrow onto Ficoll-Paque Plus and fractionate in a centrifuge at room temperature for 30 min at 1,200 × *g* (14) with no brake (see Note 3).
2. Remove the mononuclear cell layer at the interface and wash once with D-PBS and plate in a 25 cm^2 flasks in basal culture medium.
3. Maintain cultures at 37°C in a humidified atmosphere containing 5% CO_2 and provide complete medium changes twice per week.
4. Upon reaching 80–90% confluence, primary cultures were detached using trypsin, and harvested cells were replated into basal culture medium in 75 cm^2 flasks (for protein analyses) or in chamber slides (for histochemical analysis).

3.2. Flow Cytometry

1. Obtain harvested hMSCs and resuspend cells at 10^7 cells per 100 μl in D-PBS buffer.
2. Stain 100 μl aliquots of cells with Fluorescein isothiocyanate (FITC)-conjugated mouse anti-human IgGs (20 μl of 100 μg/ml CD14, 10 μl of 200 μg/ml CD45, and 10 μl for CD34) or

properly matched isotype IgG controls at 4°C for 1 h in the dark. For SH2 and SH3 staining, 100 μl aliquots of cells were incubated with 20 μl of supernatant from SH2 and SH3 hybridoma cell culture.

3. After staining, fix the cells in 4% paraformaldehyde for 5 min at room temperature.
4. Quantitative flow cytometry was performed on a FACStar flow cytometer as per the manufacturer's instructions.

3.3. Immunofluorescence Staining of SMC Markers

1. Perform immunofluorescence staining for SMC markers as described previously (14).
2. Incubate paraformaldehyde-fixed cells in chamber slides with mouse anti-human SMA (1:100) or calponin (1:50) monoclonal antibodies in 10% BSA buffer for 1 h at room temperature.
3. Wash chamber slides 3 times with PBS and incubate with the secondary antibody (FITC-conjugated goat anti-mouse IgG) in 10% BSA buffer for 30 min at room temperature.
4. Wash the chamber slides with cell culture grade water for 3 times and then remove the gaskets with a tweezer.
5. Stain nuclei with DAPI using Vectashield Mounting Medium for Fluorescence with DAPI kit with one drop of the mounting medium on top of the slide and cover with a coverslip before visualization.

3.4. Western Blot for SMC Markers

3.4.1. hMSC Cell Detachment and Cell Lysate Preparation

1. Wash a 75 cm^2 flask containing hMSCs twice with calcium-magnesium-free D-PBS and add 0.5 ml Trypsin-EDTA to each well.
2. Incubate at 37°C for no more than 5 min.
3. Shake the plate gently to detach the cells from the plate.
4. Inactivate the trypsin by adding 0.5 ml basal culture medium containing 10% FBS to each well.
5. Harvest the detached cells in a 15 ml conical tube.
6. To determine cell number, ensure cell solution is thoroughly mixed and then mix 20 μl of cell suspension with 80 μl (1:5 dilution) or 180 μl (1:10 dilution) 3% acetic acid with methylene blue in a 1 ml eppendorf tube. Load approximately 10 μl of the mixture to each side of a hemocytometer and determine cell concentration.
7. Wash cells twice with D-PBS and remove the supernatant.
8. Resuspend the cell pellet in RIPA buffer (approximately 5 times of the volume of the cell pellet) containing Triton X-100 and protease inhibitor cocktail and thoroughly homogenize by repetitive pipetting and vortexing on ice.

9. After homogenization, centrifuge the cell lysate at 14,000 × *g* in the cold room at 4°C for 30 min.
10. Transfer the supernatant to a clean eppendorf test tube and remove 25 μl of lysate for protein quantification. Mix the remaining protein lysate with ¼ volume of Laemmli's SDS-sample buffer (4×, on-reducing, Boston BioProducts) and boil at 100°C for 10 min followed by a quick chill on ice for 5 min and store at −70°C.
11. Perform protein quantification using the Quick Start Bradford Dye Reagents as per the manufacturer's instructions.

3.4.2. Western Blot Analysis of SMC-Specific Markers

1. Separate 25 microgram of protein lysate on a 10% SDS-PAGE gel and transfer to polyvinylidene difluoride membranes.
2. Develop the membrane using the SuperSignal West Pico Chemiluminescent detection system by mixing 1 ml of Reagent A and B each in a conical tube and immediately apply to the membrane. Incubate for 5 min at room temperature in the darkness.
3. Develop the film after exposing the membrane to a Kodak film on a cassette in the dark room and incubate for 1–5 min depending on the intensity of the signal.
4. After visualization, strip the membrane with 12–15 ml Restore™ Western Blot Stripping Buffer at room temperature for 5–15 min.
5. Wash the blot with TBS buffer for 5 min and incubate at 4°C overnight in blocking buffer.
6. Reprobe the blot for β-actin, which serves as an equal-loading control.
7. Quantify Western blots using Image J (National Institutes of Health, Bethesda, MD).
8. Present results as relative density after correction with β-actin.

3.5. Effect of Various Soluble Factors on hMSCs

1. Seed harvested hMSCs in 6-well plates at 5.6×10^3 cells/cm^2 in DMEM plus 5% FBS medium with one of the following supplements: 0 (control), 0.01, 0.1, 1, or 10 ng/ml transforming growth factor β1, 10 ng/ml PDGF-BB, PDGF-CC, bFGF or 50 μg/ml ascorbic acid.
2. Culture CASMCs as positive controls in SmBM supplemented with SmGM-2 SingleQuots as per the manufacturer's instructions.
3. Change culture medium on day 4 after seeding, and harvest cells after 7 days of treatment for SMC differentiation analysis.

3.6. Effect of Various Matrix Proteins on hMSCs

1. Seed harvested hMSCs at 2×10^3 cells/cm^2 in DMEM plus 10% FBS medium on 6-well plates untreated or coated with collagen type I, IV, elastin, fibronectin, and laminin.
2. After 7 days, detach the cells with trypsin and enumerate with 3% acetic acid with methylene blue on a hemocytometer (see Subheading 3.4.1, steps 1–6).
3. Lyse the remaining cells with RIPA buffer containing Triton X-100 and protease inhibitor cocktail (see Subheading 3.4.1, steps 7–10).
4. Protein lysates were quantified by the Bradford assay and stored at −70°C until needed for SMC marker western blot analysis (see Subheading 3.4.2).

3.7. Effect of Cyclic Strain on hMSCs

1. Seed hMSCs at 2×10^4 cells/ml in DMEM plus 10% FBS in a Flexcell 4000T unit and subject the seeded cells to cyclic strain in the presence or absence of 10 ng/ml PDGF-BB.
2. Subject hMSCs to equibiaxial cyclic strain for 5 days at a frequency of 0.5 Hz, resulting in 8–12% substrate elongation.
3. Prepare unstrained controls in an identical manner and culture on unstrained untreated or collagen I or fibronectin-coated flexible plates for 5 days.

3.8. Engineered Vessel Culture in a Bioreactor

3.8.1. Bioreactor Preparation the Day Before Initiating the Culture

1. Autoclave the following components: Bioreactor with a stir-bar, a lid with a feeding tube, flow system (two tubing sets – large and small), connectors, Pasteur pipettes, media cap, and injection tubes (see Figs. 1 and 2).
2. Cut PGA to 1.1 × ~8 cm (depending on bioreactor size).
3. Sew the PGA mesh around clean silicone tubing starting with three surgical knots followed by single stitches looped within each other.
4. Treat the surface of a PGA tube in 1 M NaOH for 1–2 min and rinse in three separate dH_2O baths. Pat dry with Kimwipes between each dip in water baths and dry the PGA tube in a laminar flow hood with the blower on and the UV light off.
5. Cut the Dacron into 1 cm stubs and sew the small pieces of Dacron onto each end of the PGA mesh with an overlap of 2–3 mm. Be careful not to puncture the silicone tubing.
6. Thread suture to edge of Dacron and prepare one surgical knot (two windings) and one single knot.
7. Soak the sewn mesh on silicone tubing, tools, and wire in an ethanol bath for 20–30 min.
8. Flush the tubing with ethanol by lifting one side of tube up (observe air bubbles leaving).

9. Set-up the bioreactor (see Fig. 2a) by feeding the silicone tubing through the side arms using thin wire to pull it through.
10. Pull the bioreactor out of the ethanol bath (keep mesh submerged in ethanol), and attach vessels to the glass bioreactor by tightening Dacron over the glass lip with Prolene sutures already set (surgical knot + 2 single knots).
11. Connect one side of the bioreactor side arms to connectors via silicone tubing. Pull the other side of the silicone tubing with enough tension (about one finger width) and insert remaining two connectors into silicone tubing.
12. Reinsert the bioreactor into the ethanol bath and flush with ethanol by gently pulling connectors out of the side arms.
13. Flip the bioreactor over and allow soaking for 10 min.
14. Flip bioreactor right-side up and soak in the ethanol bath for an additional 10 min.
15. Drain all ethanol.
16. Set up three large Petri dishes (tops or bottoms) in series, and place the bioreactor in the center dish.
17. Flush bioreactor (mesh included) with tissue culture water using a 5 or 10 ml pipette. Also flush the tissue culture water into the silicone tubing.
18. Thoroughly drain all excess water into the Petri dishes on either side of the bioreactor.
19. Dry the bioreactor overnight in the hood with the blower on and the UV light off.

3.8.2. Day 1 Bioreactor Culture Set-Up (see Note 4)

1. Coat each of the PGA tubes with 100 μl 0.5 mg/ml fibronectin freshly made by dissolving 5 mg fibronectin in 10 ml sterile Ca^{2+} and Mg^{2+}-free PBS. Allow to air dry for at least 45 min at room temperature.
2. Place a sterile Petri dish over the opening of the bioreactor.
3. Assemble the flow system to the bioreactor (see Fig. 3).
4. Wipe the connectors first with alcohol wipes. Attach flow system to all four arms. Close off the system as soon as possible.
5. Attach the injection port to the third, unused arm of bioreactor.
6. Remove walrus tubing and tie off the blue end as close to the Y-junction as possible. Pull the tube clamp in place to ensure no liquid transfer to this part of the tube.
7. Remove the PBS bag and attach the red end of the Walrus tubing to the far end of the PBS bag. Make sure to wipe insertion port with alcohol wipe first.

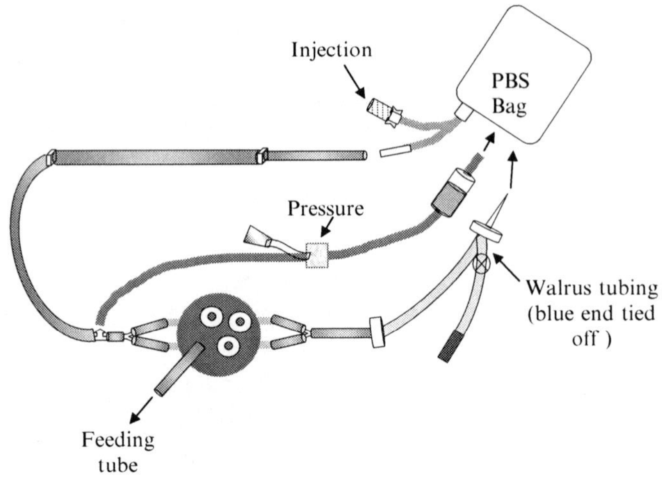

Fig. 3. Flow system set-up.

8. Attach the white end tube of the walrus tubing to one side of flow system.
9. Insert a 3-way stopcock into the flow system.
10. Remove pressure transducer from the package and connect to a 3-way stopcock.
11. Attach the other end of the pressure transducer to the middle opening of the PBS bag.
12. Parafilm everything including the connections between the tubings and between the connector and the tubings or between the connectors and the bioreactor. Spray system with ethanol.
13. Use a 60 ml syringe to inject about 350 ml of PBS with fungizone solution into the PBS bag through an open port.
14. Squeeze the bag to flush the system.
15. Move the stop cocks to allow for full movement of liquid. Check inside bioreactor to ensure there is no leaking.
16. Spray the entire system with 70% ethanol.
17. Trypsinize hMSCs, enumerate and centrifuge.
18. While waiting for the centrifugation, assemble the bioreactor lid by attaching the injection port to the feeding tube.
19. Attach the three PTFE 0.20 μm filters to each of the three air ports. Be careful not to expose or touch the bottom of the lid during this process.
20. Parafilm the injection port.
21. Resuspend one confluent 75 cm^2 flask ($5–8 \times 10^6$ cells) into 1.25 ml and seed onto one vessel. Make sure cell suspension has been introduced to the Dacron-mesh junction as well as the bottom of the mesh (see Note 5).

22. Insert the lid into the glass bioreactor, making sure that the feeding tube does not touch the seeded mesh.
23. Parafilm the lid and the lip of the glass bioreactor.
24. Place the bioreactor in the incubator on its side and rotate every 5 min for 25–30 min.
25. Fill the bioreactor with culture medium (above Dacron, below the third arm) with a pump (see Fig. 2b).

3.8.3. Day 2 Pen/Strep Feeding

1. Draw up 5 ml of pen/strep into a syringe, then attach a blue Nalgene 0.20 μm filter to the syringe.
2. Attach a green needle (21G) to a 20 ml syringe.
3. Wipe the injection port with alcohol wipe before and after every syringe injection.
4. Withdraw about 20 ml medium slowly from feeding tube with the 20 ml syringe.
5. Replace the 20 ml syringe with 10 ml syringe with pen/strep and inject the pen/strep into the bioreactor.
6. Reinject the culture medium.

3.8.4. Weekly Feeding Regimen

1. Starting with the first feeding on day 7, feed every 7 days.
2. For each medium change, remove about 200 ml (or 50% of the total volume) of medium and replace with 200 ml fresh medium.
3. Dissolve vitamin C into 5 ml sterile PBS and supplement the bioreactor 3 times per week (see Note 6) following the same protocol used for pen/strep feeding (see Subheading 3.8.2).

3.8.5. Turning on the Pump

1. After 4 weeks of culture under static conditions, change from proliferation medium to differentiation medium (14) (see Note 7). And turn on the pump (Day 29).
2. Gradually increase the speed with a target of 160 beats per minute.
3. Make sure there are no leaks.
4. Tighten all the injection ports as they tend to loosen at 37°C.
5. Record pressures daily throughout the culture.

3.8.6. Termination of the Culture

1. After 8 weeks, a complete vessel can be formed via endothelialization by seeding human endothelial or HUVEC cells into the lumen of the SMC vessel wall (14).
2. When the culture is complete (with or without endothelialization), the culture is terminated when the pump is shut down and the vessel is taken out of the bioreactor for further analysis (i.e., mechanical testing, immunohistochemical staining, or molecular analysis). See Fig. 4 as an example bioreactor in the incubator and a close-up view of an engineered vessel in the bioreactor at the completion of culture.

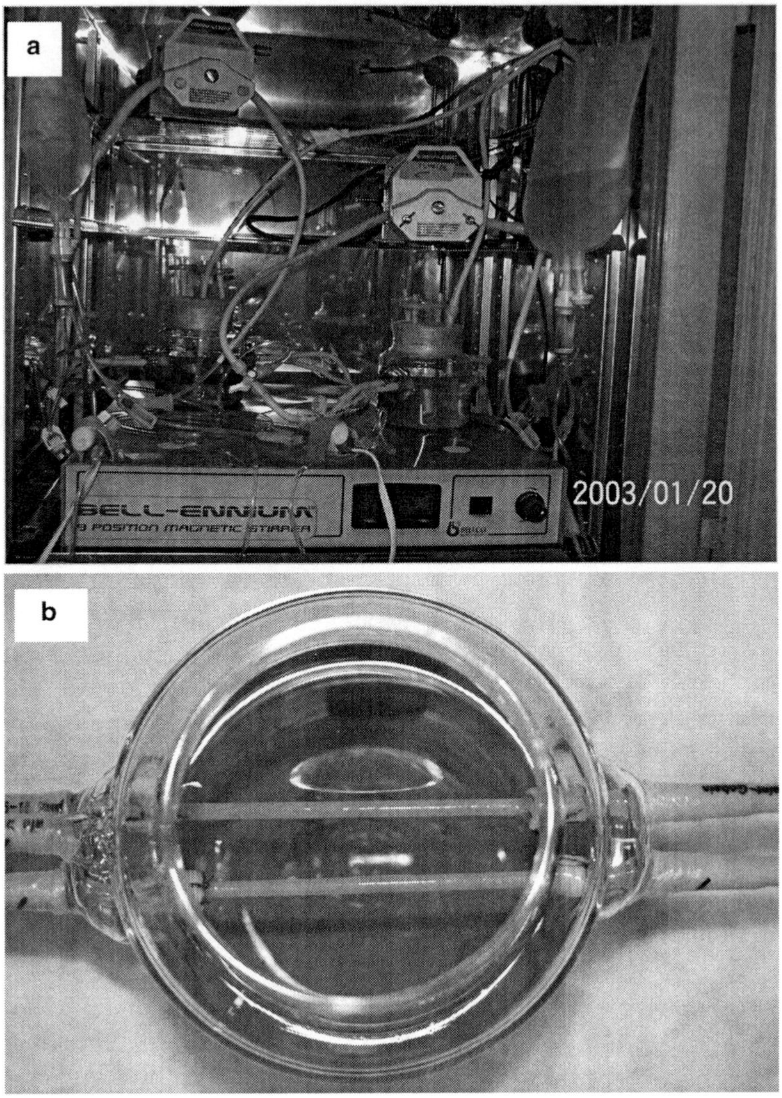

Fig. 4. Pictures of two glass bioreactors in an incubator (**a**) and two engineered vessels (**b**).

4. Notes

1. Diluted primary SMA and calponin antibodies can be recycled by collecting the solution in 15 ml conical tube and stored at 4°C for up to 1 week. Also because of different molecular weight of SMA and calponin, you can run both antibodies nicely on one blot simultaneously.
2. Preparation of growth factors: Dilute the various growth factors in 1% BSA in PBS to minimize adsorption of growth factors onto the wall of plastic eppendorf tubes.

3. Isolation of hMSC from fresh bone marrow step should be done promptly. Any delay in the processing of bone marrow will decrease the yield of hMSCs. Usually hMSC account for approximately 0.1–1 per 10^6 total bone marrow mononuclear cells obtained after the Ficoll-Paque gradient separation step.
4. On the launch day, before start, make sure the sterile stirbar is in the bioreactor. Make sure not to hover over the bioreactor to decrease risk of contamination. Cut up plenty of parafilm and soak (with paper removed) in an ethanol bath (large Petri dish works well).
5. The key to seeding the cells onto the PGA mesh is the volume of the cell suspension. When the PGA mesh is precoated with fibronectin, the total volume of cell suspension should be around 1–1.25 ml depending on the length of the PGA tube. Extra volume will cause dripping of the cell suspension off the PGA mesh to the bottom of the bioreactor.
6. Vitamin C is a very important cofactor for collagen synthesis. Because of its unstable nature, it is freshly made and added 3 times per week to the culture medium to promote collagen synthesis, which significantly affects vessel strength.
7. The whole 8-week vessel culture period is divided into 4-week of proliferation phase and 4-week of differentiation phase. During the proliferation phase, PDGF-BB, a potent mitogen for hMSCs (11), is supplemented in the culture medium to promote proliferation of hMSCs. Pulsatile flow and cyclic strain provided via the pump are not started until week 5 during the differentiation phase. In addition, PDGF-BB is replaced by TGFβ1 to promote SMC differentiation from week 5 onward.

Acknowledgments

The authors are grateful for Drs. Caroline Rhim and Shannon L. M. Dahl for their contribution to the development and optimization of the bioreactor setup protocol. This work is funded by National Institute of Health RO1HL083895 and HL063766 (both to LEN).

Disclosure

L.E.N. has a financial interest in Humacyte, Inc., a regenerative medicine company. Humacyte did not fund these studies, and Humacyte did not affect the design, interpretation, or reporting of any of the experiments herein.

References

1. Kaushal, S., Amiel, G.E., Guleserian, K.J., Sharpira, O.M., Perry, T., Sutherland, F.W., Rabkin, E., Moran, A.M., Schoen, F.J., Atala, A., Soker, S., Bischoff, J., and Mayer, J.E. (2001) Functional small-diameter neovessels created using endothelial progenitor cells expanded ex vivo. *Nat. Med.* 7, 1035–1040.
2. Kadner, A., Heorstrup, S.P., Zund, G., Eid, K., Maurus, C., Melinitchouk, S., Grunenfelder, J., and Turina, M.I. (2002) A new source for cardiovascular tissue engineering: Human bone marrow stromal cells. *Eur. J. Cardiothorac. Surg.* 21, 1055–1060.
3. Hoerstrup, S.P., Kadner, A., Melnitchouk, S., Trojan, A., Eid, K., Tracy, J., Sodian, R., Visjager, J.F., Kolb, S.A., Grunenfelder, J., Zund, G., and Turina, M.I. (2002) Tissue engineering of functional trileaflet heart valves from human marrow stromal cells. *Circulation* 106, I143–I150.
4. Perry, T.E., Kaushal, S., Sutherland, F.W., Guleserian, K.J., Bischoff, J., Sacks, M., and Mayer, J.E. (2003) Bone marrow as a cell source for tissue engineering heart valves. *Ann. Thorac. Surg.* 75, 761–767.
5. Matsumura, G., Miyagawa-Tomita, S., Shin'oka, T., Ikada, Y., and Kurosawa, H. (2003) First evidence that bone marrow cells contribute to the construction of tissue-engineered vascular autografts in vivo. *Circulation* 108, 1729–1734.
6. Wu, X., Rabkin-Aikawa, E., Guleserian, K.J., Perry, T.E., Masuda, Y., Sutherland, F.W., Schoen, F.J., Mayer, J.E., and Bischoff, J. (2004) Tissue-engineered microvessels on three-dimensional biodegradable scaffolds using human endothelial progenitor cells. *Am. J. Physiol. Heart Circ. Physiol.* 287, H480–H487.
7. Cho, S.W., Lim, S.H., Kim, I.K., Hong, Y.S., Kim, S.S., Yoo, K.J., Park, H.Y., Jang, Y., Cahng, B.C., Choi, C.Y., Hwang, K.C., and Kim, B.S. (2005) Small-diameter blood vessels engineered with bone marrow-derived cells. *Ann. Surg.* 241, 506–515.
8. Liu, J.Y., Swartz, D.D., Peng, H.F., Gugino, S.F., Russell, J.A., and Andreadis, S.T. (2007) Functional tissue-engineered blood vessels from bone marrow progenitor cells. *Cardiovasc. Res.* 75, 618–628.
9. Gong, Z., Calkins, G., Cheng, E., Krause, D., and Niklason, L.E. (2009) Influence of culture medium on smooth muscle cell differentiation from human bone marrow-derived mesenchymal stem cells. *Tissue Eng. Part A* 15, 319–330.
10. Zhang, F., Tsai, S., Kato, K., Yamanouchi, D., Wang, C., Rafii, S., Liu, B., and Kent, K.C. (2009) Transforming growth factor-beta promotes recruitment of bone marrow cells and bone marrow-derived mesenchymal stem cells through stimulation of MCP-1 production in vascular smooth muscle cells. *J. Biol. Chem.* 284, 17564–17574.
11. Kobayashi, N., Yasu, T., Ueba, H., Sata, M., Hashimoto, S., Kuroki, M., Saito, M., and Kawakami, M. (2004) Mechanical stress promotes the expression of smooth muscle-like properties in marrow stromal cells. *Exp. Hematol.* 32, 1238–1245.
12. Ball, S.G., Shuttleworth, A.C., and Kielty, C.M. (2004) Direct cell contact influences bone marrow mesenchymal stem cell fate. *Int. J. Biochem. Cell Biol.* 36, 714–727.
13. Lozito, T.P., Kuo, C.K., Taboas, J.M., and Tuan, R.S. (2009) Human mesenchymal stem cells express vascular cell phenotypes upon interaction with endothelial cell matrix. *J. Cell Biochem.* 107, 714–722.
14. Gong, Z., and Niklason, L.E. (2008) Small-diameter human vessel wall engineered from bone marrow-derived mesenchymal stem cells (hMSCs). *FASEB J.* 22, 1635–1648.
15. Niklason, L.E., Gao, J., Abbott, W.M., Hirschi, K.K., Houser, S., Marini, R., and Langer, R. (1999) Functional arteries grown in vitro. *Science* 284, 489–493.
16. Niklason, L.E., Abbott, W., Gao, J., Klagges, B., Hirschi, K.K., Ulubayram, K., Conroy, N., Jones, R., Vasanawala, A., Sanzgiri, S., and Langer, R.L. (2001) Morphologic and mechanical characteristics of bovine engineered arteries. *J. Vasc. Surg.* 33, 628–638.

Chapter 22

Dopaminergic Neuronal Differentiation Protocol for Human Mesenchymal Stem Cells

Katarzyna A. Trzaska and Pranela Rameshwar

Abstract

The generation of dopamine (DA) neurons from stem cells holds great promise for future biomedical research and in the clinical treatment of neurodegenerative diseases, such as Parkinson's disease. Mesenchymal stem cells (MSCs) derived from the adult human bone marrow (BM) can be easily isolated and expanded in culture while maintaining their immense plasticity. Here, we describe a protocol to generate DA-producing cells from adult human MSCs using a cocktail that includes sonic hedgehog (SHH), fibroblast growth factor 8 (FGF8), and basic fibroblast growth factor (bFGF). Electrophysiological functional DA neurons could be achieved by further treatment with brain-derived neurotrophic factor (BDNF). In summary, a protocol is described for the induction of primary BM-derived human MSCs to specific transdifferentiation; in this case, functional DA neurons. The MSC-derived DA cells express DA-specific markers, synthesize, and secrete dopamine. The described method could be used to generate DA cells for various model systems in which DA-producing cells are implicated in pathophysiological conditions.

Key words: Mesenchymal stem cell, Neuron, Transdifferentiation, Dopamine, Sonic hedgehog, Fibroblast growth factor, Brain-derived growth factor

1. Introduction

Mesenchymal stem cells (MSCs) have become the main focus of regenerative medicine and cellular therapeutics (1, 2). Particular attention has been brought to the treatment of a variety of neural disorders (3, 4). MSCs are easily obtained from the bone marrow (BM), can be expanded in culture, have low incidence of tumor formation, and hold immense plasticity (5). A unique property of MSCs is their ability to bypass immune rejection and could, therefore, be used for allogeneic transplantations, commonly termed as 'off-the-shelf stem cells' (6, 7). MSCs can generate various

Mohan C. Vemuri et al. (eds.), *Mesenchymal Stem Cell Assays and Applications*, Methods in Molecular Biology, vol. 698, DOI 10.1007/978-1-60761-999-4_22,

mesodermal cell types such as osteocytes, chondrocytes, adipocytes, and hematopoiesis-supporting stroma (5). Recent studies have demonstrated that MSCs can also generate cells of endodermal and ectodermal tissue (8–11). In fact, MSCs are currently in clinical trials for a number of disorders, including graft-versus-host disease, heart failure, and multiple sclerosis (3).

Dopamine (DA) neurons derived from stem cells are a valuable source to study the molecular mechanisms of DA neuron development, for screening pharmaceutical compounds that target DA disorders, and perhaps for cell replacement therapy in Parkinson's disease (PD). Stem cell therapy has been proposed as a possible treatment for PD, which is caused by the loss of DA neurons in the substantia nigra (12, 13). We have previously reported standardized in vitro conditions for the induction of adult human MSCs into DA neurons (14, 15). This induction protocol includes a cocktail of sonic hedgehog (SHH), fibroblast growth factor 8 (FGF8), basic fibroblast growth factor (bFGF), and brain-derived neurotrophic factor (BDNF). The MSC-derived cells express DA-specific markers, tyrosine hydroxylase (TH), DA transporter (DAT), Nurr1, and Pitx3. The cells also synthesize and secrete DA. The addition of BDNF at a later stage induces functional maturation based on observed postsynaptic currents and DA release upon depolarization (15). Future experiments will examine the full spectrum of excitable properties of the DA neurons generated from human MSCs. A diagram of the complete protocol described in the chapter is depicted in Fig. 1.

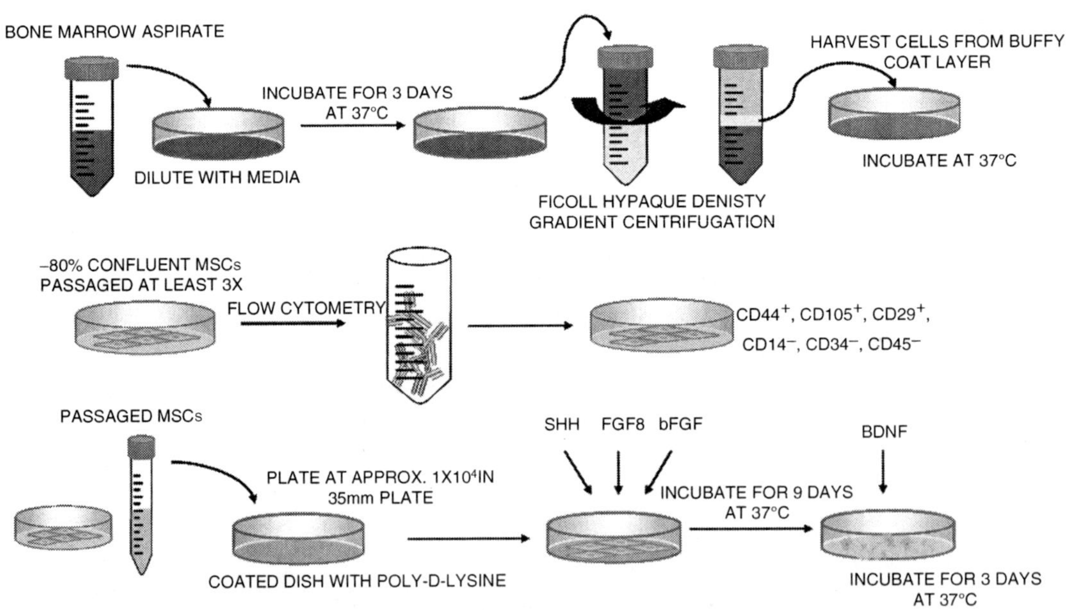

Fig. 1. Dopaminergic neuronal differentiation workflow for human mesenchymal stem cells.

2. Materials

2.1. Isolation of Human MSCs from BM Aspirates

1. Dulbecco's Modified Eagle's Medium (DMEM) with high glucose (Invitrogen/Gibco, Carlsbad, CA) supplemented with 10% fetal calf serum (Hyclone, Logan UT) and 2 mM L-glutamine (Invitrogen/Gibco).
2. Ficoll Hypaque (Sigma-Aldrich, St. Louis, MO).
3. BD Falcon 3003 vacuum gas plasma-treated 100 mm^3 petri dishes (BD Biosciences, San Diego, CA).
4. 50 ml Conical tubes.
5. Pasteur glass pipettes.

2.2. Flow Cytometry

1. Accutase Dissociation Solution (MP Biomedicals, Irvine, CA).
2. Antihuman fluorochrome-conjugated antibodies: CD44, CD29, CD14, CD34, CD45 (BD Biosciences), and CD105 (Fitzgerald Industries, Concord, MA) (see Note 1).
3. 1× PBS (Invitrogen/Gibco).
4. Bovine serum albumin (BSA) (Invitrogen/Gibco).
5. 2% Paraformaldehyde: 2 g paraformaldehyde (Sigma) dissolved in 100 ml 1× PBS, store at 4°C.

2.3. Neuronal Induction of Human MSCs

1. Neurobasal Medium (Invitrogen/Gibco).
2. 50× B27 Supplement (Invitrogen/Gibco).
3. SHH (R & D Systems, Minneapolis, MI,): Reconstitute 25 µg SHH in 500 µl PBS/BSA solution (475 µl sterile 1× PBS, 25 µl sterile 2% human or BSA (Invitrogen/Gibco)) to make 50 µg/ml. Aliquot 50 µl in 0.5 ml tubes and store at –80°C.
4. FGF8 (R & D Systems): Reconstitute 25 µg FGF8 in 500 µl PBS/BSA solution (475 µl sterile PBS, 25 µl sterile 2% human or BSA) to make 50 µg/ml. Store 250 µl of solution at –80°C as stock solution. To the remaining 250 µl, add additional 250 µl PBS/BSA solution to make a final working concentration of 25 µg/ml. Aliquot 50 µl in 0.5 ml tubes and store at –80°C.
5. bFGF (R & D Systems,): Reconstitute 25 µg bFGF in 500 µl PBS/BSA solution (475 µl sterile PBS, 25 µl sterile 2% human or BSA, 0.5 µl of 0.1 M DTT (Invitrogen)) to make 50 µg/ml. Store 250 µl of solution at –80°C as stock solution. To the remaining 250 µl, add additional 250 µl PBS/BSA solution to make a final working concentration of 25 µg/ml. Aliquot 50 µl in 0.5 ml tubes and store at –80°C.

6. BDNF (Cell Sciences, Canton, MA): Reconstitute 10 μg BDNF in 100 μl in dH_2O to make 100 μg/ml. Aliquot 10 μl in 0.5 ml tubes and store at –80°C.
7. BD poly-D-lysine (BD Biosciences): Reconstitute 20 mg poly-D-lysine in 20 ml of dH_2O to make a final working concentration of 1 mg/ml. Aliquot ml in 5 ml tubes and store at –20°C. Use within 3 months.
8. BD Falcon petri dishes: 100 mm^3 (Falcon 3003) 35 mm^3 (Falcon 3001) or 6-well plates (Falcon 3046) (BD Biosciences).

2.4. Immunocytochemistry for Neuronal and Dopaminergic Markers

1. 2–4% Paraformaldehyde: 2–4 g paraformaldehyde (Sigma) dissolved in 100 ml 1× PBS, store at 4°C.
2. Rabbit antihuman tyrosine hydroxylase antibody and mouse antihuman β III tubulin antibody (Millipore, Billerica, MA).
3. Antirabbit IgG – phycoerythrin (PE) and antimouse IgG – fluorescein (FITC) (Jackson Immunoresearch, West Grove, PA).
4. DAPI (Invitrogen/Molecular Probes).
5. 1× PBS (Invitrogen/Gibco).
6. BSA (Invitrogen/Gibco).
7. Triton X-100 (Sigma).

2.5. TH and β III Tubulin Quantitation by Flow Cytometry

1. Accutase Dissociation Solution (MP Biomedicals, Irvine, CA).
2. Rabbit antihuman tyrosine hydroxylase antibody and mouse antihuman β III tubulin antibody (Millipore, Billerica, MA).
3. Antirabbit IgG – phycoerythrin (PE) and antimouse IgG – fluorescein (FITC) (Jackson Immunoresearch, West Grove, PA).
4. 1× PBS (Invitrogen/Gibco).
5. BSA (Invitrogen/Gibco).
6. Triton X-100 (Sigma).
7. 2% Paraformaldehyde: 2 g paraformaldehyde (Sigma) dissolved in 100 ml 1× PBS, store at 4°C.

3. Methods

3.1. Isolation of Human MSCs from BM Aspirates

1. Dilute 3 ml of unfractionated BM aspirate in 3 ml of warmed DMEM containing 10% FCS and L-glutamine in Falcon 3003 vacuum gas plasma-treated 100 mm^3 petri dishes (see Note 2).
2. Incubate dishes for 3 days at 37°C.
3. At day 3, isolate the mononuclear cells by Ficoll Hypaque density gradient centrifugation. Collect media into 50 ml

conical tubes and wash dishes twice with warmed DMEM. Collect the wash into the same conical tubes. Centrifuge at 400 ×*g* for 30 min at room temperature to obtain a blood cell pellet. Aspirate off enough supernatant to obtain about 10–20 ml of the media and then gently shake tube to break up the large blood cell pellet.

4. Carefully layer a volume of 10–20 ml of the centrifuged media onto a volume of 20 ml of the Ficoll Hypaque density fluid (see Note 3). Centrifuge at 400 ×*g* for 30 min at room temperature to obtain the buffy coat layer. Harvest the cells by carefully pipetting with a glass Pasteur pipette from the buffy coat layer and place the cells into the same culture dishes used for the 3-day incubation replaced with fresh media.
5. Place dishes back at 37°C and replace 50% of the media at weekly intervals until adherent cells are approximately 80% confluent (see Note 4).
6. Passage cells at least 3 times before performing following flow cytometry analysis to ensure selection for adherent cells and removal of suspension mononuclear cells. MSCs are maintained with DMEM containing 10% FCS and L-glutamine in Falcon vacuum gas plasma-treated dishes and passaged at ~80% confluency to prevent cell growth arrest and differentiation.

3.2. Flow Cytometry

1. Detach cells with Accutase Dissociation Solution and centrifuge in a conical tube at 400 ×*g* for 10–15 min to pellet cells (see Note 5). Ensure you have at least 10^6 cells for each staining reaction.
2. Wash cells with ice-cold 1× PBS by resuspending cell pellet and centrifuge again.
3. Resuspend cells with 500 μl ice-cold 1× PBS with 3% BSA. Add 10 μl of appropriate antibody conjugated to fluorophores and incubate for at least 30 min at 4°C in the dark (see Note 6).
4. Wash cells 3 times by centrifugation at 400 ×*g* for 5–10 min and resuspend in ice-cold 1× PBS.
5. Resuspend cell pellet in 200 μl 2% paraformaldehyde to fix cells and then place the cells in 5 ml of round-bottomed tubes for analysis.

3.3. Neuronal Induction of Human MSCs

1. Coat dishes to be used for neuronal induction with BD poly-D-lysine. Let poly-D-lysine defrost at room temperature. Dilute 2 ml in 38 ml of serum-free medium (DMEM). Dilute smaller volumes if using less dishes (i.e., 500 μl in 9.5 ml). Pour enough onto dish to cover entire dish and coat for 1 h at room temperature. Aspirate off poly-D-lysine solution and wash dish 3× with 1× PBS and add fresh media for MSC culture (see Note 7).

2. At approximately 70–80% confluence, and between passages 3–10, trypsinize MSCs, centrifuge, and resuspend in 1 ml of MSC culture media. Count cells.
3. Seed approximately 10^4 cells in 35 mm^3 culture dishes that have been coated with poly-D-lysine with MSC culture media (see Note 8).
4. Allow cells to adhere to the culture surface overnight at 37°C.

   The next day, replace the media with neural induction medium. *For a 35-mm3 culture dish, proceed as follows* (see Note 8):

   (a) Change culture media to 2 ml of Neurobasal Medium

   (b) Add 0.25× (0.5%) of B27 (4 μl from the 50× stock bottle)

   (c) Add 250 ng/ml SHH (10 μl from aliquoted stock)

   (d) Add 100 ng/ml FGF8 (8 μl from aliquoted stock)

   (e) Add 50 ng/ml bFGF (4 μl from aliquoted stock)
5. Place the dishes back at 37°C and incubate for 9 days. Leave media unchanged for the culture period.
6. At the 9 day time point, add 50 ng/ml BDNF to the culture (1 μl from aliquoted stock). Replace the dishes in the 37°C incubator for an additional 3 days (total of 12 days from initial induction) (see Note 9). Cells can now be utilized for desired experiments.

3.4. Immunocytochemistry for Neuronal and Dopaminergic Markers

1. Aspirate medium from the cells and fix with 2–4% paraformaldehyde for 20 min at room temperature (see Note 10). Rinse with cold PBS.
2. Permeabilize and block with 0.1% Triton X-100 and 1% BSA in PBS for 30 min. Rinse 3 times with cold PBS.
3. Incubate cells with desired diluted antibodies in 1% BSA in PBS overnight at 4°C. Dilutions for recommended TH and β III Tubulin are 1:500 (see Note 11).
4. The next day, decant the solution and rinse cells with cold PBS 3 times.
5. Incubate with appropriate secondary antibodies at 1:1,000 in 1% BSA in PBS for 1 h at room temperature in the dark.
6. Decant the solution and rinse cells with cold PBS three times. Counterstain with DAPI to identify the nucleus. Incubate with 0.1–1 μg/ml of DAPI for 5 min. Rinse with PBS.
7. Store cells in the dark at 4°C until ready for microscopic analysis.

3.5. TH and β III Tubulin Quantitation by Flow Cytometry

1. Detach cells with Accutase Dissociation Solution and centrifuge in a conical tube at 400×*g* for 10–15 min to pellet cells (see Note 5). Ensure you have at least 10^6 cells for each staining reaction.
2. Wash cells with ice-cold 1× PBS by resuspending cell pellet and centrifuge again.
3. Fix cells with 2% paraformaldehyde for 30 min prior to labeling.
4. Centrifuge cells again and resuspend cells with 500 μl ice-cold 1× PBS with 3% BSA and 0.1% Triton X-100 for permeabilization. Wash and centrifuge again and resuspend in 1× PBS with 3% BSA and add 10 μl of appropriate TH and β III Tubulin antibodies. Incubate for 20 min at 4°C or on ice (see Note 11).
5. Wash cells at least twice by centrifugation at 400×*g* for 5–10 min and resuspend with 500 μl ice-cold 1× PBS with 3% BSA. Add 5 μl of appropriate secondary antibodies and incubate for 20 min at 4°C or on ice.
6. Centrifuge again and resuspend cell pellet in 200 μl 2% paraformaldehyde and then place the cells in 5 ml of round-bottomed tubes for analysis.

4. Notes

1. In our hands, the listed antibodies show high efficiency in flow cytometry analysis, based on the protocol. However, the studies could use other sources of antibodies.
2. Our laboratory directly acquires BM aspirates from healthy volunteers following a protocol approved by the Institutional Review Board of the University of Medicine and Dentistry of New Jersey (UMDNJ), Newark campus. We recommend obtaining fresh BM aspirates and follow the described protocol to isolate MSCs. We cannot state that aspirates or MSCs from commercial source can generate similar outcome.
3. Be sure to carefully and slowly layer the centrifuged media onto the Ficoll Hypaque. Do not apply force as some of the media may go through the density gradient and you will not get good separation. If you do not get good separation from the first centrifugation and cannot clearly see the buffy coat layer, repeat the centrifugation step for 30 min. Be careful when pipetting from the buffy coat layer as you do not want any Ficoll Hypaque to come with the cells back into the culture dish. Also, note that Ficoll Hypaque could be toxic. Therefore, avoid leaving the cells in the Ficoll for a prolonged period.

4. Do not replace media until you begin to note adherent cells. If you replace the media too early, you may take away MSCs that are still in suspension and dividing. You should start to see MSCs adhering to the bottom of the dish in about 1 week. Depending on your isolation efficiency, it may take less or more time to notice adherent cells.
5. We found that Accutase Dissociation Solution, in place of trypsin, shows consistent results for flow cytometry analysis.
6. In our hands, this method shows high efficiency for flow cytometry analysis. Alternate incubation times and amounts of antibody may be used per your specific method.
7. Unbound poly-D-lysine is toxic to the cells.
8. The described neuronal induction components are for a 35 mm^3 culture dish. Please adjust for different size culture dishes.
9. Do not replace the media at the 9-day time point when adding BDNF. Simply add 50 ng/ml BDNF to the culture dish and then return the dish in the 37°C incubator. The media should be unchanged for the entire induction period (12 days). At this time, it is unclear what is produced at various times. However, it appears that there are timeline productions of factors that contribute to the generation of functional DA neurons.
10. Any alternative fixative, such as cold acetone or methanol, can be substituted.
11. Studies can be performed with antibodies from other sources; however, these antibodies at the specified dilutions worked efficiently in our laboratory. Other neuronal markers, such as neurofilament and NeuN, can be used as alternative to β III Tubulin. TH is the best known marker for dopaminergic neurons.

Acknowledgments

This work has been supported by the F.M. Kirby Foundation.

References

1. Picinich, S.C., Mishra, P.J., Mishra, P.J., Glod, J., Banerjee, D. (2007) The therapeutic potential of mesenchymal stem cells. Cell- & tissue-based therapy. *Expert Opin Biol Ther* 7, 965–73.
2. Zipori, D. (2004) Mesenchymal stem cells: harnessing cell plasticity to tissue and organ repair. *Blood Cells Mol Dis* **33**, 211–5.
3. Giordano, A., Galderisi, U., Marino, I.R. (2007) From the laboratory bench to the patient's bedside: an update on clinical trials with mesenchymal stem cells. *J Cell Physiol* **211**, 27–35.
4. Phinney, D.G., Isakova, I. (2005) Plasticity and therapeutic potential of mesenchymal stem cells in the nervous system. *Curr Pharm Des* **11**, 1255–65.
5. Bianco, P., Riminucci, M., Gronthos, S., Robey, P.G. (2001) Bone marrow stromal stem cells: nature, biology, and potential applications. *Stem Cells* **19**, 180–92.

6. Potian, J.A., Aviv, H., Ponzio, N.M., Harrison, J.S., Rameshwar, P. (2003) Veto-like activity of mesenchymal stem cells: functional discrimination between cellular responses to alloantigens and recall antigens. *J Immunol* **171**, 3426–34.

7. Krampera, M., Pasini, A., Pizzolo, G., Cosmi, L., Romagnani, S., Annunziato, F. (2006) Regenerative and immunomodulatory potential of mesenchymal stem cells. *Curr Opin Pharmacol* **6**, 435–41.

8. Greco, S.J., Zhou, C., Ye, J.H., Rameshwar, P. (2007) An interdisciplinary approach and characterization of neuronal cells transdifferentiated from human mesenchymal stem cells. *Stem Cells Dev* **16**, 811–26.

9. Tropel, P., Platet, N., Platel, J.C., Noel, D., Albrieux, M., Benabid, A.L., Berger, F. (2006) Functional neuronal differentiation of bone marrow-derived mesenchymal stem cells. *Stem Cells* **24**, 2868–76.

10. Sato, Y., Araki, H., Kato, J., Nakamura, K., Kawano, Y., Kobune, M., Sato, T., Miyanishi, K., Takayama, T., Takahashi, M., Takimoto, R., Iyama, S., Matsunaga, T., Ohtani, S., Matsuura, A., Hamada, H., Niitsu, Y. (2005) Human mesenchymal stem cells xenografted directly to rat liver are differentiated into human hepatocytes without fusion. *Blood* **106**, 756–63.

11. Zeng, X., Cai, J., Chen, J., Luo, Y., You, Z.B., Fotter, E., Wang, Y., Harvey, B., Miura, T., Backman, C., Chen, G.J., Rao, M.S., Freed, W.J. (2004) Dopaminergic differentiation of human embryonic stem cells. *Stem Cells* **22**, 925–40.

12. Lindvall, O., Kokaia, Z., Martinez-Serrano, A. (2004) Stem cell therapy for human neurodegenerative disorders-how to make it work. *Nat Med* **10 Suppl**, S42–50.

13. Correia, A.S., Anisimov, S.V., Li, J.Y., Brundin, P. (2005) Stem cell-based therapy for Parkinson's disease. *Ann Med* **37**, 487–98.

14. Trzaska, K.A., Kuzhikandathil, E.V., Rameshwar, P. (2007) Specification of a dopaminergic phenotype from adult human mesenchymal stem cells. *Stem Cells* **25**, 2797–808.

15. Trzaska, K.A., King, C.C., Le, K.Y., Kuzhikandathil, E.V., Nowycky, M.C., Ye, J.H., Rameshwar, P. (2009) Functional maturation of mesenchymal stem cell-derived dopamine progenitors by brain-derived neurotrophic factor. *J Neurochem* **110**, 1058–69.

Chapter 23

Hepatic Differentiation of Mesenchymal Stem Cells: In Vitro Strategies

Sarah Snykers, Joery De Kock, Vanhaecke Tamara, and Vera Rogiers

Abstract

Recently, evidence has been provided that mesenchymal stem/progenitor cells (MSC) from various sources (bone marrow, adipose tissue, skin, placenta, umbilical cord) could occasionally overcome lineage borders and differentiate into endodermal (hepatocytes) and ectodermal (neural cells) cell types in vitro. Whereas unidirectional differentiation into other mesenchymal cell types, including adipocytes, chondrocytes, and osteoblasts, readily occurs in the presence of a simple cocktail of growth factors and nutrients, successful bypassing of lineage borders mainly depends on multistep processes in a coordinated signaling network. Here, we provide a reproducible basic methodology to differentiate adult MSC into functional hepatocytes in a sequential and time-dependent way. In addition, focus lies on the functional characterization of MSC-derived hepatocyte-like cells. In particular, we provide a detailed modus operandi to measure the inducible cytochrome P450 (CYP)-dependent activity of MSC-derived hepatocyte-like cells.

Key words: Adult stem cells, Mesenchymal stem cells, Hepatocytes, Differentiation, Functionality, CYP450-dependent activity, In vitro

1. Introduction

Stem cells represent a unique source of self-renewing cells within the human body. Historically, the developmental paradigm existed that adult stem cells were, in contrast to their embryonic counterparts, subjected to "cell fate determinism." New insights in stem cell potency have challenged the latter canonical developmental hierarchy (1). Still, the phenomenon of fate reprogramming and phenotypic diversification remains an enigmatic and rare process. Understanding how to control both proliferation and differentiation of stem cells and their progeny is a challenge in both preclinical drug discovery as well as clinical therapies. In this chapter, we describe an efficient technology to differentiate mesenchymal

Mohan C. Vemuri et al. (eds.), *Mesenchymal Stem Cell Assays and Applications*, Methods in Molecular Biology, vol. 698,
DOI 10.1007/978-1-60761-999-4_23, © Springer Science+Business Media, LLC 2011

stem/progenitor cells (MSC) from different origin and species into (functional) hepatocyte-like cells. In essence, the protocol includes the sequential exposure of MSC to hepatogenic factors [fibroblast growth factor (FGF)-4, hepatocyte growth factor (HGF), insulin-transferrin-sodium selenite (ITS), dexamethasone, and oncostatin M (OSM)] (2). Coexposure of MSC to epigenetic modifiers, such as the histone deacetylase inhibitor Trichostatin A (TSA), is thought to advantageously direct differentiation toward the hepatic lineage (3). Indeed, it has been revealed that in addition to the treatment of lineage-specific cytokines/growth factors (taking into account their type, concentration, mode of presentation, and order of application (4)) alterations of the epigenetic traits and chromatin code of specific gene regulatory regions play a role in bypassing cell fate determinism and reprogramming cell fate (5–9). The present methodology will elaborate on this important step.

At present, the functional characterization of MSC-derived hepatocyte-like cells is particularly focused on albumin secretion, urea metabolism, and glycogen uptake. Only little attention is being paid to xenobiotic metabolism, including cytochrome P450 (CYP)-dependent enzymatic activity and responsiveness to prototype inducers such as phenobarbital (human: CYP2B6, CYP3A4, rat CYP2B1/2), rifampicin (human CYP3A4), and 3-methylcholantrene (human and rat CYP1A1/2). Bearing in mind that inducible CYP-dependent activity is a key determinant of the functional hepatic phenotype (10, 11), characterization of stem cells-derived hepatocyte-like cells should incorporate these assays. Here we provide a detailed protocol to measure inducible CYP1A1/2 and CYP2B6-dependent activities of MSC-derived hepatocyte-like cells by means of 7-ethoxyresorufin-*O*-deethylase (EROD) and 7-penthoxy-resorufin-*O*-deethylase (PROD) micro assays, in the absence and presence of the prototype inducers 3-methylcholantrene and phenobarbital, respectively.

2. Materials

2.1. Hepatic Differentiation of MSC In Vitro

1. Benzyl Penicillin (Continental Pharma, Puurs, Belgium) (733 IU/mL) in Phosphate-buffered Saline (PBS). The solution can be stored for up to 6 months at –20°C or for 2 weeks at 4°C.
2. Dexamethasone (Sigma, Bornem, Belgium) (0.25 mM) in double-distilled water. The solution is stable for 3 months at –20°C.
3. Dulbecco's phosphate-buffered saline (DPBS), pH 7.2: 0.01325% (w/v) $CaCl_2 . 2H_2O$, 0.01% (w/v) $MgCl_2 . 6H_2O$, 0.02% (w/v) KCl, 0.02% (w/v) KH_2PO_4, 0.8% (w/v) NaCl, 0.15% (w/v) Na_2HPO_4 in Millipore-quality water. The buffer

is sterilized by passing through a 0.22-μm filter and can be stored at 4°C for 9 months.

4. Earle's Balanced Salt Solution (EBSS) without phenol red. This sterile medium can be stored for 6 months at 4°C.
5. Fetal bovine serum (FBS, Hyclone, Perbio Science, Erembodegem, Belgium). This sterile solution can be stored for up to 6 months at 4°C or –20°C.
6. Fibroblast growth factor-4 (FGF-4, R&D Systems, Minneapolis, MN) Stock Solution (25 μg/mL) in PBS. The solution can be stored for up to 12 months at –20°C.
7. Hepatocyte growth factor-4 (HGF, R&D Systems) Stock Solution (25 μg/mL) in PBS. The solution can be stored for up to 12 months at –20°C.
8. 100× Insulin/Transferrin/Selenite (ITS, Sigma) containing insulin (0.5 g/L), transferrin (0.5 g/L), and selenite (0.5 mg/L) in EBSS. This sterile solution can be stored for 12 months at 4°C.
9. 100× linoleic acid-albumin (LA-BSA, Sigma) containing bovine serum albumin (BSA) (100 mg/mL) in DPBS and two moles of linoleic acid per mole of BSA. This sterile solution can be stored for at least 3 months at 4°C.
10. L-Ascorbic Acid 2-phosphate sesquimagnesium salt (Sigma) (10 mM) in PBS. The solution is stable for 2 weeks at 4°C.
11. Nicotinamide (Sigma) (1 mg/mL) in PBS. The solution can be stored for up to 12 months at –20°C or for 2 weeks at 4°C.
12. Oncostatin M (OSM, R&D Systems) (10 μg/mL) in PBS. The solution can be stored for up to 12 months at –20°C.
13. PBS, pH 7.65: 0.28% (w/v) NaCl, 0.02% (w/v) KCl, 0.31% (w/v) $Na_2HPO_4 \cdot 12H_2O$, 0.02% (w/v) KH_2PO_4 in Millipore-quality water. The buffer is sterilized by passing through a 0.22-μm filter and can be stored at 4°C for 9 months.
14. Sodium Pyruvate (Sigma) (27.3 mg/mL) in PBS. The solution can be stored for up to 12 months at –20°C or for 2 weeks at 4°C.
15. Streptomycin Sulfate (Sigma) (5 mg/mL) in PBS. The solution can be stored for up to 6 months at –20°C or for 2 weeks at 4°C.
16. Trichostatin A (TSA, Sigma) Stock Solution (30 mM) in Ethanol p.a. The solution can be stored for up to 12 months at –20°C.
17. Williams' Medium E (Invitrogen, Merelbeke, Belgium). This sterile medium is stable for 2 weeks at 4°C.
18. Basal Medium: Williams' Medium E supplemented with Benzyl Penicillin (7.33 IU/mL), Streptomycin Sulfate (50 μg/mL), 0.1 mM L-Ascorbic Acid, LA-BSA (1 mg/mL)

Nicotinamide (4 mg/mL), and Sodium Pyruvate (27.3 mg/L). The medium is sterilized by passing through a 0.22-μm filter and can be stored at 4°C for 2 weeks.

19. Hepatic Differentiation Medium I: Basal medium supplemented with FGF-4 (10 ng/mL). The medium is sterilized by passing through a 0.22-μm filter and can be stored at 4°C for 2 weeks.
20. Hepatic Differentiation Medium II: Basal medium supplemented with FGF-4 (10 ng/mL) and HGF (20 ng/mL). The medium is sterilized by passing through a 0.22-μm filter and can be stored at 4°C for 2 weeks.
21. Hepatic Differentiation Medium III: Basal medium supplemented with FGF-4 (5 ng/mL) and HGF (30 ng/mL) and ITS (0.5×). The medium is sterilized by passing through a 0.22-μm filter and can be stored at 4°C for 2 weeks.
22. Hepatic Differentiation Medium IV: Basal medium supplemented with HGF (30 ng/mL) and ITS (0.25×), Dexamethasone (0.05 μM). The medium is sterilized by passing through a 0.22-μm filter and can be stored at 4°C for 2 weeks.
23. Hepatic Differentiation Medium V: Basal medium supplemented with HGF (20 ng/mL) and Dexamethasone (0.05 μM). The medium is sterilized by passing through a 0.22-μm filter and can be stored at 4°C for 2 weeks.
24. Hepatic Differentiation Medium VI: Basal medium supplemented with HGF (20 ng/mL) and Dexamethasone (0.05 μM) and OSM (10 ng/mL). The medium is sterilized by passing through a 0.22-μm filter and can be stored at 4°C for 2 weeks.

2.2. Collagen Thin Coating

1. Collagen, type I (BD Biosciences, Erembodegem, Belgium) (100 μg/mL) in 0.02 N Acetic Acid. The solution must be freshly prepared on ice.

2.3. EROD/PROD Micro Assays

1. β-glucuronidase/arylsulfatase solution (Roche Applied Science, Vilvoorde, Belgium) (15 Fishman units/120 Roy units) in 0.1 M Sodium Acetate Buffer, pH 4.5. This solution should be protected from light and must be freshly prepared prior to use.
2. Dexamethasone (0.25 mM) in double-distilled water. The solution is stable for 3 months at –20°C.
3. Dimethylsulfoxide (DMSO, Sigma) p.a.
4. 3,3′-Methylene-bis(4-hydroxycoumarin) (Dicoumarol, Sigma) (100 mM) in 0.5 N NaOH. The solution should be protected from light and can be stored for 1 week at room temperature.

5. 7-Ethoxyresorufine (Sigma) (51.82 mM) in DMSO. The solution should be protected from light and can be stored for up to 2 months at room temperature.
6. Assay Medium: Hepatic Differentiation medium without phenol red, containing Dicoumarol (10 μM) and Dexamethasone (0.01 μM). This medium can be stored for 1 week at 4°C.
7. EROD Assay Medium: Assay medium containing 7-ethoxy-resorufin (0–20 μM). This medium can be stored for 1 week at 4°C.
8. PBS: cf. item 2.1, 13.
9. 7-Penthoxyresorufine (Sigma) (44.14 mM) in DMSO. The solution should be protected from light and can be stored for up to 2 months at room temperature.
10. PROD Assay Medium: Assay medium containing 7-penthoxy-resorufin (0–20 μM). This medium can be stored for 1 week at 4°C.
11. Resorufin (Sigma) Stock (10 mM) in DMSO.
12. Resorufin Work Solution (100 nM) in Ethanol p.a.
13. Resorufin Standard curve (0–100 nM) in Assay Medium.
14. Sodium Acetate Buffer 0.1 M, pH4.5: 1 Volume NaOH (0.1 M)/1 Volume Acetic Acid (0.1 M). The buffer is sterilized by passing through a 0.22-μm filter and can be stored at 4°C for 6 months.
15. Williams' Medium E without phenol red (Invitrogen). This sterile medium is stable for 2 weeks at 4°C.
16. 3-Methylcholantrene (Sigma) (0.4 mM) in DMSO. This solution is sterilized by passing through a 0.22-μm filter and must be freshly prepared prior to use.
17. EROD Hepatic induction medium: Hepatic Differentiation medium containing 3-Methylcholantrene (2 μM). This medium must be freshly prepared prior to use.
18. Sodium Phenobarbital (Sigma) (75 mM) in Millipore-quality water. This solution is sterilized by passing through a 0.22-μm filter and must be freshly made prior to use.
19. PROD Hepatic induction medium: Hepatic Differentiation medium containing Sodium Phenobarbital (1 mM). This medium must be freshly prepared prior to use.

2.4. General Equipment

1. Laminar flow cabinet.
2. Aluminium foil.
3. Sterile gloves.
4. Sterile 6- and 24-well plates treated for tissue culture (BD Falcon, VWR, Heverlee, Belgium).

5. Sterile 96-well plates treated for tissue culture (Nunc, VWR).
6. Black 96-well plates for fluorescence measurements (Nunc, VWR).
7. Sterile volumetric pipets (BD Falcon, VWR).
8. Micropipets.
9. Multichannel pipettor.
10. Sterile tips.
11. Sterile Pasteur pipets.
12. Sterile 15 and 50-mL conical tubes (BD Falcon, VWR).
13. Sterile 1.5 and 2 mL conical vials (Nunc, VWR).
14. Filter (0.22 μm) (Fischer BioBlack Scientific, Doornik, Belgium).
15. Inverse-phase light microscope (Nikon).
16. Thermostated bath.
17. CO_2 incubator (HeraCell®, Hereaus).
18. Vortex.
19. Victor^3 Multiwell plate reader (filters: Excitation λ: 530 nm and Emission λ: 590 nm) (Perkin Elmer, Zaventem, Belgium).

3. Methods

3.1. Collagen Thin Coating

1. Add 50, 250 or 1,200 μL of 100 μg/mL Collagen Type I per well of a 96-well plate (0.3 cm^2), a 24-well plate (2 cm^2) or a 6-well plate (9.6 cm^2), respectively.
2. Swirl well to cover the whole area.
3. Incubate plates for 1 h at room temperature in a laminar airflow cabinet.
4. Aspirate the Collagen solution carefully.
5. Rinse 3× with PBS to remove acid.
6. The plates can be used immediately or may be air dried; they can be stored at 4°C for up to 1 week under sterile conditions.

3.2. Hepatic Differentiation of MSC In Vitro

1. Plate MSC at a cell density of 2.13×10^4 cells/cm^2 on collagen type I precoated wells of a 96-well plate (0.3 cm^2), a 24-well plate (2 cm^2) or a 6-well plate (9.6 cm^2).
2. Incubate MSC at 37°C in Basal Medium including 2% FBS.
3. Refresh media every 2 days until MSC have reached 90–100% confluence.

4. At 90–100% confluence, remove Basal Medium and start differentiation (Day 0) by exposing cells to basal medium supplemented with hepatogenic factors. Change media every 2–3 days unless specified otherwise (see Note 1).
   (a) Day 0: Incubate cells with Hepatic Differentiation Medium I (see Note 2).
   (b) Days 1–2: Incubate cells with Hepatic Differentiation Medium II.
   (c) Days 3–5: Incubate cells with Hepatic Differentiation Medium III.
   (d) Days 6–8: Incubate cells with Hepatic Differentiation Medium IV supplemented with 1 μM TSA (see Note 3).
   (e) Days 9–11: Incubate cells with Hepatic Differentiation Medium V supplemented with 1 μM TSA.
   (f) Day 12 onward: Incubate cells with Hepatic Differentiation Medium VI supplemented with 1 μM TSA.

3.3. Functional Characterization of MSC-Derived Hepatocyte-Like Cells by EROD/PROD Micro Assays

The optimal concentration and incubation time of 7-ethoxyresorufin and 7-penthoxyresorufin must be determined (see Note 4).

3.3.1. Optimization of EROD/PROD Micro Assays

1. For CYP inducibility measurements: Incubate cells during differentiation at least for 6 days with EROD Hepatic induction medium or PROD Hepatic induction medium. Refeed with fresh media daily.
2. Upon differentiation, wash cells 2× with PBS at 37°C.
3. Incubate cultured cells with EROD/PROD Assay medium at 37°C for 24 h in multiwell plate reader. Protect from light (see Note 5).
4. Measure the formation of fluorescent resorufin in supernatants at an Excitation λ of 530 nm and Emission λ of 590 nm during this time period.
5. Measure the fluorescence of a Resorufin Standard curve (0–100 nM) (see Note 6).
6. Determine the most optimal concentration and the incubation time of 7-ethoxyresorufin and 7-penthoxyresorufin based on the Michaelis–Menten kinetics (see Note 7).

3.3.2. Optimized EROD/PROD Micro Assays

1. For CYP inducibility measurements: incubate cells during differentiation at least for 6 days with either EROD Hepatic induction medium or PROD Hepatic induction medium. Refresh media daily.
2. Upon differentiation, wash cells 2× with PBS at 37°C.
3. Incubate cultured cells with optimized EROD/PROD Assay medium at 37°C for an optimized time period (mostly between 45 min and 3 h). Protect from light (see Notes 4 and 5).

4. Upon incubation time, transfer 150 μL supernatant to a black 96-well plate for fluorescence measurements (=assay plate). Protect from light.
5. Add 50 μL β-glucuronidase/arylsulfatase solution, homogenize, and protect from light.
6. Incubate assay plate for 30 min at 37°C while shaking. Protect from light.
7. After 30 min add 100 μL ethanol p.a. and homogenize. Protect from light.
8. Measure fluorescence of supernatants at an excitation λ of 530 nm and emission λ of 590 nm.
9. Measure the fluorescence of a Resorufin Standard curve (0–100 nM) (see Note 6).
10. Determine CYP-dependent activity in nanomoles resorufin formed/h × 10^3 cells (see Note 7).

4. Notes

1. Slight modifications to the hepatic differentiation protocol might be needed depending on the origin, species, and maturation state of the postnatal stem/progenitor cells. For example, liver epithelial cells, which are more hepatic committed cells, may prefer immediate exposed to HGF in order to progress along the hepatic lineage (12).
2. During the first 2 days of differentiation, incubation with 2% FBS might be preferred in order to prevent cell death.
3. The concentration and incubation time of TSA might vary dependent on the stem/progenitor cell type used. Usually a concentration between 100 nM and 2 μM TSA is applied. Fine tuning should be performed by measuring the DNA synthesis by means of e.g., [3H] thymidine incorporation. It is recommended to search for those conditions where the cells are in cell cycle arrest without the occurrence of massive cell death. Note that apoptosis of the nondifferentiating cells might occur. For bone marrow MSC, a concentration of 1 μM TSA, added from day 6 of differentiation onward, was found to be particularly efficient for stimulating the hepatic differentiation process (3).
4. The most optimal concentration and incubation time of 7-ethoxyresorufin and 7-penthoxyresorufin depend on the cell type used. For bone marrow MSC and freshly isolated hepatocytes, the optimized incubation time is between 45 min and 3 h for 7.5 μM 7-ethoxyresorufin and 15 μM 7-penthoxyresorufin.

5. The EROD/PROD micro assay must be performed on living cells.
6. The concentration range of the Resorufin Standard Curve depends on the actual enzymatic activities of CYP1A1/2 and CYP2B6 enzymes. For freshly isolated hepatocytes, use a Resorufin standard curve in the μM range.
7. Include appropriate controls. As a positive control, include freshly isolated hepatocytes incubated in EROD/PROD assay medium. As negative controls, include MSC-derived hepatocyte-like cells incubated in Assay medium, undifferentiated MSC incubated in EROD/PROD assay medium, and boiled hepatocytes incubated in EROD/PROD assay medium.

Acknowledgments

This work was supported by the Fund for Scientific Research – Flanders (FWO), the Research Council (OZR) of the Vrije Universiteit Brussel (Belgium), European Community's Sixth Framework Program (FP6/2005-2011) under grant agreement n°037499, the European Community's Seventh Framework Program (FP7/2007-2013) under grant agreement n°20161, ISRIB (Brustem), and BELSPO (IAP HEPRO).

Valorization

Patent submission PCT/EP2004/0012134 Differentiation of stem cells and stabilization of phenotypical properties of primary cells. Applicant: VUB – Inventors: Rogiers V, Vanhaecke T, Snykers S, Papeleu P, Vinken M, Henkens T, and Elaut G.

Patent submission PCT/EP2007/055754 Differentiation of cells. Applicant: VUB – Inventors: Rogiers V, Vanhaecke T, Snykers S, Papeleu P, Vinken M, Henkens T, and Elaut G.

References

1. Zech, N. H., Shkumatov, A. and Koestenbauer, S. (2007) The magic behind stem cells. *J Assist Reprod Genet.* **24**(6), 208–214.
2. Snykers, S., Vanhaecke, T., Papeleu, P., Luttun, A., Jiang, Y., Vander Heyden, Y., Verfaillie, C. and Rogiers, V. (2006) Sequential exposure to cytokines reflecting embryogenesis: the key for in vitro differentiation of adult bone marrow stem cells into functional hepatocyte-like cells. *Toxicol Sci.* **94**, 330–341.
3. Snykers, S., Vanhaecke, T., De Becker, A., Papeleu, P., Vinken, M., Van Riet, I. and Rogiers, V. (2007) Chromatin remodelling agent trichostatin A: a key-factor in the hepatic differentiation of human mesenchymal stem cells derived of adult bone marrow. *BMC Dev Biol.* 7, 24.
4. Moore, K. A. and Lemischka, I. R. (2006) Stem cells and their niches. *Science.* **311**, 1880–1885.

5. Kondo, T. (2006) Epigenetic alchemy for cell fate conversion. *Curr Opin Genet Dev.* **16**(5), 502–507.

6. Zardo, G., Cimino, G. and Nervi, C. (2008) Epigenetic plasticity of chromatin in embryonic and hematopoietic stem/progenitor cells: therapeutic potential of cell reprogramming. *Leukemia.* **22**(8), 1503–1518.

7. Collas, P. (2008) Epigenetic states in stem cells. *Biochim Biophys Acta.* ahead of print.

8. Reik, W. (2007) Stability and flexibility of epigenetic gene regulation in mammalian development. *Nature.* **447**(7143), 425–432.

9. Perry, P., Sauer, S., Billon, N., Richardson, W. D., Spivakov, M., Warnes, G., Livesey, F. J., Merkenschlager, M., Fisher, A. G. and Azuara, V. (2004) A dynamic switch in the replication timing of key regulator genes in embryonic stem cells upon neural induction. *Cell Cycle.* **3**(12), 1645–1650.

10. Vanhaecke, T. and Rogiers, V. (2006) Hepatocyte cultures in drug metabolism and toxicological research and testing. In: Phillips IR and Shephard EA, editors. Cytochrome P450 Protocols, 2nd ed. Series in: Methods in Molecular Biology. Totowa: *Humana.* p. 209–227.

11. Rozga, J. (2006) Liver support technology – an update. *Xenotransplantation.* **13**(5), 380–389.

12. Snykers, S., De Kock, J., Vanhaecke, T. and Rogiers, V. (2007) Differentiation of neonatal rat epithelial cells from biliary origin into immature hepatic cells by sequential exposure to hepatogenic cytokines and growth factors reflecting liver development. *Toxicol In Vitro.* **21**(7), 1325–1331.

Chapter 24

Hepatic Transplantation of Mesenchymal Stem Cells in Rodent Animal Models

Bruno Christ, Sandra Brückner, and Peggy Stock

Abstract

The hepatocyte is the smallest functional entity of the liver and executes the majority of this organ's metabolic functions. Hence, hepatocyte transplantation has become a versatile alternative to whole organ liver transplantation. This novel treatment option is based on the assumption that transplanted hepatocytes integrate into the host liver, proliferate at the site of tissue damage, take over the long-term hepatic synthetic capacity, and thus substitute for the diseased host tissue. However, clinical success is still waiting for a breakthrough, likely because of two major reasons including (1) the scarcity of cadaveric donor livers and (2) the largely poor quality of cells isolated from marginal quality donor organs. Therefore, alternative cell sources have to be established to further prompt the clinical success of hepatocyte transplantation. Due to their multiple differentiation potential and nearly unlimited availability, stem cells are an attractive alternate resource. Because of both clinical and ethical objections, adult stem cells are often preferred over embryonic stem cells as a starting material. Recent studies have demonstrated the ability of mesenchymal stem cells derived from various tissues to differentiate into hepatocyte-like cells in vitro as well as showing specific hepatocyte functions in vivo after transplantation into the livers of mice or rats.

Key words: Cell transplantation, Mesenchymal stem cell, Hepatocyte differentiation

1. Introduction

Mesenchymal stem cells (MSC) have become an easily accessible and readily available cell resource for clinical applications in regenerative medicine. This utility is based on their proliferation potential in vitro and their capacity for multiple lineage differentiation (1, 2). The clinical feasibility of MSC has been proven through the transplantation of these cells into patients suffering from a number of various diseases (3–6). MSC reside in multiple tissues and organs. Although they share gross biological and biochemical properties, they feature distinct growth and

Mohan C. Vemuri et al. (eds.), *Mesenchymal Stem Cell Assays and Applications*, Methods in Molecular Biology, vol. 698, DOI 10.1007/978-1-60761-999-4_24, © Springer Science+Business Media, LLC 2011

differentiation kinetics depending on the tissue source they originate from (7–9).

Studies in rodents (10, 11) and humans (12–16) have demonstrated that MSC from bone marrow, umbilical cord blood, or adipose tissue may differentiate into hepatocyte-like cells in vitro by applying specific inductive growth conditions. After hepatic transplantation in a xenogeneic mouse model, differentiation and integration of MSC from human bone marrow was less efficient than that of hepatocyte-like cells predifferentiated from MSCs in vitro prior to transplantation. This enhanced differentiation and integration was shown by immunohistochemical staining of human hepatocyte-specific markers and FACS analyses quantifying the amount of human cells in the chimeric mouse livers (17). The use of xenogeneic rodent models is the gold standard for the transplantation of human cells. Identifying transplanted human cells in the host liver is a major challenge for these types of therapeutic studies. In general, functionally integrated human donor-derived cells are identified in the recipient liver by the expression of human hepatocyte-specific antigens. An elegant example of such a model is a syngeneic rat model that takes advantage of the naturally occurring mutation of the CD26 gene in the Fischer F344 rat (18, 19). This model allows the identification of wild-type CD26 expressing MSC-derived hepatocyte-like cells in the otherwise negative host liver while undifferentiated MSC do not express CD26. Hence, in these models the expression of specific antigens by transplanted cells can allow the evaluation of both topological and functional host cell integration.

2. Materials

2.1. Isolation of Human MSC from Femur Bone Marrow and Cell Culture

1. 0.106 M Trisodium citrate dihydrate, pH 7.4.
2. Collagenase NB 4G (Serva, Heidelberg, Germany).
3. Phosphate-buffered saline (Biochrom AG, Berlin, Germany).
4. Biocoll separation solution ($\delta = 1.077$ g/ml) (Biochrom AG).
5. Stem cell maintenance medium: Prepare D-MEM (1.0 g/L D-Glucose; w L-Glutamine, w/o $NaHCO_3$) (Biochrom AG) by dissolving the powder in 1 L of Aqua dest. To this medium add 3.7 g $NaHCO_3$, adjust the pH to 7.2, and pass through a 0.22-μm sterile filter. Prepare MCDB 201 (Sigma-Aldrich, Steinheim, Germany) by dissolving the powder in 1 L of Aqua dest. Adjust the pH to 7.2 and pass through a sterile filter. Next, combine three parts of D-MEM with two parts of MCDB. Mix the solution well and add 5 ng/ml selenious acid (98%), 5 μg/ml apo-transferrin, 4.7 μg/ml linoleic acid (liquid, 99%), 1 nM dexamethasone, 5 μg/ml insulin, 0.1 mM

L-ascorbic acid 2-phosphate (all from Sigma), and 1% Penicillin/Streptomycin (c-c-pro, Oberdorla, Germany). Finally, supplement the medium with 15% heat-inactivated FBS (Invitrogen, Karlsruhe, Germany). Substances are added from frozen aliquots just prior to use (see Note 1).

6. Dimethylsulfoxide (Roth, Karlsruhe, Germany).

2.2. Isolation of Rat MSC from Adipose Tissue and Cell Culture

1. Washing buffer: Made of 20 mM Hepes (Serva), 120 mM NaCl (Roth), 4.8 mM KCl (Merck, Darmstadt, Germany), 1.2 mM $MgSO_4 \times 7H_2O$ (Sigma-Aldrich), 1.2 mM KH_2PO_4 (Merck), 4% BSA (Paesel+Lorei, Duisburg, Germany), 50 mg/L DNAse (Sigma-Aldrich).
2. DMEM-HG (PAA, Pasching, Austria).
3. Animals: Wild-type Fischer F344 rats serve as MSC donors and may be purchased from various commercial suppliers.

2.3. Transplantation of MSC into Mouse and Rat Liver

1. Trypsin (Invitrogen).
2. Phosphate-buffered saline (Biochrom AG).
3. Syringe and needle 25G (rat) or 26G (mouse).
4. Sterile swab (Fuhrmann Verbandstoffe GmbH, Bövingen, Germany).
5. Set of instruments: Scissors, tissue retractor 5 cm (rat) or 3 cm (mouse), surgical forceps, needle holder.
6. Sterile gauze compress (Fuhrmann Verbandstoffe GmbH).
7. Iodine solution for disinfection (B. Braun Melsungen AG, Melsungen, Germany).
8. Electric shaver.
9. Surgical suture material (3/0) (Catgut, Markneukirchen, Germany).
10. Ethyl alcohol (70%) (Sigma-Aldrich).
11. Sterile absorbable gelatin sponge (Chauvin ankerpharm GmbH, Berlin, Germany).
12. Isoflurane (Abbott, Wiesbaden, Germany).
13. Oxygen cylinder (Linde AG, Leuna, Germany).
14. Anesthetics inhalator (Völker GmbH, Kaltenkirchen, Germany).
15. Electric heating pad (Beurer, Ulm, Germany).
16. Retrorsine (Sigma-Aldrich).
17. Propranolol (Schwarz Pharma, Monheim, Germany).
18. Animals: All experiments including animals must conform to national and institutional regulations and be filed at the federal state authorities.

Recipient rats: Fischer rats defective in the CD26 protein are used. The animals are not available commercially and are grown in our own animal breeding facilities at the University of Halle-Wittenberg (19) (see Note 2). Animals are kept under the normal daily light cycle and are fed a standard diet with free access to drinking water. They should have a body weight of 200–250 g when being used in the experiments. By blocking the regenerating capacity of the host hepatocytes in the recipient liver, transplanted hepatocytes are provided with a growth advantage in order to promote the repopulation of the host liver. This is achieved by i.p. injection of 30 mg/kg of the pyrrolizidine alkaloid retrorsine 6 and 4 weeks prior to the cell transplantation.

Recipient mice: Pfp/Rag2$^{-/-}$ immunodeficient mice lacking functional NK as well as mature T and B cells (Taconic, Ejby, Denmark) (see Note 3). Animals are kept under the normal daily light cycle and are fed a standard diet with free access to drinking water. They have a body weight of 25 g when being used in the experiments. The regenerating capacity of the recipient liver is temporarily blocked by propranolol (final dose: 60 mg/kg) given in the drinking water over a time period of 3 days prior to cell transplantation.

2.4. Immunohistochemical Detection of Transplanted Human MSC in the Mouse Liver

1. 4% Histofix (Roth).
2. Silan (Sigma-Aldrich).
3. Glass slides (Schütt24, Göttingen, Germany).
4. Microtome (Microm, Walldorf, Germany).
5. Xylol (Roth).
6. Ethyl alcohol (96, 80, 70, 50%) (Sigma-Aldrich).
7. 0.01 M citrate buffer pH 6.0 (Merck).
8. Phosphate-buffered saline (Biochrom AG).
9. Blocking solution: Made of 5% BSA (Paesel + Lorei) and 0.5% Tween 20 (Merck) in PBS.
10. 3% H_2O_2.
11. DAB Substrate Kit for Peroxidase (Linaris, Wertheim-Bettingen, Germany).
12. Hemalaun (Merck).
13. Entellan (Merck).
14. Primary antibody: Monoclonal mouse antihuman hepatocyte antigen, clone OCH1E5, specifically detecting the human HepPar1 antigen (Dako, Hamburg, Germany).
15. Secondary antibody: Goat antimouse Ig (HRP) (BD, Erembodegem, Belgium).

2.5. Histochemical Detection of CD26-Expressing Rat MSC in the Rat Liver

1. Acetone (Sigma-Aldrich).
2. Ethyl alcohol (96%) (Sigma-Aldrich).
3. Substrate solution: To 10 ml of 100 mmol/L Tris maleate buffer (Roth) (pH 6.5) add 58 mg of NaCl (AppliChem, Darmstadt, Germany). In 1 ml of this buffer, dissolve 0.5 mg Gly-pro-methoxy-β-naphthylamide (Sigma-Aldrich) and 1 mg of Fast Blue (Sigma-Aldrich).
4. Phosphate-buffered saline (Biochrom AG).
5. Hemalaun (Merck).
6. Formalin (10%) (Roth).
7. Glycerol-gelatine (Merck).
8. Mr. Frosty™ Cryo Freezing Container (Nalgene, Roskilde, Denmark).
9. MEV cryostat (SLEE Technical GmbH, Mainz, Germany).

2.6. Determination of the Repopulation Rate by Flow Cytometry

1. Set of instruments: scissors, surgical forceps.
2. Permanent venous catheter (Johnson&Johnson, Neuss, Germany).
3. Peristaltic pump (IKA®, Staufen, Germany).
4. Perfusion tubings (Roth).
5. Bubble trap.
6. Glass heat exchanger.
7. Water bath (Lauda, Königshofen, Germany).
8. Narcorene i.p. at a dose of 700 μl/kg (Merial GmbH, Hallbergmoos, Germany).
9. Heparin 25,000 I.E./5 ml, i.p. at a dose of 1,000 μl/kg (ROTEXMEDICA, Trittau, Germany).
10. Krebs-Ringer-buffer (KRP): Made of 120 mM NaCl (Roth), 4.8 mM KCl (Merck), 1.2 mM $MgSO_4 \times H_2O$ (Sigma-Aldrich), 1.2 mM KH_2PO_4 (Merck), 24.4 mM $NaHCO_3$ (Merck).
11. First perfusion buffer (PPP): Dissolve 30 mg EGTA (Roth) in 300 ml KRP and equilibrate with carbogen for 30 min. Adjust the pH to 7.35, filter the buffer sterile into a bottle, and preheat to 37°C.
12. Second perfusion buffer (CPP): Dissolve 892.5 mg Hepes (Serva) and 147.5 mg $CaCl_2 \times 2H_2O$ (Merck) in 250 ml of KRP and equilibrate with carbogen for 30 min. Adjust the pH to 7.5, sterile filter the buffer into a bottle, and preheat to 37°C. Immediately before use, dissolve 0.12 U/ml collagenase NB 4G (Serva) in a small volume of this CPP and filter sterile into the buffer.
13. Washing buffer (see Subheading 2.2, item 1).

14. FACS tubes (BD).
15. Mouse antirat CD26 antibody (BD).
16. Isotype control (BD).
17. PBS containing 1% FBS.
18. FACS scan (BD).
19. Cell Quest software (BD).

3. Methods

3.1. Preparation of Human Bone Marrow-Derived MSC for Transplantation

1. In conical tubes, add 10-times a volume of sodium citrate to the bone marrow (usually 4–10 ml) immediately after harvest to prevent blood clotting. Keep sample on ice and quickly start preparation, otherwise efficiency decreases dramatically (see Note 4).
2. Add 0.075% (w/v) (15 mg) collagenase in DMEM and adjust the volume to 20 ml with DMEM.
3. Mix thoroughly and incubate for 25–30 min at 37°C with gentle agitation of the tubes in horizontal position.
4. Stop digestion by adding 5 ml of FBS, pass mixture through a Nylon mesh (125 μm) to remove undigested tissue and bone fragments. Spin for 7 min at 150×*g* and 4°C.
5. Suck off the supernatant with rotating movements to remove residual fat thoroughly.
6. Resuspend the pellet in 20–40 ml cold PBS and spin for 7 min at 150×*g* and 4°C.
7. Repeat the previous step 2–3 times and resuspend the pellet in 1 ml PBS after the final washing step.
8. Prepare a density gradient by layering the cell suspension onto a 10-ml Percoll cushion in a 15-ml tube. Carefully avoid mixing of the layers.
9. Spin for 20 min at 150×*g* and 4°C (brake off).
10. The interphase containing the mononuclear cell fraction is then carefully recovered by using a sterile Pasteur pipet. If red blood cells are still visible, repeat steps 8 and 9.
11. Finally add 10 ml PBS to the cell suspension and spin for 5 min at 150×*g* and 4°C.
12. Repeat the previous step twice.
13. After the final centrifugation, resuspend cells in 5 ml of stem cell maintenance medium and determine cell number.
14. Plate cells onto uncoated plastic dishes or bottles at a cell density of 200 cells/cm^2 and culture for 24–48 h under a gas atmosphere of 5% CO_2 at 37°C (see Note 5).

15. After the initial medium change (after 24–48 h), subsequent medium changes are made every 4 days. After 1 week of culture, small colonies should appear. Do not allow colonies to grow to more than 60% confluence to avoid spontaneous differentiation. At this stage, cells are trypsinized and either processed for downstream applications or stored in undiluted FBS supplemented with 7.5% DMSO at −196°C. For the controlled freezing process, use the Mr. Frosty™ Cryo Freezing Container. Stocks might contain between 25,000 and 250,000 cells.

3.2. Preparation of Rat Adipose Tissue-Derived MSC for Transplantation

1. Open the peritoneal cavity of a wild-type Fischer 344 rat and excise peritoneal adipose tissue with a scissors and forceps. Place the tissue immediately into cold washing buffer in a conical tube. Determine the weight of the tissue by weighing the tube without and with the tissue and start preparation of cells as soon as possible to avoid decreased cell recovery.
2. Suck off the washing buffer and add a small volume of DMEM-HG. The following procedure is described for the use of not more than 5 g of tissue. If more is required, run the protocol appropriately in parallel at the same scale.
3. Cut the tissue into pieces and add 0.075% (w/v) (15 mg) collagenase in DMEM-HG. Adjust the volume to 20 ml with DMEM-HG.
4. Continue this protocol as described in steps 3–15 under Subheading 3.1 (see Note 6).

3.3. Hepatic Transplantation of Human MSC in Immunodeficient Mice via Intrasplenic Application

1. Detach MSCs from the culture dishes by incubation with a trypsin solution (50%) for 5–10 min at 37°C. Wash cells 3 times with phosphate-buffered saline and test cell viability by trypan blue exclusion. Aliquots of the cell suspension (50–100 μl) containing $0.5–1 \times 10^6$ cells are stored on ice until intrasplenic application. The cell suspension is pulled into the syringe just before the injection (see Subheading 3.3, step 7).
2. After isoflurane (2.0%) and oxygen (2 L per min) inhalation anesthesia, shave the abdomen and the region underneath the left costal arch of the recipient mouse and disinfect with ethyl alcohol (70%) and iodine solution.
3. Uncover the abdominal musculature with surgical forceps and scissors by a 1.5-cm incision of the skin and open the peritoneal cavity with a 1-cm long incision. Fix with a small retractor in an optimal position.
4. For the 1/3 hepatectomy, a proximal ligature of the left lateral liver lobe is followed by a sharp resection of the liver lobe and shortening of the sutural material (see Note 7).

5. After suture of the abdominal musculature and the skin, shift the mouse to expose the region underneath the left costal.
6. Carefully set a small (0.5 cm) incision without damaging the costal arch. Pull out the spleen attached to adipose tissue with the help of surgical forceps and a sterile swab.
7. Fix the spleen by a median ligature, which also should fix the 26-gauge needle pricked into the apical pole of the spleen to prevent back retraction during the slow injection of the cell suspension (Fig. 1).

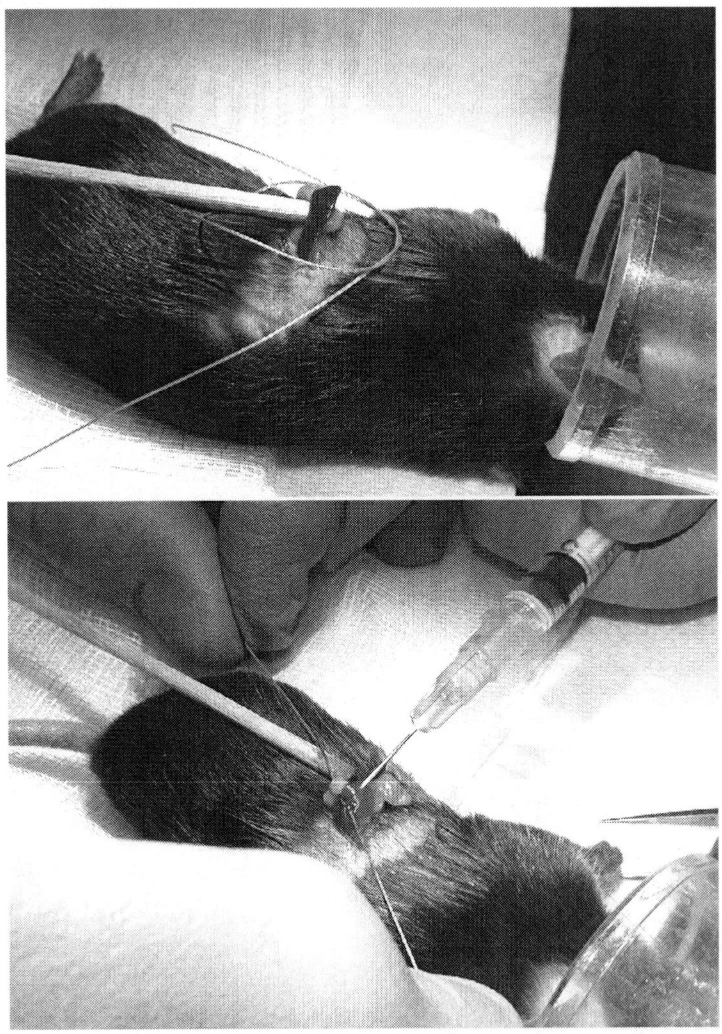

Fig. 1. Mice are placed on the right body side and a small incision is carried out beneath the left costal arch to expose the spleen with the aid of a sterile swab. A ligature is prepared (*upper panel*), which later will fix the needle at the time of infusion of the cells (*lower panel*). After completing injection, the needle is retracted and the ligation tightened to prevent leakage of injected cells. On the right, a tube for the application of inhalation anesthesia is visible.

8. After the infusion of the cells, the ligature is tightened and the needle removed. The spleen is slipped back, and the musculature and the skin are sutured.
9. Recover the animals from anesthesia by warming over an electric heating pad (see Note 8).

3.4. Hepatic Transplantation of MSC in the Rat via Portal Vein Application

1. Detach the MSCs from the culture dishes and wash as described under Subheading 3.3, step 1. Aliquots of the cell suspension (500 μl) containing $1.5–2.5 \times 10^6$ cells are stored on ice until portal vein application (step 4). The cell suspension is pulled into the syringe just before the injection.
2. Perform anesthesia as described under Subheading 3.3, step 2. Shave the abdomen and disinfect with ethyl alcohol (70%) and iodine solution. Completely cover the animal except the abdomen with sterile gauze compresses.
3. Uncover the abdominal musculature with surgical forceps and scissors by a 6-cm incision of the skin and open the peritoneal cavity with a 5-cm long incision. Fix with a small retractor in an optimal position.
4. For the 1/3 hepatectomy, a proximal ligature of the left lateral liver lobe is followed by a sharp resection of the liver lobe (see Note 7). Dislocate the omentum majus and the intestine to expose the portal vein as shown in Fig. 2. Puncture the vein

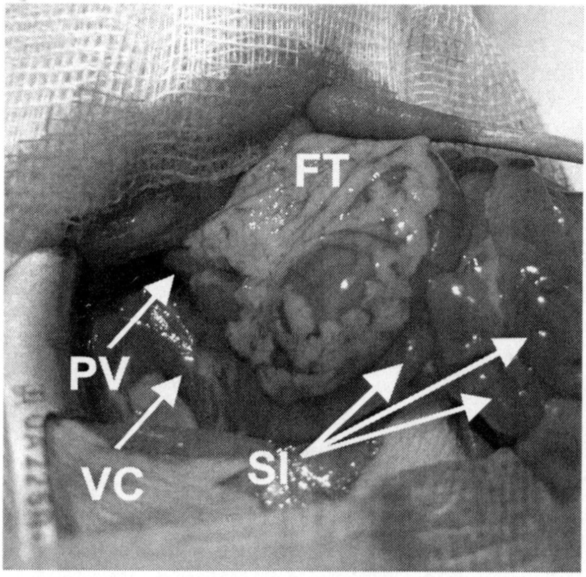

Fig. 2. The abdomen of an anesthetized rat is opened and the portal vein (PV) exposed by dislocation of the small intestine (SI). The portal vein must be distinguished from the vena cava inferior (VC), which is located dorsal from the portal vein and in general is much larger than the latter. Occasionally, the portal vein is hardly visible because fat tissue (FT) is incorporated in the omentum majus. Cells are slowly injected with a syringe and a needle into the portal vein.

and slowly inject the cells (prepared as described in step 1) using a syringe and a needle (25G). After injection is completed, stop bleeding after retraction of the needle with an absorbable gelatin sponge.

5. Put the intestinal loop back into the peritoneal cavity and suture the abdominal musculature and the skin.
6. Place the animals on an electric heating pad to recover from anesthesia (see Note 8).

3.5. Preparation of Liver Tissue Sections for the Detection of Transplanted Cells

3.5.1. Preparation of Paraffin Sections from Mouse Livers (see Note 9)

1. Administer (i.p.) a lethal dose of narcorene to the animals, open the peritoneal cavity, and explant the whole liver.
2. Split the liver into pieces of 1 × 1 cm size and place them into tubes filled with 4% histofix. The pieces are paraffinized in an automated procedure and sectioned into 2 μm slices using a microtome. Slices are mounted onto silan-coated glass slides.

3.5.2. Preparation of Cryo-Sections from Rat Livers (see Note 10)

1. Administer (i.p.) a lethal dose of narcorene to the animals, open the peritoneal cavity and explant the whole liver.
2. Split the liver into pieces of 1 × 1 cm size and place them into cryotubes. To achieve a controlled freezing process, a cooling rate of −1°C/min is adjusted using the Mr. Frosty™ Cryo Freezing Container.
3. Prepare 5 μm thick tissue sections of the liver using a cryostat and place them onto silan-coated glass slides. Until histochemical detection of CD26, store the slides at −80°C.

3.6. Immunohistochemical Detection of Transplanted Human Cells in the Mouse Liver

1. First dewax the slices (prepared as described under Subheading 3.5, step 2) in a descending ethanol series (xylol, 96, 80, 70, 50%).
2. Unmask the slices by incubation in a pressure cooker with 0.01 M citrate buffer for 20 min.
3. After washing with PBS, incubate with 3% H_2O_2 at 4°C for 30 min to block endogenous peroxidases. Wash again in PBS and cover the slices with blocking solution for 1 h.
4. Dilute the primary antibody to 1:50 in PBS + 1% BSA and incubate the slices with antibody solution for 2 h in a humid chamber. After washing 3 times with PBS, incubate with the secondary antibody diluted to 1:200. Wash again 3 times.
5. To detect the secondary antibody-coupled peroxidase, use the DAB substrate kit following the instruction manual.
6. Counterstain the slices with hemalaun and dehydrate in an ascending ethanol series (50, 70, 80, 96%, isopropanol, xylol).

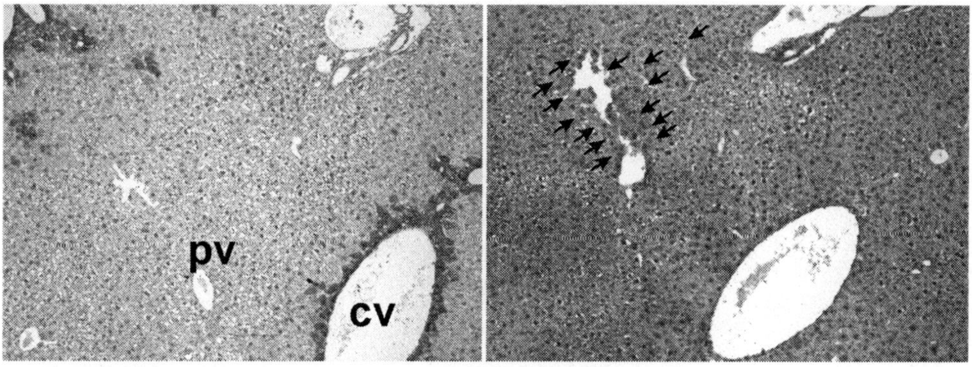

Fig. 3. Perivenous and periportal areas in the mouse liver lobule may be distinguished by the detection of glutamine synthase, which is specifically located in 1–3 layers of hepatocytes surrounding the branches of the central vein (cv). Hepatocytes surrounding branches of the portal vein (pv) are negative (*left panel*). This typical staining pattern allows for the topological orientation of human MSCs in the mouse liver in serial sections. Human cells are identified by staining of human-specific antigens. Here, we used the antihuman-specific hepatocyte antigen HepPar1, which is expressed in human hepatocytes. Seven weeks after transplantation, clusters of human cells are predominantly visible in periportal areas (arrows in *right panel*). They are differentiated into hepatocyte-like cells. This can be concluded from the expression of HepPar1, which is not expressed in undifferentiated MSC. MSC used in the experiment shown have been predifferentiated in vitro into hepatocyte-like cells prior to transplantation as described in ref. (17). This procedure augments repopulation of the host liver by the transplanted donor cells significantly.

7. Finally cover the slices with entellan and analyze under a microscope (Fig. 3).

3.7. Histochemical Detection of CD 26 in the Rat Liver

1. Fix the slices (prepared as described under Subheading 3.5, step 5) with acetone for 5 min at –20°C.
2. Wash the slices with 96% ethyl alcohol, dry under air and incubate for 20 min with substrate solution.
3. After washing with PBS, stain with hemalaun and fix in formalin (10%).
4. Finally cover the slices with glycerol-gelatine and analyze under a microscope (Fig. 4).

3.8. Preparation of Isolated Rat Hepatocytes for Flow Cytometry by Liver Perfusion (see Note 11)

1. Inject animals i.p. with heparin and narcorene at a lethal dose. Open the peritoneal cavity and dislocate the intestinal loop to expose the portal vein.
2. Ligate the portal vein and insert the permanent venous catheter, fix the catheter with the ligature.
3. Connect the perfusion tubing to the catheter and start perfusion with PPP (20–25 ml/min) using a peristaltic pump. Immediately puncture the vena cava inferior to allow blood and buffer outflow. For the setup of the perfusion equipment see Fig. 5.
4. After 10 min, the liver should be free of blood, exchange PPP for CPP and continue with the perfusion.

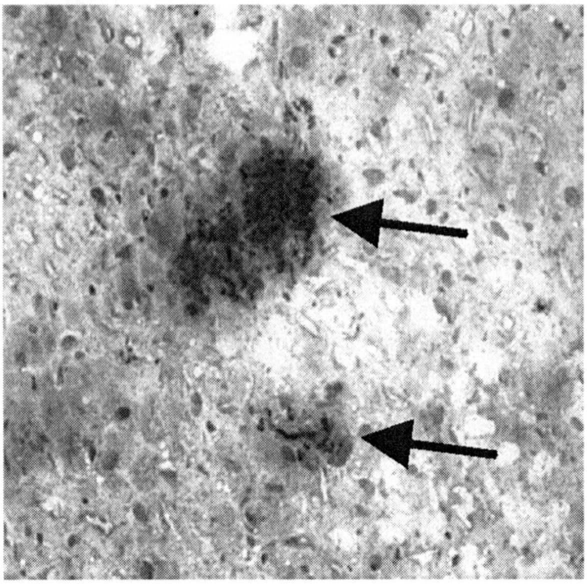

Fig. 4. Ten weeks after hepatic transplantation, clusters of CD26-expressing donor cells (*arrows*) are visible in the negative host liver. Histochemical staining reveals a red color, which indicates that the transplanted MSC at least in part differentiated into hepatocyte-like cells, because undifferentiated MSC do not express CD26. MSC used in the experiment shown have been predifferentiated in vitro into hepatocyte-like cells prior to transplantation as described in ref. (20). This procedure augments repopulation of the host liver by the transplanted donor cells significantly.

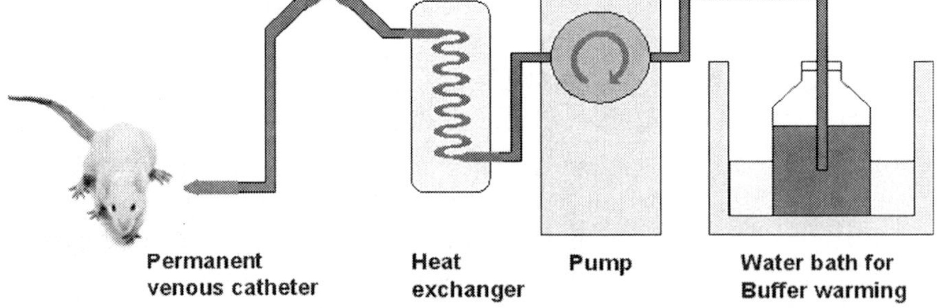

Fig. 5. Set-up of the equipment for the isolation of hepatocytes from the rat liver. After perfusion with a washing buffer to flush out the blood from the liver, buffer containing collagenase is perfused to digest the liver tissue. During perfusion media are prewarmed to 37°C.

5. The digestion is optimal when dark fields of digested liver tissue become visible. Stop the perfusion and excise the liver.
6. Place the liver into washing buffer and open the liver capsule by gentle scratching with a needle. Shake the liver to wash out the hepatocytes. Pass the cell suspension through a sterile gauze, compress and spin for 5 min at 70×g and 4°C. Discard the supernatant and wash the cells twice with washing buffer.

3.9. Flow Cytometry

1. Wash aliquots (200,000 cells) of the isolated hepatocytes prepared as described under Subheading 3.8 with PBS+1% FBS in FACS tubes. Spin for 5 min at 70×g and 4°C and discard the supernatant.
2. Dilute the CD26 antibody 1:50 in PBS+1% FBS and incubate the cells with 75 µl of diluted antibody for 1 h at room temperature. The isotype control is diluted 1:7 in PBS+1% FBS (see Note12).
3. Wash the cells twice with PBS+1% FBS and finally resuspend in 500 µl PBS+1% FBS.
4. Analyze the cells by using a FACSscan and the Cell Quest software. Ten weeks after transplantation, the number of CD26-expressing cells in a host hepatocyte preparation amounts to about 1–2% of the analyzed population. If MSC predifferentiated into hepatocyte-like cells prior to transplantation are used, the repopulation rate increases by a factor of 10 as shown in Fig. 6.

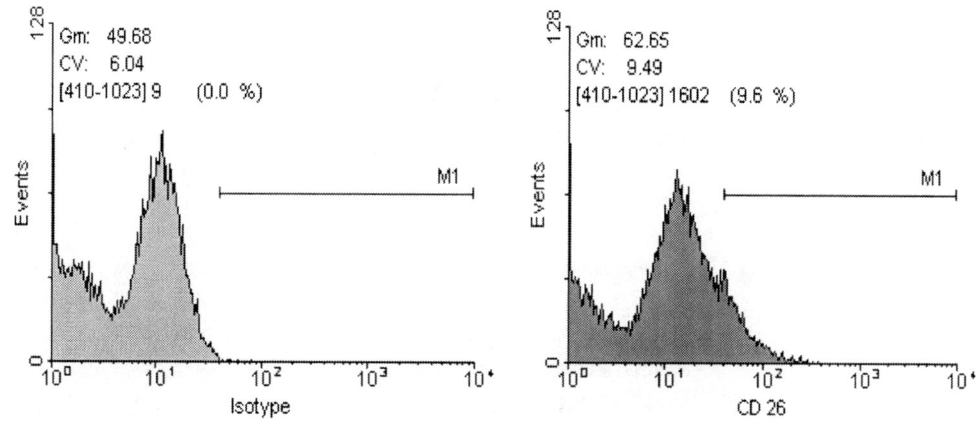

Fig. 6. Histograms typically showing the isotype control (*left*) and the amount of CD26-expressing cells in a whole hepatocyte preparation from a CD26-deficient Fischer rat liver (*right*) after transplantation of MSC derived from adipose tissue of wild-type Fischer rats. MSC used in the experiment shown have been predifferentiated in vitro into hepatocyte-like cells prior to transplantation as described in ref. (20). This procedure augments repopulation of the host liver by the transplanted donor cells significantly, in the case shown up to around 10% of the hepatocytes.

4. Notes

1. For individual components, stock solutions are prepared at appropriate concentrations and stored in aliquots at –20°C. Aliquots are used only once and discarded after thawing.
2. The rat model used here takes advantage of the naturally occurring mutant CD26 Fischer rat, in which CD26 is expressed as a nonfunctional protein. Therefore, hepatic transplantation of wild-type MSC into a mutant animal allows for both the histochemical and the immunohistochemical detection of CD26 enzyme activity and protein, respectively, in the transplanted cells in the otherwise negative host liver background.
3. Xenotransplantation of human MSC into immunodeficient mice enables the identification of transplanted cells by the expression of human proteins in the negative mouse liver background. Here we describe the detection of the antihuman hepatocyte antigen HepPar1. In addition to the identification of the human cells in the recipient mice, this marker is not expressed in undifferentiated MSC and ultimately indicates the differentiation of MSC to hepatocyte-like cells.
4. The isolation of MSC from human tissue requires the approval by the Ethics Committee and a patients' informed written consent. MSC may be isolated from human bone marrow aspirates from the iliac crest donated by volunteers or from femur bone marrow gained during artificial hip or knee joint implantation. Peritoneal or subcutaneous adipose tissue is acquired during elective surgery.
5. MSC might be enriched by growth on uncoated plastic surfaces. This enables the removal of nonadherent cells and cell debris after 24–48 h with the first medium change.
6. The procedure for the isolation of MSC from human adipose tissue is essentially the same as described for rat adipose tissue. Rat bone marrow MSCs are isolated from tibia and femur bones. The marrow is flushed out of the bones with 0.075% collagenase (w/v) (15 mg) in DMEM-HG supplemented with 2 ml sodium citrate. The total volume is adjusted to 20 ml with DMEM-HG and MSC selection continued as described in Subheading 3.1.
7. The hepatectomy is needed to provide a proliferative challenge, which in conjunction with the attenuation of host hepatocyte proliferation by previous retrorsine treatment (see Subheading 2.3, item 18) fosters amplification of transplanted cells in the host liver.
8. Transplanted cells might be detected in the host liver directly after transplantation. However, cells are initially trapped in the

blood vessels and it takes hours to days to allow penetration of the cells through the endothelia. The kinetics of integration of donor cells in the recipient parenchyma is hard to follow short-term, because only a small part of the transplanted cells invades the parenchyma so that only single cells are visible. The majority of cells are cleared from the hepatic vasculature by still unknown mechanisms. Ideally, cells proliferate at the site of integration forming clusters of MSC-derived hepatocyte-like cells 10 weeks after transplantation. The periportal areas of the liver lobule are the preferred sites of integration because after infusion cells are entrapped in the proximal parts of the liver sinusoids.

9. Antihuman antigen-specific antibodies are used to detect human cells in the otherwise negative mouse liver background. This might be done either in cryo- or paraffin sections.
10. In the rat model transplanted wild-type cells are identified by histochemical detection of CD26 in the negative host liver using cryosections.
11. The rat model used here allows for the quantification of donor-derived MSCs in the recipient liver by determining the number of CD26-expressing cells in a whole hepatocyte population isolated from a CD26 deficient host liver. CD26 is labeled by an antibody and cells quantified by FACS.
12. For an optimal labeling, it is necessary to shake the tubes gently during antibody incubation.

Acknowledgments

The authors are greatly indebted to Sabine Ebensing and Madlen Hempel, who carefully assisted throughout the methods setup. Work was supported by grants to BC from The German Ministry of Education and Research (NBL3-NG4 and BMBF, PtJ-Bio, 0313909, 1106SF) as well as the German Research Council (Ch 109/15-1).

References

1. Barry, F. P. and Murphy, J. M. (2004) Mesenchymal stem cells: clinical applications and biological characterization. *Int J Biochem Cell Biol* **36**, 568–584.
2. Kassem, M., Kristiansen, M. and Abdallah, B. M. (2004) Mesenchymal stem cells: cell biology and potential use in therapy. *Basic Clin Pharmacol Toxicol* **95**, 209–214.
3. Bunnell, B. A., Deng, W., Robinson, C. M., Waldron, P. R., Bivalacqua, T. J., Baber, S. R., Hyman, A. L. and Kadowitz, P. J. (2005) Potential application for mesenchymal stem cells in the treatment of cardiovascular diseases. *Can J Physiol Pharmacol* **83**, 529–539.
4. Christopeit, M., Schendel, M., Foll, J., Muller, L. P., Keysser, G. and Behre, G. (2008) Marked improvement of severe progressive systemic sclerosis after transplantation of mesenchymal stem cells from an allogeneic haploidentical-related donor mediated by ligation of CD137L. *Leukemia* **22**, 1062–1064.

5. Ohnishi, S. and Nagaya, N. (2007) Prepare cells to repair the heart: mesenchymal stem cells for the treatment of heart failure. *Am J Nephrol* **27**, 301–307.
6. Ringdén, O., Uzunel, M., Rasmusson, I., Remberger, M., Sundberg, B., Lönnies, H., Marschall, H.-U., Dlugosz, A., Szakos, A., Hassan, Z., Omazic, B., Aschan, J., Barkholt, L. and Le Blanc, K. (2006) Mesenchymal stem cells for treatment of therapy-resistant graft-versus-host disease. *Transplantation* **81**, 1390–1397.
7. Taléns-Visconti, R., Bonora, A., Jover, R., Mirabet, V., Carbonell, F., Castell, J. V. and Gómez-Lechón, M. J. (2006) Hepatogenic differentiation of human mesenchymal stem cells from adipose tissue in comparison with bone marrow mesenchymal stem cells. *World J Gastroenterol* **12**, 5834–5845.
8. Izadpanah, R., Trygg, C., Patel, B., Kriedt, C., Dufour, J., Gimble, J. M. and Bunnell, B. A. (2006) Biologic properties of mesenchymal stem cells derived from bone marrow and adipose tissue. *J Cell Biochem* **99**, 1285–1297.
9. Wagner, W., Wein, F., Seckinger, A., Frankhauser, M., Wirkner, U., Krause, U., Blake, J., Schwager, C., Eckstein, V., Ansorge, W. and Ho, A. D. (2005) Comparative characteristics of mesenchymal stem cells from human bone marrow, adipose tissue, and umbilical cord blood. *Exp Hematol* **33**, 1402–1416.
10. Lange, C., Bassler, P., Lioznov, M. V., Bruns, H., Kluth, D., Zander, A. R. and Fiegel, H. C. (2005) Hepatocytic gene expression in cultured rat mesenchymal stem cells. *Transplant Proc* **37**, 276–279.
11. Wang, P. P., Wang, J. H., Yan, Z. P., Hu, M. Y., Lau, G. K., Fan, S. T. and Luk, J. M. (2004) Expression of hepatocyte-like phenotypes in bone marrow stromal cells after HGF induction. *Biochem Biophys Res Commun* **320**, 712–716.
12. Lee, K. D., Kuo, T. K., Whang-Peng, J., Chung, Y. F., Lin, C. T., Chou, S. H., Chen, J. R., Chen, Y. P. and Lee, O. K. (2004) In vitro hepatic differentiation of human mesenchymal stem cells. *Hepatology* **40**, 1275–1284.
13. Hong, S. H., Gang, E. J., Jeong, J. A., Ahn, C., Hwang, S. H., Yang, I. H., Park, H. K., Han, H. and Kim, H. (2005) In vitro differentiation of human umbilical cord blood-derived mesenchymal stem cells into hepatocyte-like cells. *Biochem Biophys Res Commun* **330**, 1153–1161.
14. Seo, M. J., Suh, S. Y., Bae, Y. C. and Jung, J. S. (2005) Differentiation of human adipose stromal cells into hepatic lineage in vitro and in vivo. *Biochem Biophys Res Commun* **328**, 258–264.
15. Banas, A., Teratani, T., Yamamoto, Y., Tokuhara, M., Takeshita, F., Quinn, G., Okochi, H. and Ochiya, T. (2007) Adipose tissue-derived mesenchymal stem cells as a source of human hepatocytes. *Hepatology* **46**, 219–228.
16. Aurich, I., Mueller, L. P., Aurich, H., Luetzkendorf, J., Tisljar, K., Dollinger, M. M., Schormann, W., Walldorf, J., Hengstler, J. G., Fleig, W. E. and Christ, B. (2007) Functional integration of hepatocytes derived from human mesenchymal stem cells into mouse livers. *Gut* **56**, 405–415.
17. Aurich, H., Sgodda, M., Kaltwaßer, P., Vetter, M., Weise, A., Liehr, T., Brulport, M., Hengstler, J. G., Dollinger, M. M., Fleig, W. E. and Christ, B. (2009) Hepatocyte differentiation of mesenchymal stem cells from human adipose tissue in vitro promotes hepatic integration in vivo. *Gut* **58**, 570–581.
18. Laconi, E., Oren, R., Mukhopadhyay, D. K., Hurston, E., Laconi, S., Pani, P., Dabeva, M. D. and Shafritz, D. A. (1998) Long-term, near-total liver replacement by transplantation of isolated hepatocytes in rats treated with retrorsine. *Am J Pathol* **153**, 319–329.
19. Laconi, S., Curreli, F., Diana, S., Pasciu, D., Filippo, G. D., Sarma, D. S. R., Pani, P. and Laconi, E. (1999) Liver regeneration in response to partial hepatectomy in rats treated with retrorsine: a kinetic study. *J Hepatol* **31**, 1069–1074.
20. Sgodda, M., Aurich, H., Kleist, S., Aurich, I., Konig, S., Dollinger, M. M., Fleig, W. E. and Christ, B. (2007) Hepatocyte differentiation of mesenchymal stem cells from rat peritoneal adipose tissue in vitro and in vivo. *Exp Cell Res* **313**, 2875–2886.

Chapter 25

Phenotypic Analysis and Differentiation of Murine Mesenchymal Stem Cells

Lindolfo da Silva Meirelles and Dimas Tadeu Covas

Abstract

Mesenchymal stem cells (MSCs) hold great promise as therapeutic tools to treat different types of disease and the use of preclinical animal models is mandatory for the development of novel MSC-based therapies. The mouse is one of the most important species used for preclinical experiments and, by extension, so is the isolation and characterization of murine MSCs. This chapter presents methods for the phenotypic analysis of cultured murine MSCs.

Key words: Mesenchymal stem cell, Mouse, Characterization, Cell culture, Flow cytometry

1. Introduction

Mesenchymal stem cells (MSCs) can be defined as nonhematopoietic, plastic-adherent cells able to proliferate in culture and to possess a phenotype characteristic of cells from connective tissues such as bone, cartilage, adipose, and others (1). Differentiation can be induced by culture in specific medium composition, or by implantation in vivo (2). In addition to their differentiation capabilities, MSCs are attractive for therapeutic purposes because they secrete a multitude of bioactive factors which, combined with their ability to dock at injured sites when systemically infused, contribute to the generation of local regenerative microenvironments (3). Consequently, clinical trials are currently underway to verify if these properties are effective for the treatment of some types of disease (reviewed in (4)).

The use of animal models is necessary to assess the suitability of different therapeutic strategies for clinical trials. The mouse is the ideal entry-level species for preclinical experiments for a number of reasons. First, their genetics and other aspects of murine

Mohan C. Vemuri et al. (eds.), *Mesenchymal Stem Cell Assays and Applications*, Methods in Molecular Biology, vol. 698, DOI 10.1007/978-1-60761-999-4_25,

physiology are well known, and many parallels can be established between mice and humans. Second, due to the availability of methods that allow genetic manipulation of murine embryonic stem cells, transgenic, knock-out, or knock-in mice can be generated in order to study specific biological problems. In addition, animal husbandry techniques can be used to generate genetically homogeneous or heterogeneous animals. Finally, and not rarely a decisive point for many laboratories around the globe, the cost of maintenance of mice is relatively low.

The isolation and characterization of mouse mesenchymal stem cells (mMSCs) represents a starting point to explore different MSC-based therapeutic strategies using murine models. We have previously isolated and characterized murine MSCs from bone marrow (5) and other organs such as brain, lungs, liver, spleen, kidneys, and skeletal muscle (6). This chapter describes some of the methods used for the phenotypic analysis of murine MSCs, along with notes that emphasize details deemed important. These methods are described below as basic protocols that can be used as a starting point for working with murine MSCs.

2. Materials

Specific materials for the protocols listed in this chapter are listed below. Equipments and common materials such as laminar flow hood, CO_2 incubator, inverted microscope with phase-contrast optics (see Note 1), centrifuge, vortex, pH meter, analytical scale, water bath, hotplate/magnetic stirrer, mortar with pestle, hemocytometer (Neubauer chamber), surgical instruments (scissors, forceps), syringes, and needles are expected to be available.

2.1. *In Vitro* Osteogenesis

1. Standard Culture Medium: Low-glucose Dulbecco's modified Eagle's Medium (DMEM-LG; Sigma-Aldrich, St. Louis, MO, USA, cat. # D5523) containing 3.7 g/l sodium bicarbonate (Sigma), 10 mM 4-(2-hydroxyethyl)-1-piperazineethanesulfonic acid (HEPES, free acid, Sigma), and 10% (v/v) fetal bovine serum (Hyclone, Logan, UT, USA) (see Note 2).
2. Ascorbic Acid 2-phosphate (AAP) Solution: Dissolve 50 mg ascorbic acid-2 phosphate (Sigma) in 10 ml DMEM-LG containing 15 mM HEPES. In a tissue culture hood, filter through a 0.22-μm pore size membrane (Millipore, Billerica, MA, USA). Store the solution in a closed flask at 4°C. It should be stable for several months. Use this solution as a 1,000× supplement.
3. β-glycerophosphate (BGP) Solution: Dissolve 630 mg BGP (Sigma) in 20 ml DMEM-LG containing 15 mM HEPES,

free acid (Sigma). In the tissue culture hood, filter through a 0.22-μm pore size membrane. Store the solution in a closed flask at 4°C. It should be stable for at least 2 months (see Note 3).

4. Dexamethasone (Dex) Solution: In the tissue culture hood, dissolve 1.2 mg dexamethasone (Sigma) in 1.223 ml ethanol (Sigma) to obtain a 2.5×10^{-3} M solution. This solution can be stored at −20°C for future use. Transfer 10 μl of the Dex solution into 2.5 ml sterile culture medium (or any buffered saline solution) to obtain a 1×10^{-5} solution, which will be used as 1,000× supplement. Prepare single-use aliquots and store at −20°C until use (see Note 4).

5. Osteogenic Differentiation Medium: Prepare 10 ml Osteogenic Differentiation Medium by combining 8 ml DMEM-LG containing 10 mM HEPES, 1 ml BGP solution (final concentration: 10 mM), 1 ml FBS, 10 μl AAP solution (final concentration: 5 μg/ml), and 10 μl Dex solution (final concentration: 1×10^{-8} M) (see Note 5).

2.2. Detection of In Vitro Osteogenesis

1. Paraformaldehyde Solution: Add 40 g paraformaldehyde (Sigma) to 100 ml phosphate-buffered saline (PBS; Sigma). Mix on a hotplate/magnetic stirrer at 60°C (see Note 6) until it dissolves (it will take some hours) (see Note 7). To avoid contamination, filter the solution through a 0.22-μm pore size membrane in a tissue culture hood and store it under sterile conditions. To avoid paraformaldehyde deterioration, store in aliquots at −20°C. Prepare a fresh working solution by thawing frozen aliquots and diluting 10× with PBS.

2. Alizarin Red S Solution: Dissolve 2 g Alizarin Red S (Sigma) in 90 ml deionized water. Adjust the pH to 4.1 with the addition of ammonium hydroxide (Sigma) (see Note 8). Raise the volume to 100 ml with deionized water. Filter through filter paper to remove precipitates.

2.3. In Vitro Adipogenesis

1. Insulin Solution: Add hydrochloric acid (Sigma) to ultrapure water until a pH of 2.5 is achieved. Dissolve 50 mg of insulin from bovine pancreas (Sigma) in 10 ml of the acid solution. Filter through a 0.22-μm pore membrane. This solution is stable for up to a year when stored at 4°C. Alternatively, aliquots can be stored for several years at −20°C.

2. Indomethacin Solution: Dissolve 0.36 g of indomethacin (Sigma) in 10 ml DMSO. This solution is stable for a year at room temperature. Use as a 1,000× supplement.

3. Rosiglitazone Solution: Grind two 8 mg Avandia® (GlaxoSmithKline Brasil Ltda., Rio de Janeiro, RJ, Brazil) tablets into a fine powder using a mortar with pestle.

Add 8.95 ml DMSO (Sigma) to the powder, rinsing the pestle to retrieve any remnants. Transfer the suspension to a 15 ml conical centrifuge tube and centrifuge at 1,000 × *g* for 5 min. Collect the supernatant and filter through a 0.22-μm pore membrane. Store this solution in small aliquots at –20°C until use. Use as a 1,000× supplement.

4. 3-isobutyl-1-methylxanthine (IBMX) Solution: Dissolve 111 mg of IBMX (Sigma) in 1 ml of DMSO. This solution is stable for at least 3 months when stored at –20°C. Use as a 1,000× supplement.
5. Caprylic acid solution: Dissolve 2.01 g of sodium octanoate (Sigma) in 50 ml of DMEM-LG and filter through a 0.22-μm pore membrane. Use this as a 50× supplement.
6. Citrated Human Platelet-poor Plasma: Draw 18 ml of peripheral blood with a 20-ml syringe containing 2 ml of 3.8% sodium citrate (Sigma). Transfer the citrated blood to a 50-ml centrifuge tube. Centrifuge at 3,600 × *g* or higher speed for 30 min at 22°C. Transfer the supernatant to a fresh centrifuge tube and allow it to stand at –70°C overnight. Thaw the plasma at 4°C. Centrifuge at 1,000 × *g* for 10 min at 4°C. Collect the supernatant and prefilter it using a 0.45-μm pore membrane (Millipore, Billerica, MA, USA). Filter the prefiltered plasma through a 0.22-μm pore membrane.
7. Adipogenic Differentiation Medium A: Prepare 10 ml Adipogenic Differentiation Medium A by combining 9 ml DMEM-LG containing 10 mM HEPES (HEPES is optional), 1 ml FBS, 10 μl Dex solution (final concentration: 1×10^{-8} M), and 10 μl insulin solution (final concentration: 5 μg/ml).
8. Adipogenic Differentiation Medium B: Prepare 10 ml Adipogenic Differentiation Medium B by combining 9 ml DMEM-LG or DMEM-HG containing 10 mM HEPES (HEPES is optional), 1 ml FBS, 10 μl Dex solution (final concentration: 1×10^{-8} M), 10 μl insulin solution (final concentration: 5 μg/ml), and 10 μl indomethacin (final concentration: 100 μM).
9. Adipogenic Differentiation Medium C: Prepare 10 ml Adipogenic Differentiation Medium C by combining 9 ml DMEM-LG or DMEM-HG containing 15 mM HEPES (HEPES is optional), 1 ml FBS, 10 μl Dex solution (final concentration: 1×10^{-8} M), 5 μl insulin solution (final concentration: 2.5 μg/ml), 10 μl indomethacin solution (final concentration: 100 μM), and 10 μl rosiglitazone solution (final concentration: 5 μM).
10. Adipogenic Differentiation Medium D: Prepare 10 ml Adipogenic Differentiation Medium D by combining 9 ml DMEM-LG or DMEM-HG containing 10 mM HEPES

(HEPES is optional), 1 ml FBS, 10 μl Dex solution (final concentration: 1×10^{-8} M), 5 μl insulin solution (final concentration: 2.5 μg/ml), 10 μl indomethacin solution (final concentration: 100 μM), and 10 μl IBMX solution (final concentration: 500 μM).

11. Adipogenic Differentiation Medium E: Prepare 10 ml Adipogenic Differentiation Medium E by combining 9.5 ml DMEM-LG or DMEM-HG containing 10 mM HEPES (HEPES is optional), 100 μl FBS, 250 μl Dex solution (final concentration: 2.5×10^{-7} M), 5 μl insulin solution (final concentration: 2.5 μg/ml), and 200 μl caprylic acid solution (final concentration: 5 mM).
12. Adipogenic Differentiation Medium F: Prepare 10 ml Adipogenic Differentiation Medium F by combining 7.5 ml DMEM-HG, 2.5 ml citrated human platelet-poor plasma, and 15 μl sodium heparin (Liquemine® for subcutaneous administration; Produtos Roche Químicos e Farmacêuticos S.A., Rio de Janeiro, RJ, Brazil; final concentration: 30 units/ml).

2.4. Detection of In Vitro Adipogenesis

1. Oil Red O Solution: Add 2 g of Oil Red O (Sigma) to a beaker containing 100 ml isopropanol (Sigma). Stir the solution while warming up in a water bath at 37°C for a few minutes to improve dissolution of the stain. Add 66.5 ml of deionized water and mix. Filter the resulting solution through filter paper to remove undissolved stain. Store in a closed flask at room temperature. This solution is stable for several months at room temperature. If precipitates form, refilter the solution.
2. Sudan Black B Solution: Add 2 g of Sudan Black B (Sigma) to a beaker containing 100 ml isopropanol. Stir the solution while warming up in a water bath at 37°C for a few minutes to improve dissolution of the stain. Add 66.5 ml of deionized water and mix. Filter the resulting solution through filter paper to remove undissolved stain. Store in a closed flask at room temperature.
3. Harris Hematoxylin (ready-to-use solution, Sigma).

2.5. Reagents and Equipment for Flow Cytometric Methods

1. FACSCalibur Flow Cytometer with CellQuest Software (BD Biosciences, San Jose, CA, USA).
2. Antibodies: Fluorescein isothiocyanate (FITC)-conjugated rat anti-mouse IgG, R-phycoerythrin (PE)-conjugated rat anti-mouse CD19, FITC-conjugated rat anti-mouse CD29, PE-conjugated rat anti-mouse CD44, FITC-conjugated rat anti-mouse CD45, PE-conjugated rat anti-mouse CD11b (all from BD Biosciences), and unconjugated anti-murine alpha smooth muscle actin mouse IgG (Chemicon, Temecula, CA, USA). For unconjugated antibody, dissolve a small

amount of the stock antibody in PBS containing 1% bovine serum albumin (Sigma) to achieve a concentration of 0.2 μg/μl. For all antibodies, use 5 μl antibody per 5×10^5 cells in a 100-μl staining volume.

3. Permeabilization solution: Using a 1-ml syringe, transfer 0.2 ml Triton® X-100 (Sigma) to 100 ml PBS. To minimize contamination, store at 4°C, and warm it up to room temperature prior to use.
4. Propidium iodide solution: Dissolve 25 mg propidium iodide (Calbiochem, San Diego, CA, USA) in 5 ml ultrapure water. Store at 4°C protected from light until use.
5. RNase A Solution: Add 100 μl of a 20 mg/ml stock RNAse A solution (Invitrogen Brasil Ltda., São Paulo, SP, Brazil) to 1.9 ml PBS. Store at 4°C until use.

2.6. CFU-F Assay

1. Giemsa Stain Stock and Working Solutions: Dissolve 3.8 g of Giemsa powder (Sigma) in 250 ml of methanol in a 1-l beaker. Stir this solution in a magnetic stirrer with warm heat for 10 min. Add 250 ml of glycerin. Filter the solution with filter paper to remove undissolved stain. Allow this solution to sit at room temperature for 2 months. Prepare working solution by mixing one volume of the stock solution with one volume of methanol and eight volumes of distilled water.

3. Methods

3.1. In Vitro Osteogenic Differentiation

Osteogenic differentiation can be attained by augmentation of the standard culture medium with dexamethasone, which upregulates alkaline phosphatase expression; ascorbic acid, which is a cofactor for prolyl hydroxylase (an enzyme that adds hydroxyl groups to proline lateral residues of procollagen, allowing collagen fiber assembly); and inorganic phosphate, which is a building block for hydroxyapatite. Ascorbic acid is highly labile in aqueous solutions, which makes the use of a more stable compound, AAP, more appropriate for the purpose of in vitro osteogenic differentiation.

1. Reserve a nearly confluent MSC culture to perform the differentiation.
2. The osteogenic induction consists of switching from complete expansion medium to osteogenic medium (OM) and performing medium changes twice a week for 3–4 weeks (see Note 9).
3. Generally, mineralized extracellular matrix deposition can be observed as early as 1 week after the onset of differentiation and becomes more evident with time (see Note 10).

3.2. Detection of Mineralized Matrix During Osteogenesis

1. Gently rinse the cell layer with PBS.
2. Fix with paraformaldehyde solution for 20 min at room temperature.
3. Remove the paraformaldehyde solution and wash once with deionized water.
4. Cover the cell layer with Alizarin Red S solution and wait for 5 min.
5. Remove the stain solution, and wash several times with deionized water.
6. Calcium-rich extracellular matrix is stained red and can be observed macroscopically or microscopically (Fig. 1), depending on the amount deposited.

3.3. In Vitro Adipogenic Differentiation

Adipogenic differentiation protocols often employ high-glucose media containing dexamethasone (a synthetic steroid that differs from hydrocortisone due to one single atom substitution) and insulin. Depending on the cell population to be differentiated, other molecules such as IBMX, indomethacin, and PPAR-γ ligands (e.g., rosiglitazone) can be added in order to increase differentiation efficiency. In some instances, substituting human platelet-poor plasma for FBS, or combination of these two types of supplement, may be required to induce adipogenesis. Two protocols are listed below. Protocol #1 may be employed using different medium compositions to best suit specific cell populations' requirements for adipogenic differentiation.

3.3.1. Adipogenic Differentiation – Protocol #1

1. Reserve a culture that is just about to reach confluence.
2. Replace standard culture medium with the respective adipogenic induction medium.
3. Incubate for 3 days.
4. There are two alternatives for subsequent medium changes:
   (a) Switching to adipogenic maintenance medium – complete expansion medium plus insulin and dexamethasone only – and performing medium changes twice a week until there is a satisfactory degree of differentiation as identified through lipid vesicle formation (Fig. 1).
   (b) Keep using the whole differentiation medium and performing medium changes twice a week until there is a satisfactory degree of differentiation (see Notes 11–19).

3.3.2. Adipogenic Differentiation – Protocol #2 (see Note 20)

1. Lift cells from the dish using trypsin-EDTA.
2. Collect cells by centrifugation, and resuspend them in enough standard culture medium to allow comfortable cell count using a hemocytometer.

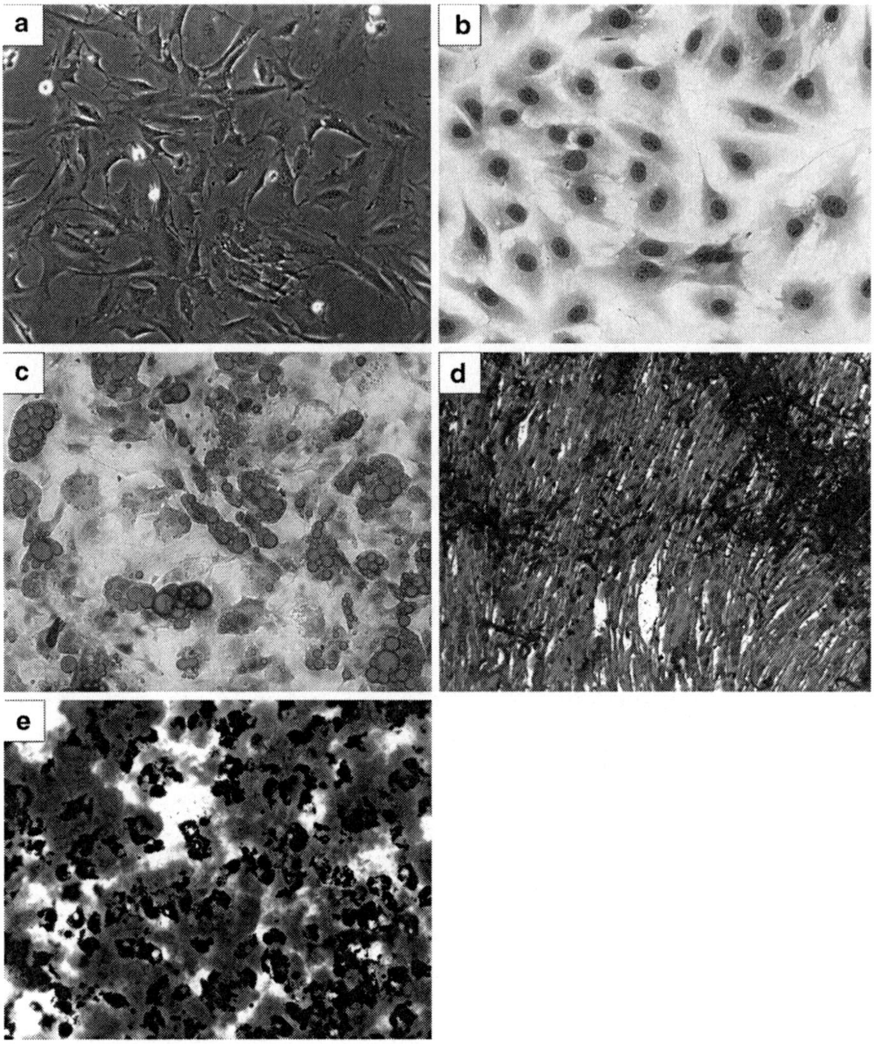

Fig. 1. In vitro differentiation of murine MSCs. (**a**) Undifferentiated vena cava-derived MSCs as observed with phase contrast. (**b**) Undifferentiated bone marrow-derived MSCs stained with Giemsa. (**c**) Aorta-derived MSCs subjected to adipogenic differentiation using adipogenic differentiation protocol #1 with adipogenic medium C, stained with Oil Red O, and counterstained with Harris hematoxylin. (**d**) Pancreas-derived MSCs after osteogenic differentiation. Calcium-rich extracellular matrix is revealed by staining with Alizarin Red S. (**e**) Bone marrow-derived MSCs after osteogenic induction. Adipocytes that developed over the course of osteogenic induction are stained in black with Sudan Black IV to allow for visualization after mineralized extracellular matrix staining with Alizarin Red S. Original magnifications: 100× (**a**), 200× (**b**), 200× (**c**), 50× (**d**), and 100× (**e**).

3. Resuspend the cells in standard culture medium to achieve a final cell density of 5×10^3 cells/cm^2.
4. Incubate for 2 days, or until the culture reaches ~25% confluence.
5. Switch to adipogenic medium composition F and incubate for additional 4 days.

3.4. Detection of Fat Vacuole Formation During Adipogenesis

1. Gently rinse the cell monolayer with PBS.
2. Fix with paraformaldehyde solution for 1 h at room temperature.
3. Remove the paraformaldehyde solution and wash once with deionized water.
4. Cover the monolayer with either Oil Red O or Sudan Black B solution (see Note 21) and wait for 5 min.
5. Remove the stain solution and wash 5 times with deionized water or until no precipitate is visible.
6. Cover the cell layer with Harris hematoxylin and incubate at room temperature for 5 min.
7. Remove the stain solution and wash 3 times with deionized water or until no precipitate is visible.
8. Observe the result on the inverted microscope. Fat vacuoles stained with Oil Red O and Sudan Black B are orange and black, respectively. Nuclei are stained pale blue (Fig. 1).

3.5. Two-Color Flow Cytometric Detection of Surface Antigens

1. Harvest cells using trypsin-EDTA, and resuspend cells in PBS at a concentration of 1×10^6 cells/ml.
2. Dispense 100 μl-aliquots into 12 × 75 mm, round-bottom polystyrene tubes (Becton Dickinson, Franklin Lakes, NJ, USA).
3. Label tubes as follows: CD19-PE/Ig-FITC (control), CD44-PE/CD29-FITC, CD11b-PE/CD45-FITC.
4. Add 3 μl of each antibody to the corresponding tubes and incubate for 15 min at room temperature or 30 min at 4°C (see Note 22). Protect tubes from light.
5. Add 1 ml of PBS to each tube, swirl and centrifuge at $400 \times g$ for 5 min.
6. Resuspend the contents of each tube in 250 μl of PBS.
7. Before running the samples in the cytometer, open CellQuest and draw an acquisition FSC × SSC dot plot and an acquisition FL1-H × FL2-H dot plot.
8. Run the control tube through the flow cytometer.
9. Bring the cells to the center of the FSC-H × SSC-H plot by adjusting the voltage and the amplification gain of the FSC and SSC detectors, and adjust the FSC-H threshold to rule out debris. An example of FSC-H × SSC-H dot plot is shown in Fig. 2a.
10. Draw a gate around the main population in the FSC-H × SSC-H plot, and set the FL1-H × FL2-H plot to use this gate.
11. Adjust the voltage and amplification gain of the FL1 and FL2 detectors so as to place most of the events between the 10^1 and 10^2 notches on each axis.

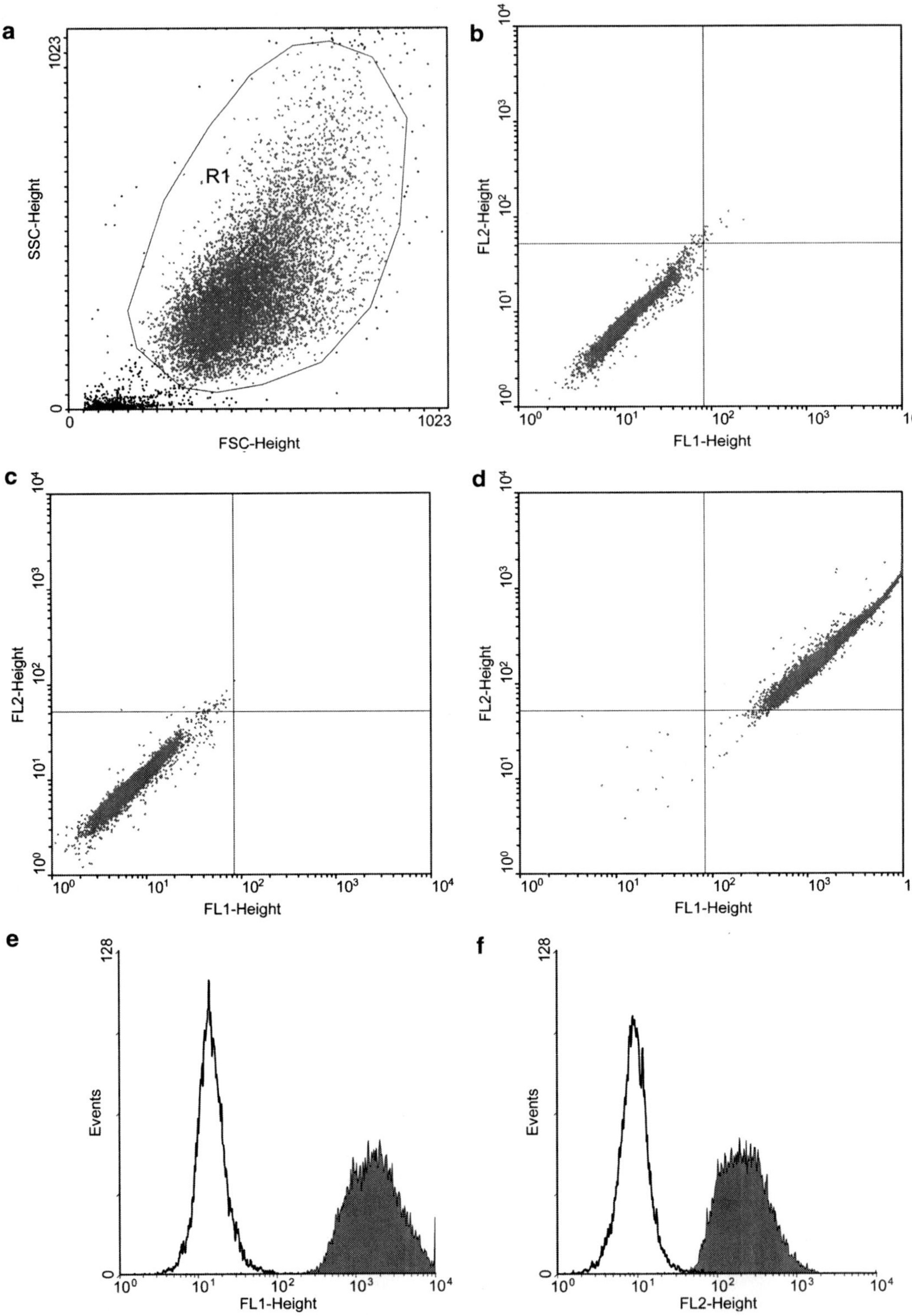

Fig. 2. Surface marker analysis of bone marrow-derived murine MSCs at the 15th passage. (**a**) FSC-H × SSC-H dot plot showing the gate used for the analysis. *Dots* in the lower left portion of the plot represent cellular debris. (**b**) Control FL1-H × FL2-H dot plot. (**c**) FL1-H × FL2-H dot plot showing absence of expression of CD45 (FL1-H) and CD11b (FL2-H). (**d**) FL1-H × FL2-H dot plot showing coexpression of CD29 (FL1-H) and CD44 (FL2-H) in nearly all cells. (**e**) Histogram depicting expression of CD29 (*solid fill*) overlaid with control. (**f**) Histogram depicting expression of CD44 (*solid fill*) overlaid with control.

12. Use quadrant lines to confine the cells to the lower left quadrant as shown in Fig. 2b.
13. Run the CD44-PE/CD29-FITC tube and use it to adjust the compensation so that the main cell population is displayed as shown in Fig. 2d (see Note 23).
14. Run the CD11b-PE/CD45-FITC tube (Fig. 2c).
15. Run the control tube. Make sure cells remain confined to the lower left quadrant after adjusting the compensation. Otherwise, adjust the quadrant accordingly.
16. Acquire at least 10,000 events and save to a file.
17. Repeat step 16 for the other tubes.
18. After completion of data acquisition, data files can be analyzed using histograms depicting fluorescence for each marker overlaid with histograms of the controls as shown in Fig. 2. Data analyses can be carried out using various computer programs including WinMDI 2.8, which is freely available at http://facs.scripps.edu/software.html and was used to build the graphics shown here.

3.6. Detection of Intracellular Antigens by Flow Cytometry

1. Harvest cells by trypsinization.
2. Wash twice by resuspending in PBS and centrifuging at 400 × *g* for 10 min at room temperature.
3. Count cells, prepare two tubes containing 5×10^5 cells in 100 μl PBS, and label one of the tubes as "control."
4. Add 1 ml paraformaldehyde solution to each tube.
5. Incubate for 1 h at room temperature.
6. Add 9 ml PBS to each tube and centrifuge at 400 × *g* for 1 min.
7. Resuspend the pellet in 100 μl PBS.
8. Resuspend the cells in 5 ml permeabilization solution.
9. Incubate for 15 min at room temperature.
10. Wash once with PBS.
11. Resuspend each pellet in 100 μl PBS.
12. Add 5 μl of the anti-alpha smooth muscle actin antibody to the unlabeled tube.
13. Incubate both tubes overnight at 4°C.
14. Wash cells once with PBS.
15. Resuspend cells in 100 μl PBS.
16. Add 10 μl of FITC-conjugated anti-murine IgG antibody to each of the tubes.
17. Incubate for 1 h at room temperature, protecting the tubes from light.

18. Wash cells once with PBS and resuspend each pellet in 300 μl PBS.
19. Read the fluorescence of the samples as described above (see Subheading 3.5), but use a FL1-H × SSC-H dot plot instead of an FL1-H × FL2-H dot plot.

3.7. Flow Cytometric Assessment of DNA Content and Cell Cycle Status

3.7.1. Cell Cycle Analysis

1. To obtain murine spleen cells for use as diploid control, euthanize a mouse of the same sex of the animal used to derive murine MSC cultures using procedures approved by your institution.
2. Dissect the animal and transfer the spleen to a 100-mm petri dish containing 5 ml PBS.
3. Cut one of the ends of the spleen using scissors and proceed with the "toothpaste tube" method: hold the opposite end of the spleen with a straight pair of forceps and slide a curved pair of forceps toward the open end of the spleen, pressing it against the petri dish.
4. Filter the cell suspension through a 40-μm cell strainer.
5. Wash the cells twice with PBS and count using a hemocytometer.

3.7.2. DNA Staining

1. Obtain murine spleen cells or harvest cultured MSCs. Make sure cell clumping is reduced to the minimum possible.
2. Wash cells twice with PBS.
3. Resuspend 2×10^6 cells in 0.5 ml PBS and chill in ice.
4. While vortexing, add 2 ml of ice-cold methanol drop by drop to minimize cell clumping.
5. Store cells at 4°C overnight.
6. Wash cells twice with PBS.
7. Resuspend the cells in 0.9 ml PBS.
8. Add 100 μl of RNase A solution and incubate at 37°C for 25 min (see Note 24).
9. Add 30 μl of propidium iodide solution and incubate at room temperature for 30 min in the dark (see Note 25).
10. Wash cells twice with PBS and resuspend the cell pellet in 0.5 ml PBS.

3.7.3. Data Acquisition

1. Set up an acquisition panel with the following graphics: FSC-H × SSC-H dot plot, FL2-W × FL2-A dot plot, and an FL2-A histogram.
2. Run the murine spleen cells through the cytometer.
3. Adjust the amplification and/or gain of the FSC and SSC detectors to position the main populations within the limits of the graph (see Note 26).

4. Adjust the threshold of the FSC axis to rule out most debris.
5. Adjust the voltage and/or amplification gain of the FL2 detectors so that G0/G1 cells (see Fig. 3b) sit at around the top of the 200 mark on the FL2-A axis. G2/M cells should sit on the 400 mark.
6. Draw a gate that runs in parallel with the cells in G0/G1, S, and G2/M phases to rule out cell clumps as shown in Fig. 3b (see Note 27).
7. Use the FL2-A histogram set to display only cells inside the gate as an aid to fine-tune the positioning of the cells while increasing/decreasing the FL2-A amplification gain.
8. Acquire at least 10,000 events in the gate and save to a file.
9. Run the cultured mMSCs through the cytometer. There should be no need to readjust the parameters. If slight changes are required, data from the spleen cells should be acquired again.
10. Analyze the data using your favorite flow cytometry software. A simple analysis is shown in Fig. 3. For further comparison between the cultured mMSCs and the diploid control, FL2-W × FL2-A graphics can be overlaid using graphic manipulation software.

3.8. CFU-F Assay

The CFU-F assay described below has been used to score the number of fibroblastic colonies in murine BM to provide an estimated frequency of MSCs (5). This assay is also useful for the selection of serum batches that best suit the requirements of murine MSCs. This protocol can be used to estimate the frequency of MSCs in other murine tissues as long as the number of input cells is adjusted.

1. Euthanize a mouse using a procedure approved by your institution.
2. Immerse the animal in 70% ethanol and transfer to the tissue culture hood. Place it on a clean Styrofoam block and pin the paws using needles.
3. Using sterile forceps and scissors, remove one femur and dissect away connective tissue.
4. Take a 5-ml syringe filled with standard culture medium coupled to a 22G needle. Insert the needle through one end of the femur and cut the opposite end. Alternatively, both epiphyses can be cut, and the needle inserted through one of the extremities.
5. Flush marrow out into a 100-mm bacteriological petri dish, holding it in an angle.
6. Dissociate total bone marrow cells by flushing it in and out the syringe (~5 times).

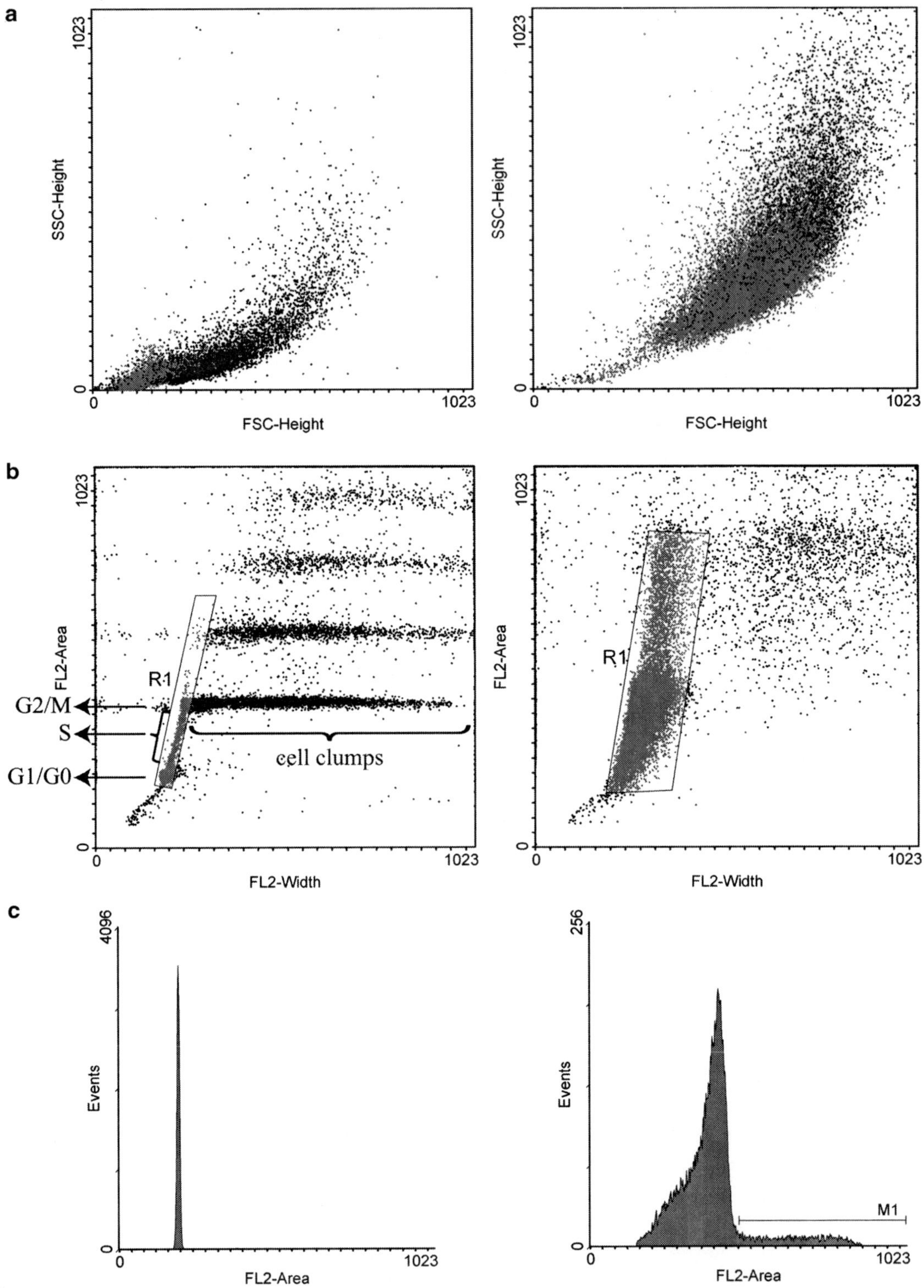

Fig. 3. DNA content/cell cycle analysis of muscle-derived murine MSCs at the 63rd passage as compared to murine spleen cells. (**a**) FSC-H × SSC-H plots of murine spleen cells (*left*) and MSCs (*right*), with *dots* corresponding to single cells shown in a *lighter color*. (**b**) FL2-W × FL2-A plots of murine spleen cells (*left*) and MSCs (*right*). Gates (R1) define single cells, whereas *dots* situated on the *right* of the gates represent clumped cells. Cells in G0/G1, S, and G2/M phases of the cell cycle are indicated. (**c**) FL2-A histograms, representative of DNA content, of murine spleen cells (*left*) and MSCs (*right*).

7. Take a sample of flushed cells, mix with an equal volume of 0.4% Trypan blue solution (Sigma), and count the number of viable nucleated cells using a hemocytometer.
8. Prepare 13 ml of a cell suspension containing 2.25×10^5 viable cells per milliliter in standard culture medium (see Note 28).
9. Dispense 2 ml of the cell suspension per well in a 6-well plate (TPP, Trasadingen, Switzerland) and place in a humidified incubator at 37°C, with 5% CO_2 in air.
10. 72 h postplating, simply aspirate the entire old medium, and replace it with 2 ml of fresh standard culture medium.
11. Keep the plate in the incubator for an additional 5 days and change medium (see Note 29).
12. Incubate for additional 5 days (13 days total).
13. Remove the medium, rinse with PBS, and fix cells with 96–100% ethanol for 2 min at room temperature.
14. Stain with Giemsa for 2.5 min at room temperature.
15. Score colonies with 5 or more MSC-like cells on the inverted microscope, excluding non-MSC-like colonies (see Fig. 4 for an example of an MSC-like colony), and calculate the mean colony number per well (see Note 30).

4. Notes

1. Phase contrast is essential for the observation of live flat adherent cells (which is the case for murine MSCs).
2. If possible, try different batches of FBS from different suppliers using the CFU-F assay described in this chapter. Although the use of heat-inactivated serum is very popular, heat-inactivation may decrease FBS performance (7).
3. In this case, BGP is assumed to have water content around 30%. If your reagent has different water content, make the calculations to achieve a final concentration of 100 mM.
4. Solutions in absolute ethanol or DMSO are intrinsically sterile, and usually there is no need to filter through 0.22 μm in order to sterilize them. If contamination problems occur, a filtration step can be included.

These plots indicate that, while the majority of spleen cells are in G0/G1 phase, most MSCs of this particular population are actively cycling. Marker M1 delimits around 20% of the gated population, which comprises cells that contain around twice the normal DNA content and are in S and G2/M phases. Consequently, the peak of the histogram includes diploid cells in G2/M phase in addition to cells with a nearly tetraploid DNA content in G0/G1 phase.

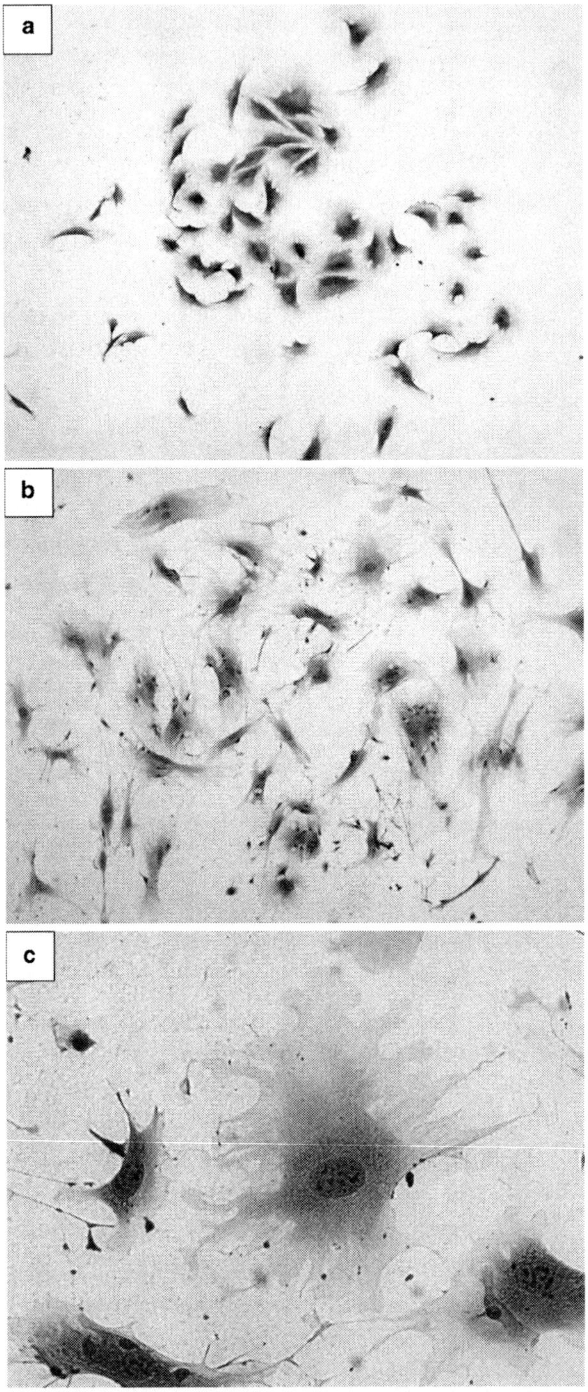

Fig. 4. Fibroblastic colonies. (**a**) A fibroblastic colony observed after plating cells from an established bone marrow-derived MSC population at low density (50×). (**b**) An MSC-like fibroblastic colony obtained by means of the CFU-F assay (50×). (**c**) A higher magnification (200×) of the fibroblastic colony shown in (**b**). All cells were stained with Giemsa.

5. Dex concentration may require some adjustment as murine MSCs may differentially respond to it according to their tissue of origin and, also, cortisol levels in FBS may vary depending on the supplier or batch. 1×10^{-8} M Dex works well for the osteogenic differentiation of bone marrow-derived murine MSCs and is a good starting concentration for MSCs from other tissues.
6. It is important that the temperature does not exceed 60°C during dissolution of paraformaldehyde, as it decreases performance of the resulting solution.
7. Dissolution of paraformaldehyde can be accelerated by adding some drops of a 1 M NaOH solution.
8. Since this solution is not buffered, ammonium hydroxide must be added very carefully to reach the desired pH.
9. OM is stable for 1 month at 4°C, the length of time required for matrix mineralization to occur. If OM is to be used over the course of a longer extent of time, it is recommended to prepare it without Dex, and add Dex at each medium change.
10. Human MSCs can be induced to differentiate along the osteogenic pathway using a two-step induction system (culture in DMEM/10% FBS supplemented with Dex and AAP for 10 days, then culture in DMEM/10% FBS + Dex + AAP + BGP) (2). Murine MSCs usually become overconfluent if cultured for extended time in complete medium supplemented with Dex and AAP, and some populations become contractile and detach from the dish. Addition of BGP at the beginning of the osteogenic induction provides conditions for early extracellular matrix mineralization, which minimizes cell detachment problems.
11. In some murine MSC populations, the use of insulin may favor cell proliferation rather than differentiation when added at the beginning of adipogenic differentiation induction. In this case, it is advisable to suppress insulin from the differentiation medium, and add it only when preadipocytes are visible.
12. Protocol #1 with adipogenic medium A works well for bone marrow-derived murine MSCs, which are usually proadipogenic, but may be ineffective for MSCs from some organs.
13. Depending on the cell population used, cell monolayers may detach from the dish when using protocol #1/A. In this case, other adipogenic medium compositions should be used.
14. Protocol #1/D is usually employed with the option "a."
15. Although differentiation time usually ranges from 1 to 3 weeks, it may extend to up to 2 months, if the cell population studied has a poor adipogenic differentiation potential.

16. Rosiglitazone is labile and for this reason it is recommended that it be stored in single-use aliquots and added to the adipogenic medium at each medium change. Although insulin is stable at a pH between 2 and 3 when stored at 4°C, it will degrade quickly if the culture medium becomes alkaline, which is very common for a medium such as DMEM; hence, it is best to add it to the adipogenic medium at each medium change.
17. Although IBMX solutions in DMSO are considered to be stable for some months when stored at –20°C, the best results are obtained by preparing a small amount of fresh solution right before addition to the differentiation medium.
18. Stable reagents such as indomethacin can be stored in stock solutions at room temperature, or incorporated into the base medium at will.
19. The final concentration of additives' diluents such as DMSO or ethanol should not exceed 0.1% (v/v).
20. This protocol is well suited for proadipogenic cell populations.
21. Oil Red O and Sudan Black B are commonly used stains to evidence fat vacuoles in adipocytes. Although Oil Red O is more esthetically attractive, Sudan Black B can be used to counterstain adipocyte-containing osteogenic cultures whose mineralized matrix has been stained with Alizarin Red S as shown in Fig. 1.
22. Incubation at room temperature makes staining faster, but can increase background fluorescence. Background fluorescence can be reduced by incubating the cells at 4°C, but incubation time needs to be increased to 30 min to ensure effective staining.
23. Murine MSCs are positive for both CD29 and CD44; however, CD44 is sensitive to the action of trypsin, which many times renders the apparent fluorescence intensity medium to low. For this reason, it is important that cells be harvested with the lowest possible trypsin concentration and that incubation with the enzyme be limited to just long enough to lift the cells from the dish. A trypsin concentration of 0.05% in Ca^{2+}- and Mg^{2+}-free saline solution containing 0.5 mM EDTA is effective for this purpose. For optimal results, the cells should not be more than 90% confluent before trypsinization.
24. This step is required because binding of propidium iodide to RNA makes readings nonstoichiometric.
25. Propidium iodide is sticky and increases the chance of clogging inside the fluidic components of the flow cytometer.

Make sure to carefully perform thorough cleaning procedures after running your samples through the cytometer.

26. Murine MSCs are much larger than spleen cells, and this difference is reflected in both FSC-H × SSC-H and FL2-W × FL2-A dot plots. Make sure to use voltage and amplification gain values that allow both cell populations to be contained within the limits of the plots.
27. If the presence of cells with higher DNA content is suspected, the gate on the FL2-W × FL2-A dot plot should be extended toward the top of the dot plot without changing the angle (see Fig. 3b).
28. Cell concentration and density used in this protocol have been optimized for C57Bl/6 mice. Although the same protocol has been used for BALB/c mice, optimal cell densities/concentrations may be different for other mouse strains.
29. The CFU-F assay can be used as a basis for simple osteoblast colony-forming unit (CFU-O) and adipocyte colony-forming unit (CFU-A) assays. These can be performed by applying the osteogenic and adipogenic differentiation protocols described above starting at this point (eighth day) of the CFU-F assay.
30. Each 6-well plate provides a mean and a SEM or SD that may be used for statistical purposes.

Acknowledgments

The authors are indebted to Conselho Nacional de Desenvolvimento Científico e Tecnológico (CNPq), Fundação de Amparo à Pesquisa do Estado de São Paulo (FAPESP), and Financiadora de Estudos e Projetos (FINEP) for funding. Data showed in this chapter were acquired at Laboratório de Imunogenética, Department of Genetics of Universidade Federal do Rio Grande do Sul (Porto Alegre, RS, Brazil). L. da S. M. holds a postdoctoral fellowship from FAPESP.

References

1. Caplan, A. I. (1991) Mesenchymal stem cells, *J Orthop Res 9*, 641–650.
2. Lennon, D. P., and Caplan, A. I. (2006) Mesenchymal Stem Cells for Tissue Engineering, in *Culture of Cells for Tissue Engineering*. (Vunjak-Novakovic, G., and Freshney, R. I., Eds.), pp 23–59, Wiley, Hoboken.
3. da Silva Meirelles, L., Caplan, A. I., and Nardi, N. B. (2008) In search of the in vivo identity of mesenchymal stem cells, *Stem Cells 26*, 2287–2299.
4. Meirelles Lda, S., and Nardi, N. B. (2009) Methodology, biology and clinical applications of mesenchymal stem cells, *Front Biosci 14*, 4281–4298.

5. Meirelles Lda, S., and Nardi, N. B. (2003) Murine marrow-derived mesenchymal stem cell: isolation, in vitro expansion, and characterization, *Br J Haematol 123*, 702–711.
6. da Silva Meirelles, L., Chagastelles, P. C., and Nardi, N. B. (2006) Mesenchymal stem cells reside in virtually all post-natal organs and tissues, *J. Cell Sci 119*, 2204–2213.
7. Kuznetsov, S. A., Mankani, M. H., Bianco, P., and Robey, P. G. (2009) Enumeration of the colony-forming units-fibroblast from mouse and human bone marrow in normal and pathological conditions, *Stem Cell Res 2*, 83–94.

Part IV

MSC Phenotypic Characterization and Extended Applications

Chapter 26

Immunohistochemical Analysis of Human Mesenchymal Stem Cells Differentiating into Chondrogenic, Osteogenic, and Adipogenic Lineages

Zheng Yang, Jacqueline Frida Schmitt, and Eng Hin Lee

Abstract

Mesenchymal stem cells (MSCs) are multipotent cells that have the potential to differentiate into various mesenchymal lineages in vitro and in vivo. Due to their availability from tissues such as bone marrow, synovium, fat, and muscle, and their highly proliferative capacity, MSCs have evoked interest as a potential cell source for repair and regeneration of various types of tissues. Characterization by the expression of a panel of surface markers and the ability of MSCs to undergo multilineage differentiation is the benchmark for identifying this stem cell population. In this chapter, the protocols for the differentiation of MSC to chondrogenic, osteogenic, and adipogenic lineages and histological and immunostaining protocols for confirming trilineage differentiation of the MSC cells are described.

Key words: Mesenchymal stem cells, Differentiation, Chondrogenic, Osteogenic and adipogenic, Immunohistochemical analysis

1. Introduction

Mesenchymal stem cells (MSCs) are multipotent cells that are present in various adult tissues of mesenchymal origin. MSCs are a potential cell source for tissue engineering applications as they are readily availability from various tissues and are easily expanded in culture. Under appropriate conditions, MSCs are able to differentiate into a variety of tissues such as bone, cartilage, ligament, tendon, muscle, and fat (1–4). MSCs delivered in vivo have been found to differentiate in a site-specific manner and contribute to tissue regeneration (2, 5–8). In addition, animal studies have shown that MSCs injected systemically migrate to regions of injured tissue and contribute to the healing process (9).

Mohan C. Vemuri et al. (eds.), *Mesenchymal Stem Cell Assays and Applications*, Methods in Molecular Biology, vol. 698, DOI 10.1007/978-1-60761-999-4_26, © Springer Science+Business Media, LLC 2011

More recently, the trophic effect of MSCs has been recognized (10). Cultured MSCs are found to secrete various bioactive molecules that display antiapoptotic, immunomodulatory, angiogenic, antiscarring, and chemoattractant properties (11, 12), providing a basis for their use as tools to create local regenerative environments in vivo. Another striking feature of MSCs is their low inherent immunogenicity as they induce little, if any, proliferation of allogeneic lymphocytes and antigen-presenting cells (9, 13). Instead, MSCs appear to be immunosuppressive in vitro (14). MSCs multilineage differentiation potential and trophic effects, coupled with their immunoprivileged properties, are being widely employed in autologous and allogeneic cell replacement strategies (15, 16).

MSCs are frequently isolated from the heterogeneous mononuclear cell population of bone marrow by their propensity to adhere to the plastic surface of tissue culture containers. Apart from characterization of MSCs by their expression of a panel of surface markers (CD29, CD44, CD73, CD90, CD105, CD166), the ability of these MSCs to differentiate to three of the mesenchymal lineages (osteogenic, chondrogenic, and adipogenic) has been used as a benchmark to ascertain the functionality of the ex vivo expanded MSCs. In this chapter, we present the protocols for the differentiation of MSCs to chondrogenic, osteogenic, and adipogenic lineages and the immunohistochemical analytical methods for confirmation of development into these three lineages.

2. Materials

2.1. Cell Culture

2.1.1. Bone Marrow MSC Isolation and Expansion

1. Dulbecco's Modified Eagle's Medium (DMEM) (Gibco/Invitrogen, Carlsbad, CA).
2. Fetal bovine serum (FBS) (Gibco/Invitrogen).
3. Dulbecco's Ca^{2+}- and Mg^{2+}-free phosphate-buffered saline (10× 1 L; first Base, Singapore).
4. TrypLE™ Express Stable Trypsin-Like Enzyme with Phenol Red (Gibco/Invitrogen).
5. RossetteSep® Human Mesenchymal Stem Cell Enrichment Cocktail (STEMCELL Technologies, Vancouver, Canada).

2.1.2. Chondrogenic Differentiation of MSC

1. DMEM high-glucose (4.5 g/L; Sigma).
2. GlutaMax-I Supplement, 200 mM; 100× (Gibco/Invitrogen).
3. Penicillin/Streptomycin (P/S) 10,000 U/10,000 µg (Gibco/Invitrogen).

4. Sodium pyruvate, 100 mM; 100× (Gibco/Invitrogen).
5. ITS[+1] (10 μg/ml insulin, 5.5 μg/ml transferrin, 5 ng/ml selenium, 0.5 mg/ml bovine serum albumin, 4.7 μg/ml linoleic acid, BD Bioscience, San Jose, CA).
6. Ascorbic acid 2-phosphate (AA2P) (Sigma, St. Louis, MO) a 5 mg/ml stock is prepared in Dulbecco's Ca^{2+}/Mg^{2+}-free phosphate-buffered saline, aliquoted, and stored at −20°C.
7. L-Proline (Sigma), a 0.4-M stock is prepared in Dulbecco's Ca^{2+}- and Mg^{2+}-free phosphate-buffered saline, aliquoted, and stored at −20°C.
8. Dexamethasone (Sigma) a 1 mM stock is prepared in absolute ethanol (analytical reagent grade, Cat No: E/0650DF/17, Fisher Scientific), aliquoted, and stored at −20°C.
9. Recombinant human transforming growth factor-beta3 (TGFβ3) (R&D Systems, Minneapolis, MN). Prepare stock solution of 10 ng/μl by reconstituting in filter-sterilized 4 mM HCl containing 0.1% (w/v) bovine serum albumin (BSA, Sigma).

2.1.3. Osteogenic Differentiation of MSC

1. DMEM (Gibco/Invitrogen).
2. FBS (Gibco/Invitrogen).
3. GlutaMax-I Supplement, 200 mM; 100× (Gibco/Invitrogen).
4. Penicillin/Streptomycin (P/S) 10,000 U/10,000 μg (Gibco/Invitrogen).
5. Sodium pyruvate, 100 mM; 100× (Gibco/Invitrogen).
6. AA2P (Sigma) a 5 mg/ml stock is prepared in Dulbecco's Ca^{2+}/Mg^{2+}-free phosphate-buffered saline, aliquoted, and stored at −20°C.
7. Dexamethasone (Sigma) a 1 mM stock is prepared in absolute ethanol (analytical reagent grade, Cat No: E/0650DF/17, Fisher Scientific), aliquoted, and stored at −20°C.
8. β-glycerol phosphate disodium salt pentahydrate (Sigma) a 1 M stock is prepared in Dulbecco's Ca^{2+}/Mg^{2+}-free phosphate-buffered saline, aliquoted, and stored at −20°C.

2.1.4. Adipogenic Differentiation of MSC

1. DMEM high-glucose (4.5 g/L; Sigma).
2. FBS (Gibco/Invitrogen).
3. GlutaMax-I Supplement, 200 mM; 100× (Gibco/Invitrogen).
4. Penicillin/Streptomycin (P/S) 10,000 U/10,000 μg (Gibco/Invitrogen).
5. Dexamethasone (Sigma) a 1 mM stock is prepared in absolute ethanol (analytical reagent grade), aliquoted, and stored at −20°C.

6. Insulin (Gibco/Invitrogen) a 4 mg/ml stock is prepared in Dulbecco's Ca^{2+}/Mg^{2+}-free phosphate-buffered saline, aliquoted, and stored at –20°C.
7. Isobutylmethylxanthine (IBMX) (GMCN; Calbiochem/Merck, San Diego, CA) a 0.5 M stock is prepared in dimethylsulphoxide (DMSO, Sigma), aliquoted, and stored at –20°C.
8. Indomethacin (Sigma) a 66.6 mM stock is prepared in absolute ethanol (analytical reagent grade), aliquoted, and stored at –20°C.

2.2. Analysis of Matrix Protein Synthesis in Chondrogenesis

2.2.1. Fixation and Paraffin-Processing of Chondrogenic Samples

1. 10% Buffered formalin (Sigma).
2. Premium ethanol absolute 99%. Prepare graded dilutions of 70, 80, 95% v/v of ethanol in milliQ water.
3. Eosin Y (Sigma).
4. Xylene (VWR, International S.A.S, Briare, France).
5. Paraffin wax (Leica Biosystems, Nussloch GmbH Heidelberger, Germany).
6. An embedding machine for making paraffin blocks, Leica, EG1160 (Leica Biosystems).
7. A microtome for cutting paraffin sections, Leica RM2135 (Leica Biosystems).
8. Mounting media for coverslipping (VWR International Ltd, England).
9. Superfrost® Plus microscope slides (Menzel-Glaser, Braunschweig, Germany).

2.2.2. Hematoxylin and Eosin Staining

1. Xylene and graded dilutions of ethanol as for Subheading 2.2.1.
2. Hematoxylin (Sigma).
3. Differentiating solution consisting of 10 ml concentrated hydrochloride acid diluted in 1,000 ml of 70% ethanol.
4. Scott's tap water (Sigma).

2.2.3. Proteoglycan Synthesis/Deposition

2.2.3.1. Alcian Blue Staining

1. 0.5% (w/v) Alcian Blue staining solution. Dissolve 0.25 g Alcian blue 8GX (Sigma) with 0.1 M hydrochloride acid and top up to 50 ml after adjusting to pH 1.0. Filter before use.
2. Nuclear Fast Red Solution (Sigma).

2.2.3.2. Safranin O

1. Fast green (Sigma).
2. Safranin O (Acros Organics, Geel, Belgium).
3. Immedge Hydrophobic Barrier pen: (Vector Laboratories, Burlingame, CA).

2.2.4. Immunohistochemical Detection of Collagen Deposition

1. UltraVision HRP Detection System (Lab Vision Inc., Fremont, CA) containing Ultra V Block, Biotinylated Goat Antimouse IgG, Streptavidin Peroxidase, Diaminobenzidine (DAB) substrate, and DAB Chromogen (see Note 1).
2. Pepsin (Lab Vision).
3. Hydrogen peroxide block (Lab Vision).
4. Monoclonal anticollagen type I, clone COL-1 (Sigma).
5. Monoclonal anticollagen type II, clone II-II6B3 (Chemicon Inc., Temecuela, CA).
6. Monoclonal anticollagen type X, clone X-53 (Quartett Immunodiagnostika GmBH, Berlin, Germany).
7. Control mouse IgG isotype (Invitrogen).
8. Dulbecco's Ca^{2+}- and Mg^{2+}-free phosphate-buffered saline (10× 1 L; first Base).

2.3. Analysis of Mineralization in Osteogenesis

2.3.1. Alizarin Red Staining

1. 10% Formalin (Sigma).
2. Alizarin red S (Sigma); 2 g alizarin red power dissolved in 100 ml distilled water with pH adjusted to 4.1–4.3 using 0.5% ammonium hydroxide. The pH is critical. The solution can be kept in dark for up to 6 months. Filter using filter paper before use.
3. 0.5% Ammonium hydroxide (Sigma).
4. Dulbecco's Ca^{2+}- and Mg^{2+}-free phosphate-buffered saline (first Base).
5. Filter paper (70 mm) (Whatman).

2.4. Analysis of Oil Droplet Formation in Adipogenesis

2.4.1. Oil Red O Staining

1. 70% Ethanol (make from 100% ethanol with distilled water).
2. Oil Red O (Sigma) stock of 0.5% (w/v) is prepared freshly before use by dissolving 0.05 g Oil Red O in 10 ml propanol alcohol (Singapore). The Oil Red O solution is filtered with filter paper.
3. Hematoxylin (Sigma).
4. Dulbecco's Ca^{2+}- and Mg^{2+}-free phosphate-buffered saline (10× 1 L; first Base).
5. Filter paper (70 mm) (Whatman).

3. Methods

3.1. Isolation and Culture of Bone Marrow-Derived MSC

1. Human bone marrow mononuclear cells from iliac crest bone marrow aspirates are isolated using RosetteSep® Human Mesenchymal Stem Cell Enrichment Cocktail (see Note 2). Isolated mononuclear cells are resuspended in MSC expansion

media (DMEM supplemented with 10% FBS (see Note 3) and 100 U/100 μg P/S), seeded in T75 tissue culture flasks and incubated in a 37°C incubator. Nonadherent cells are removed 4–5 days after plating, and adherent MSCs allowed to proliferate in expansion media until they reach 90% confluence. The medium is replaced every 3–4 days.

2. To further expand the MSCs, the media is aspirated, and the cells treated with 2 ml of TrypLE at 37°C for 5 min. The tissue culture flask is lightly tapped on the side to help lift off the cells. The enzyme is inactivated with addition of 8 ml of expansion media. Cells are counted and seeded at 4×10^3 cells per cm^2.

3.2. Chondrogenic Differentiation of MSC

Chondrogenic differentiation of MSC is carried out in a three-dimensional culture system where the cells are pelleted to form an aggregate (17, 18).

1. Basic serum-free chondrogenic medium consisting of DMEM high-glucose supplemented with ITS^{+1} (10 μg/mL insulin, 5.5 μg/mL transferrin, 5 ng/mL selenium, 0.5 mg/mL bovine serum albumin, 4.7 μg/mL linoleic acid), 4 mM L-proline, 50 μg/mL AA2P, 1% sodium pyruvate, 10^{-7} M dexamethasone, 2 mM Glutamax, and 100 U/100 μg P/S was prepared from stock reagents (see Subheading 2.1.2). Chondrogenic medium can be kept for up to 1 month at 4°C. Warm to 37°C in a waterbath before use.
2. MSCs at confluency of less than 90% are harvested by TrypLE treatment as described in Subheading 3.1. Cells are collected by centrifugation at $250 \times g$ for 5 min and resuspended in chondrogenic medium to achieve a cell density of 5.0×10^5 cells per ml.
3. The control pellets are prepared by aliquoting 0.5 ml of cell suspension into 15 ml polypropylene screw-cap tubes. Prepare at least two control pellets for each time point. The control pellets are cultured in the absence of any growth factors (see Note 4).
4. For the remaining cell suspension, add TGFβ3 at 1 μl of TGFβ3 stock solution (10 ng/μl) to each ml of the cell suspension, giving the final TGFβ3 concentration of 10 ng/ml. Aliquot 0.5 ml of TGFβ3-cell suspension into 15 ml polypropylene screw-cap tubes.
5. The pellets are formed by centrifugation at $300 \times g$ for 5 min. Loosen the caps of the tubes to allow air exchange and incubate in a 37°C incubator with 5% CO_2.
6. Media are replaced every 2–3 days up to a period of 21 days with fresh 0.5 ml chondrogenic medium with TGFβ3 (1 μl TGFβ3 stock solution/ml medium) or without TGFβ3

(control samples). For even differentiation of the pellet, ensure the pellet does not stick to the bottom of the tube.

3.3. Osteogenic Differentiation of MSC

1. MSCs are plated at a density of 3×10^4 cells/cm^2 on a 24-well tissue culture plate in MSC expansion media and allowed to adhere overnight at 37°C.
2. Osteogenic differentiation is induced by replacing the MSC expansion medium with osteogenic media consisting of DMEM with 10% FBS, 1 mM sodium pyruvate, and 100 U/100 μg P/S supplemented with 10^{-7} M dexamethasone, 50 μg/ml ascorbic acid, and 10 mM β-glycerophosphate.
3. Parallel undifferentiated MSC are plated and maintained in MSC expansion media.
4. Media are changed every 3 days up to a period of 21 days.

3.4. Adipogenic Differentiation of MSC

1. MSCs are plated at a density of 6×10^4 cells/cm^2 on a 24-well plate in MSC expansion media and allowed to adhere overnight at 37°C.
2. Adipogenic differentiation is induced by replacing the MSC expansion medium with adipogenic media consisting of DMEM with 10% FBS, 1 mM sodium pyruvate, and 100 U/100 μg P/S supplemented with 10^{-6} M dexamethasone, 10 μg/ml insulin and 0.5 mM IBMX, and 200 μM indomethacin.
3. Parallel undifferentiated MSCs are plated and maintained in MSC expansion media.
4. Media are changed every 3 days up to a period of 21 days.

3.5. Analysis of Chondrogenic Matrix Protein Synthesis

3.5.1. Preparation of Chondrogenic Sections for Histology and Immunohistochemistry

1. After completion of chondrogenesis, the medium is removed from the pellets and replaced with 2 ml 10% buffered formalin for 1 h at room temperature. The formalin is then replaced with 2 ml of fresh formalin and the pellets maintained at room temperature for a further 3–5 h, or kept at 4°C overnight (see Note 5).
2. Remove the formalin and dehydrate the pellets with serial washes in ethanol for 2×15 min in 70% ethanol, then 2×15 min in 80% ethanol, then 2×15 min in 95% ethanol.
3. To facilitate handling of the pellets, stain for 20 s in a drop of eosin Y, before washing 3 times in 100% ethanol. Then wash 2×15 min in 100% ethanol (see Note 6).
4. Wash the pellets 2×20 min in xylene to remove the ethanol.
5. Incubate the pellets 3×20 min in melted paraffin wax at 56°C.
6. Embed the pellets in a block of paraffin.

7. Cut 5 μm sections using a microtome and transfer them to Superfrost® Plus microscope slides. Dry on a 37°C hotplate, then leave at 60°C overnight to facilitate adherence of the section to the slide.
8. Immediately before staining, dewax and rehydrate the sections by washing 2 × 10 min in xylene, followed by 2 × 2 min in 100% ethanol, 2 × 2 min in 95% ethanol, 2 × 2 min in 70% ethanol. Finish with 2 × 3 min washes in PBS.

3.5.2. Morphological Analysis of Chondrogenic Sections

To examine cell morphology, sections of chondrogenic pellets are stained with hematoxylin and eosin.

1. Rehydrated sections on microscope slides are placed in hematoxylin for 5 min, and then rinsed thoroughly in several changes of tap water.
2. Dip the sections briefly (no more than 5 s) in differentiating solution, then rinse in tap water.
3. Dip the sections briefly in Scott's Tap water, and then wash twice for 5 min each time in tap water.
4. Stain in eosin for 5 min. Rinse briefly in two changes of 95% ethanol followed by two changes of 100% ethanol.
5. Rinse twice in xylene for 2 min each time. Air-dry the slides for 5 min.
6. Apply coverslips using a permanent (nonaqueous) mounting medium. An example of a hematoxylin and eosin-stained chondrogenic pellet section is shown in Fig. 1a.

3.5.3. Histological Staining for Proteoglycan

3.5.3.1. Alcian Blue Staining

1. Rehydrated sections on microscope slides are stained with 0.5% alcian blue solution for 30 min. Then rinsed in distilled water 3 × 3 min.
2. The nuclei of the cells are stained with nuclear fast red for 5 min, then rinsed briefly in two changes of 95% ethanol followed by two changes of 100% ethanol.
3. Wash in xylene, air-dry, and apply coverslips as outlined in steps 5 and 6 of Subheading 3.5.2.
4. An example of an alcian blue-stained chondrogenic pellet section is shown in Fig. 1b.

3.5.3.2. Safranin O Staining

1. Rehydrated sections on microscope slides are stained with hematoxylin following steps 1–3 as outlined in Subheading 3.5.2 above.
2. Stain sections with fast green solution for 10 min, and rinse in 1% acetic acid for no more than 5 s.
3. Stain in 0.1% safranin O solution for 3 min, then rinse in 95% ethanol twice followed by two rinses in 100% ethanol.

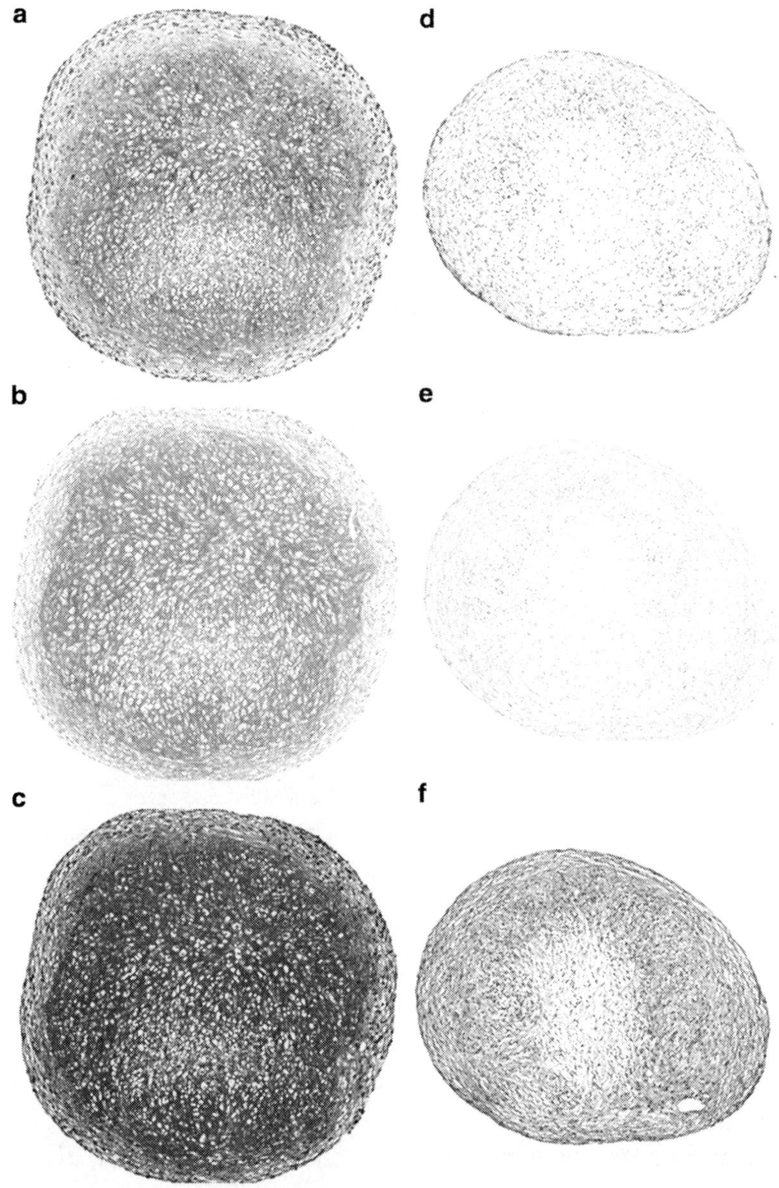

Fig. 1. MSCs were differentiated into the chondrogenic lineage for 21 days in the presence (**a–c**) or absence (**d–f**) of TGFβ3. The morphology of the chondrogenic samples was analyzed by hematoxylin and eosin staining (**a**, **d**). Proteoglycan deposition was analyzed by alcian blue (**b**, **e**) and Safranin O staining (**c**, **f**). Images are taken at ×100 magnification.

4. Clear slides by washing 2 × 2 min in xylene, then air-dry slides for 5 min before coverslipping with mounting medium.
5. An example of the safranin O-stained chondrogenic pellet section is shown in Fig. 1c.

3.5.4. Immunostaining for Collagen Proteins

Collagen Type I, II, and X expression is detected by immunohistochemistry using the UltraVision HRP Detection System (see Note 1).

1. Using a wax pen, draw a circle on the microscope slide to surround the rehydrated sections. Add a drop of hydrogen peroxide block to the section within the wax circle and incubate for 15 min. Wash 2 × 5 min with PBS.
2. Incubate sections under drops of pepsin at 37°C for 15 min and then wash 2 × 5 min with PBS.
3. Incubate in Ultra V Block for 5 min (see Note 7).
4. Drain excess Ultra V Block and then add the PBS-diluted primary antibody or matching isotype control. See Table 1 for the antibody and isotype concentrations and the incubation times. After incubation is complete, wash the sections 4 × 2 min in PBS.
5. Incubate the sections with biotinylated goat antimouse for 30 min and then wash the section 4 × 2 min in PBS.
6. Incubate the sections with strepavidin peroxidase for 45 min. and then wash 4 × 2 min in PBS.
7. Prepare color development substrate by mixing 1 drop DAB chromagen (see Note 8) with 1 ml of DAB substrate and mix by inverting tube.
8. Add the color development substrate to the sections and incubate until staining is evident (1–3 min). Remove the color solution and stop the reaction by immersing the slides in distilled water.
9. Counterstain the sections with hematoxylin. Follow the procedure outlined in Subheading 3.5.2 for hematoxylin and eosin staining omitting the eosin staining step.

Table 1
Primary antibodies incubation condition for detection of collagen protein

Antibody or control	Concentration	Isotype	Dilution to use	Incubation time (mins)	Incubation temperature
Collagen I	Ascites fluid	IgG_1	1/500	60	Room temperature
Collagen II	0.2 mg/ml	IgG_1	1/500	60	Room temperature
Collagen X	Ascites fluid	IgG_1	1/25	60	Room temperature
IgG_1 control	1.0 mg/ml		1/2,500	60	Room temperature

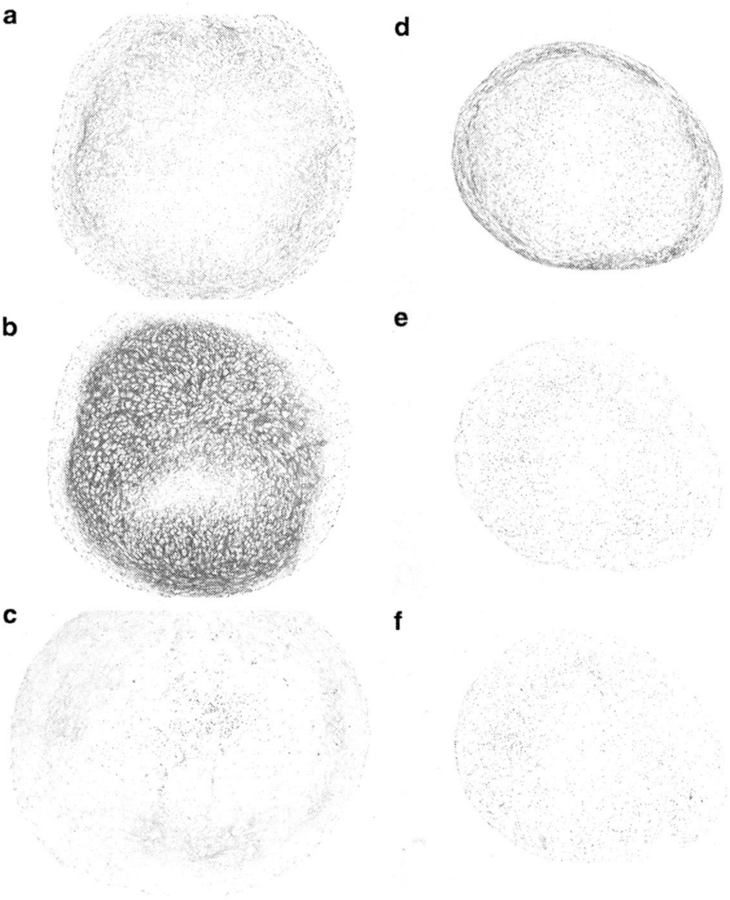

Fig. 2. MSCs were differentiated into the chondrogenic lineage for 21 days in the presence (**a–c**) or absence (**d–f**) of TGFβ3. Collagen deposition of the chondrogenic pellets was analyzed by immunohistochemical staining for collagen type I (**a**, **d**), type II (**b**, **e**), and type X (**c**, **f**). Images are taken at ×100 magnification.

10. Examples of collagen-stained chondrogenic pellet sections are shown in Fig. 2.

3.6. Analysis of Osteogenic Mineralization (Alizarin Red Staining)

1. Media is removed after 21 days of differentiation and cells are washed once with PBS.
2. Cells are fixed with 10% formalin for 30 min at room temperature, followed by three washes with PBS.
3. Cells are stained with Alizarin Red S solution for 5 min.
4. The staining solution is removed, and cells are washed with distilled water with gentle agitation until nonspecific staining is removed.
5. MSCs that have undergone 21 days of differentiation would have mineralization deposition detected as a brick-red color as shown in Fig. 3.

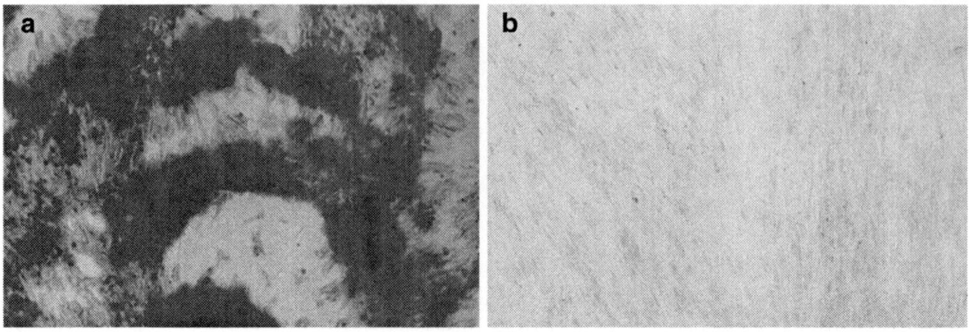

Fig. 3. MSCs were differentiated into the osteogenic lineage for 21 days. Calcium deposition analyzed by alizarin red staining shows brick-red staining of the osteogenic cells (**a**), compared to negative staining in the control (**b**). Images are taken at ×100 magnification.

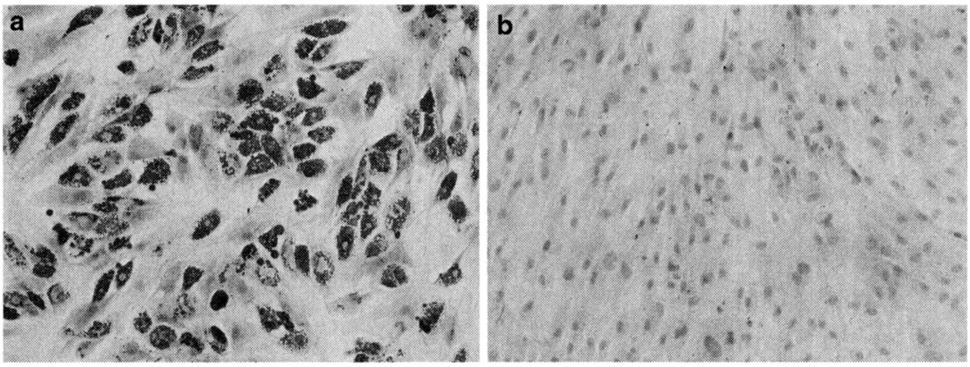

Fig. 4. MSCs were differentiated into the adipogenic lineage for 21 days. Oil deposition analyzed by Oil Red staining shows oil droplet accumulation in the adipocytes of the adipogenic sample (**a**), compared to negative staining in the control (**b**). Images are taken at ×200 magnification.

3.7. Analysis of Oil Droplet (Oil Red Staining)

1. Media is removed after 21 days of differentiation and cells are washed twice with PBS.
2. Cells are fixed with 70% ethanol for 20 s.
3. Cells are stained with Oil Red O working solution (prepared with 6 ml freshly prepared and filtered 0.5% Oil Red O stock solution in 4 ml distilled water) for 15 min.
4. The staining solution is removed, and cells washed with 70% ethanol for a few seconds.
5. Cells are then washed again with distilled water.
6. Cells are counterstained with hematoxylin for 30 s.
7. The counterstaining solution is removed, and cells are washed with distilled water.
8. MSCs that have undergone 21 days of differentiation would have oil droplets stained in red as shown in Fig. 4.

4. Notes

1. Alternatively, individual components of the kit (i.e., Ultra V Block, Biotinylated Goat Antimouse IgG, Streptavidin Peroxidase, DAB substrate, and DAB Chromogen) can be obtained individually from other sources. Alternative for Ultra V Block, see Note 7.
2. Isolation of mononuclear cells from bone marrow aspirates using RosetteSep® Human Mesenchymal Stem Cell Enrichment Cocktail follows exactly the recommendation of the manufacturers' protocol (http://www.stemcell.com/technical/15128_15168-PIS.pdf).
3. The source of FBS affects MSCs due to variation of, yet to be identified, components in the serum. To maintain good performance of the expanded MSC, i.e., high proliferation rate and multilineage differentiation capacity, it is recommended that different batches of FBS be tested for their ability to support the expansion and clonal enumeration (colony forming unit fibroblast assay) of MSCs, or that commercially available MSC-grade FBS be used.
4. Minus TGFβ3 control pellets are included to serve as negative controls to the chondrogenic differentiation pellet in the presence of TGFβ3.
5. Fixation times of 4–16 h are preferable.
6. After the first 15 min wash in 100% ethanol, the pellets can be left at 4°C for a few days if necessary. Then the pellets are warmed to room temperature and the final 15 min ethanol wash is carried out before proceeding to step 4.
7. Incubations >10 min may reduce the signal. Alternatively, normal goat serum (normal serum from the host of the secondary antibody) diluted to 1:5 (final concentration 20%), incubated for 1 h room temperature, can be used to block nonspecific sites.
8. DAB is a potential carcinogen and needs to be handled with care and disposed off according to safety regulations.

References

1. Wakitani S., Goto T., Pineda S.J., Young R.G., Mansour J.M., Caplan A.I., Goldberg V.M. (1994) Mesenchymal cell-based repair of large, full-thickness defects of articular cartilage. *J Bone Joint Surg Am* **76**, 579–92.
2. Lee K.B., Hui J.H., Song I.C., Ardany L., Lee E.H. (2007) Injectable mesenchymal stem cell therapy for large cartilage defects–a porcine model. *Stem Cells* **25**, 2964–71.
3. Chong A.K., Ang A.D., Goh J.C., Hui J.H., Lim A.Y., Lee E.H., Lim B.H. (2007) Bone marrow-derived mesenchymal stem cells influence early tendon-healing in a rabbit achilles tendon model. *J Bone Joint Surg Am* **89**, 74–81.

4. Lee E.H., Hui J.H. (2006) The potential of stem cells in orthopaedic surgery. *J Bone Joint Surg Br* **88**, 841–51.
5. Koc O.N., Gerson S.L, Cooper B.W., Dyhouse S.M., Haynesworth S.E., Caplan A.I., Lazarus H.M. (2000) Rapid hematopoietic recovery after co-infusion of autologous blood stem cells and culture expanded marrow mesenchymal stem cells in advanced breast cancer patients receiving high dose chemotherapy. *J Clin Oncol* **8**, 307–16.
6. Horwitz E.M., Gordon P.L., Koo W.K., Marx J.C., Neel M.D., McNall R.Y., Muul L., Hofmann T. (2002) Isolated allogeneic bone marrow-derived mesenchymal cells engraft and stimulate growth in children with osteogenesis imperfecta: Implications for cell therapy of bone. *Proc Natl Acad Sci U S A* **99**, 8932–7.
7. Orlic D., Kajstura J., Chimenti S., Limana F., Jakoniuk I., Quaini F., Nadal-Ginard B., Bodine D.M., Leri A., Anversa P. (2001) Mobilized bone marrow cells repair the infarcted heart, improving function and survival. *Proc Natl Acad Sci USA* **98**, 10344–49.
8. Jackson K.A., Majka S.M., Wang H., Pocius J., Hartley C.J., Majesky M.W., Entman M.L., Michael L.H., Hirschi K.K., Goodell M.A. (2001) Regeneration of ischemic cardiac muscle and vascular endothelium by adult stem cells. *J Clin Invest* **107**, 1395–402.
9. Chamberlain G., Fox J., Ashton B., Middleton J. (2007) Concise review: mesenchymal stem cells: their phenotype, differentiation capacity, immunological features, and potential for homing. *Stem Cells* **25**, 2739–49. Review
10. Caplan A.I. (2009) Why are MSCs therapeutic? New data: new insight. *J Pathol* **217**, 318–24. Review.
11. Majumdar M.K., Thiede M.A., Mosca J.D., Moorman M., Gerson S.L. (1998) Phenotypic and functional comparison of cultures of marrow derived mesenchymal stem cells (MSCs) and stromal cells. *J Cell Physiol* **176**, 57–66.
12. Haynesworth S.E., Baber M.A., Caplan A.I. (1996) Cytokine expression by human marrow-derived mesenchymal progenitor cells in vitro: effects of dexamethasone and IL-1α. *J Cell Physiol* **166**, 585–92.
13. Aggarwal S., Pittenger M.F. (2005) Human mesenchymal stem cells modulate allogeneic immune cell responses. *Blood* **105**, 1815–22.
14. Di Nicola M., Carlo-Stella C., Magni M., Milanesi M., Longoni P.D., Matteucci P., Grisanti S., Gianni A.M. (2002) Human bone marrow stromal cells suppress T-lymphocyte proliferation induced by cellular or nonspecific mitogenic stimuli. *Blood* **99**, 3838–43.
15. González M.A., Gonzalez-Rey E., Rico L., Büscher D., Delgado M. (2009) Treatment of experimental arthritis by inducing immune tolerance with human adipose-derived mesenchymal stem cells. *Arthritis Rheum* **60**, 1006–19.
16. Oh J.Y., Kim M.K., Shin M.S., Lee H.J., Ko J.H., Wee W.R., Lee J.H. (2008) The anti-inflammatory and anti-angiogenic role of mesenchymal stem cells in corneal wound healing following chemical injury. *Stem Cells* **26**, 1047–55.
17. Yoo J.U., Barthel T.S., Nishimura K., Solchaga L., Caplan A.I., Goldberg V.M., Johnstone B. (1998) The chondrogenic potential of human bone-marrow-derived mesenchymal progenitor cells. *J Bone Joint Surg Am* **80**, 1745–57.
18. Johnstone B., Hering T.M., Caplan A.I., Goldberg V.M., Yoo J.U. (1998) In vitro chondrogensis of bone marrow-derived mesenchymal progenitor cells. *Exp Cell Res* **238**, 265–72.

Chapter 27

Panel Development for Multicolor Flow-Cytometry Testing of Proliferation and Immunophenotype in hMSCs

Jolene A. Bradford and Scott T. Clarke

Abstract

Adult human mesenchymal stem cells (hMSC) are rare fibroblast-like cells capable of differentiation into a variety of cell tissues which include bone, cartilage, muscle, ligament, tendon, and adipose. Normal adult bone marrow and adipose tissue are the most common sources of these cells. The International Society for Cellular Therapy (ISCT) has proposed a set of standards to define hMSC for laboratory investigations and preclinical studies: adherence to plastic in standard culture conditions; in vitro differentiation into osteoblasts, adipocytes, and chondroblasts; and specific surface antigen expression. Direct measurement of proliferation combined with simultaneous detection of the ISCT-consensus immunophenotypic profile provides data that is used to determine the differentiation status and health of the cells. Flow cytometry provides a powerful technology that is routinely used to simultaneously and rapidly measure multiple parameters in a single sample. This chapter describes a flow cytometric panel for the simultaneous detection of immunophenotypic profile, proliferative capacity, and DNA content measurement in hMSC. Because a relatively small number of cells are needed with this approach, measurements can be made with minimal impact on expansion potential. The ability to assess antigen expression and proliferative status enables the investigator to make informed decisions on expansion and harvesting.

Key words: hMSC, Mesenchymal stem cells, Flow cytometry, Immunophenotype, Proliferation, EdU, Click chemistry, Multicolor, Expansion, ISCT

1. Introduction

Adult human mesenchymal stem cells (hMSC) are rare fibroblast-like cells capable of differentiation into, and contributing to the regeneration of, a variety of cell tissues with limited capacity for self-repair. These tissues include bone, cartilage, muscle, ligament, tendon, and adipose (1–3). The therapeutic potential of these cells has drawn interest across a wide range of disciplines, as the use of hMSC may offer a mechanism to repair and/or regenerate these tissues (2, 3). Normal adult bone marrow and adipose

Mohan C. Vemuri et al. (eds.), *Mesenchymal Stem Cell Assays and Applications*, Methods in Molecular Biology, vol. 698, DOI 10.1007/978-1-60761-999-4_27,

tissues are the most common sources of these cells (3). Until now, the characteristics used to define hMSC have been inconsistently reported. To address this issue, the International Society for Cellular Therapy (ISCT) proposed a set of standards to define hMSC for laboratory investigations and preclinical studies. These studies include: adherence to plastic in standard culture conditions; in vitro differentiation into osteoblasts, adipocytes and chondroblasts; and specific surface antigen expression in which 95% of the cells express the antigens recognized by CD105, CD73, and CD90 with the same cells lacking (2% positive) the antigens recognized by CD45, CD34, CD14 or CD11b, CD79a or CD19, and HLA-DR (2). Recent studies have shown that CD34 antigen may be present, but its expression is transient and present only in early passages from cells derived from some isolates (3–5).

Conventional hMSC cultures exhibit decreased proliferation with increasing passage. Direct measurement of proliferation combined with simultaneous detection of the ISCT-recommended immunophenotypic profile provides information-rich data that is used to determine the differentiation status and health of the cells. Flow cytometry is a powerful technology that is routinely used to rapidly and simultaneously measure multiple parameters in a single sample (6–8). This chapter describes a flow cytometric panel for the simultaneous detection of the ISCT-consensus immunophenotypic profile combined with percent proliferation and DNA content measurements in a single sample. Because relatively few cells are needed with this approach, evaluation of cells at very low passage numbers can be made with minimal impact on yield following cell expansion. With this approach, the immunophenotype of the cells can be confirmed while information gathered on proliferative status enables the investigator to make informed decisions on expansion and harvesting.

Proliferation measurements, which detect the incorporation of a thymidine analog in newly synthesized DNA, provide the most direct measurement of proliferation within cells. The traditional method of pulsing cell cultures with the thymidine analog bromodeoxyuridine (BrdU) has limitations. In particular, multiparameter panels are not readily compatible with the BrdU methodology because of the requirement for harsh treatments in order to adequately denature the DNA for antibody-based BrdU detection. An alternative thymidine analog, ethynyldeoxyuridine (EdU), uses small molecule chemistry based detection, and allows treatment methods compatible with the immunophenotyping panel (9, 10). The basis of this method employs the chemistry of the Husigen reaction (also know as click chemistry), a copper (I) catalyzed cycloaddition between an azide and an alkyne. In this application, a dye azide conjugate is covalently bonded to the alkyne group found on the EdU molecule incorporated in the DNA (11, 12).

Detection of the incorporated EdU in combination with a stoichiometric DNA-binding dye allows for accurate DNA content cell cycle analysis, including direct S-phase measurements (9, 10).

Reagent selection for the flow cytometric panel begins with an understanding of the configuration of the instrument to be used. Laser excitation wavelengths and detector numbers will guide the use of particular fluorochromes and their combinations. The selection of antigens for the immunophenotyping panel follows the recommendation of the ISCT position paper (2). The selection of the fluorochromes is based upon excitation spectra, emission spectra, background staining, performance, and stability of individual fluorochromes, spectral overlap between fluorochromes, and availability of direct antibody conjugates or conjugation kits and services (7, 8, 13–17). The fluorescent reagents used to measure DNA content and EdU content are also considered in this panel design. Once the fluorochromes are chosen, optimal collection filters are selected to minimize spectral overlap between fluorochromes while maintaining sensitivity in each detection channel (14, 16, 18). Each antibody-fluorochrome conjugate requires validation on a cellular target known to be positive, and the individual antibody-fluorochrome conjugates require titration to determine optimal concentration giving the best signal-to-background ratio (6, 13, 14, 16, 18). The treatment of hMSC with EdU likewise requires optimization with regards to reagent concentration and pulse time (19). Performance of individual markers also requires validation in the full panel (8, 14, 16, 18). Since this procedure uses cells that are fixed, each antibody target requires testing before and after target fixation to validate that the antibody will detect the antigen of interest after fixation. In the experimental design, several controls are needed to ensure accurate results: compensation, fluorescence-minus-one (FMO), and biological (6–8, 14, 16, 20–22). In multicolor flow cytometry, spectral overlap between the fluorochromes needs to be eliminated using a mathematical process known as compensation. This is done using a series of control samples labeled individually with each of the fluorochromes to be used in the experiment. Compensation can be performed using cells, antibody-capture beads, or a combination of both and requires the use of the same reagents used in the panel (6, 7, 14, 20, 21, 23, 24). Compensation is best set using software as opposed to setting compensation manually using graphics (7, 14, 21, 24). Compensated data may result in an unexpected distribution (spread) of data that precludes the adjustment of signal to the level of the auto fluorescence found in unlabeled cells. The use of staining controls that include all reagents except for the one of interest, FMO controls, allows a precise definition for the delineation of positive and negative populations. This method also adjusts for the spread of the data (6, 14, 15, 20, 21, 23, 24).

Finally, biological controls that provide biologically relevant information should be included to ensure that the sample preparation, staining, compensation, and gating are performing as expected (8, 14, 16).

2. Materials

2.1. Cells and Culture Supplies

1. Adipose derived stem cells (ADSCs) (Invitrogen, Carlsbad, California) or bone marrow derived mesenchymal stem cells (BMSCs) (Lonza, Allendale, NJ), store in liquid nitrogen.
2. CELLstart™ (Invitrogen), substrate for coating flasks, store at 2–8°C.
3. Serum-free medium (SFM): Complete SFM (StemPro® MSC SFM; Invitrogen) consists of a basal medium (store at 4°C) and a frozen supplement (store at −20°C). To prepare complete SFM, thaw supplement at 2–8°C overnight and, to 84 mL basal medium, add 15 mL supplement and 1 mL of 200mM L-Glutamine or GlutaMAX™ (2 mM final; Invitrogen). Store complete SFM protected from light at 2–8°C for up to 1 month.
4. TrypLE™ Express, Stable Trypsin-Like Replacement Enzyme (Invitrogen), store at 2–8°C.
5. Dulbecco's phosphate buffered saline (DPBS) with Ca^{2+}/Mg^{2+} (1×) (Invitrogen). Store at room temperature.
6. DPBS without Ca^{2+}/Mg^{2+} (1×) (Invitrogen), store at room temperature.
7. T75 culture flasks (Greiner Bio One, Monroe, North Carolina).
8. CO_2 incubator: humidified, 37°C, and 5% CO_2.

2.2. Proliferation Detection

1. Click-iT® EdU Alexa Fluor® 647 Flow Cytometry Assay Kit (Invitrogen): The kit consists of the fixative, EdU (thymidine analog), permeabilization buffer, and reagents for detection of incorporated EdU using click chemistry. Follow package insert instructions for reagent preparation with the exception of EdU (component A). Resuspend the EdU in 1/6 of the recommended volume of DMSO, making a 60 mM stock solution. Reagents can all be stored at 4°C. Reconstituted Alexa Fluor® 647 azide and the buffer additive should be stored at −20°C for increased stability (see Notes 1–3).
2. FxCycle™ Violet stain (Invitrogen), for DNA content cell cycle analysis: To make a 1 mg/mL stock solution of FxCycle™ Violet stain, add 100 μL deionized water to one vial (containing

100 μg) of the stain. Mix well. Store this solution at 2–8°C, protected from light, do not freeze. The FxCycle™ Violet stock solution is stable for at least 6 months (see Note 4).

3. 1% bovine serum albumin (BSA) in PBS wash solution: To make 1 L of solution, add 33 mL of 30% BSA (Sigma-Aldrich, St. Louis, Missouri) to 967 mL phosphate buffered saline, pH 7.4 (PBS; Invitrogen). Store at 2–8°C for 1 month.

2.3. Immunostaining

1. CD14, Mouse anti-human, R-PE direct conjugate (Invitrogen), clone TüK4, isotype IgG2a, suspended in PBS with 0.1% sodium azide. Store at 2–8°C until listed expiration date, do not freeze (see Notes 5–8).
2. CD19, Mouse anti-human, R-PE conjugate (Invitrogen), clone SJ25-C1, isotype IgG1, suspended in PBS with 0.1% sodium azide. Store at 2–8°C until listed expiration date, do not freeze.
3. CD45, Mouse anti-human, R-PE conjugate (Invitrogen), clone HI30, isotype IgG1, suspended in PBS with 0.1% sodium azide. Store at 2–8°C until listed expiration date, do not freeze.
4. HLA-DR, Mouse anti-human, R-PE direct conjugate (Invitrogen), clone Tü36, isotype IgG2b, suspended in PBS with 0.1% sodium azide. Store at 2–8°C until listed expiration date, do not freeze.
5. CD34, Mouse anti-human, Qdot® 800 conjugate (Invitrogen custom conjugate), clone 581 (class III), isotype IgG1, 1 μM in 50 mM borate, 1 M betaine, pH 8.3 with 0.05% sodium azide. Store at 2–8°C until listed expiration date, do not freeze (see Notes 9–12).
6. CD73, Mouse anti-human, PerCP conjugate (Invitrogen custom conjugate), clone 7G2, isotype IgG2b, suspended in PBS with 0.1% sodium azide. Store at 2–8°C until listed expiration date, do not freeze.
7. CD90, Mouse anti-human, FITC conjugate (Invitrogen), clone 5E10, isotype IgG1, suspended in PBS with 0.1% sodium azide. Store at 2–8°C until listed expiration date, do not freeze.
8. CD105, Mouse anti-human, Alexa Fluor® 700 conjugate (Invitrogen custom conjugate), clone SN6, isotype IgG1, suspended in PBS with 0.1% sodium azide. Store at 2–8°C until listed expiration date, do not freeze.
9. Mouse IgG, 10 mg/mL in PBS with 0.1% sodium azide, (Sigma) for use as an Fc blocking agent. Can be stored at 2–8°C for short periods. For longer storage, divide solution into aliquots and freeze at ≤–20°C, avoid multiple freeze thaw cycles.
10. AbC™ anti-Mouse Bead Kit (Invitrogen), for use in setting compensation. Store at 2–8°C, do not freeze.

3. Methods

3.1. Human MSC Cell Culture

1. Precoat T75 flasks plastic flasks with the substrate, CELLstart™, at least 1 h before use.
2. Dilute CELLstart™ 1:100 in DPBS with Ca^{2+}/Mg^{2+}. Add 10 mL of diluted CELLstart™ to each flask and place in the CO_2 incubator for 1 h (see Note 13).
3. Aseptically remove the CELLstart™ from the flask, and replace with 10 mL of complete medium. Place flasks in the CO_2 incubator for sufficient time to reach equilibrium (see Note 14).
4. Source hMSCs from the liquid nitrogen. Release any trapped liquid nitrogen.
5. Rapidly thaw the frozen vial in a 37°C water bath.
6. Transfer the contents of the vial into a sterile 50 mL plastic tube.
7. While gently mixing, slowly add the 10 mL of the preequilibrated SFM (from T75 flask) drop-wise to the tube. Transfer the contents of the tube back into the treated flask and place horizontally in the CO_2 incubator.
8. Replace complete medium every 2–3 days and periodically monitor cell confluence.
9. Upon reaching ~70–90% confluence, cells are ready to be passaged and further expanded, if so desired.

3.2. Passaging the Culture

1. Prewarm DPBS (without Ca^{2+}/Mg^{2+}), TrpLE™ Express and complete SFM medium in a water bath to 37°C.
2. Aspirate the medium in the flask and wash twice with 10 mL of DPBS. Replace with 1.5 mL of TrpLE™. Rock the flask from side to side such that the enzyme coats the entire surface of the flask. Incubate at 37°C for 3 min (see Note 15).
3. Gently tap the side of the flask to detach the cells. Examine flasks under the microscope to insure complete cell release.
4. Add 8.5 mL of prewarmed complete medium and transfer the entire contents into a sterile 50 mL centrifuge tube. Wash the flask with 10 mL SFM and collect all remaining cells in the same tube.
5. Spin tube at 250×*g* for 5 min. Aseptically remove the medium. Dislodge the pellet and add 10 mL of prewarmed complete SFM.
6. Wash once more to completely remove TrpLE™ from the medium before resuspending the seed stock in 5–10 mL of complete medium.
7. Gently vortex and remove a small aliquot to count by hemocytometer or automated counting instrument.

8. Inoculate to a density of 3×10^3 cells/cm^2 (2.25×10^5 cells/T75 flask) to a new set of flasks pretreated with CELLstart™.
9. Place flasks in the CO_2 incubator.

3.3. EdU Treatment

1. Pulse label cell cultures grown in SFM with 60 μM EdU for the purpose of labeling the newly synthesized DNA (see Notes 16 and 17).
2. Prepare a working stock of EdU by first diluting 10.5 μL of 60 mM EdU into 0.5 mL of prewarmed medium for each flask to be labeled. Mix with gentle vortexing and then add 0.5 mL of diluted EdU to each flask containing 10 mL of culture medium. Mix by rocking, and replace the flask back in the incubator.
3. After 4 h of EdU labeling, harvest the flask, as described in Subheading 3.1. Briefly: aspirate medium, wash twice with DPBS without Ca^{2+}/Mg^{2+}, then TrypLE™ treat for 3 min, add medium or prewarmed 1% BSA/DPBS. Repeatedly pipet to remove the attached cells from the bottom of the flask. Transfer the entire culture to a 50 mL centrifuge tube and spin at $250 \times g$ for 5 min.
4. After removing the supernatant, dislodge the pellet with brief, gentle vortex or by raking the tube across a rack. Repeat the wash step with 10 mL of 1% BSA/DPBS.
5. Aspirate the wash buffer, dislodge the pellet, and add between 0.2 and 1.0 mL of 1% BSA/DPBS, depending on the cell density of the flask.
6. Determine the cell density with a hemocytometer or automated cell counting instrument.
7. Add an equal volume of Click-iT® EdU kit fixative (component D) or either 4% formaldehyde or paraformaldehyde/PBS to the cells.
8. Vortex (or vigorously mix) briefly and incubate for 15 min at room temperature.
9. Add 10 mL of 1% BSA/PBS wash buffer and spin at $250 \times g$ for 5 min.
10. Remove the supernatant, dislodge pellet, and resuspend the cells to a sufficient density. To conserve cells used for analysis, 100 μL of 2×10^6 cells/mL is sufficient for performing the multiparametric flow analysis, however up to 1×10^7 cells/mL can be used. At this stage the fixed cells can be safely stored at 4°C until beginning the staining procedure (see Note 18).

3.4. Immunostaining

Using the markers recommended for confirmation of hMSC immunophenotype, this multicolor flow cytometry panel is designed to include the three positive markers CD73, CD90, and CD105, as well as the CD34 marker in individual detection channels, while including the remaining negative markers CD14, CD19, CD45, and HLA-DR combined into a single detection channel, a so-called lineage channel (see Table 1). The emission profile of each fluorochrome and emission bandpass used for collection is shown in Figs. 1–3.

1. See Table 2 for sample set-up. The test samples will have every marker included. Single-color compensation controls will have only one marker per sample. Some compensation samples use cells while other compensation samples use antibody-capture beads. FMO controls that label cells will have all markers expect one. A negative cell control without EdU treatment that is stained with the Click-iT® reaction mixture is an optional gating control. A single labeled sample with the DNA content dye to be used for EdU gating is another optional control. Label a 12 × 75 mm tube for each control and sample being tested.
2. Add 100 μL fixed hMSC suspended in 1% BSA in PBS to each 12 × 17 mm tube. Ensure that the cell concentration is in the range of 5×10^5 to 1×10^7 cells/mL, and that the cell concentration is the same for every sample within the experiment. If using another cell type for positive and negative controls for the EdU detection, these samples should also be included in this staining procedure.
3. Add 100 μL Click-iT® EdU kit Triton® X-100-based permeabilization reagent to samples and mix well.
4. Incubate samples for 15 min at room temperature.

Table 1
Listing of markers and fluorochromes used in the flow-cytometry panel including the excitation laser and emission band pass filters

Marker	Fluorochrome	Excitation laser (nm)	Emission bandpass
CD73	PerCP	488	675/20
CD90	FITC	488	530/30
CD105	Alexa Fluor® 700	633	730/30
CD14 CD19 CD45 HLA-DR	R-PE	488	575/26
CD34	Qdot® 800 nanocrystal	488	780/60
Cell proliferation	Click-iT® EdU Alexa Fluor® 647 azide	633	660/20
DNA content	FxCycle™ Violet stain	405	450/50

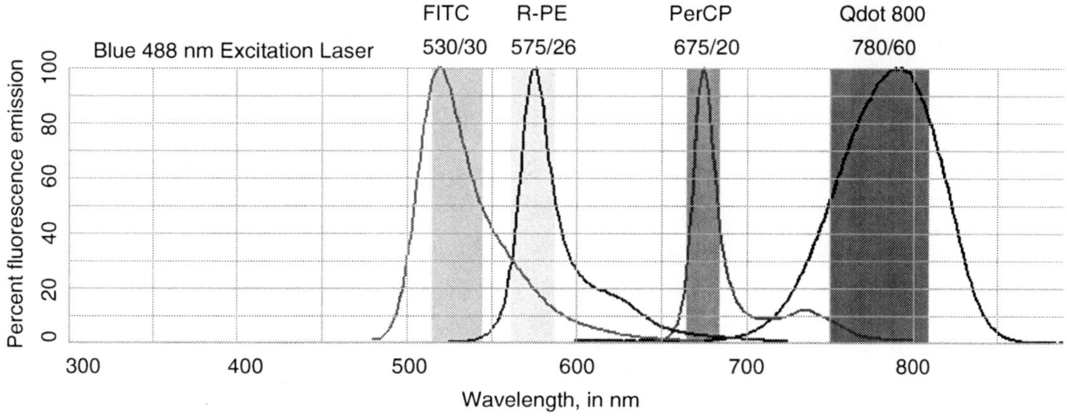

Fig. 1. Fluorescence emission spectra of FITC, R-PE, PerCP, and Qdot® 800 antibody conjugates including the bandpass filters used for collection used with the 488 nm excitation laser.

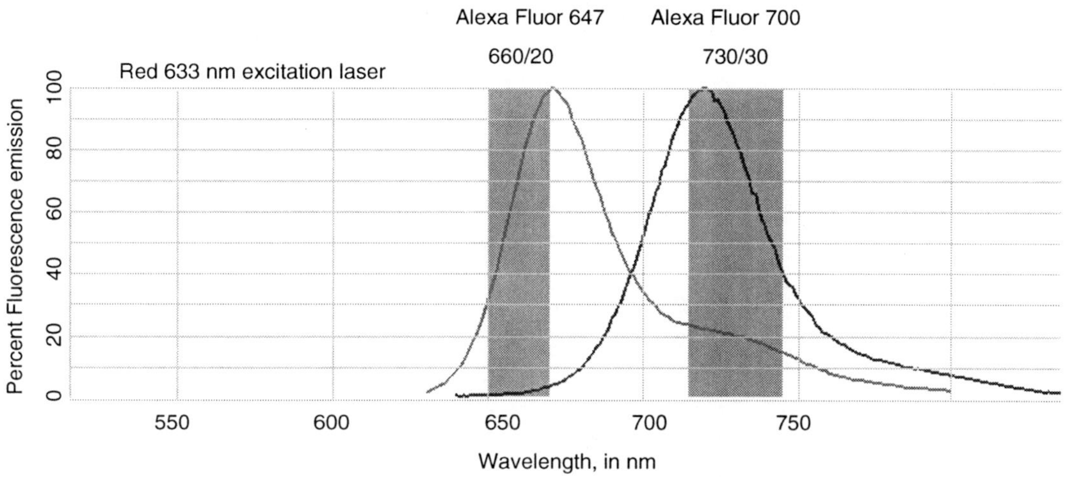

Fig. 2. Fluorescence emission spectra of Alexa Fluor® 647 and Alexa Fluor® 700 conjugates including the bandpass filters used for collection with the 633 nm excitation laser.

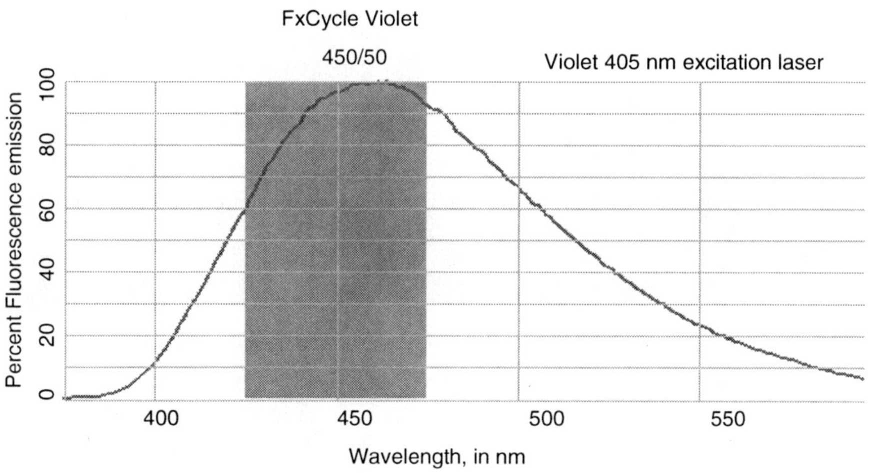

Fig. 3. Fluorescence emission spectra of FxCycle™ Violet stain including the bandpass filter used for collection with the 405 nm excitation laser.

Table 2
Sample set-up including unstained cells, single-color compensation samples, FMO gating samples, control samples, and test samples

Single-color compensation controls (capture beads or cells)	CD90 FITC	Negative R-PE	CD73 PerCP	CD105 Alexa Fluor® 700	CD34 QDot® 800	Click-iT® Alexa Fluor® 647	FxCycle™ Violet
hMSC cells-unstained	–	–	–	–	–	–	–
Beads CD90-FITC	Add	–	–	–	–	–	–
Beads negative markers-R-PE	–	Add	–	–	–	–	–
Beads-CD73-PerCP	–	–	Add	–	–	–	–
Beads-CD105-Alexa Fluor®700	–	–	–	Add	–	–	–
Beads CD34-Qdot® 800	–	–	–	–	Add	–	–
hMSC cells Click-iT® EdU+	–	–	–	–	–	Add	Add
hMSC cells FxCycle™ Violet Stain	–	–	–	–	–	–	
FMO gating controls (all markers except one)							
FMO, no CD90-FITC	–	Add	Add	Add	Add	Add	Add
FMO, no R-PE	Add	–	Add	Add	Add	Add	Add
FMO, no CD73-PerCP	Add	Add	–	Add	Add	Add	Add
FMO, no CD105-Alexa Fluor® 700	Add	Add	Add	–	Add	Add	Add
FMO, no CD34-Qdot® 800	Add	Add	Add	Add	–	Add	Add
FMO, no Click-iT® reaction	Add	Add	Add	Add	Add	–	Add
FMO, no FxCycle™ Violet stain	Add	Add	Add	Add	add	Add	–
Negative EdU controls							
hMSC (EdU-negative)	–	–	–	–	–	Add	–
hMSC-FxCycle™ Violet stain	–	–	–	–	–	–	Add
Test samples							
Etc. all markers	Add	Add	Add	Add	Add	Add	Add

5. Prepare Click-iT® reaction mixture during the previous incubation step. Prepare adequate Click-iT® reaction mixture for 0.5 mL usage for each sample requiring EdU detection. Use the reaction mixture within 15 min of preparation.
6. Add 0.5 mL Click-iT® reaction mixture to designated samples and mix well. It is not necessary to remove the permeabilization buffer before adding the Click-iT® reaction mixture.
7. Incubate the samples for 30 min at room temperature or on ice, protected from light.
8. Wash samples once with 3 mL 1% BSA in PBS, pellet cells by centrifugation at 500×*g* for 5 min, remove supernatant.
9. Dislodge the cell pellet and mix well to ensure a homogenous sample.
10. Add 20 μL mouse IgG to samples for Fc blocking. This is a final concentration of 200 μg/mL. Alternatively, goat IgG may be used. Incubate for 10 min at room temperature.
11. Vortex (or vigorously mix) the AbC™ capture beads (Component A in the AbC™ anti-mouse bead kit) to mix the bead sample.
12. Add one drop capture beads to designated 12×75 mm test tubes, ensure the bead sample is in the bottom of the tube.
13. Add pretitered amounts of each antibody in the panel to designated tubes (containing either cells or beads) and mix well. Make sure the antibody is deposited directly to the bead or cell suspension.
14. Incubate samples for 30 min at room temperature or on ice, protected from light.
15. Wash all samples once with 3 mL 1% BSA in PBS, pellet cells by centrifugation at 500×*g* for 5 min, remove supernatant.
16. Dislodge the cell pellet and mix well to ensure a homogenous sample.
17. Resuspend samples in 0.5 mL 1% BSA in PBS.
18. Add 0.5 μL FxCycle™ Violet stain to designated tubes. Do not wash samples (see Note 19).
19. Incubate samples for 15–30 min at room temperature, protected from light.
20. If data acquisition will be delayed, store samples at 2–8°C, protected from light, until acquisition. Collect data within 18 h of staining.

3.5. Instrument Set-up and Data Acquisition

1. Lasers and detection parameters: The instrument for this particular panel will require three lasers for excitation. Detection will require two scatter parameters and seven fluorescent channels: Violet (405 nm) excitation with one fluorescent detection channel, Blue (488 nm) excitation with two scatter

detectors and four fluorescent detection channels, and Red (633/5 nm) excitation with two fluorescent detection channels. See Table 1 (see Notes 20 and 21).

2. Select the optimal optic filters to minimize spectral overlap between fluorochromes while maintaining sensitivity in each detection channel. The particular instrument in use will in part determine the actual optical filters used based on the optical path (6, 8, 14). The excitation lasers, reagents, and bandpass filters used in this particular panel are listed in Table 1. Equivalent or similar bandpass filters may be used in place of these filters. Online spectra viewers are invaluable in determining fit of fluorochromes, bandpass filters, and excitation lasers for experimental design (25) (see Notes 22 and 23).
3. Verify the performance of the instrument on the day of use. The performance of the instrument can have a significant impact on the measurements in this panel, especially the DNA content measurement. Instrument performance is typically monitored with fluorescent bead suspensions as per manufacturer recommendations. The fluidics and laser alignment need to be checked; additionally the instrument should have periodic check of laser delay, window extension, linearity, and sensitivity (6, 20, 26).
4. Select the parameters for analysis. Linear parameters include forward scatter (FCS), side scatter (SSC), DNA content Area, and DNA content Width. The area and width DNA signal parameters are used for gating of singlet cells to exclude cell aggregates (26–28). Logarithmic parameters include the detectors for CD90-FITC, negative marker mixture in R-PE, CD73-PerCP, CD34-Qdot® 800 nanocrystal, Alexa Fluor® 647 azide for EdU detection, and CD105-Alexa Fluor® 700 conjugate. Select the log/linear transformed plot if available on the specific instrument in use to more accurately display compensated data (6, 13, 22, 29, 30).
5. Set optimal voltage for each parameter. A key element in the experiment is setting optimal voltage for each detector to achieve maximal signal-to-background reading. Set each detector voltage high enough to bring the autofluorescence of the cells above background fluorescence of the instrument. Ideally each detector voltage is set with a sample or samples containing the brightest and dimmest signals expected for the assay to ensure the positive and negative signals appear on scale (8, 14). Once the fluorescent voltage has been set, it is important not to change it throughout the entire data collection (6) (see Note 24).
6. Compensation. In multicolor flow cytometric analysis, spectral overlap between the fluorochromes needs to be eliminated using compensation. The spillover is a result of fluorescent dyes that are measurable in more than one detector and it is something that can be corrected (14, 18, 20, 24). Compensation is considered an experiment-related rather than instrument-associated setting,

and should be determined for each multicolor experiment (20). This panel uses a combination of hMSC and antibody-capture beads to set compensation. Using a FSC vs. SSC scatter plot, gate on either the cells or the beads to collect the single-color compensation controls for each fluorescent parameter. A minimum of 5,000 gated events should be used for collection of compensation controls. Use the positive signal generated from the beads for antibody compensation, the positive azide dye signal on EdU-incorporated hMSCs for the EdU compensation, and the entire DNA content signal (in linear scaling) for the DNA compensation. Once all the single-color compensation samples have been collected, use software to calculate the compensation matrix. These compensation settings are used to analyze the remainder of the experiment (see Notes 25–28).

7. A low flow rate should be used for sample acquisition when using instruments that use hydrodynamic focusing. When performing DNA content analysis, using a low flow rate of 200–400 events/second for acquisition provides lower percent coefficient of variation (CV) and thus more accurate DNA content analysis (26, 28).
8. The minimum number of events to collect is 10,000 per sample, gated on the DNA content width vs. area singlet gate. This minimum number is necessary for reproducible DNA content measurements. Higher numbers of events may produce better results, with 20–30,000 events usually providing adequate numbers. Set the instrument to stop data collection after a minimum of 10,000 singlet-gated events (26).

3.6. Analysis and Gating

1. FMO gating and immunophenotype evaluation: First create a region on the main population of cells on the FSC vs. SSC dual parameter plot to exclude debris. Gated on this main population of cells, create a second dual parameter plot of DNA Width vs. Area. Create a region on DNA width vs. area plot around the singlet cell population, the singlet gate. All of the fluorescent parameters are gated on this singlet gate. Include the following two parameter plots: FITC vs. R-PE, Qdot® 800 vs. R-PE, PerCP vs. R-PE, Alexa Fluor® 700 vs. R-PE, Qdot® 800 vs. FITC, PerCP vs. FITC, Alexa Fluor® 700 vs. PerCP. By using these plots, considerably fewer than all possible dual plot combinations, allows each antibody to be displayed against a positive and a negative marker. Create flexible gating quadrants based on the FMO controls for gating. Create statistics measurements for each quadrant in each plot to include percent of parent. Evaluation of Immunophenotype of each sample can be performed on these plots (see Fig. 4).
2. Proliferation Evaluation: Create two single parameter histograms, each gated on the singlet gate, one linear histogram for DNA content and one log histogram for the Click-iT® azide fluorescence. Create a marker on the azide-positive cells on the

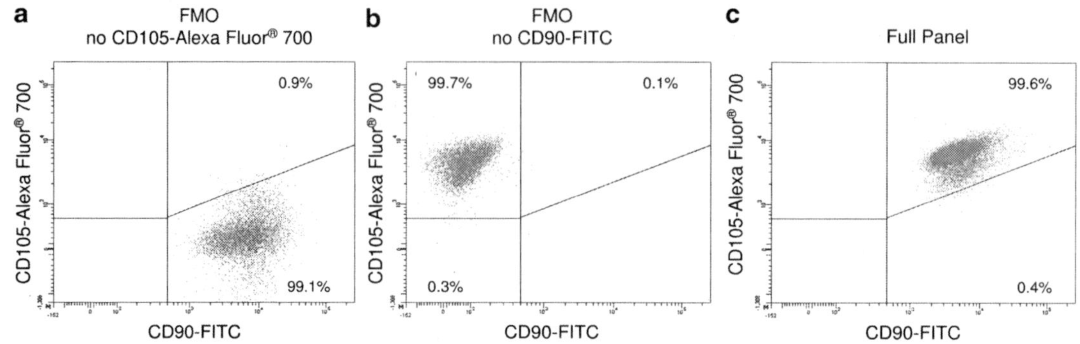

Fig. 4. Example of fluorescence-minus-one (FMO) gating. All three plots show dual parameter dot plots of CD90-FITC vs. CD105-Alexa Fluor® 700 fluorescence; both markers are positive in hMSCs. Plot (**a**) displays data from the FMO sample containing all markers except CD105-Alexa Fluor® 700; the flexible quad maker is drawn to exclude any positive signal from CD105 in the CD90-FITC positive cells. Plot (**b**) displays data from the FMO sample containing all markers except CD90-FITC; the quad marker is drawn to exclude any positive signal from CD90. Plot (**c**) displays data for the sample containing all markers in the panel, using the gating determined from the FMO samples in plots (**a**) and (**b**). The cells show the immunophenotype of CD90 and CD105 copositive expected in hMSCs.

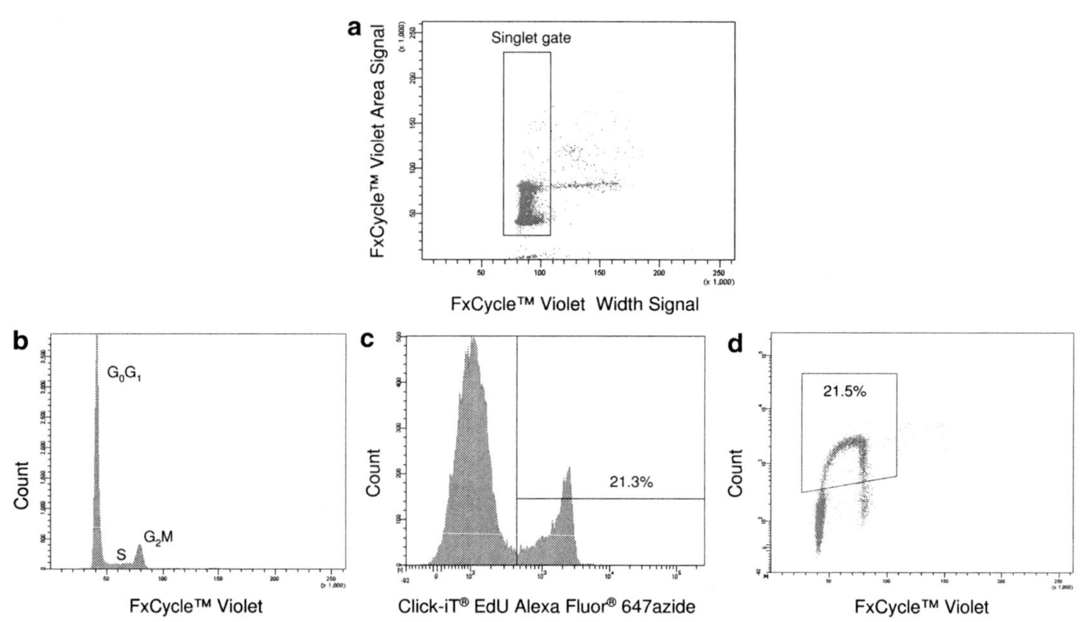

Fig. 5. Example of gating for analysis. Plot (**a**) displays the DNA width vs. area signal with region gated around the singlet cell population to exclude any cell aggregates. Plots (**b**–**d**) are all gated on this singlet region. Plots (**b**) displays a DNA content histogram, with cells distributed among three major phases of the cell cycle: G0/G1 phase, S-phase, and G2/M phase. Plot (**c**) displays the Alexa Fluor® 647 azide detection of EdU with a marker drawn on the positive cells showing 21.3% of the cells are proliferating. Plot (**d**) is a dual parameter plots of DNA content vs. Alexa Fluor® 647 azide detection of EdU, with a region drawn around the EdU positive cells showing 21.5% of the cells are proliferating.

Click-iT® azide. Create a dual parameter plot gated on the singlet gate for DNA content vs. Click-iT® azide fluorescence; create a region on the azide positive cells, indicating cells which are proliferating. These plots are used for evaluation of DNA content and percent proliferation. See Fig. 5 (see Note 29).

4. Notes

1. DMSO (Component C) provided as a solvent in this kit is known to facilitate the entry of organic molecules into tissues. Handle reagents containing DMSO using equipment and practices appropriate for the hazards posed by such materials.
2. IC Fixation Buffer (Invitrogen) or other 4% formaldehyde or paraformaldehyde in PBS may be used in place of the Click-iT® kit fixative.
3. 0.1% Triton®X-100 in DPBS permeabilization reagent may be used in place of the Click-iT™ Triton®-X100-based permeabilization reagent (Component F) in the Click-iT® kit permeabilization buffer, may offer increased access of reagents to the nuclear components.
4. The stain is a known mutagen and may cause sensitization by inhalation and skin contact, and is irritating to eyes, respiratory system, and skin. Do not breathe dust. In case of contact with eyes, rinse immediately with plenty of water and seek medical advice. Wear suitable protective clothing, safety glasses, and gloves. Avoid contact with skin and eyes.
5. Sodium azide is an extremely toxic and dangerous compound particularly when combined with acids or metals. Solutions containing sodium azide should be disposed off properly.
6. Exposure to light should be avoided with fluorochromes-conjugated antibodies. Use dim light during handling, incubation, and analysis. It is recommended that data acquisition be completed within 18 h of staining. If the conjugate is being diluted, it is recommended that only the quantity to be used within 1 week be diluted.
7. Antibodies should be titered for optimal staining. CD14, CD19, CD45, and HLD-DR antibodies can be titered using human mononuclear cells; KG1a culture cells or other CD34 positive cells may be used for CD34 antibody titration; hMSCs should be used for titration of the markers CD73, CD90, and CD105.
8. The use of fluorochrome-conjugated isotype-matched controls may be useful in determining nonspecific staining for each antibody being tested. The isotype controls should not, however, be used for defining positive signals (20).
9. Several custom conjugates were used in this panel, many other companies also offer custom conjugation services. It is also possible to label IgG antibody using kits for monoclonal labeling or protein labeling kits (17).

10. Because Qdot® nanocrystals are conjugated to biological materials, some loss of activity may be observed with prolonged storage. The Qdot® conjugate contains cadmium and selenium in an inorganic crystalline form. Dispose of the material in compliance with all applicable local, state, and federal regulations for disposal of these classes of material
11. Qdot® nanocrystals are nanometer-scale atom clusters of semiconductor material which exhibit narrow and symmetrical emission bandwidths with very long Stokes shifts. The nanocrystals can be excited with any wavelength below their emission maximum but are best excited by UV or violet light. Qdot® nanocrystals demonstrate an intrinsic brightness and photostability that can be many times greater than observed with other classes of fluorophores (30–32).
12. Qdot® nanocrystals are excited optimally with UV or 405 nm light, although excitation can be obtained with any wavelength below the emission wavelength of a given nanocrystal. Qdot® nanocrystals can be used on cytometers that do not have UV or violet excitation sources as long as they have appropriate emission filters. Be sure to check for Qdot® nanocrystals emission in any channel that can capture nanocrystals emission off of other lasers on the cytometer (32).
13. Alternatively, flasks can be coated in advance by placing horizontally in 4°C. They can be stored for several weeks before use.
14. To achieve increased viability from the frozen seed vial, the medium should be equilibrated to temperature and CO_2 prior to inoculation.
15. Do not overdigest by incubating longer than 7 min at 37°C.
16. This is a longer pulse at a higher concentration than required for most mammalian culture conditions.
17. To avoid possible DMSO toxicity, add no more than 1/1,000 volume to mammalian cells.
18. For prolonged storage, 0.1% sodium azide should be added to the wash buffer as a preservative.
19. Nucleic acid dyes stain in equilibrium with the suspension buffer and therefore must be present in the external media to maintain fluorescent intensity.
20. If the available instrument does not have a Violet (405 nm) excitation laser, the DNA content measurement may be eliminated. DNA content cell cycle data will not be obtained, however the percent proliferation will be available from the EdU detection.
21. If the available instrument has only three detection channels with the Blue (488 nm) excitation laser, the CD34 may be combined with the negative marker mixture in the R-PE channel instead of using a dedicated channel for CD34 detection.

22. Spectra viewers are available on the web sites of Invitrogen, BD Biosciences, and Beckman Coulter.

23. Qdot® nanocrystals can be excited by wavelengths below their emission wavelength, but with decreasing efficiency at longer excitation wavelengths. Nanocrystals can be used efficiently with 488 nm excitation sources. In this panel, Qdot® 800 nanocrystal is used in place of an R-PE tandem dye as the fourth fluorochrome using the 488 nm excitation laser. Its use provides two distinct advantages: the Qdot® nanocrystal does not exhibit the light sensitivity issues that are problematic with tandem dyes. In addition, there will be no residual fluorescence in the R-PE channel, also a problem when using the tandem dyes. Residual R-PE fluorescence from an R-PE tandem dye could be measured as a false positive of the negative marker mixture measured in the R-PE channel (15, 16, 32).

24. Many instruments perform fluorescence optimization as part of an automated instrument standardization using microspheres (6, 16). If the instrument to be used does not have this automated standardization, the procedure may be performed manually (8, 20).

25. The use of antibody-capture beads for compensation has several advantages. The antibody-capture beads can replace cells in the compensation set-up, an advantage when cells are scarce, as is the case with hMSCs. Antibodies are captured on the beads without regard to specificity, and so provide a bright and uniform signal regardless of how bright or dim the antibody stains cells. In this panel, the beads are able to set the compensation with the negative markers in the R-PE and Qdot® channels, even though these antibodies are expected not to have expression on the hMSCs. The same antibody-fluorochrome conjugate used in the panel is used for compensation calculation using beads, allowing the same lot used in the actual testing (14, 20).

26. Log/linear transformed plots are extremely valuable for correct visualization of compensated data. The name used for this type of data presentation is variable, and includes Bi-Exponential, HyperLog™ (Verity House), Lin/Log, and Logicle. Dual parameter plot display using log/linear scaling requires flexible quadrant gating (20–22, 29).

27. All data may be collected uncompensated, including all controls, and then compensated post-acquisition in analysis software.

28. Compensation is best set using software vs. any manual procedures using visual graphics (7, 14, 21, 24).

29. The DNA content data may be evaluated using cell cycle modeling software analysis. Although this is not a direct measurement of S-phase like that obtained with Click-iT® EdU detection, it can give approximate S-phase percentage.

Acknowledgments

The authors would like to thank Drs. Lucas Chase and Shayne Boucher for helpful discussions and technical expertise. In addition, we would like to thank Tami Nyberg, Shulamit Jaron, Kristi Haataja, Judie Berlier, and John Ivanovitch for their technical assistance.

References

1. Chamberlain, G., Fox, J., Ashton, B., and Middleton, J. (2007) Concise review: mesenchymal stem cells: their phenotype, differentiation capacity, immunological features, and potential for homing, *Stem Cells* 25, 2739–2749.
2. Dominici, M., Le Blanc, K., Mueller, I., Slaper-Cortenbach, I., Marini, F., Krause, D., Deans, R., Keating, A., Prockop, D., and Horwitz, E. (2006) Minimal criteria for defining multipotent mesenchymal stromal cells. The International Society for Cellular Therapy position statement, *Cytotherapy* 8, 315–317.
3. Roche, S., Delorme, B., Oostendorp, R. A., Barbet, R., Caton, D., Noel, D., Boumediene, K., Papadaki, H. A., Cousin, B., Crozet, C., Milhavet, O., Casteilla, L., Hatzfeld, J., Jorgensen, C., Charbord, P., and Lehmann, S. (2009) Comparative proteomic analysis of human mesenchymal and embryonic stem cells: towards the definition of a mesenchymal stem cell proteomic signature, *Proteomics* 9, 223–232.
4. Lin, G., Garcia, M., Ning, H., Banie, L., Guo, Y. L., Lue, T. F., and Lin, C. S. (2008) Defining stem and progenitor cells within adipose tissue, *Stem Cells Dev* 17, 1053–1063.
5. Gimble, J. M., Katz, A. J., and Bunnell, B. A. (2007) Adipose-derived stem cells for regenerative medicine, *Circ Res* 100, 1249–1260.
6. Tung, J. W., Parks, D. R., Moore, W. A., and Herzenberg, L. A. (2004) Identification of B-cell subsets: an exposition of 11-color (Hi-D) FACS methods, *Methods Mol Biol* 271, 37–58.
7. Perfetto, S. P., Chattopadhyay, P. K., and Roederer, M. (2004) Seventeen-colour flow cytometry: unravelling the immune system, *Nat Rev Immunol* 4, 648–655.
8. Wood, B. (2006) 9-color and 10-color flow cytometry in the clinical laboratory, *Arch Pathol Lab Med* 130, 680–690.
9. Buck, S. B., Bradford, J., Gee, K. R., Agnew, B. J., Clarke, S. T., and Salic, A. (2008) Detection of S-phase cell cycle progression using 5-ethynyl-2′-deoxyuridine incorporation with click chemistry, an alternative to using 5-bromo-2′-deoxyuridine antibodies, *Biotechniques* 44, 927–929.
10. Salic, A., and Mitchison, T. J. (2008) A chemical method for fast and sensitive detection of DNA synthesis in vivo, *Proc Natl Acad Sci U S A* 105, 2415–2420.
11. Meldal, M., and Tornoe, C. W. (2008) Cu-catalyzed azide-alkyne cycloaddition, *Chem Rev* 108, 2952–3015.
12. Tornoe, C. W., Christensen, C., and Meldal, M. (2002) Peptidotriazoles on solid phase: [1,2,3]-triazoles by regiospecific copper(i)-catalyzed 1,3-dipolar cycloadditions of terminal alkynes to azides, *J Org Chem* 67, 3057–3064.
13. Ormerod, M. G. (2008) Flow cytometry – a basic introduction, at http:flowbook.denovo-software.com
14. Baumgarth, N., and Roederer, M. (2000) A practical approach to multicolor flow cytometry for immunophenotyping, *J Immunol Methods* 243, 77–97.
15. Maecker, H. T., Frey, T., Nomura, L. E., and Trotter, J. (2004) Selecting fluorochrome conjugates for maximum sensitivity, *Cytometry A* 62, 169–173.
16. Maecker, H. T., and Trotter, J. (2008) Selecting reagents for multicolor flow cytometry, BD Biosciences Application Note.
17. Roederer, M. (1996) Conjugation of monoclonal antibodies, at http://www.drmr.com/abcon/index.html
18. Roederer, M., De Rosa, S., Gerstein, R., Anderson, M., Bigos, M., Stovel, R., Nozaki, T., Parks, D., and Herzenberg, L. (1997) 8 color, 10-parameter flow cytometry to elucidate complex leukocyte heterogeneity, *Cytometry* 29, 328–339.
19. Diermeier-Daucher, S., Clarke, S. T., Hill, D., Vollmann-Zwerenz, A., Bradford, J. A., and Brockhoff, G. (2009) Cell type specific applicability of 5-ethynyl-2′-deoxyuridine (EdU) for dynamic proliferation assessment in flow cytometry, *Cytometry A* 75, 535–546.

20. Maecker, H. T., and Trotter, J. (2006) Flow cytometry controls, instrument setup, and the determination of positivity, *Cytometry A* 69, 1037–1042.

21. Bayer, J., Grunwald, D., Lambert, C., Mayol, J. F., and Maynadie, M. (2007) Thematic workshop on fluorescence compensation settings in multicolor flow cytometry, *Cytometry B Clin Cytom* 72, 8–13.

22. Herzenberg, L. A., Tung, J., Moore, W. A., and Parks, D. R. (2006) Interpreting flow cytometry data: a guide for the perplexed, *Nat Immunol* 7, 681–685.

23. McLaughlin, B. E., Baumgarth, N., Bigos, M., Roederer, M., De Rosa, S. C., Altman, J. D., Nixon, D. F., Ottinger, J., Oxford, C., Evans, T. G., and Asmuth, D. M. (2008) Nine-color flow cytometry for accurate measurement of T cell subsets and cytokine responses. Part I: Panel design by an empiric approach, *Cytometry A* 73, 400–410.

24. Roederer, M. (2001) Spectral compensation for flow cytometry: visualization artifacts, limitations, and caveats, *Cytometry* 45, 194–205.

25. Invitrogen. Spectra Viewer, at http://www.invitrogen.com/site/us/en/home/support/research-tools/fluorescence-SpectraViewer.html

26. Shankey, T. V., Rabinovitch, P. S., Bagwell, B., Bauer, K. D., Duque, R. E., Hedley, D. W., Mayall, B. H., Wheeless, L., and Cox, C. (1993) Guidelines for implementation of clinical DNA cytometry. International Society for Analytical Cytology, *Cytometry* 14, 472–477.

27. Wersto, R. P., Chrest, F. J., Leary, J. F., Morris, C., Stetler-Stevenson, M. A., and Gabrielson, E. (2001) Doublet discrimination in DNA cell-cycle analysis, *Cytometry* 46, 296–306.

28. Givan, A. L. (2001) *Flow Cytometry: First Principles*, Wiley, New York.

29. Bagwell, C. B. (2004) DNA histogram analysis for node-negative breast cancer, *Cytometry A* 58, 76–78.

30. Telford, W. G. (2004) Analysis of UV-excited fluorochromes by flow cytometry using near-ultraviolet laser diodes, *Cytometry A* 61, 9–17.

31. Shapiro, H. (2003) *Practical Flow Cytometry*, Fourth Edition, John, Hobboken.

32. Godfrey, W. L., Zhang, Y., Jaron, S., and Buller, G. (2008) Use of Qdot® nanocrystal primary antibody conjugates in flow cytometry, Invitrogen Application Note, at http://www.probes.invitrogen.com/products/qdot/images/O-073210%20QDot%20AppNote_HRF.pdf

Chapter 28

Simplified PCR Assay for Detecting Early Stages of Multipotent Mesenchymal Stromal Cell Differentiation

Shayne E. Boucher

Abstract

With increased demand for standardized stem cell-based assays in basic and clinical research, there is a concerted effort to develop and share quick, robust validated assays for tracking multipotent mesenchymal stromal cell (MSC) status and multipotency retention. With respect to determining differentiation capacity, classical method is to perform time-consuming histological stain assays to detect mature differentiated cell types, which can take up to 1 month or more. To accelerate identification of MSC lineage commitment, we developed and validated a simple PCR-based growth and differentiation assay to routinely detect MSC lineage commitment within 7 days. By establishing a standardized PCR assay system, critical attributes can be temporally tracked in cultured MSC. In addition to meeting the reference criteria for MSC identification, this approach is also utilized in quality testing and lot release of stem cell media products.

Key words: Induction, Multipotency, Adipogenesis, Osteogenesis, Chondrogenesis

1. Introduction

There has been a rapid increase in the level of understanding the genomic basis of multipotent mesenchymal stromal cell (MSC) function and regulation since it was first described (1–4). One area of particular interest is the impact of media systems on MSC behavior where precise control is critical for finely delineating stem cell function (5, 6). However, standardized expansion and characterization methods are not in common use for generating results that can be readily compared with results from other labs. Thus, a premium is placed on streamlining expansion and characterization assay workflows that can generate results in a relatively

Mohan C. Vemuri et al. (eds.), *Mesenchymal Stem Cell Assays and Applications*, Methods in Molecular Biology, vol. 698, DOI 10.1007/978-1-60761-999-4_28,

Table 1
Extended criteria for identifying MSC and its differentiated progeny

Extended criteria for identifying mesenchymal stromal stem cells

Adherence to plastic in standard culture conditions	
Phenotype:	
Positive (≥95% +)	*Negative* (≤2% +)
CD105	CD45
CD73	CD34
CD90	CD14 or CD11b
FOSB (*PCR*)	CD79a or CD19
	HLA-DR
Early differentiation (identified by gene microarray, verified by PCR)	
Cell type	*7 days*
Osteoblasts/osteocytes	RGC32/FKBP5
Preadipocytes/adipocytes	FABP4/HP
Chondroblasts/chondrocytes	SPP1/TNFAIP6

short period of time, especially for demonstrating commitment of stem cells to specific differentiated lineages (3, 4, 7–9).

To focus on generating unambiguous and consistent differentiation results for recently released osteogenesis, adipogenesis, and chondrogenesis differentiation kits, we designed, optimized, and validated culture- and genomic-based assays for qualifying these kits. Part of the workflow simplification process was to use a common template protocol for expansion and duplexed gel-based PCR assays on uninduced and differentiating cell types within 7 days (10). By focusing on a simplified culture- and genomic-based approach, this allows MSC researchers to work with standardized characterization methods that facilitate data sharing among stem cell research labs (Table 1).

2. Materials

2.1. MSC Expansion

1. MSC Expansion Medium: Prepare by adding 450 mL D-MEM low glucose with GlutaMAX-I (Invitrogen, Carlsbad, CA) with 50 mL MSC-Qualified FBS (Invitrogen), and 250 μL 10 mg/mL gentamicin solution (optional) (Invitrogen). Store at 4°C in the dark for up to 1 month.
2. TrypLE Express cell dissociation solution (Invitrogen).
3. Tissue culture-treated 75 and 225 cm^2 cell culture flasks with vented cap (BD Biosciences, San Jose, CA).

2.2. MSC Differentiation

1. Osteogenic Differentiation Medium: Prepare StemPro Osteogenesis Differentiation medium (Invitrogen), according to the manufacturer's protocol.
2. Adipogenic Differentiation Medium: Prepare StemPro Adipogenesis Differentiation medium (Invitrogen) according to the manufacturer's protocol.
3. Chondrogenic Differentiation Medium: Prepare StemPro Chondrogenesis Differentiation medium (Invitrogen) according to the manufacturer's protocol.
4. Disposable reagent reservoir (VWR, West Chester, PA).
5. 10 μL 8-channel pipetter (Rainin Instrument, Oakland, CA).
6. Tissue culture-treated 100 mm dishes (BD Biosciences).
7. Cell scrapers (BD Biosciences).
8. TrypLE Express cell dissociation solution (Invitrogen).
9. 0.2% Collagenase Type II solution: Prepare by adding 0.1 g collagenase type II (Invitrogen) and 50 mL HBSS with Ca^{2+} and Mg^{2+} (Invitrogen) into a 50 mL conical tube. Mix well and filter sterilize. Aliquot into 1 mL samples and store at −20°C.

2.3. RNA Isolation and Quality Check

1. 1× TAE buffer: add 20 mL UltraPure DNA Typing Grade 50× TAE buffer (Invitrogen) to 980 mL distilled water (Invitrogen). Store at room temperature.
2. 125 mL Erlenmeyer glass flask (VWR).
3. RNase AWAY reagent (Invitrogen).
4. 100% ethanol.
5. Chloroform.
6. UltraPure Agarose (Invitrogen).
7. 10 mg/mL ethidium bromide solution (Invitrogen).
8. PureLink Micro-to-Midi Total RNA Purification System (Invitrogen).
9. DNAse (Invitrogen).
10. Mini-Sub-Cell GT with casting stand, gel tray and comb (Bio-Rad, Hercules, CA).

2.4. Order Primer Sets from DNA Oligonucleotide Synthesizer Source

1. B2M (beta-2-microglobin) – housekeeping gene, 314 bp product

   B2M primer forward : GCGTACTCCAAAGATTCAG

   B2M primer reverse: CAAACCTCCATGATGCTG
2. CD73 (5′ nucleotidase, ecto) – MSC cell surface marker, 414 bp product

CD73 primer forward: CAATTGTCTATCTGGATGGC

CD73 primer reverse: GACACTTGGTGCAAAGAAC

3. RGC32 (response gene to complement 32) – early osteocyte cell marker, 166 bp product

   RGC32 primer forward: GCCACTTCCACTACGAGGAG

   RGC32 primer reverse: GCTGGGGTAGAGTCTGTTGG

4. FABP4 (Fatty acid-binding protein 4) – early adipocyte cell marker, 215 bp product

   FABP4 primer forward: TCATACTGGGCCAGGAAT

   FABP4 primer reverse: TCCCTTGGCTTATGCTCT

5. SPP1 (Bone sialoprotein 1) – early chondrocyte cell marker, 229 bp product

   SPP1 primer forward: CTCCATTGACTCGAACGACTC

   SPP1 primer reverse: CAGGTCTGCGAAACTTCTTAGAT

2.5. Prepare 100 μM Primer Stock Solution

1. Note nanomoles value for each primer on the Certificate of Analysis. Calculate amount of RNAse, DNase-free water required to make a 100 μM stock solution (e.g., if a primer sample contains 50 nmol, 500 μL RNAse, DNase-free water is needed to generate a 100 μM stock solution; a minimum of 250 μL solution should be prepared for each primer).
2. Add calculated amount of water to primer vial and pipette mix thoroughly including the sides of tube. Proceed to next step; store unused portion at −20°C.

2.6. Prepare 10 μM Duplex Primer Working Solution

1. Label a 2 mL microfuge tube with B2M and single gene of interest (GOI), where GOI is RGC32, FABP4, or SPP1. Pipette 40 μL 100 μM each of B2M forward and reverse primers in microfuge tube. Pipette 80 μL 100 μM each of GOI forward and reverse primers in same tube.
2. Pipette 1,480 μL RNAse, DNAse-free water in tube and pipette mix thoroughly to generate 1,600 μL of duplex primer solution.
3. Aliquot 125 μL duplex primer working solution into labeled 12 × 200 μL PCR tubes. Store at −20°C.
4. Repeat process for remaining GOIs.

2.7. cDNA Synthesis and PCR Analysis

1. SuperScript III Platinum Two-Step qRT-PCR Kit with SYBR Green (Invitrogen).
2. 1× TAE buffer: add 20 mL UltraPure DNA Typing Grade 50× TAE buffer (Invitrogen) to 980 mL distilled water (Invitrogen). Store at room temperature.

3. UltraPure Agarose (Invitrogen).
4. 10 mg/mL ethidium bromide solution (Invitrogen).
5. Sub-Cell GT with casting stand, gel tray, and comb (Bio-Rad).
6. 96-well skirted or semi-skirted white PCR plates (VWR).
7. 8-well PCR strip caps (VWR).
8. Strip-cap tool (Bio-Rad).
9. DNA Engine Thermal Cycler PTC-0200 (Bio-Rad).

3. Methods

The following protocol has been established to allow researchers to perform a PCR analysis comparing uninduced and differentiated MSC. The provided protocol has been broken down into three phases: (1) cell expansion and differentiation, (2) RNA isolation and quality check, and (3) cDNA and PCR. There are two protocols for performing MSC Differentiation. Osteogenic and adipogenic differentiation follows the same protocol as outlined in Subheading 3.1, steps 4–6. By contrast, chondrogenic differentiation requires special handling procedures that are covered in Subheading 3.1, steps 7–10. For all other steps, the protocol is the same for all three differentiation methods.

3.1. Cell Expansion and Differentiation

1. Day 0: Recover and plate MSC in T75 flasks (see Fig. 1 for schematic workflow). Pipette 30–40 mL MSC Expansion Medium into a 50 mL conical tube and warm in 37°C water bath for at least 15 min. Remove vial of frozen MSC from liquid nitrogen storage and thaw vial with rapid mixing in a 37°C water bath. Aseptically transfer the entire contents of the vial into an empty 50 mL conical tube. Slowly add 10 mL prewarmed MSC Expansion Medium to the tube containing cells following the steps detailed below (see Note 1). Add 5 mL of medium at a rate of approximately 2–3 drops per 10 s while gently swirling the tube. Add the final 5 mL of medium at a rate of approximately 5 drops per 10 s while gently swirling the tube. Split the 10 mL of cell solution evenly between two 75 cm^2 flasks. For analysis of chondrogenesis, cell solution is split evenly between two 225 cm^2 flasks to allow a longer expansion period and ultimately a higher cell number output as required for plating micromass cultures. Add 10–15 mL prewarmed MSC Expansion Medium to each 75 cm^2 flask for a total of 15–20 mL medium per flask. These flasks are for osteogenesis and adipogenesis set up

Typical PCR Assay Process Workflow

Day 0: Recover & Plate MSCs in Flasks (e.g. Thursday)

↓

Day 1: Fluid Change Flasks with Expansion Media (e.g. Friday)

↓

Day 4: Fluid Change Flasks with Expansion Media (e.g. Monday)

↓

Day 5: Differentiate MSCs in Flasks (e.g. Tuesday)

↓

Day 8: Fluid Change Flasks with Differentiation Media (e.g. Friday)

↓

Day 12: Harvest Flasks with Differentiation Media (e.g. Tuesday)

↓

Day 13: Isolate total RNA & Synthesize First Strand cDNA (e.g. Wednesday)

↓

Day 14: PCR Run & Gel Analysis (e.g. Thursday)

Fig. 1. PCR assay process workflow for testing osteogenesis and adipogenesis kits.

(see Subheading 3.1, step 4). For 225 cm^2 flasks, adjust the volume accordingly to achieve a total of 30–40 mL medium per flask. These flasks are for chondrogenesis set up (see Subheading 3.1, step 7). Incubate the flasks overnight at 37°C, 5% CO_2 in humidified air.

2. Day 1: Fluid change flasks with MSC Expansion Medium. Warm MSC Expansion Medium in a 37°C water bath for at least 15 min. Remove the flasks from the incubator and visually confirm sufficient MSC recovery and attachment (>30–40%). If poor cell recovery and attachment is observed, discard flasks and initiate new cultures. Aspirate and discard spent medium. Add the appropriate volume of prewarmed medium to each flask. Add 15–20 mL for each 75 cm^2 flask and 30–40 mL for each 225 cm^2 flask. Return the flasks to the incubator.
3. Day 4: Fluid change flasks with MSC Expansion Medium. Warm MSC Expansion Medium in a 37°C water bath for at least 15 min. Remove the flasks from the incubator and visually confirm sufficient expansion of MSC. If cells are expanding poorly, discard flasks and initiate new cultures. Aspirate and

discard spent medium. Add the appropriate volume of prewarmed MSC Expansion Medium to each flask (see Subheading 3.1, step 2) and return flasks to the incubator.

4. Day 5 (osteogenesis and adipogenesis set-up): Harvest uninduced MSC cultures and initiate osteogenic or adipogenic differentiation (see Subheading 3.1, step 7 for chondrogenesis differentiation): Warm 15–20 mL Osteogenic and Adipogenic Differentiation Medium in a 37°C water bath for at least 15 min. Harvest and Trizol treat uninduced MSC cultures (see Note 2). Remove a T75 flask from the incubator and visually observe under the microscope for cell attachment, distribution, and greater than 50% confluence. Aspirate and discard medium from the flask and rinse with 10 mL D-PBS. Discard D-PBS, add 5 mL TrypLE to the flask, and incubate at 37°C for 4–6 min. Firmly rap the flask to facilitate cell detachment and transfer the entire solution to a labeled 15 mL conical tube. Centrifuge the tube at 300 × *g* for 5 min at room temperature. Remove the supernatant from the tube and resuspend the cell pellet in 0.5 mL of Trizol in a fume hood (see Note 3). Store the Trizol samples in a –70°C freezer until ready to perform RNA isolation (see Subheading 3.2). Initiate osteogenic or adipogenic differentiation of MSC. Remove T75 flasks from the incubator and visually observe under the microscope for cell attachment, distribution, and greater than 50% confluence. Aspirate and discard the medium from each flask and add 15–20 mL prewarmed Osteogenic or Adipogenic Differentiation Medium to respective flasks. Return the flasks to the incubator.
5. Days 8 and 11: Fluid change flasks with Osteogenic or Adipogenic Differentiation Medium. Warm 15–20 mL Osteogenic and Adipogenic Differentiation Medium in 37°C water bath for at least 15 min. Aspirate and discard the medium from each flask and add 15–20 mL of the respective Differentiation Medium to each flask. Return flasks to the incubator.
6. Day 12: Harvest osteogenic or adipogenic differentiation cultures. Remove a T75 flask from incubator and visually observe under the microscope for cell attachment and even distribution. For osteogenic cultures, a granular phenotype may appear (Fig. 2a). For adipogenic cultures, spherical lipid vesicles will be present in some cells (Fig. 2b). Aspirate and discard the medium from each flask and rinse with 10 mL D-PBS. Discard D-PBS, add 5 mL TrypLE to each flask, and incubate at 37°C for 10–15 min. Firmly rap each flask to facilitate cell detachment and transfer the entire solution to a labeled 15-mL conical tube (see Note 4). Centrifuge the tubes at 300 × *g* for 5 min at room temperature. Remove the supernatant

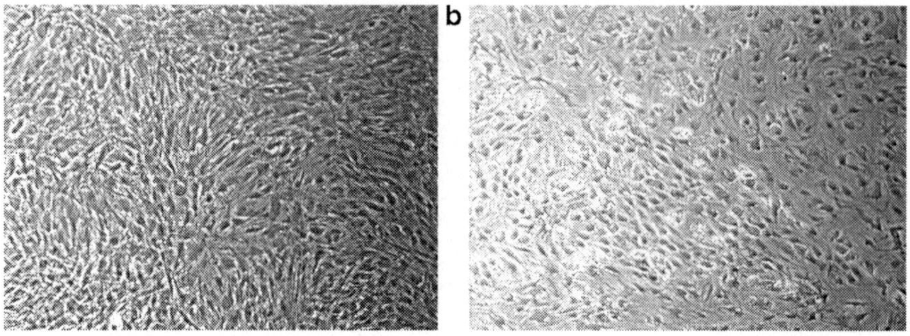

Fig. 2. Morphology of MSC cultured under osteogenic and adipogenic differentiation conditions. (**a**) For osteogenic cultures, induced MSC become tightly packed and will exhibit initial signs of granulation that will lead to formation of calcium matrix. (**b**) For adipogenic cultures, induced MSC will be near confluency and begin to display lipid vesicles within the cytoplasmic compartment.

from each tube and resuspend the cell pellet in 0.5 mL of Trizol in a fume hood (see Note 5). Store Trizol samples in a –70°C freezer until ready to perform RNA isolation (see Subheading 3.2).

7. Day 6 (chondrogenesis set-up): Fluid change flasks with MSC Expansion Medium. Warm 60–80 mL MSC Expansion Medium in a 37°C water bath for at least 15 min. Remove T225 flasks from the incubator and visually confirm sufficient expansion of MSC. If the cells are expanding poorly, discard the flasks and initiate new cultures. Aspirate and discard spent medium and add prewarmed MSC Expansion Medium to each flask. Return the flasks to the incubator.
8. Day 7: Harvest uninduced MSC cultures and initiate chondrogenic differentiation of MSC. Warm 40 mL Chondrogenic Differentiation Medium in a 37°C water bath for at least 15 min. Harvest and Trizol treat uninduced MSC cultures (see Note 2). Remove one flask from the incubator and visually observe under the microscope for cell attachment, distribution, and greater than 70% confluence. Aspirate and discard medium from the flask and rinse with 10 mL D-PBS. Remove D-PBS, add 5 mL TrypLE to the flask, and incubate at 37°C for 4–6 min. Firmly rap the flask to facilitate cell detachment and transfer the entire solution to a labeled 15 mL conical tube. Centrifuge the tube at $300\times g$ for 5 min at room temperature. Remove the supernatant from each tube and resuspend cell pellet in 0.5 mL of Trizol in fume hood (see Note 3). Store the Trizol sample in a –70°C freezer until ready to perform RNA isolation (see Subheading 3.2). Initiate chondrogenic differentiation of MSC. Remove the flask from incubator and visually observe under the microscope for cell attachment, distribution, and greater than 70% confluence.

Aspirate and discard the medium from the flask and rinse with 10 mL D-PBS. Discard the D-PBS, add 5 mL TrypLE to flask, and incubate at 37°C for 4–6 min. Firmly rap the flask to facilitate cell detachment. Mix the detached cells to generate a homogenous solution and transfer the entire solution to a labeled 15 mL conical tube. Take a 20 μL sample from the tube and mix with 20 μL Trypan Blue solution. Determine the viable cell count using a hemocytometer. Centrifuge the tube at $300\times g$ for 5 min at room temperature. Remove the supernatant from the tube and resuspend the cell pellet in MSC Expansion Medium at a concentration of 1.6×10^7 cells/mL. Load the entire resuspended cell solution into a disposable reagent reservoir and attach seven sterile tips to a 10 μL 8-channel pipetter. Generate a 7×7 grid of 5 μL droplets on multiple (2 or 3) 100 mm tissue culture-treated dishes depending on total cell yield (Fig. 3) (see Note 6). Incubate the plates in a 37°C, 5% CO_2 humidified incubator for 2 h. After 2 h incubation, carefully add 12 mL prewarmed Chondrogenic Differentiation Medium slowly down the side of the dish to cover the micromass cultures. Return the dishes to the incubator.

9. Days 9, 11, and 13: Fluid change plates with Chondrogenic Differentiation Medium. Prewarm Chondrogenic

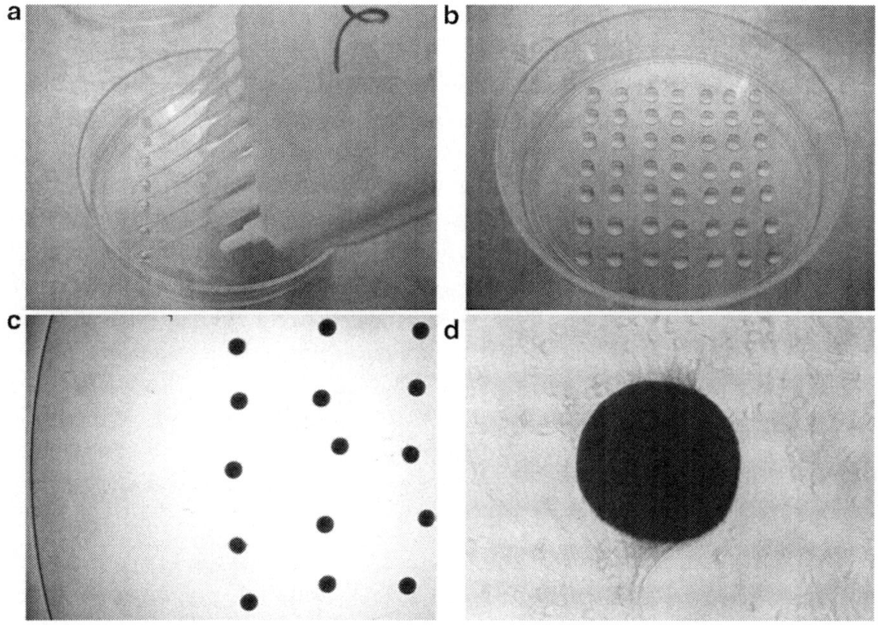

Fig. 3. Plate layout and morphology of chondrogenic micromass culture. The *top row* images show a suggested protocol for generating 7×7 grid pattern of micromass cultures. The *bottom row* images show low and high magnification of chondrogenic pellets morphology after 7 days induction.

Differentiation Medium in a 37°C water bath for at least 15 min. Carefully aspirate and discard the medium without removing chondrogenic pellets. Slowly add 12 mL Chondrogenic Differentiation Medium to the dishes and return to the incubator (see Note 7).

10. Day 14: Harvest chondrogenic differentiation cultures. Remove plates from the incubator and observe under inverted microscope (see Note 8). Using a cell scraper, detach the pellets from the culture dishes. Pipette the entire medium, including detached chondrogenic pellets, into 15 mL conical tubes. Centrifuge the tubes at 300 × *g* for 5 min at room temperature. Remove supernatant from each conical tube and add 1 mL 0.2% collagenase type II solution. Mix well by gentle finger vortexing and incubate the tubes in a 37°C, 5% CO_2 humidified incubator for 1 h. After incubation, centrifuge the tubes at 300 × *g* for 5 min at room temperature. Remove the supernatant from each tube and resuspend the pellets in 0.5 mL of Trizol in a fume hood (see Notes 3 and 9). Store resuspended Trizol samples in a –70°C freezer until ready to perform RNA isolation (see Subheading 3.2).

3.2. RNA Isolation and Quality Check

1. Prepare Wash Buffer II according to PureLink Micro-to-Midi Total RNA Purification kit's instruction.
2. Prepare 1× TAE buffer.
3. Prepare 1.5% Agarose for RNA Quality Check. Add 50 mL 1× TAE buffer and 0.75 g agarose into a 125 mL Erlenmeyer glass flask, swirl contents of flask, and heat solution in a microwave until agarose has melted into a homogenous solution. With heat protective gloves or mittens, remove the melted agarose solution from the microwave and transfer to a chemical hood. Allow the solution to cool until the flask can be comfortably handled. As the agarose solution is cooling, add 5 μL of 10 mg/mL ethidium bromide (EtBr) directly to the agarose (see Note 10). Prepare a gel tray with an 8-well comb. After the agarose solution has cooled (~5 min), pour the solution into the gel tray (see Note 11). Allow the gel to harden at room temperature until it becomes translucent (~15 min). Unlock the clamp and remove the gel tray from the gel loading tray. Transfer the gel out of the hood and place into a gel box containing 1× TAE solution (see Note 12).
4. Perform Total RNA Extraction. Wipe down the working surface with RNase AWAY reagent. Fill an ice container with crushed ice and place frozen 10× DNAse buffer and RNAse-, DNAse-free water on the ice and let thaw. Remove Trizol samples from the –70°C freezer and thaw in a chemical hood (see Note 2). After thaw, transfer 0.5 mL Trizol samples with

a P1000 pipetter to labeled 1.5 mL microcentrifuge tubes for chloroform extraction step. Using a P200 pipetter, add 100 μL chloroform to the sample tubes. Shake the tubes vigorously by hand for 15 s and incubate at room temperature for 2–3 min (see Note 13). Centrifuge the samples in a microcentrifuge at 12,000 × *g* for 15 min at 4°C. During centrifugation, prepare DNAse I and 70% EtOH solutions. After centrifugation, remove the tubes from the microcentrifuge, taking care to keep all tubes in an upright position and not disturbing the resultant layers. Visual observation should yield an ~700 μL lower red phase, a proteinaceous interface, and an ~300 μL upper aqueous phase. Using a P200 pipetter, slowly extract 150 μL of aqueous phase in 50 μL increments from a sample and place into a correspondingly labeled 1.5 mL tube. Repeat this step for all samples. Add 150 μL freshly prepared 70% EtOH solution to all tubes. Tightly seal the caps and vortex for 30 s. Follow RNA Wash Step instruction from the RNA Purification Kit for all samples. After isolating RNA, load a 1 μL sample of RNA onto the stage of a Nanodrop spectrophotometer. Read the sample, verify that the A260/A280 ratio is at least 1.7 or higher, and repeat for all samples. Based on RNA concentration, determine the volume of total RNA required to obtain 1 μg for RNA purity analysis and cDNA preparation.

5. Perform RNA Agarose Quality Check. For each sample, mix 1 μg RNA with 1 μL Blue Juice in a 200 μL microfuge tube. Add a remaining volume of RNAse-, DNAse-free water needed for a total volume of 10 μL. Load 5 μL of 100 bp and 1 Kp MW marker in the outside lanes (see Note 14). Load 10 μL of each RNA sample in the inner lanes. Run electrophoresis with the power supply on a program of 90 V constant voltage for at least 30 min. Transfer the gel to a UV imaging system and capture an image of the gel. If distinct 28S and 18S bands are present, the RNA quality is good and cDNA synthesis can be performed.

3.3. cDNA Preparation

1. Thaw the RT Enzyme Mix and 2× RT Reaction Mix from the SuperScript III Platinum Two-Step qRT-PCR Kit with SYBR Green kit. Upon thawing, place reagents on ice until needed.
2. Calculate the volumes needed to prepare 2 μg of sample RNA in RNAse-, DNAse-free water for a total volume of 16 μL.
3. Pipette the required volume of RNAse-, DNAse-free water into a 200 μL PCR tube, followed by the RNA sample. Mix thoroughly.
4. Place the 200 μL PCR tubes into a DNA thermal cycler and run a 70°C preheat program for 5 min.

5. During the preheat step, prepare cDNA master mix by adding 10 μL RT Enzyme Mix and 50 μL 2× RT Reaction Mix to a microfuge tube.
6. After preheating, immediately place the cDNA tubes on ice.
7. Add 24 μL cDNA master mix to each PCR tubes and pipette mix up and down thoroughly.
8. Place the sample PCR tubes in a DNA thermal cycler and run a cDNA synthesis program in the following order: (1) 25°C for 10 min, (2) 42°C for 50 min, (3) 85°C for 5 min, (4) hold at 4°C.
9. After chilling sample tubes, add 2 μL RNase H to the PCR tubes. Pipette mix up and down and place the tubes in a DNA thermal cycler and run an RNase treatment program by incubating the samples at 37°C for 20 min.
10. Place the RNAse treated samples immediately on ice.

3.4. PCR Amplification

1. Prepare 2% Agarose Gel and buffers. Set up a Sub-Cell gel tray in a casting stand with two 20-well combs. Add 150 mL 1× TAE buffer and 3 g pure agarose into a 250 mL glass flask. Swirl the contents of the flask and heat the solution in a microwave with occasional swirling until the agarose has melted into a homogenous solution. Transfer the solution to a chemical hood. As the agarose solution is cooling, add 15 μL of 10 mg/mL ethidium bromide (EtBr) solution (see Note 10). After the solution has cooled (~5 min), pour the agarose solution into a gel tray (see Note 11). Let the gel harden at room temperature until the gel becomes translucent (~15 min). Transfer the gel into a Sub-Cell GT gel box with 1× TAE solution and store until ready for PCR endpoint gel analysis.
2. Prepare PCR samples. To prepare samples for PCR analysis, combine cDNA, primer mix and PCR master mix together such that triplicate reactions for each sample can be made. Place the following items in an ice bucket: (1) tissue positive control (POS) cDNA, (2) uninduced sample (D0) cDNA, (3) differentiated sample (D7) cDNA, (4) RNAse- and DNAse-free water, (5) duplex primer mixes, and (6) SYBR Super Mix. Prepare a set of tubes for primer and PCR:cDNA master mix preparation. We routinely use colored PCR 8-well strip 500 μL tubes to track samples.
3. Prepare SYBR:cDNA Mix for each cDNA sample (Fig. 4). Set up four 500 μL tubes for cDNA negative control (NEG), tissue positive control (POS), day 0 uninduced (D0), and day 7 differentiated (D7). Pipette 187.5 μL SYBR SuperMix to each 500 μL tube for each sample. Pipette 15 μL RNAse- and DNAse-free water (Blank Control) to NEG. Pipette 15 μL

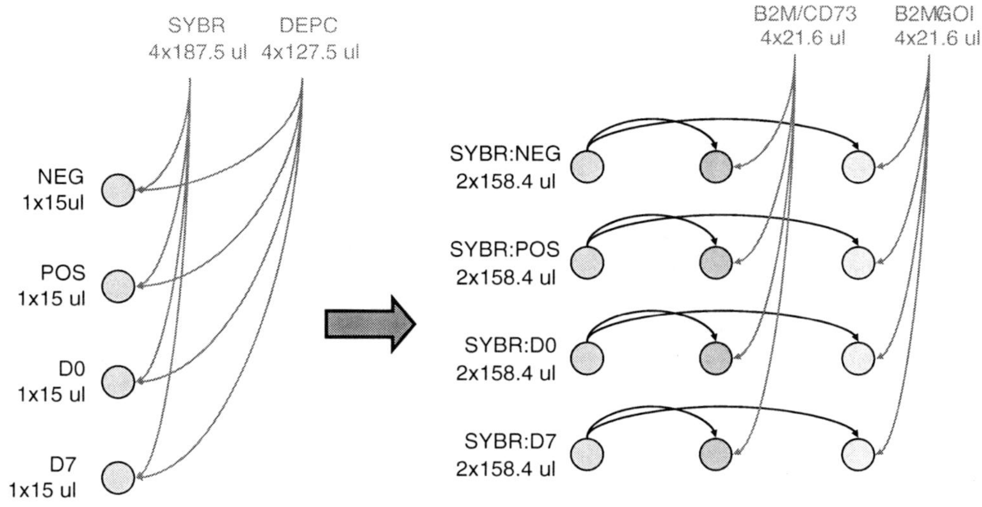

Fig. 4. Process flow for combining together SYBR:cDNA: mix with duplex primer solutions.

POS, D0, and D7 cDNA sample to each corresponding tube. Pipette and mix thoroughly 127.5 μL RNAse- and DNAse-free water for each individual PCR tube.

4. Prepare PCR master mixes. Pipette 158.4 μL SYBR:NEG mix to each of the two top row tubes (Fig. 4). Repeat step accordingly for SYBR:POS, SYBR:D0, and SYBR:D7 (Fig. 4). For the first column, pipette 21.6 μL of B2M/CD73 duplex primer mix to each tube. This mix will detect for the presence of housekeeping and MSC-specific genes expressed in tissue positive sample, uninduced sample, and differentiated sample. For the second column, pipette 21.6 μL of B2M/GOI duplex primer mix to each tube. This mix will detect for the presence of housekeeping gene expressed in tissue positive sample, uninduced sample, and differentiated sample, but lineage-specific GOI will only be detected in tissue positive and differentiated samples. After preparing PCR master mixes, pipette 50 μL of each mix into triplicate wells in the plate (Fig. 5). Triplicate wells are done for each sample to minimize potential loss of PCR product; readout and verify that the technician is performing the assay satisfactorily. Confirm that all strip caps are sealed evenly into place on the PCR plate and follow a matrix pattern observed in (Fig. 5). Transfer the PCR plate to a DNA thermal cycler and run the PCR gene amplification program for osteogenesis, adipogenesis, or chondrogenesis analysis. Following is an example of a PCR program for adipogenesis analysis: Step 1 – 95°C for 2 min (initial activation). Step 2 – 95°C for 30 s (denaturation).

a

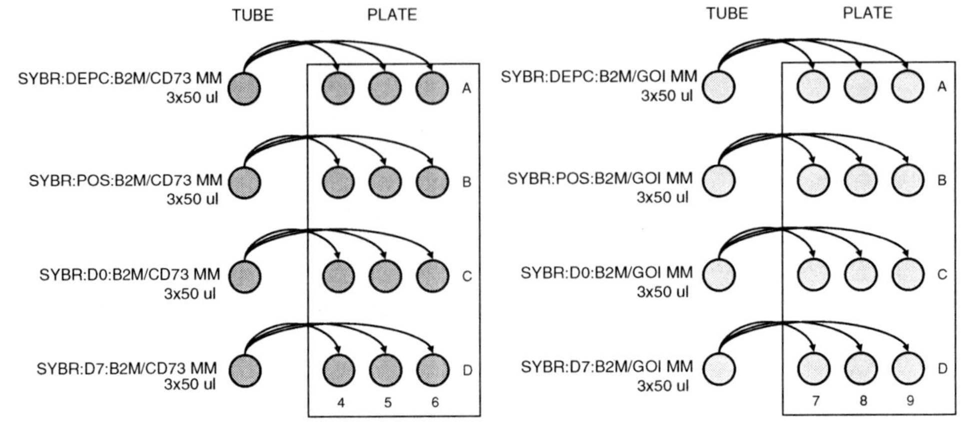

b

				B2M/CD73			B2M/GOI					
	1	2	3	4	5	6	7	8	9	10	11	12
A				NEG	NEG	NEG	NEG	NEG	NEG			
B				POS	POS	POS	POS	POS	POS			
C				D0	D0	D0	D0	D0	D0			
D				D7	D7	D7	D7	D7	D7			
E												
F												
G												
H												

Fig. 5. Plating PCR plate. *Top* figure shows process workflow of transferring PCR MasterMixes into PCR plate. *Bottom* figure shows layout of reaction triplicates for each sample in PCR plate.

Step 3 – 58°C for 30 s (annealing). Step 4 – 72°C for 30 s (extension). Step 5 – repeat steps 2–4, 34 times. Step 6 – 72°C for 5 min (final extension). Step 7 – hold at 4°C. PCR program for osteogenesis and chondrogenesis analysis differs only by going through the amplification steps 30 and 28 cycles respectively, instead of 35 cycles for adipogenesis. After the PCR run, immediately place the 96-well plate on ice and proceed to next step.

5. Perform PCR gel analysis. Carefully remove the strip cap from the 96-well plate one column at a time and pipette mix 2.5 μL 10× BlueJuice to each well in the 96-well plate. Load 10 μL of each sample into wells of 2% agarose gel. Load 5 μL of the 100 bp MW marker in first lane of the gel and pipette 5 μL of the 1 Kb MW marker in last lane. Run the gel at 90 V constant voltage for 1 h. Transfer the gel to a gel documentation system and capture images. As an example, in Fig. 6 CD73 and B2M are expressed in all samples except negative control (Water). However, the FABP4 gene is expressed only in adipocyte tissue positive control (Adipo) and day 7 adipogenic differentiated MSC (AD7). FABP4 is not expressed in uninduced MSC (P0). This result confirms that the differentiated MSC specifically expressed FABP4, an early adipocyte cell marker.

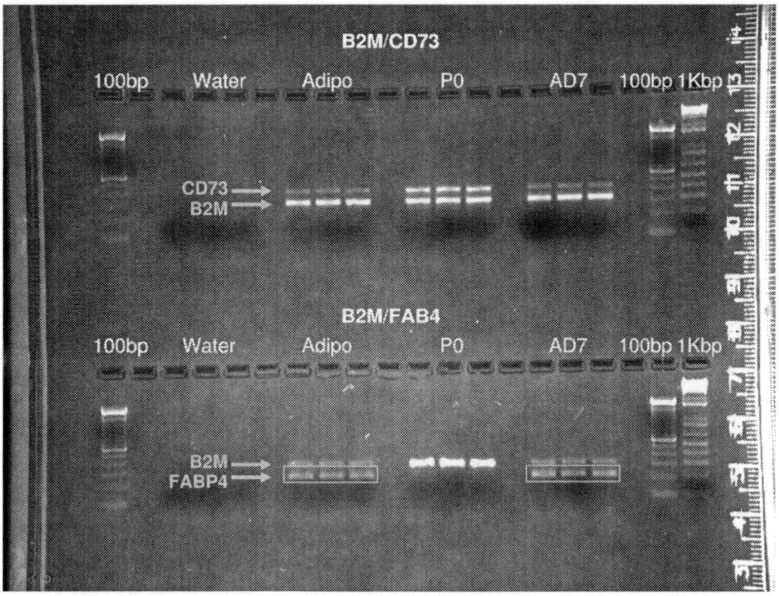

Fig. 6. Captured image of PCR gel showing PCR products generated and loaded into 2% gel. Bands represent expression of specific genes in tissue positive control (Adipo), uninduced MSC on day 0 (P0) and differentiated MSC on day 7 (AD7). Detection of adipogenic-specific FABP4 bands was observed only in tissue positive control and day 7 differentiated MSC (see *gold boxes*). No bands are expressed in negative control (Water).

4. Notes

1. Slow media addition is critical for optimal cell viability and subsequent growth. If media is added too quickly, it can dramatically stunt cell growth.
2. It is strongly recommended that Trizol be handled in a fume hood. Trizol is toxic in contact with skin and if swallowed. Trizol can cause burns. If material comes in contact with your gloves, replace with new gloves. If material contacts your skin, wash immediately with plenty of detergent and water. If you feel unwell, seek medical advice. See MSDS for proper handling of Trizol.
3. It is critical to work quickly and not disturb cell pellet as you aspirate supernatant. If too much time is elapsed, then the cell pellet will become loose and there is an increased risk of cells becoming aspirated.
4. For osteogenic cultures, the cells tend to come off culture surface as a sheet and form a viscous solution. Osteocyte cultures may need to be treated with TrypLE for longer periods

if the cells do not initially come off after 10 min. For adipogenic cultures, the cells will come off as a single cell suspension but may need more than 10 min enzymatic digestion to completely detach from the culture surface. Tap the flasks and observe under microscope to verify that adipocytes are completely detached.

5. For osteoblast pellets, you may need to break up the pellet into small pieces with a P1000 tip by pushing the tip into the pellet in a cookie cutter fashion while simultaneously pipetting up and down. Once the pellet is broken up into smaller pieces, resuspension of the cells in Trizol will be easier.
6. These droplets will form micromass cultures that lead to generation of chondrogenic pellets needed for chondrogenesis analysis.
7. It is acceptable to have pellets detach and float around the dish during chondrogenesis. These floating pellets are still undergoing proper chondrogenesis.
8. You should observe many chondrogenic pellets which will typically appear spherical in shape and many will still be attached to the culture surface.
9. The chondrogenic pellets may not completely dissolve in Trizol. The freezing step will ensure sufficient breakdown of the pellets.
10. *Warning*: Ethidium bromide is a mutagen. Ethidium bromide solution must be handled in the chemical hood. When Ethidium bromide is added into melted Agarose solution, it must remain in the hood until the Agarose has solidified into a gel. Refer to product MSDS on proper handling of ethidium bromide.
11. Agarose solution that is cooled too much before pouring will lead to poor and inconsistent gel hardening.
12. The gel can be prepared ahead of time, wrapped in plastic wrap, and stored overnight in 4°C refrigerator until ready to use. Allow the gel warm to room temperature before placing into gel box and loading samples.
13. *Warning*: Chloroform is a potentially lethal chemical if not done in chemical fume hood. Ensure caps are well-sealed prior to shaking. Replace gloves immediately, if chloroform comes in contact. See MSDS for proper handling of chloroform.
14. Use caution in loading the lanes, avoiding puncture of the gel and resultant spillover of contents into adjacent lanes.

Acknowledgments

I thank Uma Lakshmipathy, Lucas Chase, and Mohan Vemuri for early discussions and insight in generating helpful comments on the manuscript.

References

1. Friedenstein, A.J., I.I. Piatetzky-Shapiro, and K.V. Petrakova. 1966. Osteogenesis in transplants of bone marrow cells. *J Embryol Exp Morphol* 16:381–390.
2. Friedenstein, A.J., R.K. Chailakhjan, and K.S. Lalykina. 1970. The development of fibroblast colonies in monolayer cultures of guinea-pig bone marrow and spleen cells. *Cell Tissue Kinet* 3:393–403.
3. Jaiswal, N., S.E. Haynesworth, A.I. Caplan, and S.P. Bruder. 1997. Osteogenic differentiation of purified, culture-expanded human mesenchymal stem cells in vitro. *J Cell Biochem* 64:295–312.
4. Pittenger M.F., A.M. Mackay, S.C. Beck, R.K. Jaiswal, R. Douglas, J.D. Mosca, M.A. Moorman, D.W. Simonetti, et al. 1999. Multilineage potential of adult human mesenchymal stem cells. *Science* 284:143–147.
5. Ng, F., S. Boucher, S. Koh, K.S. Sastry, L. Chase, U. Lakshmipathy, C. Choong, Z. Yang, et al. 2008. PDGF, TGF-b and FGF signaling is important for differentiation and growth of mesenchymal stem cells (MSCs): transcriptional profiling can identify markers and signaling pathways important in differentiation of MSC into adipogenic, chondrogenic and osteogenic lineages. *Blood* 112:295–307.
6. Goff, L.A., S. Boucher, C.L. Ricupero, S. Fenstermacher, M. Swerdel, L.G. Chase, C.C. Adams, J. Chesnut, U. Lakshmipathy, and R.P. Hart. 2008. Differentiating human multipotent mesenchymal stromal cells regulate microRNAs: prediction of microRNA regulation by PDGF during osteogenesis. *Exp Hematol* 36:1354–1369.
7. Dominici, M., K. Le Blanc, I. Mueller, I. Slaper-Cortenbach, F. Marini, D. Krause, R. Deans, A. Keating, et al. 2006. Minimal criteria for defining multipotent mesenchymal stromal cells. The International Society for Cellular Therapy position statement. *Cytotherapy* 8:315–317.
8. Mackay, A.M., S.C. Beck, J.M. Murphy, F.P. Barry, C.O. Chichester, and M.F. Pittenger. 1998. Chondrogenic differentiation of cultured human mesenchymal stem cells from marrow. *Tissue Eng* 4:415–428.
9. Nakamura, T., S. Shiojima, Y. Hirai, T. Iwama, N. Tsuruzoe, A. Hirasawa, S. Katsuma and G. Tsujimoto. 2003. Temporal gene expression changes during adipogenesis in human mesenchymal stem cells. *Biochem Biophys Res Commun* 303:306–312.
10. S. Boucher, U. Lakshmipathy, M. Vemuri. 2009. A simplified culture and polymerase chain reaction identification assay for quality control performance testing of stem cell media products. *Cytotherapy* 11:761–767.

Chapter 29

Transcriptome Analysis of Common Gene Expression in Human Mesenchymal Stem Cells Derived from Four Different Origins

Tzu-Hao Wang, Yun-Shien Lee, and Shiaw-Min Hwang

Abstract

We have used Affymetrix oligonucleotide microarrays to analyze common transcriptomes and thereby learn about the core gene expression profile in human mesenchymal stem cells (MSC) from different tissues, including fetal amniotic fluid-derived MSC, term pregnancy amniotic membrane-derived MSC, term pregnancy umbilical cord blood-derived MSC, and adult bone marrow-derived MSC. The beauty of microarray analysis of gene expression (MAGE) is that it can be used to discover associating genes that were previously thought to be unrelated to a physiological or pathological event. However, interpreting complex biological processes from gene expression profiles often requires extensive knowledge mining in biomedical literature. In this chapter, we describe, step-by-step, how to use a commercially available biological database and software program, MetaCore (GeneGo Inc.), for functional network analysis.

Key words: Mesenchymal stem cells, Amniotic fluid, Amniotic membrane, Cord blood, Bone marrow, Functional network analysis, Metacore (GeneGo)

1. Introduction

Multipotent mesenchymal stem cells (MSC) found in many adult and fetal tissues are a promising source of cells for tissue engineering and cell-based therapeutics due to their extensive capability of self-renewal and multi-lineage differentiation potential. However, the characteristics of MSC are still not equivocal, especially in their differentiation tendencies and plasticity (1, 2). The increasing interest in MSC for biological and clinical application has prompted the International Society for Cellular Therapy (ISCT) to propose the minimal criteria for defining human MSC as follows: MSC should (a) be plastic-adherent, (b) express CD73, CD90, and CD105, and lack expression of CD45, CD34, CD14 or

Mohan C. Vemuri et al. (eds.), *Mesenchymal Stem Cell Assays and Applications*, Methods in Molecular Biology, vol. 698, DOI 10.1007/978-1-60761-999-4_29,

CD11bm CD79a or CD19, and HLA-DR surface molecules, and (c) differentiate to osteoblasts, adipocytes, and chondroblasts in vitro (3). The molecular level of characterization for MSC is in need of further clarification. In this study, we focused on the core gene expression profiles of MSC from different tissues, including fetal amniotic fluid-derived MSC, termed amniotic membrane-derived MSC, termed umbilical cord blood-derived MSC, and adult bone marrow-derived MSC.

One major benefit of microarray analysis of gene expression (MAGE) is that it can be used to discover associating genes that were previously thought to be unrelated to a physiological or pathological event (4). On the other hand, it is a daunting challenge to interpret complex biological processes from such a long list of genes, of which their known functions may be seemingly unrelated to one another. To deal with these difficulties, knowledge mining throughout biomedical literature is essential and database-dependent bioinformatics research has proved instrumental (5–7). More recently, we have used this type of approach to gain insight into individual and common gene expression profiles of various human mesenchymal stem cells derived from amniotic fluid, amniotic membrane, cord blood, and bone marrow (8) as well as gene expression response in cervical cancer patients undergoing concurrent chemoradiation therapy (9).

In the following sections, we describe, step-by-step, how to use a commercially available biological database and software, MetaCore (GeneGo, Inc.), for functional network analysis. In the demonstrated example (8), we have learned that a set of core gene expression profiles was preserved in four kinds of MSCs. The core signature transcriptomes of all MSCs, when compared with those of 6-week-old fetal organs, included genes that were mainly involved in the regulation of extracellular matrix and adhesion, whereas the genes common to all MSC were the least involved in cell division and system development.

2. Materials

2.1. Culture of Mesenchymal Stem Cells

1. Mesenchymal stem cells were obtained from the Bioresource Collection and Research Center (BCRC, http://www.bcrc.firdi.org.tw/wwwbcrc/index.jsp), Taiwan. The samples included human amniotic fluid-derived (AF) MSC, amniotic membrane-derived (AM) MSC, umbilical cord blood-derived (CB) MSC, and bone marrow-derived (BM) MSC.
2. Culture medium: α-modified minimum essential medium (α-MEM, Hyclone, Logan, UT, USA) supplemented with 20% fetal bovine serum (FBS, Hyclone) and 4 ng/ml basic

fibroblast growth factor (bFGF, R&D System, Minneapolis, MN, USA).

3. Trypsin–EDTA (Invitrogen, Carlsbad, CA, USA).
4. Tissue culture flasks, 75 cm^2 (Corning, Corning, NY, USA).

2.2. RNA Extraction, Quality Analysis, and Microarray Analysis

1. TRIZOL reagent (Invitrogen, Carlsbad, CA).
2. RNeasy purification kit (Qiagen, Valencia, CA).
3. Total RNA from human fetal tissues (ViroGen, Watertown, MA).
4. Bioanalyzer 2100 (Agilent Technologies, Palo Alto, CA).
5. Human U133A GeneChip (Affymetrix, Santa Clara, CA).

2.3. MetaCore (http://www.genego.com/metacore.php)

1. Download (http://www.genego.com/productTrials.php) and complete the trial paperwork agreement to obtain a free 2-week trial.
2. List MetaCore as the one to be tried and fax the paperwork to the United States phone number: (760) 479 2059.
3. Once GeneGo, Inc., receives your paperwork, it will be processed and an URL and access passwords will be emailed to you. All trials are fully supported by customer support.

3. Methods

3.1. Culture of Multipotent Mesenchymal Stem Cells

1. Frozen MSC of four different tissues were obtained from the cell bank of BCRC, including AF MSC, AM MSC, CB MSC, and BM MSC. These cells were characterized as MSC following the criteria of ISCT (1, 8).
2. All MSC were cultured in 75 cm^2 flasks with α-MEM plus 20% FBS, 4 ng/ml bFGF, and incubated at 37°C in a humidified atmosphere with 5% CO_2. Cells were kept at 90% confluence before passage or RNA extraction.

3.2. RNA Extraction and Affymetrix Microarray Analysis for Gene Expression

1. At 90% confluence, the MSCs were briefly rinsed with ice-cold PBS and lysed in TRIZOL reagent for RNA extraction.
2. Total RNA was further purified using the RNeasy purification kit.
3. Total RNA specimens of human fetal brain, heart, lung, liver, kidney, and muscle at the sixth gestational week were obtained from ViroGen, Inc.
4. RNA quality and quantity were confirmed using the Bioanalyzer 2100.

5. Gene expression profiles in MSCs and early fetal tissues were analyzed with the human U133A GeneChip and the manufacturer's protocol was strictly followed.
6. The GeneChip data were analyzed using the GeneChip Operating Software (GCOS) version 4.1.

3.3. Identification of MSC Unique Gene Expression Profiles

1. Normalize all Affymetrix data and save the results in Excel files. In this example, we compared twenty MSC samples and six 6-week-old fetal organs, as previously reported (see Note 1) (8).
2. Do an appropriate statistical analysis to compare the two groups of genes (see Note 2).
3. Apply a nonstringent *P* value ($P < 0.05$) to select the differently expressed genes between two groups.
4. Rank the list of genes according to fold changes, from large to small.
5. Select the top ranked gene number of appropriate size (see Note 3). For example, select and copy 100 top-ranked Affymetrix probe sets, displayed as "202627_s_at," to be pasted into the MetaCore text data space (see Subheading 3.4, step 3).
6. Similarly, select the bottom ranked genes of appropriate size, representing those were least expressed in MSC than in differentiated fetal organs (see Note 4).

3.4. Database-Dependent Analysis of Potential Functional Networks

1. Following the instructions provided by GeneGo's technical support, connect to the MetaCore website.
2. Activate "Signaling networks" in MetaCore. In the example, we used MetaCore (version 4.5, build 11165) for functional network visualization and analysis.
3. Select "Draw from the list," and paste the copied list of Affymetrix probe sets into the "Create list manually" text data space, and select "Parse."
4. Check to make sure that all Affymetrix probe sets correspond to a gene (network object), then activate "Add selected objects."
5. Activate "Select all" of network objects, and then "Build network for selected objects."
6. Select "Analyze network," and "Build network."
7. On the page of Network list, examine how many networks are derived.
8. Double click "Target," and the list will be ranked with the network that contains the highest number of targets (i.e., the genes in your input list) on the top.
9. Activate "Open" the first network on top of the list.
10. Click the "Show" tab, select "Root Object" to show target objects in the network and "Loc line."

11. Click "View Options" and select "Loc mode" to rearrange the network according to the localization of each genes in extracellular membrane, cytoplasm, nucleus, or unspecified compartments.
12. Drag the genes around to visualize all genes clearly. Be careful not to drag them out of their original compartment.
13. When "Hint" is selected under the "Show" tab, place the cursor at the gene of interest to show its description in a balloon.
14. When "Effect" is selected under the "Show" tab, just place the cursor at the effect marker to show the interactions (such as activation or inhibition) between two objects in a balloon.
15. Click an "Effect" icon of specific interest, a new window of references (Link) appears. This can be used to directly connect to the PubMed database for viewing summaries of reference papers.
16. After examining the interaction using the "Effect" option, deselect "Effect" for a clearer view of the network.
17. Click "Processes" to see the main functions that genes in this network are involved in.

4. Notes

1. To avoid generating biased weight by the same MSC sample, data from three samples of the same MSC at different passage times were averaged and considered as one, resulting in a total of 12 MSC samples from four different origin (8). Nonparametric Mann–Whitney *U* test was performed between 12 MSC samples and six fetal organ samples (8).
2. A fold-change ranking method with a *P*-value cutoff <0.05 has been shown to be reproducible in selecting the signature gene list from results using different microarray platforms (10). These selecting criteria have been shown to be more reproducible than *t*-test *P* value or significance analysis of microarrays (SAM) (10). We applied a nonstringent *P* value ($P<0.05$) to select genes of significant different expression levels between MSC and fetal organs. Then we ranked the list of genes, from large to small, by the fold changes between MSC and fetal organs (8).
3. In the demonstration example (8), we selected the top 100 ranked genes, which represented the upregulated expression in MSC, to be analyzed using MetaCore. Among 100 probe sets, 72 genes (see Table 1) were found to participate in signaling networks of that build by MetaCore. All of them were selected

Table 1
Upregulated genes in mesenchymal stem cells compared with fetal tissues

5′-NTD	ENO1	PBR
		PHLDA1
Activin	FHL2	PLAT (TPA)
Activin A	Fibronectin	PLAU (UPA)
Activin beta A	FoxD1	PLAUR (uPAR)
ADAM9		PLOD2
Alpha-actinin 1	Galectin-1	PTRF
AMIGO2	GLIPR	PTX3
Annexin I	Gremlin	
Annexin II		RGS4
	IEX1	
Beta-IG-H3	IL-6	Serglycin
	IL-8	SERPINE2
Calcyclin		SH3BGRL3
Caldesmon	KDELR	Stanniocalcin 2
Calgizzarin	LIF	TAGLN2
Calumenin	LOXL1	TFPI-2
Caveolin-1	LOXL2	Thrombospondin 1
Caveolin-2	Lysyl oxidase	TIMP1
CCL2		TM4SF1 (TAAL6)
CD44	MafF	Transgelin
CD59	MCFD2	
CLIC4	MICAL2	Tropomyosin-2
CTGF	MMP-1	Tropomyosin-4
Cyr61	Myoferlin	TSSC3
DKK1	NNMT	UDP-*N*-acetylhexosamine pyrophosphorylase
	NQO1	
EMP3		VEGF-C
ENO	PAI1	

to build networks. Using the "Analyze network" function, 25 networks were built. The top network, represented by fibronectin, MMP-1, CD44, TIMP1, CCL2, etc., contained 15 target objects (root genes) in a 50-gene network, with the least *P* value being 4.71×10^{-27}, and the highest *z*-score of 38.78 (see Note 5). Notably, 13 genes from the original list were localized in the extracellular compartment of this network (see Fig. 1). The main processes of the top network were as follows: response to external stimuli, chemotaxis, response to wounding, biological adhesion, cell adhesion, cell motility, and localization of cell (see Table 2). The second network, containing 13 root objects (i.e., genes from the input list), was represented with PAI1, Gremlin, activin beta A, PLAU, and ENO1 (see Fig. 2). This network, which contained 5 genes in the extracellular compartment and 5 genes in the cytosolic compartment, was involved in biological processes such as signal transduction, differentiation, and development (see Table 3).

4. To detect what genes were least expressed in mesenchymal stem cells, we selected bottom 100 ranked genes to be analyzed using MetaCore. Among the 100 Affymetrix probe sets, 75 genes (see Table 4) were found in signaling networks of

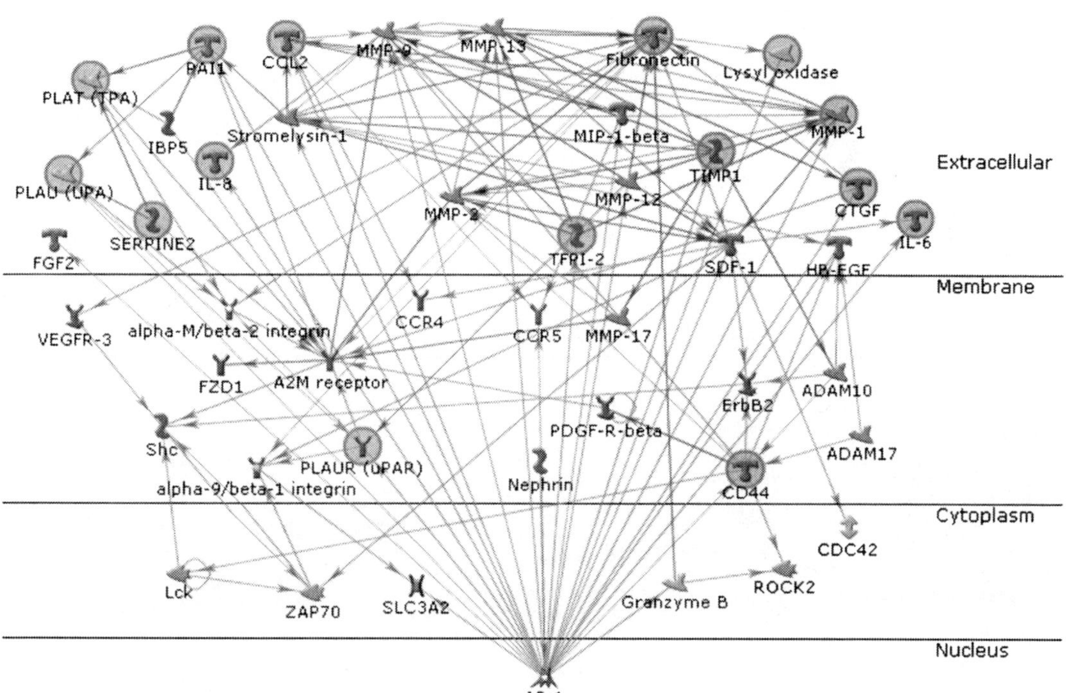

Fig. 1. Functional network of 15 genes (marked with *solid grey circles*) that were upregulated in all four types of MSCs, when compared with six 6-week-old fetal organs.

Table 2
Biological processes that might be related to the 15 genes displayed in Fig. 1

Process	%	*P* value (↓)
Response to external stimulus	41.30	9.066e-10
Taxis	21.74	4.844e-09
Chemotaxis	21.74	4.844e-09
Response to wounding	32.61	8.064e-09
Biological adhesion	32.61	3.989e-08
Cell adhesion	32.61	3.989e-08
Cell mobility	32.61	4.515e-08
Localization of cell	32.61	4.515e-08
Proteolysis	26.09	1.283e-07
Blood vessel development	21.74	3.100e-07
Locomotory behavior	21.74	3.479e-07
Vasculature development	21.74	3.614e-07

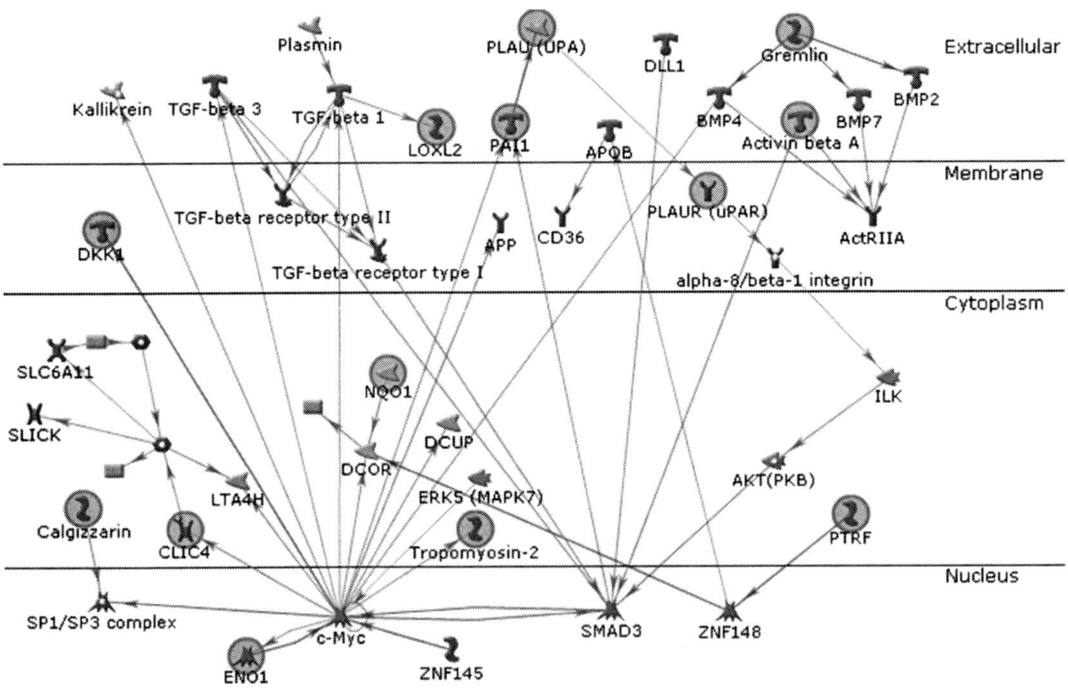

Fig. 2. Functional network of 13 genes (marked with *solid grey circles*) that were upregulated in all four types of MSCs, when compared with six 6-week-old fetal organs.

Table 3
Biological processes that might be related to the 13 genes displayed in Fig. 2

Process	%	*P* value (↓)
Transmembrane receptor protein serine/threonine kinase signaling pathway	25.64	1.049e-11
Organ development	58.97	1.198e-08
Negative regulation of biological process	51.28	3.066e-08
Regulation of cell differentiation	25.64	3.096e-08
Negative regulation of cellular process	48.72	7.671e-08
Regulation of developmental differentiation	28.21	9.284e-08
Tissue development	30.77	9.946e-08
Enzyme linked receptor protein signaling pathway	28.21	1.189e-07
Anatomical structure development	66.67	1.298e-07
System development	61.54	2.216e-07
Anatomical structure morphogenesis	48.72	2.412e-07
Transforming growth factor beta receptor signaling pathway	15.38	3.683e-07

the build 11165 of MetaCore version 4.5. Using "Analyze network" function, 17 networks were built. The top network (see Fig. 3), represented by DAPK1, IGF-2, Reelin, Glypican-3, selenoprotein-P, etc., contained nine target objects (root genes) in a 50-gene network, with the lowest *P* value being 4.10×10^{-15}, and the highest *z*-score of 25.28 (see Note 5). The main processes of this network were as follows: multicellular organismal development, cell division, system development, and others (see Fig. 3).

5. The *P* value is used to initially rank the subnetworks. The *P* values throughout MetaCore – for maps, networks, and processes – are all calculated using the same basic formula: a hypergeometric distribution where the *P* value essentially represents the probability of any particular mapping arising by chance, given the numbers of genes in the set of all genes on maps/networks/processes, genes on a particular map/network/process and genes in your experiment. This function uses the same variables as the *z*-score.

$$P \text{ value} = \frac{R!n!(N-R)!(N-n)!}{N!} \cdot i$$

$$\sum_{\max(r,R+n=N)}^{\min(n,R)} \frac{1}{i!(R-i)!(n-!)(N-R-n+i)!}.$$

Table 4
Downregulated genes in mesenchymal stem cells compared with fetal tissues

A2M	DLK	NCAM1
ABLIM1		NFIB
Adult hemoglobin	E-cadherin	Non-A beta of AD amyloid
AFP		
AFP-cyclopeptide	FGF13	OGT (GlcNAc transferase)
Albumin	FGFR2	Osteoglycin
Alpha1-globin	FGFR3	
Alpha-synuclein	FXYD6	P311
Alpha-synuclein filaments		p57
Ankyrin-B	GABT	Peg10
Asporin	Glypican-3	PEG3
	Golgin-67	Pleiotrophin
		PLGF
BChE	HBB	PTCH1
BCKD-E2	HBE	PTPR-delta
BEX1	HBG1	
	HBGA	
C1qRp	HBZ	RBP1
C6orf111	HMGCS1	Reelin
Calgranulin A	HNRPA3	Reep1
Carboxypeptidase H		RNase 1
CD24		
COL9A3	IGF-2	
Collagen II	ITM2A	Selenoprotein P
Collagen IX		SPARCL
Collagen XI	KLC1	SULT1E1
Collagen XXI, alpha 1	M6B	TK2
CXADR	MEG3	TSLC1
Cyclin D2	MTSS1	WSB1
	MYL4	
DAPK1	Myosin II	XIST

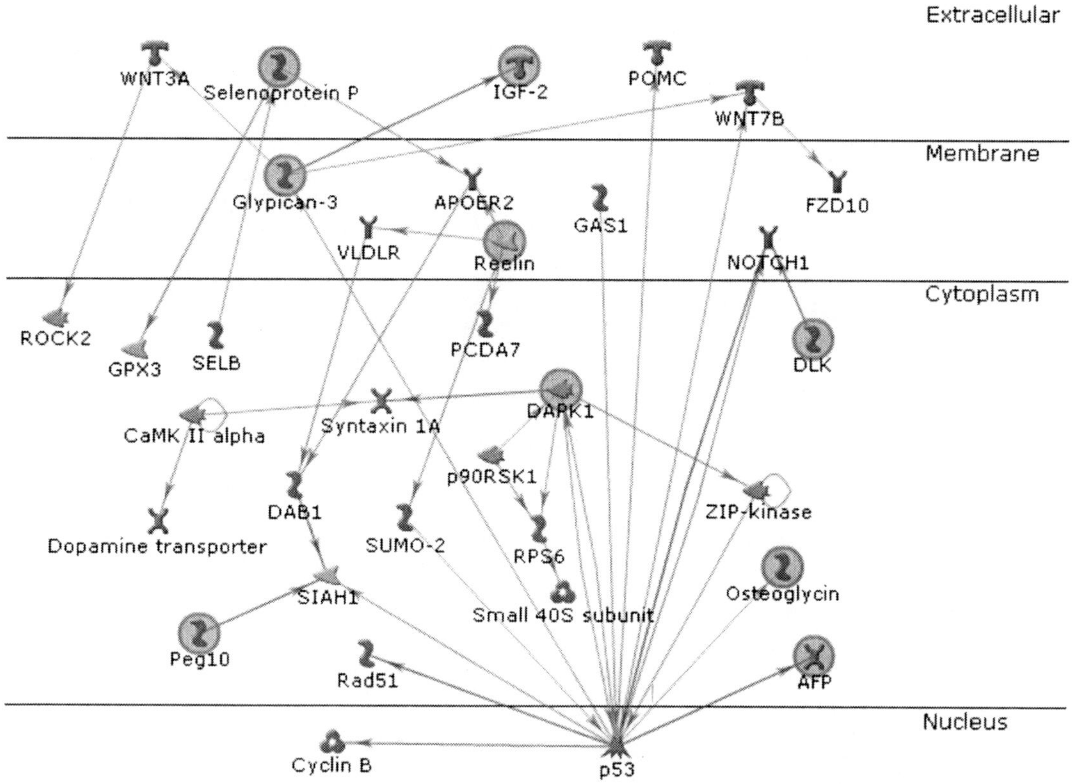

Fig. 3. Functional networks of nine genes (marked with *solid grey circles*) that were downregulated in all four types of MSCs, when compared with six 6-week-old fetal organs.

Variables:

- N – total number of nodes in MetaCore database.
- R – number of the network's objects corresponding to the genes and proteins in your list.
- n – total number of nodes in each small network generated from your list.
- r – number of nodes with data in each small network generated from your list.

The *z*-score ranks the Analyze Network algorithm's subnetworks with regard to their saturation with genes from your experiments (Table 5). A high *z*-score means the network is highly saturated with genes from your experiments.

$$z \text{ score} = \frac{r - n\dfrac{R}{N}}{\sqrt{n\left(\dfrac{R}{N}\right)\left(1-\dfrac{R}{N}\right)\left(1-\dfrac{n-1}{N-1}\right)}}.$$

Table 5
Biological processes that might be related to the nine genes displayed in Fig. 3

Process	%	*P* value (↓)
Multicellular organismal process	68.75	2.698e-04
Somatic stem cell division	6.25	3.503e-04
Generation of neurons	21.88	3.610e-04
Neurogenesis	21.88	5.444e-04
Postembryonic development	6.25	5.604e-04
Nervous system development	31.25	5.641e-04
System development	50.00	6.699e-04
Stem cell division	6.25	6.834e-04
Multicellular organismal development	53.12	8.215e-04
Cell division	9.38	1.280e-03
Protein oligomerization	9.38	2.166e-03
Cell communication	59.38	2.395e-03

Acknowledgments

The authors like to thank Li-Wen Hsu (BCRC), Ching-Ling Wang, Hsiao-Wen Lu, Yu-Chi Lin (Chang Gung Memorial Hospital), and George Hsu (Tri-I Biotech, Inc. Taiwan) for their excellent technical help, and Shih-Yee Mimi Wang (University of Illinois College of Medicine) for editing. This study was supported in part by NSC95-2314-B-182A-156 (to Wang TH), NSC 95-2311-B-130-001 (to Lee YS), and 98-EC-17-A-17-R7-0525 (to Hwang SM).

References

1. Chang, Y.J., Shih, D., Tseng, C.P., Hseih, T.B., Lee, D.C., and Hwang, S.M. (2006) Disparate mesenchyme-lineage tendencies in mesenchymal stem cells from human bone marrow and umbilical cord blood. *Stem Cells* **24**, 679–685.
2. Wagner, W., Wein, F., Seckinger, A., Schwager, C., Wirkner, U., Krause, U., et al. (2005) Comparative characteristics of mesenchymal stem cells from human bone marrow, adipose tissue, and umbilical cord blood. *Exp. Hematol.* **33**, 1402–1416.
3. Dominici, M., Le Blanc, K., Mueller, I., Slaper-Cortenbach, I., Marini, F.C., Krause, D.S., et al. (2006) Minimal criteria for defining multipotent mesenchymal stromal cells. The International Society for Cellular Therapy position statement. *Cytotherapy* **8**, 315–317.
4. Wang, T.H. and Chao, A. (2007) Microarray analysis of gene expression of cancer to guide

the use of chemotherapeutics. *Taiwan J. Obstet. Gynecol.* **46**, 222–229.

5. Nikolsky, Y., Ekins, S., Nikolskaya, T., and Bugrim, A. (2005) A novel method for generation of signature networks as biomarkers from complex high throughput data. *Toxicol. Lett.* **158**, 20–29.
6. Nikolsky, Y., Nikolskaya, T., and Bugrim, A. (2005) Biological networks and analysis of experimental data in drug discovery. *Drug Discov. Today* **10**, 653–662.
7. Mason, C.W., Swaan, P.W., and Weiner, C.P. (2006) Identification of interactive gene networks: a novel approach in gene array profiling of myometrial events during guinea pig pregnancy. *Am. J. Obstet. Gynecol.* **194**, 1513–1523.
8. Tsai, M.S., Hwang, S.M., Chen, K.D., Lee, Y.S., Hsu, L.W., Chang, Y.J., et al. (2007) Functional network analysis on the transcriptomes of mesenchymal stem cells derived from amniotic fluid, amniotic membrane, cord blood, and bone marrow. *Stem Cells* **25**, 2511–2523.
9. Chao, A., Wang, T.H., Lee, Y.S., Hong, J.H., Tsai, C.N., Chen, M.H., et al. (2008) analysis of functional groups of differentially expressed genes in the peripheral blood of patients with cervical cancer undergoing concurrent chemoradiation. *Radiat. Res.* **169**, 76–86.
10. Guo, L., Lobenhofer, E.K., Wang, C., Shippy, R., Harris, S.C., Zhang, L., et al. (2006) Rat toxicogenomic study reveals analytical consistency across microarray platforms. *Nat. Biotechnol.* **24**, 1162–1169.

Chapter 30

Comparison of Microarray and Quantitative Real-Time PCR Methods for Measuring MicroRNA Levels in MSC Cultures

Cynthia Camarillo, Mavis Swerdel, and Ronald P. Hart

Abstract

The capacity for self-renewal and the multilineage potential of mesenchymal stromal cells (MSC) offer a therapeutic promise for regenerative medicine. MicroRNAs (miRNAs) are small noncoding RNAs that play a key regulatory role during differentiation both at the level of posttranslational modulation and epigenetic control. Studies on MSCs have just begun to identify miRNA profiles in MSC and differentiated MSC. While several methods are available for miRNA exploration, microarrays and quantitative real-time PCR (qPCR) are the most common. Since there are several microarray and qPCR platforms available for miRNA detection, it is valuable to explore how these methods compare. We used the NCode Multi-Species miRNA microarray (Invitrogen) and the TaqMan Human microRNA array (Applied Biosystems) to compare microRNA expression in undifferentiated MSCs and MSCs differentiated into early osteoblasts. We show that while there is a somewhat low correlation between these two methods, there is a subset of miRNA measurements that did correlate.

Key words: Mesenchymal stem cell, MicroRNA, Microarray, MSC differentiation, Quantitative PCR

1. Introduction

Adult mesenchymal stromal cells (MSC) are undifferentiated multipotent cells isolated from various tissues including bone marrow, fetal tissue, peripheral blood, and adipose tissue (1–5). The capacity of MSC for self-renewal and their ability to differentiate into osteoblasts, chondrocytes, adipocytes, or fibroblasts make these cells attractive for therapeutic transplantation. However, the multiple derivations of MSCs and their ability to generate multiple lineages will require characterizing these MSC subpopulations by their unique genetic profiles. This has been done with mRNA microarrays (6, 7), but miRNA profiles are often simpler and exhibit more drastic differences upon differentiation (8).

Mohan C. Vemuri et al. (eds.), *Mesenchymal Stem Cell Assays and Applications*, Methods in Molecular Biology, vol. 698, DOI 10.1007/978-1-60761-999-4_30,

MicroRNAs (miRNA) are 17–24 nucleotide long, single-stranded, noncoding RNAs that are genome-encoded and regulate multiple cellular processes. MiRNAs can modulate gene expression by affecting transcription rate (9), mRNA stability (10), or mRNA translation into functional proteins (11). Several studies have identified populations of miRNAs that are unique to mouse embryonic stem (ES) cells (12, 13), mouse or human embryonic carcinoma cells (14), and human ES cells (13, 15, 16). Likewise, MSCs exhibit unique miRNA expression profiles (8, 17, 18). What these studies reveal is that miRNAs may be used as diagnostic tools to discriminate between cell types as well as to uncover differentiation mechanisms. As diagnostic indicators, groups or individual miRNAs regulate cardiac differentiation (13), ESC self-renewal (19), osteogenic differentiation (20), myogenesis (21, 22), neural differentiation (14, 23, 24), and cancer (25). Such groups of miRNAs, as in the case of ESC self-renewal serve as regulators of cell cycle (26, 27). Also, the presence of miRNAs in stem cells and their disappearance during differentiation suggest roles in restricting cell differentiation. In MSCs, specific populations of miRNAs are regulated during osteogenic differentiation and these miRNAs have been shown to be responsive to extracellular signals (8). Thus, there is a need for further examination of miRNA expression profiles in MSC populations.

Methods used to investigate miRNA expression include microarrays, northern blotting, small RNA library sequencing, in situ hybridization, deep sequencing, and quantitative real-time PCR (qPCR). The decision of which method to choose depends on the level of data one needs to extract or the number of samples to be processed. For identifying which groups of miRNAs are being expressed, microarrays are probably the most logical method, while qPCR is often used to validate those results or to screen fewer miRNAs across larger numbers of culture conditions. However, improvements to qPCR have made it possible to detect the expression of multiple miRNAs using a microfluidics-based array of qPCR reactions. The two methods outlined here use either microarrays or qPCR-based arrays to investigate miRNA expression in undifferentiated and differentiated MSCs. For the microarray platform, we used the NCode Multi-Species miRNA microarray system (Invitrogen Corporation) which requires poly(A) tailing of the miRNAs followed by tagging the tail with a DNA polymer attached to fluorescence dye molecules. Although there are disadvantages of using microarrays, such as high variability in the signal-to-noise ratio (likely due to differences in labeling and the difficulty of specifically hybridizing such short probes), there are several advantages of using the NCode platform including increased sensitivity of the labeling reaction due to the multiple Alexa Fluor® molecules present on the DNA polymer and the inclusion of positive control probes.

To minimize dye bias effects, it is also recommended to perform dye-swap hybridization, a technique used in the below described protocol. An alternative to microarrays is a PCR-based approach. To demonstrate such an approach, we employed the TaqMan Human microRNA Array (Applied Biosystems). This technology requires reverse transcription of miRNAs from total RNA with up to 380 stem-loop reverse transcription primers followed by the option to pre-amplify the cDNA and PCR amplification by real-time PCR analysis. Utilizing both the NCode platform and the TaqMan platform, we demonstrate the ability to identify miRNAs present in undifferentiated and differentiated MSCs and compare the microarray-detected miRNA expression to that of qPCR-detected miRNA expression (see Fig. 1).

2. Materials

2.1. MSC Cell Cultures and Differentiation

1. MSC medium: DMEM/F12 low osmolality (Invitrogen, Carlsbad, CA) supplemented with 10% FBS (HyClone, Odgen, UT), 1× GlutaMAX (Invitrogen), 1× MEM nonessential amino acids (Invitrogen), 10 mM HEPES (Invitrogen), and 0.5 μg/ml Gentamycin (Invitrogen). MSC medium can be stored at 4°C for up to 4 weeks.
2. Osteoblast medium: MSC medium supplemented with 10 mM Glycerol-2-phosphate (Sigma-Aldrich, St. Louis, MO), 50 μM L-ascorbic acid-2-phosphate (Sigma-Aldrich), and 100 nM Dexamethasone (Sigma Aldrich). Osteoblast medium can be stored at 4°C for up to 4 weeks.
3. Trizol (Invitrogen).
4. NanoDrop ND-1000 Spectrophotometer (Thermo Scientific, Waltham, MA).

2.2. MicroRNA Microarrays

1. Total RNA sample containing small RNA fraction.
2. NCode™ Multi-species miRNA microarray (Invitrogen).
3. Glass Coverslips or LifterSlips™ (Thermo Scientific).
4. Wash buffers: 2× SSC/0.2% SDS, 2× SSC, and 0.2× SSC.
5. 1 mM Tris–HCl, pH 8.0.
6. Hybex microarray incubation system (SciGene, Sunnyvale, CA) or alternative hybridization chamber.
7. NCode™ Rapid miRNA labeling System (Invitrogen).
8. Microarray scanner such as the GenePix 4000B (Molecular Devices, Sunnyvale, CA).
9. Wash containers for submerging slide rack.

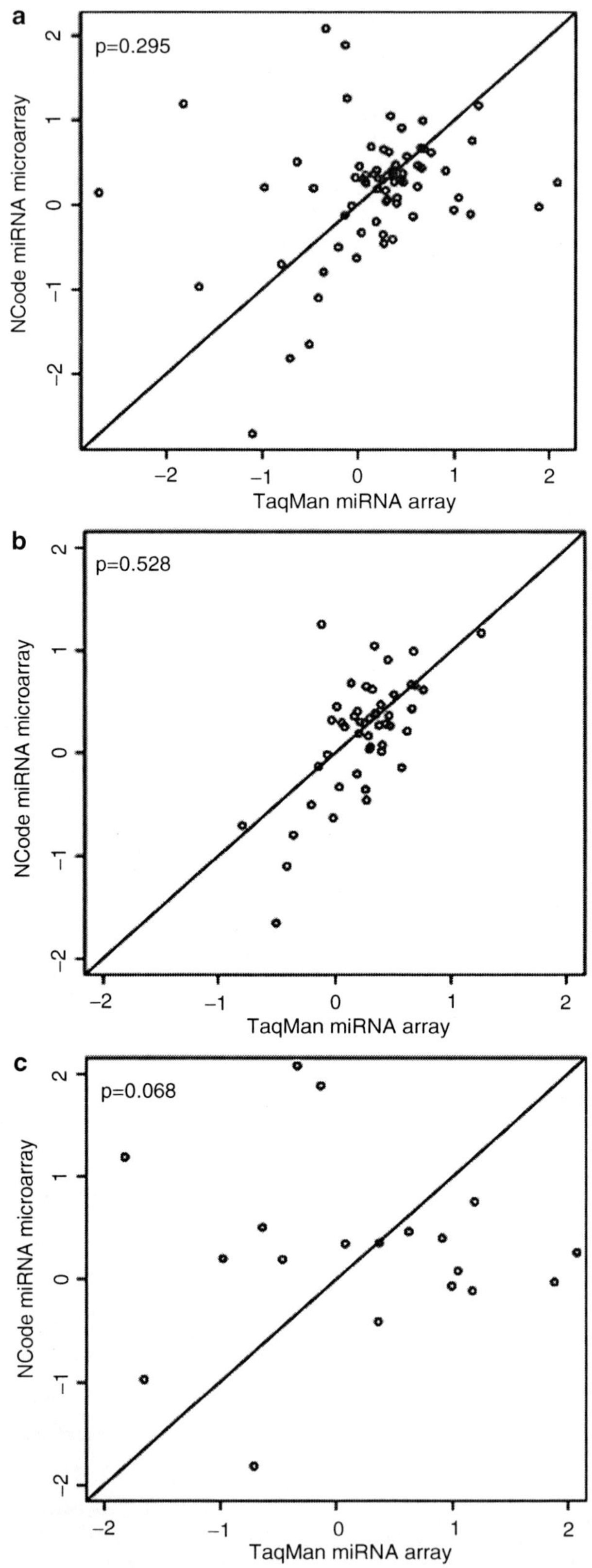
a
p=0.295
NCode miRNA microarray
TaqMan miRNA array
b
p=0.528
NCode miRNA microarray
TaqMan miRNA array
c
p=0.068
NCode miRNA microarray
TaqMan miRNA array

10. Vacuum concentrator.
11. Slide box.

2.3. TaqMan® MicroRNA qPCR

1. Total RNA sample containing small RNA fraction.
2. Megaplex™ RT primers for Human Pool A and Human Pool B (Applied Biosystems, Carlsbad, CA).
3. TaqMan® Human MicroRNA Array A and Array B (Applied Biosystems).
4. TaqMan® MicroRNA Reverse Transcription kit (Applied Biosystems).
5. TaqMan® Universal PCR Master mix, No AmpErase® UNG, 2× (Applied Biosystems).
6. 7900HT Fast Real-Time PCR system with the 7900HT TaqMan® Low Density Array unit (Applied Biosystems).

3. Methods

3.1. MSC Differentiation and Total RNA Isolation

1. MSCs (8) were first expanded in MSC medium on 10 cm plasma-treated culture plates and replenished with fresh media every 2 days. After reaching 80% confluence, cells were passaged onto two, 6-well plates at 10,000 cells/cm^2. Following a recovery period of 24 h, three wells were harvested with 1 ml of Trizol and stored at −80°C while a second set of three wells was selected for differentiation. The medium was changed to osteoblast differentiation medium and replenished every 3 days. On the seventh day, osteoblasts were harvested with 1 ml of Trizol and the cell lysis was stored at −80°C.
2. Total RNA was isolated following the Trizol method (Invitrogen) and quantified using a NanoDrop ND-1000 Spectrophotometer (see Note 1).

Fig. 1. Correlation of NCode™ miRNA microarray and TaqMan® miRNA array measurements for the 70 miRNAs detected by both assays is dependent on expression level. Out of 454 miRNA probes present on the NCode arrays, 93 miRNA measurements showed a positive mean intensity above background. Results are plotted as the $\log_2$ fold change between MSC and nascent osteoblasts. Out of these 93 miRNAs, 70 were present in both the NCode and TaqMan arrays. (**a**) Comparison between the two miRNA platforms using all 70 probes indicated poor correlation between the two methods (Spearman's rho = 0.295). (**b**) However, using only the qPCR results filtered for a higher level of expression (a C_t or cycle threshold value of less than 30) produced increased correlation (Spearman's rho = 0.528). (**c**) By contrast, when qPCR results were filtered for low levels of expression (C_t value of 30 or more cycles), there was less of an association between the two methods (Spearman's rho = 0.068). Results indicate that fold-change measurements are somewhat similar between the two methods at higher levels of expression but that the methods disagree at lower expression levels. This is likely due to the greater sensitivity of qPCR.

3.2. NCode™ Microarray Hybridization

1. A total of 1 μg of total RNA in a volume of no more than 9 μl is required for each experimental condition (two undifferentiated MSC samples and two differentiated MSC samples) (see Notes 2 and 3).
2. Poly(A) tailing of the RNA is carried out using the NCode™ Rapid miRNA Labeling System. Prepare the ATP dilution with 1 mM Tris–HCl, pH 8.0 for using total RNA according to manufacturer's recommendations prior to beginning the poly(A) tailing procedure but do not add it to the total RNA until the procedure calls for the diluted ATP.
3. Dilute the NCode™ multi-species miRNA microarray control according to the manufacturer's recommendations. Add 1 μl of the diluted NCode™ multi-species miRNA microarray control to each of the four total RNA samples and bring up the volume to 10 μl.
4. Add the 10× miRNA reaction buffer, 25 mM $MnCl_2$, diluted ATP, and poly(A) polymerase to the 10 μl RNA as indicated in the manufacturer's protocol.
5. Mix, centrifuge, and incubate the four reactions at 37°C for 15 min (see Note 4).
6. Briefly centrifuge the 15 μl of poly(A) tailed RNA and add 4 μl of the ligation fluorescent mix to the corresponding RNA (see Note 5).
7. Add 2 μl of T4 DNA ligase to the reaction.
8. Mix and briefly centrifuge the reaction (see Note 6).
9. Incubate the reactions for 30 min at room temperature under low-light conditions (see Note 7).
10. Add 2.5 μl of Stop solution to each reaction. Mix and briefly centrifuge the reactions.
11. Combine the appropriate control sample (labeled with the first fluorescent dye) with its corresponding test sample (labeled with the second fluorescent dye) to yield two ligation reactions with a total volume of 47 μl each.
12. Concentrate the two reactions with a vacuum concentrator to half the volume (~23.5 μl) (see Note 8).
13. Add 5 μl of 10% BSA to each concentrated ligation reaction.
14. Add 28.5 μl of 2× prewarmed hybridization buffer (see Note 9).
15. Incubate this hybridization mix for 10 min at 65°C (see Note 10). Protect the reactions from light.
16. Add 57 μl of the hybridization mix to the array, cover with a glass coverslip, and hybridize overnight at 53°C in a dark, humidified chamber (see Notes 11 and 12).

17. Warm up 2× SSC/0.2% SDS wash buffer to 52°C.
18. Remove the coverslip by immersing the slide in prewarmed 2× SSC/0.2% SDS for 2 min or until the coverslip floats off (see Note 13).
19. Wash slide in prewarmed 2× SSC/0.2% SDS at 52°C for 15 min.
20. Wash slide in 2× SSC for 15 min at room temperature.
21. Wash slide in 0.2× SSC for 15 min at room temperature.
22. Quickly transfer the slide to a centrifuge and spin at 1,100 rpm for 3 min (see Note 14).
23. Scan and analyze the array on a microarray scanner (see Note 15).

3.3. TaqMan® qPCR

24. Set up two 0.2 ml tubes with undifferentiated MSC total RNA and two 0.2 ml tubes with differentiated MSC total RNA. Each tube should contain 3 μl of total RNA (350–1,000 ng).
25. Set up one RT master mix with Megaplex™ RT Primers Pool A, enough for two samples and one RT master mix with Megaplex™ RT Primers Pool B for two samples according to manufacturer's protocol (see Note 16).
26. Add 4.5 μl of the appropriate RT Master Mix to each 3 μl of total RNA.
27. Mix and briefly centrifuge contents.
28. Incubate each reaction on ice for 5 min.
29. Perform reverse transcription by setting up a thermal cycling run according to the manufacturer's protocol.
30. If necessary, thaw the Megaplex™ RT product on ice. Add 444 μl of nuclease-free water to each RT product. Mix and briefly centrifuge each reaction.
31. Prior to setting up the PCR reactions, mix the 2× TaqMan® Universal PCR Master Mix, No AmpErase® UNG. To each Megaplex™ RT product, add 450 μl of the Universal PCR Master Mix, mix and briefly centrifuge.
32. Allow each TaqMan® MicroRNA array to reach room temperature (two of Array A and two of Array B). Aliquot 100 μl of RT product into each fill reservoir of the array. Repeat this step for the remaining three arrays.
33. Centrifuge the arrays to distribute the RT product into the reaction wells of the array according to the manufacturer's instructions.
34. Seal the array according to the manufacturer's instructions and trim the fill reservoirs from the array.

35. Load and run each array using the relative quantification mode and the 384 well TaqMan® Low Density Array thermal cycling conditions (see Note 17).

4. Notes

1. If RNA yield is of concern (≤100 ng), RNA may be precipitated by adding Linear Acrylamide (Applied Biosystems, Carlsbad, CA) to a final concentration of 10 μg/μl prior to isopropanol precipitation. A ratio of the absorbance at 260 and 280 ($A_{260/A280}$) and a ratio of the absorbance at 260 and 230 ($A_{260/A230}$) need to be ≥2.0 when measured at pH 8.0. If either ratio is less than 1.8, additional cleanup steps are required to remove any contaminants that remain from the RNA isolation. To determine the integrity of total RNA, the samples may be analyzed by gel electrophoresis or on the 2100 Bioanalyzer (Agilent Technologies, Santa Clara, CA). If the Bioanalyzer is used, a RNA Integrity Number (RIN) of ≥8.0 is recommended.
2. Since environmental ozone levels can have an effect on fluorescence intensity, ensure ozone levels are as low as 2 ppb prior to beginning hybridization unless the microarray laboratory has been installed with an adequate air filtering system (28, 29). A dye-flip experiment is performed to minimize any dye effects. Thus, two arrays and four samples (two control samples and two test samples) are required for each hybridization reaction. For experiments with additional replicates, linear modeling may be performed (30).
3. If the total RNA is too dilute, the RNA may be concentrated using a vacuum concentrator. However, be careful not to over-dry the RNA or it will not dissolve properly.
4. While the poly(A) tailing reaction is incubating, store all poly(A) tailing components and begin to thaw the materials required for ligation. Perform the ligation reaction under low light conditions.
5. One group of samples, a control (undifferentiated MSC) and a test sample (differentiated MSC), is labeled with one fluorescent dye. The second group of samples is labeled with the second fluorescent dye.
6. The NCode™ dye normalization control is not necessary since a dye-flip experiment is being conducted.
7. Inspect the slide to ensure it is free of dust or lint particles. If necessary, gently clean the slide with a stream of air.

8. Preheat the 2× hybridization buffer at 53°C until the buffer is completely dissolved.
9. The 2× hybridization buffer is viscous, so pipette it slowly.
10. Maintain the ligation reactions at 65°C until ready to load the reactions onto the array slides.
11. To load a slide there are two options, one option is to place the glass coverslip on the array and then align the pipette tip at the center of the glass coverslip and slowly pipette the liquid by allowing it to flow throughout the slide. The second option is to apply the hybridization mix first and then gently lower the coverslip onto the slide.
12. If using the Hybex microarray incubation system (SciGene), only one chamber (which fits 4 slides) is required. Unscrew the lid of the microarray incubation chamber from its base, soak the filter paper within the chamber with 1 ml of water, allow the water to soak throughout the filter paper, and remove any excess water from the chamber. After loading the slide, close the chamber tightly. Load the chambers into the incubation system according to the manufacturer's recommendations.
13. If the slide dried during hybridization, it will take longer than 2 min for the coverslip to float off. Do not allow the slides to dry in between the wash steps.
14. The slide may be dried in a 50 ml centrifuge tube by orienting the slide so that any label is at the bottom of the tube. If a swinging bucket centrifuge is available, orient the slides so that they are parallel to the swinging bucket. The slide may be stored in a slide box until ready to analyze the array.
15. Set PMT settings accordingly to control spots available on the array and then proceed with gridding of the array.
16. One sample of undifferentiated MSC and one sample of differentiated MSC are labeled with Megaplex™ RT Primers Pool A. The second sample of undifferentiated MSC and sample of differentiated MSC are labeled with Megaplex™ RT Primers Pool B.
17. To analyze the data, load the SDS files for the samples labeled with the Megaplex™ RT primers Pool A into the SDS RQ manager software provided by Applied Biosystems. In the analysis settings, set the detector settings to a manual C_t and a threshold of 0.2 and select an automatic baseline. Also select the mammalian U6 probe as the endogenous control detector and the control sample (undifferentiated MSC) as the calibrator. Analyze all probes and export data. Repeat these steps for Pool B.

Acknowledgments

Supported by grants from the New Jersey Commission on Science & Technology (Stem Cell Program), NIH 5R21 NS054028, and Invitrogen, Inc. CC was supported by a fellowship from the New Jersey Commission on Spinal Cord Research.

References

1. Wagner, W., Wein, F., Seckinger, A., Frankhauser, M., Wirkner, U., Krause, U., Blake, J., Schwager, C., Eckstein, V., Ansorge, W., and Ho, A. D. (2005) Comparative characteristics of mesenchymal stem cells from human bone marrow, adipose tissue, and umbilical cord blood, *Exp Hematol 33*, 1402–1416.
2. Mansilla, E., Marin, G. H., Drago, H., Sturla, F., Salas, E., Gardiner, C., Bossi, S., Lamonega, R., Guzman, A., Nunez, A., Gil, M. A., Piccinelli, G., Ibar, R., and Soratti, C. (2006) Bloodstream cells phenotypically identical to human mesenchymal bone marrow stem cells circulate in large amounts under the influence of acute large skin damage: new evidence for their use in regenerative medicine, *Transplant Proc 38*, 967–969.
3. Igura, K., Zhang, X., Takahashi, K., Mitsuru, A., Yamaguchi, S., and Takashi, T. A. (2004) Isolation and characterization of mesenchymal progenitor cells from chorionic villi of human placenta, *Cytotherapy 6*, 543–553.
4. In't Anker, P. S., Scherjon, S. A., Kleijburg-van der Keur, C., Noort, W. A., Claas, F. H., Willemze, R., Fibbe, W. E., and Kanhai, H. H. (2003) Amniotic fluid as a novel source of mesenchymal stem cells for therapeutic transplantation, *Blood 102*, 1548–1549.
5. In't Anker, P. S., Noort, W. A., Scherjon, S. A., Kleijburg-van der Keur, C., Kruisselbrink, A. B., van Bezooijen, R. L., Beekhuizen, W., Willemze, R., Kanhai, H. H., and Fibbe, W. E. (2003) Mesenchymal stem cells in human second-trimester bone marrow, liver, lung, and spleen exhibit a similar immunophenotype but a heterogeneous multilineage differentiation potential, *Haematologica 88*, 845–852.
6. Ng, F., Boucher, S., Koh, S., Sastry, K. S., Chase, L., Lakshmipathy, U., Choong, C., Yang, Z., Vemuri, M., Rao, M. S., and Tanavde, V. (2008) PDGF, TGF-b and FGF signaling is important for differentiation and growth of mesenchymal stem cells (MSCs): transcriptional profiling can identify markers and signaling pathways important in differentiation of MSC into adipogenic, chondrogenic and ostoegenic lineages, *Blood 112*, 295–307.
7. Kulterer, B., Friedl, G., Jandrositz, A., Sanchez-Cabo, F., Prokesch, A., Paar, C., Scheideler, M., Windhager, R., Preisegger, K. H., and Trajanoski, Z. (2007) Gene expression profiling of human mesenchymal stem cells derived from bone marrow during expansion and osteoblast differentiation, *BMC Genomics 8*, 70.
8. Goff, L. A., Boucher, S., Ricupero, C. L., Fenstermacher, S., Swerdel, M., Chase, L. G., Adams, C. C., Chesnut, J., Lakshmipathy, U., and Hart, R. P. (2008) Differentiating human multipotent mesenchymal stromal cells regulate microRNAs: prediction of microRNA regulation by PDGF during osteogenesis, *Exp Hematol 36*, 1354–1369.
9. Kuwabara, T., Hsieh, J., Nakashima, K., Taira, K., and Gage, F. H. (2004) A small modulatory dsRNA specifies the fate of adult neural stem cells, *Cell 116*, 779–793.
10. Zhou, L., Qi, X., Potashkin, J. A., Abdul-Karim, F. W., and Gorodeski, G. I. (2008) MicroRNAs miR-186 and miR-150 down-regulate expression of the pro-apoptotic purinergic P2X7 receptor by activation of instability sites at the 3′-untranslated region of the gene that decrease steady-state levels of the transcript, *J Biol Chem 283*, 28274–28286.
11. Filipowicz, W., Bhattacharyya, S. N., and Sonenberg, N. (2008) Mechanisms of post-transcriptional regulation by microRNAs: are the answers in sight?, *Nat Rev Genet 9*, 102–114.
12. Houbaviy, H. B., Murray, M. F., and Sharp, P. A. (2003) Embryonic stem cell-specific MicroRNAs, *Dev Cell 5*, 351.
13. Ivey, K. N., Muth, A., Arnold, J., King, F. W., Yeh, R. F., Fish, J. E., Hsiao, E. C., Schwartz, R. J., Conklin, B. R., Bernstein, H. S., and Srivastava, D. (2008) MicroRNA regulation

of cell lineages in mouse and human embryonic stem cells, *Cell Stem Cell 2*, 219–229.

14. Sempere, L. F., Freemantle, S., Pitha-Rowe, I., Moss, E., Dmitrovsky, E., and Ambros, V. (2004) Expression profiling of mammalian microRNAs uncovers a subset of brain-expressed microRNAs with possible roles in murine and human neuronal differentiation, *Genome Biol 5*, R13.
15. Suh, M. R., Lee, Y., Kim, J. Y., Kim, S. K., Moon, S. H., Lee, J. Y., Cha, K. Y., Chung, H. M., Yoon, H. S., Moon, S. Y., Kim, V. N., and Kim, K. S. (2004) Human embryonic stem cells express a unique set of microRNAs, *Dev Biol 270*, 488.
16. Lakshmipathy, U., Love, B., Goff, L. A., Jornsten, R., Graichen, R., Hart, R. P., and Chesnut, J. D. (2007) MicroRNA expression pattern of undifferentiated and differentiated human embryonic stem cells, *Stem Cells Dev 16*, 1003–1016.
17. Greco, S. J., and Rameshwar, P. (2007) MicroRNAs regulate synthesis of the neurotransmitter substance P in human mesenchymal stem cell-derived neuronal cells, *Proc Natl Acad Sci USA 104*, 15484–15489.
18. Bae, S., Ahn, J. H., Park, C. W., Son, H. K., Kim, K. S., Lim, N. K., Jeon, C. J., and Kim, H. (2009) Gene and microRNA expression signatures of human mesenchymal stromal cells in comparison to fibroblasts, *Cell Tissue Res 335*, 565–573.
19. Xu, N., Papagiannakopoulos, T., Pan, G., Thomson, J. A., and Kosik, K. S. (2009) MicroRNA-145 regulates OCT4, SOX2, and KLF4 and represses pluripotency in human embryonic stem cells, *Cell 137*, 647–658.
20. Kim, Y. J., Bae, S. W., Yu, S. S., Bae, Y. C., and Jung, J. S. (2009) miR-196a regulates proliferation and osteogenic differentiation in mesenchymal stem cells derived from human adipose tissue, *J Bone Miner Res 24*, 816–825.
21. Rao, P. K., Kumar, R. M., Farkhondeh, M., Baskerville, S., and Lodish, H. F. (2006) Myogenic factors that regulate expression of muscle-specific microRNAs, *Proc Natl Acad Sci USA 103*, 8721–8726.
22. Kim, H. K., Lee, Y. S., Sivaprasad, U., Malhotra, A., and Dutta, A. (2006) Muscle-specific microRNA miR-206 promotes muscle differentiation, *J Cell Biol 174*, 677–687.
23. Schratt, G. M., Tuebing, F., Nigh, E. A., Kane, C. G., Sabatini, M. E., Kiebler, M., and Greenberg, M. E. (2006) A brain-specific microRNA regulates dendritic spine development, *Nature 439*, 283–289.
24. Krichevsky, A. M., Sonntag, K.-C., Isacson, O., and Kosik, K. S. (2006) Specific microRNAs modulate embryonic stem cell-derived neurogenesis, *Stem Cells 24*, 857–864.
25. Calin, G. A., and Croce, C. M. (2006) MicroRNA signatures in human cancers, *Nat Rev Cancer 6*, 857–866.
26. Greer Card, D. A., Hebbar, P. B., Li, L., Trotter, K. W., Komatsu, Y., Mishina, Y., and Archer, T. K. (2008) Oct4/Sox2-regulated miR-302 Targets Cyclin D1 in human embryonic stem cells, *Mol Cell Biol 28*, 6426–6438.
27. Wang, Y., Baskerville, S., Shenoy, A., Babiarz, J. E., Baehner, L., and Blelloch, R. (2008) Embryonic stem cell-specific microRNAs regulate the G1-S transition and promote rapid proliferation, *Nat Genet 40*, 1478–1483.
28. Branham, W., Melvin, C., Han, T., Desai, V., Moland, C., Scully, A., and Fuscoe, J. (2007) Elimination of laboratory ozone leads to a dramatic improvement in the reproducibility of microarray gene expression measurements, *BMC Biotechnol 7*, 8.
29. Fare, T. L., Coffey, E. M., Dai, H., He, Y. D., Kessler, D. A., Kilian, K. A., Koch, J. E., LeProust, E., Marton, M. J., Meyer, M. R., Stoughton, R. B., Tokiwa, G. Y., and Wang, Y. (2003) Effects of atmospheric ozone on microarray data quality, *Anal Chem 75*, 4672–4675.
30. Goff, L. A., Lakshmipathy, U., and Hart, R. P. (2008) The analysis of microRNAs in stem cells, in *Stem Cell Research and Therapeutics* (Shi, Y., Ed.), pp 141–167, Springer, The Netherlands.

Chapter 31

Two Dimensional Gel Electrophoresis Analysis of Mesenchymal Stem Cells

Monique Provansal, Christian Jorgensen, Sylvain Lehmann, and Stéphane Roche

Abstract

Proteomic analysis is a powerful tool to follow physiological modifications and phenotypes of mesenchymal stem cells (MSC). This approach generates informative data on expression and post-translational modifications of proteins which are of interest to assess the true potential of MSC in regenerative medicine. No matter the technologies used, proteomic analysis is always a challenge as the proteome is extremely diverse (in terms of constituents and concentrations), is changing with time, and is highly sensitive to pre-analytical conditions. In the framework of a European project (GENOSTEM http://www.genostem.org/), we have set up a multisite two dimensional gel electrophoresis (2DE) proteomic comparison of MSC. The goal is to compare cells from different origins, to follow their differentiation and to ultimately define a specific MSC proteomic signature. One important initial task is the optimization of 2DE protocols such that they are robust enough to be used in a multisite project. In this chapter, we detail these protocols which can be used not only for MSC but also for other cells in culture.

Key words: Mesenchymal stem cell, Proteomics, 2DE electrophoresis, Western blotting, Silver staining

1. Introduction

Stem cells and in particular mesenchymal stem cells (MSC) represent a major hope for regenerative medicine. While their existence in various tissues is undisputed, many questions remain regarding their physiology and their capability to give rise to various functional cell types. To define and exploit their true potential, additional studies on the molecular determinants of MSC isolation, renewal, and differentiation are needed.

Proteomic analysis relies on a set of powerful technologies to study cell physiology (1). This approach has access to expression

Mohan C. Vemuri et al. (eds.), *Mesenchymal Stem Cell Assays and Applications*, Methods in Molecular Biology, vol. 698, DOI 10.1007/978-1-60761-999-4_31,

levels and post-translational modifications of proteins and adds to protein profiling as changes in the mRNA level do not always correlate with protein expression (2).

Two dimensional gel electrophoresis (2DE) is a core technology for proteomics. It involves the separation of proteins from complex mixtures based on two dimensions. First, proteins are separated by their isoelectric point on a pH gradient gel under denaturing condition. Second, the proteins in the pH gradient are further separated based on their molecular size by SDS-polyacrylamide gel electrophoresis (Fig. 1).

2DE has been used by us and others on MSC cultures to address the important questions mentioned above and to define a MSC-specific proteomic signature (3). However, it has been difficult to compare and combine the different published works which differ in terms of paradigms, protein extraction and 2DE protocols (1). In the present chapter, we detail the methods and

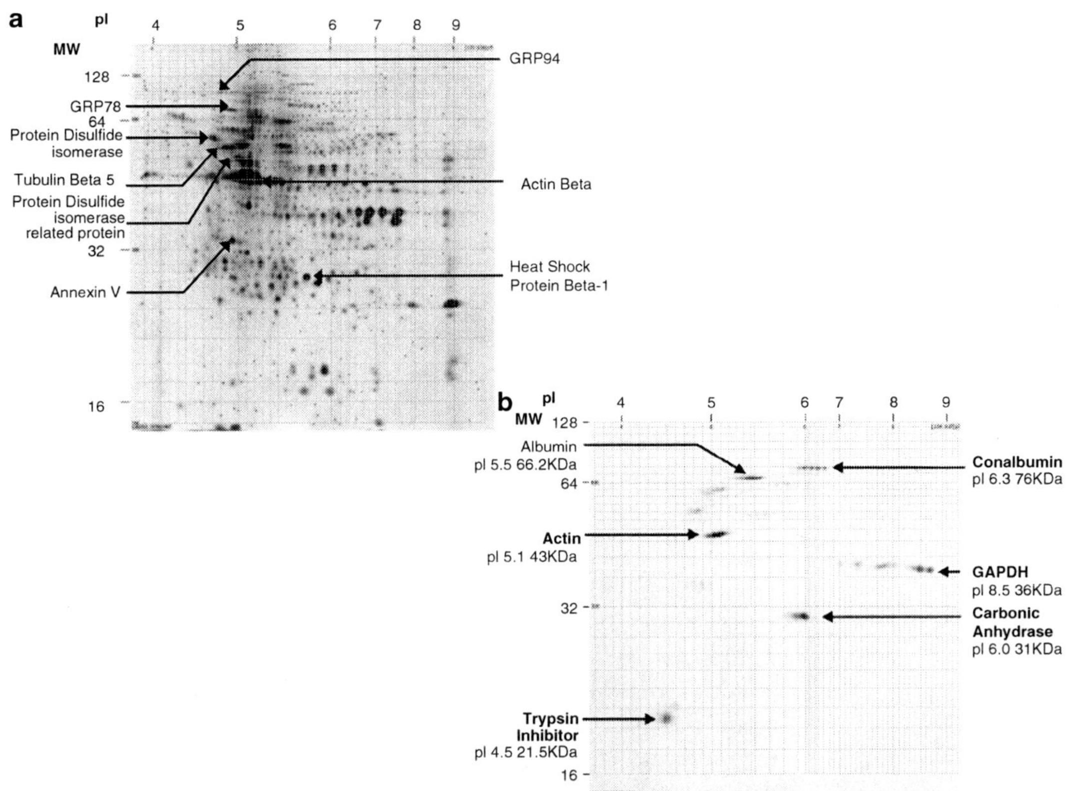

Fig. 1. (**a**) 2DE silver stained gel with the localization of identified proteins. Lysate of human primary bone marrow mesenchymal stem cells (MSC) was analysed by 2DE electrophoresis using a pH 3–10 IEF strip for the first dimension, a 12% SDS-PAGE for the second dimension. The identification of proteins was performed by peptide mass fingerprints after trypsin digestion and MALDI-TOF. (**b**) 2DE gel of IEF protein standard. The 2DE protein standard is first analysed alone, then mixed with a sample (not shown). This comigration is important for the calibration in mass and pHi. The analytical software computes automatically corresponding mass and pHi of others spots.

protocols for the 2DE analysis of MSC which was designed through a program that included a multisite comparison of MSC (3–5).

2. Materials

2.1. Cell Lysis

1. Orthovandate solution: to prepare 100 mL of 10 mM Na_4VO_3 (Sigma) in ddH_2O, boil the solution during 5 min until complete dissolution. The solution will appear yellow to uncoloured. Cool it and adjust the volume to 100 mL with ddH_2O. The solution can be stored in a glass bottle for a few months at room temperature (RT) or at –20°C aliquoted in polypropylene tubes.
2. Wash solution: add 500 µL of orthovanadate solution and 50 µL of protease inhibitor (Sigma) to 50 mL of phosphate buffer saline without Mg/Ca (PBS).
3. Lysis buffer (6, 7): 8 M Urea (GE Healthcare), 1 M Thiourea (GE Healthcare), 4.8% CHAPS (Sigma), 50 mM DTT (Sigma) (see Note 1). When urea is put in solution, the volume increases significantly. Therefore, to prepare 20 mL of lysis buffer, dissolve the components first in 10 mL of water and adjust to 20 mL with ddH_2O afterward (see Note 2). Store lysis buffer at –20 or –80°C (see Note 3). Avoid repeated freeze–thaw cycles.

2.2. Removal of DNA and Protein Quantification

1. Spermine solution: 2.4 M Spermine dihydrate (Sigma). Prepare 5 mL of solution in ddH_2O. Aliquot by 20 µL in eppendorfs, store at –20°C.
2. PlusOne 2-D Quant Kit (GE Healthcare).

2.3. Isoelectric Focusing

1. Rehydration buffer: 9.8 M Urea, 4.8% CHAPS, 50 mM DTT. Store in 250 µL aliquot at –20 or –80°C. To prepare 20 mL of buffer, follow the same recommendations as for the lysis buffer.
2. Low viscosity paraffin oil with a density from 0.827 to 0.89 (Sigma).

2.4. SDS-Polyacrylamide Gel Electrophoresis (SDS-PAGE)

1. SDS-equilibration buffer: 6 M urea, 50 mM Tris–HCl pH8.8, 30% Glycerol, 2% SDS, trace of bromophenol blue. Prepare 200 mL, aliquot by 10 mL and store at –20°C.
2. Gel buffer: 1.5 M Tris–HCl, pH8.8, 0.4% SDS. Prepare 100 mL, store at 4°C.
3. Ammonium persulfate: prepare 10 mL at 10% in water and make 500 µL aliquots stored at –20°C for single use.
4. Running buffer 10×: 250 mM Tris, 1.92 M Glycine, 1% SDS. Prepare 3 L, store at 4°C.

5. Agarose: prepare 50 mL of 1% low melting agarose in 1× running buffer, add a few drops of bromophenol blue to colour the solution. The solution needs to be heated at 90°C to dissolve the agarose. Can be stored at 4°C and reheated at 90°C before use.
6. Unstained molecular weight standard (Bio-Rad).
7. MultiTemp III cooling device (GE Healthcare).

2.5. Gel Staining

2.5.1. Silver Staining

1. All solutions are made fresh in ddH_2O; all references are from Sigma. Prepare 250 mL of each solution per gel.
2. Buffer A: 50% ethanol, 10% acetic acid.
3. Buffer B: 5% ethanol.
4. Buffer C: 0.02% sodium thiosulfate anhydrous.
5. Buffer D: 0.2% $AgNO_3$.
6. Buffer E: 3% sodium carbonate decahydrate, 0.0005% Formaldehyde, 2% buffer C.
7. Buffer F: 1.4% EDTA-Na.

2.5.2. Coomassie Staining

1. Buffer A: 50% ethanol, 10% acetic acid.
2. Buffer B: 5% ethanol.
3. Colloidal Coomassie stain (Fermantas).

2.5.3. Western Blotting

1. The procedure for Western blotting is adapted to the NuPAGE system and gel (Bis/Tris 12% gel); all reagents and associated buffers are from Invitrogen.
2. The MOPS running buffer is diluted to 1× as recommended by the manufacturer.
3. The transfer buffer is prepared as follows: 5% transfer buffer stock solution and 20% ethanol in ddH_2O.
4. Before transfer, the PVDF membrane is incubated in 100% ethanol for 10 s and rinsed in the transfer buffer for 5 min on a rocking platform.
5. PBS-T: Stock solution: 1% Tween-20 in 10× PBS (without Ca and Mg). Dilute at 1/10 in ddH_2O for final solution.
6. Blocking buffer: 5% (w/v) non fat dry milk in PBS-T.
7. Primary and secondary antibody dilutions are performed in the blocking buffer. A multiplex solution of primary antibodies is prepared using anti-14-3-3 (1:1,000, Santa Cruz, Clone K19), anti-Actin (1:500, Santa Cruz), and Sox-9 (1:1,000, kind gift from F. Poulat, IGH, CNRS) antibodies. Secondary antibody multiplex solution is prepared using anti-rabbit and anti-mouse antibodies (Jackson ImmunoResearch) (both at 1:10,000).
8. Enhanced chemiluminescent (ECL) (SuperSignal West Pico) (Pierce Biotechnology). ECL prepared as described by manufacturer by mixing 2 mL of enhancer and 2 mL of

peroxide solution. The ECL camera system is ChemiDoc XRS from Bio-Rad.

2.6. Scan of the 2DE Gels, Detection and Spot Analysis

1. Epson 1680PRO scanner.
2. Greyscale chart (Kodak).
3. Labscan 3 software (GE Healthcare).
4. ImageMaster 2DE Platinium version 6 (GE Healthcare).
5. Progenesis SameSpot version 2 (Non Linear Dynamics).
6. Hierarchical Clustering Explorer v3.5 (8).

2.7. Molecular Weight and Isoelectric Point Calibration

1. 2DE protein standard (Bio-Rad).

2.8. Protein Identification

1. Mass spectrometry qualified tube, 1.5 mL polypropylene tube (TreffLab).
2. 5%, 20% Formic acid, and 100 mM NH_4CO_3 are made fresh.
3. Sequencing grade tryspin (Promega). Reconstitute to a concentration of 0.1 μg/μL using the buffer supplied by the manufacturer. Stored in 10 μL aliquot at −20°C. The working solution is obtained by mixing 2 μL of trypsin to 8 μL of 100 mM NH_4CO_3.
4. ZipTip C18 column (Millipore).
5. Benchtop speepvac "CentriVac" (Labconco).
6. Ultraflex Mass Spectrometer (Bruker).

3. Methods

3.1. Cell Lysis

1. Wash twice cultured cells with the wash solution at RT.
2. Remove all residual buffer. The cell layer needs to be near dry (see Note 4).
3. Put the lysis buffer directly on the cells in the culture dish. For a 100 mm Petri dish, use 200 μL of lysis buffer. Do the lysis at RT (see Note 5).
4. Using a clean cell scraper, collect the lysate in the dish, and transfer it to a 1.5 mL eppendorf tube.
5. Do not spin or cool the tube and freeze it as soon as possible at −80°C.

3.2. Removal of DNA and Protein Quantification

1. Thaw samples at RT.
2. Add 1 volume of spermine dihydrate to 100 volumes of sample and incubate at RT during 1 h. DNA that is charged negatively will associate with spermine.

3. Centrifuge at 21,000 × *g*, 1 h, at 20°C to pellet spermine and DNA. Use a refrigerated centrifuge to control the temperature which otherwise increases during centrifugation. Do not set up the temperature below 20°C, otherwise the urea will precipitate.
4. Collect and aliquot supernatant in 1.5 mL eppendorf tubes.
5. Use a fraction of the lysate to measure the protein concentration with the PlusOne 2-D Quant Kit following the supplier recommendations. Briefly, 10 μL of samples (in duplicates) is used and quantify using a standard curve with 2, 5, 10, 20, and 40 μg of proteins.

3.3. Isoelectric Focusing

1. Thaw samples (after spermine treatment) and rehydration buffer to RT.
2. For 13 cm isoelectric focusing (IEF) dry strips, a loading volume of 250 μL is optimal. The concentration of the protein in the samples will indicate if a dilution with the rehydration buffer is needed in order to have 30 μg of protein in 250 μL for silver staining and 300 μg of protein in 250 μL for Coomassie staining.
3. Add 1 μL of IPG ampholine buffer to the 250 μL diluted samples. Mix the samples well and spin briefly to remove bubbles. The IPG buffer is a mix of soluble ampholines which affect sample solubility and focalisation (see Note 6).
4. Deposit each sample along an IPGphor "sarcophagus," avoiding bubbles. Cover the sample with the IEF dry strip. The acrylamide gel of the strip must be in direct contact with the samples (the acrylamide face is at the bottom). The "plus" or acidic part of the IEF dry strip must be at the head part of the sarcophagus. Cover samples and strips using low viscosity paraffin oil.
5. Place the sarcophagus in the IPGphor apparatus. Set the temperature to 20°C and the current to 60 μA. The program should include passive rehydration, active rehydration and focalisation as follows: 7 h without current, 3 h at 30 V, 2 h at 200 V, 1 h at 1,000 V, 1 h at 3,000 V, then a gradient to reach 8,000 V in 2 h and a final phase of 7 h at 8,000 V.
6. At the end of the focalisation, we expect to have reached 8,000 V with less than 50 μA and a total close to 76,000 V × h (see Note 7).
7. The IPGphor stops automatically at the end of the program. Recover the IEF strips rapidly. Put them in a plastic tube and store them at −20°C. There is no need to rinse the strips.

3.4. SDS-Polyacrylamide Gel Electrophoresis (SDS-PAGE)

1. The SDS-PAGE gels (1.5 mm thick, 12% acrylamide) for the second dimension of the 2DE need to be prepared in advance. The following protocol is adapted to the Hoeffer SE-600 apparatus. One important factor to obtain optimal gels is to

use clean and unscratched glass plates. Avoid using detergents to clean the plates, prefer ddH_2O and ethanol.

2. For one gel, prepare an acrylamide solution as follows: mix 15 mL of acrylamide/bisacrylamide solution (37.5:1, T40%), 12.5 mL of gel buffer and 22.25 mL of ddH_2O. Add 25 μL of TEMED and 250 μL of 10% ammonium persulfate. Pour the solution between the glass plates, leaving 5 mm on the top empty for the IEF strip. No stacking gel is needed. Pour isopropanol on the top of the gel to flatten the surface. To reduce the amount of unpolymerized acrylamide monomers, let the gel polymerize at least 2 h. The gels can be prepared 24 h in advance and kept at 4°C in a wet atmosphere.
3. Thaw the IEF strips and incubate them for 30 min on a rocking platform in 5 mL of SDS-equilibration buffer supplemented with 0.25% DTT. Remove the buffer and replace it with 5 mL of SDS-equilibration buffer supplemented with 4.5% iodoacetamide for another 30 min on the rocking platform.
4. Before using the SDS gel, verify the absence of bubbles. Pour off isopropanol and rinse the top of the gel with ddH_2O.
5. Using small tweezers, place the IEF strip on the top of the gel. Usually, the "plus" end of the strip is positioned on the left of the gel. On the right part, insert a small 1.5 mm thick piece of plastic to create a molecular weight (MW) well. To secure the strip to the gel, pour warm agarose on the top paying attention not to leave any bubbles between them. Place the gel at 4°C for 5 min to solidify the agarose.
6. Remove the above piece of plastic to create a small well in the agarose. Add to this well 10 μL of unstained molecular weight standard and seal the well with a drop of warm agarose.
7. Mount the gel on the apparatus and connect it to the power supply. Run the gel for 17 h at 70 V. To obtain an optimal migration, the electrophoresis apparatus needs to be associated to a MultiTemp III cooling device, maintaining the system at 10°C.

3.5. Gel Staining

3.5.1. Silver Staining

1. At the end of the SDS-PAGE, carefully disassemble the plates and put the gel in the buffer A in order to fix proteins. Incubate for a minimum of 2 h on a rocking platform (Fig. 1a).
2. Rehydrate the gel in buffer B for 15 min on the rocking platform and rinse 3 times for 5 min with ddH_2O.
3. Incubate the gel in buffer C for 2 min on the rocking platform and rinse 3 times for 30 s with ddH_2O (9).
4. Incubate the gel in buffer D for 25 min on the rocking platform and rinse twice for 1 min with ddH_2O (9).
5. Stain proteins using buffer E (9). The staining takes place from 2 to 10 min. It needs to be stopped at the optimal time

in order to obtain the best signal/background ratio (see Note 8).

6. To stop the staining, replace the buffer E with the buffer F for 10 min in a rocking platform.
7. Dry gels or conserve them in 10% glycerol in a sealed plastic pouch.

3.5.2. Coomassie Staining

1. At the end of the SDS-PAGE, carefully disassemble the plates and put the gel in buffer A in order to fix proteins. Incubate for a minimum of 2 h on a rocking platform.
2. Rehydrate the gel in buffer B for 15 min on the rocking platform and rinse 3 times for 5 min with ddH_2O.
3. Incubate the gel in colloidal Coomassie stain for at least 1 h and remove excess stain by rinsing 2–3 times in ddH_2O.
4. The gels can be dried or conserved in 10% glycerol in a sealed plastic pouch.

3.5.3. Western Blotting

1. For Western blotting, the first dimension of the 2DE is performed using a 7 cm IEF strip. The only modification in the protocol (when compared to the 13 cm IEF strip) is the lower amount of samples (125 μL) and IPG buffer (0.5 μL) used (Fig. 2).
2. The second dimension is performed with the NuPAGE system with one well Bis/Tris 12% gels.
3. After the final incubation in the SDS-equilibration buffer, carefully position the IEF strip on the top of the gel. Usually, the "plus" end of the strip is positioned on the left of the gel. To secure the strip to the gel, pour warm agarose prepared in 1× MOPS buffer on the top, paying attention not to leave any bubbles between them. Place the gel at 4°C for 5 min to allow the agarose to solidify. The NuPAGE migration at 70 V will take 2 h. No cooling system is necessary for this migration.

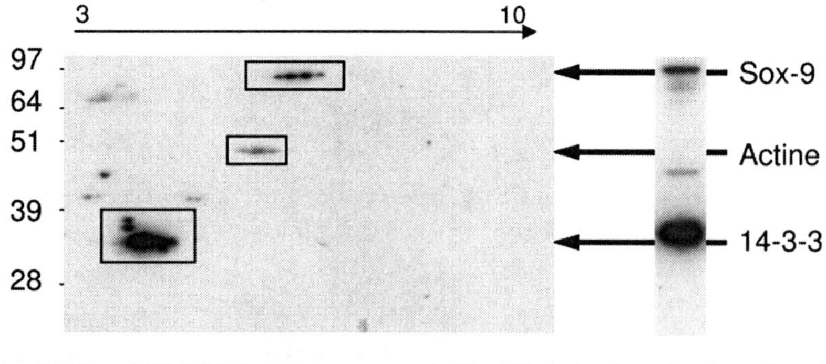

Fig. 2. 2DE Western blotting of BM-MSC. Proteins were separated by 2DE electrophoresis using a 7 cm dry strip pH 3–10 for the first dimension and a Bis/Tris 12% NuPAGE for the second dimension. After transfer on PVDF membrane, targeted proteins are detected by multiplex Western blotting using antibodies against 14-3-3, actin, and Sox-9 proteins. Note the different spots of 14-3-3 and Sox-9 indicating the presence of post-translational modified isoforms.

4. At the end of the migration, open the gel cassette, discard the IEF strip and the agarose, then incubate the gel a few minutes in the transfer buffer. The transfer, using pads and Whatman paper sheets on a PVDF membrane, is performed as per the manufacturer instruction. The transfer is performed at 30 V for 1 h at RT.
5. After transfer, rinse the membrane for 30 s in ddH_2O and incubate it for 30 min in the blocking buffer on a rocking platform.
6. Discard the blocking buffer and incubate the membrane in the primary antibodies for 1 h at RT.
7. Wash 4 times for 5 min with PBS-T and incubate the membrane with the secondary antibodies for 1 h at RT.
8. Wash 4 times for 5 min with PBS-T, incubate with the ECL mix for a few minutes and detect the signal using the CCD camera. Save the digital file. Include in the file name the gel number, date, nature of sample and nature of antibodies.

3.6. 2DE Gel Scan, Detection and Protein Spot Analysis

1. Scan the stained gels at 300 dots per inch using the Labscan 3 software after a calibration procedure using the greyscale chart.
2. For spot detection and quantitation of the silver stained gels, use the ImageMaster 2DE Platinum or the Progenesis SameSpot software. A value for each spot is obtained by an integration of the grey level of each spot and normalisation with the total spot volume. In this method, the volume of each spot is divided by the total volume of all of the spots in the gel.
3. Run three to five replicates with independent samples to ensure reproducibility of the results. The software includes many functions for selected spots and can perform statistical comparison between experimental conditions. Select a gel with the maximum number of spots to serve as a reference.
4. As a threshold for significant differential spot expression, use a fold increase/decrease ratio of 1.7 and a *p*-value of 0.05 as calculated by Progenesis SameSpot.
5. A global analysis of the data can be performed using the Hierarchical Clustering Explorer 3.0 software (8). This software relies on an average linkage hierarchical clustering algorithm based on the centred Manhattan Distance coefficient as the similarity metric.

3.7. Molecular Weight and Isoelectric Point Calibration

1. Run a 2DE gel with a mixture of 10 μg of whole cell extract and 5 μL of 2DE standard (Fig. 1b).
2. These standard spots are associated with their respective MW and pI. After a comparison step with this gel, the software calibrates all homologous gels with pHi and MW.

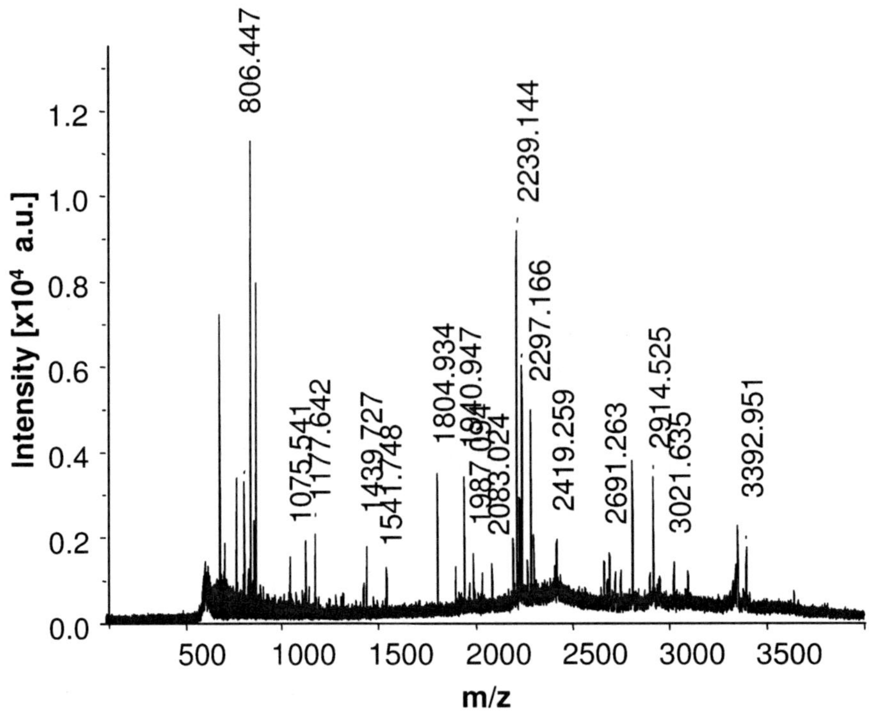

Fig. 3. Mass spectrum of tryspin in gel digestion of candidate protein. A selected spot is excised from the gel, dehydrated, and rehydrated with a sequencing grad trypsin. After digestion, the peptides are extracted, purified, and submitted to MALDI-TOF mass spectrometry. After database query, the protein can be identified with a good probability.

3.8. Protein Identification

1. Excise spots from fresh Coomassie stained 2DE gels run with 300 μg of proteins. Put them separately in 1.5 mL specific tubes (Fig. 3).
2. Incubate pieces of gels with 15 μL of 100 mM NH_4HCO_3 for 10 min. Add 15 μL of acetonitrile for 10 min. Remove and discard supernatant, repeat the procedure.
3. Use a speedvac to evaporate the remaining buffer. Rehydrate in 10 μL of trypsin working solution and digest overnight at 25°C.
4. Extract the digested peptides as follows: add 10 μL of 100 mM NH_4HCO_3 for 10 min then 10 μL of acetonitrile for 10 min. Collect the supernatant, repeat the procedure twice, and combine the supernatants. Add 10 μL of 5% formic acid for 10 min then 10 μL of acetonitrile for 10 additional min. Collect the supernatant, repeat the procedure twice, and combine all the supernatants.
5. Dry the supernatants using the speedvac and put back in solution with 10 μL of 20% formic acid. Desalt using a ZipTip C18 column as recommended by the manufacturer.

6. Determine peptide masses in the positive-ion reflector mode using the Ultraflex mass spectrometer. Compare peptide mass fingerprints to mammalian databases (SWISS PROT and Trembl) using MASCOT (http://www.matrixscience.com/cgi/search_form.pl?FORMVER=2&SEARCH=PMF) and Aldente (http://www.expasy.org/tools/aldente/) algorithms with one miss trypsin cleavage site and a mass deviation smaller than 20 ppm. The probabilistic score of MASCOT/Aldente is required for the identification of an unnamed protein.

4. Notes

1. 2DE results are affected by the composition of the lysis buffer used. This is in relation with differences in protein solubility vs. extraction and with the negative effect of salt on protein focalisation in 2DE. Other types of lysis exist for example design to solubilised membrane proteins (7).
2. Do not heat the lysis buffer in order to speed up dissolution as it will alter urea lysis properties.
3. To optimize the use of the lysis buffer it is recommended to store it in 250 μL aliquots.
4. Do one dish at a time to reduce the risk of complete drying of the cells. Dry cells disrupt and fix the proteins and limit extraction efficacy.
5. Low temperatures induce urea crystallisation and reduce extraction efficacy.
6. It is recommended to use the same pH range of ampholines in the IPG buffer and in the IEF strip. Increasing the volume of ampholines may alter focalisation performance.
7. If the power supply do not reach 8,000 V and/or 76,000 V × h, the focalisation might be poor. The decrease of final voltage or V × h value is due to the interferences of salts and amino acids originate from culture medium and PBS.
8. Above 10 min, no additional spots will generally appear. The staining is the most critical step of 2DE electrophoresis.

Acknowledgments

This work was supported by the European Commission through the FP6 project GENOSTEM (LHSB-CT-2003-503161).

References

1. Roche, S., Provansal, M., Tiers, L., Jorgensen, C., and Lehmann, S. (2006) Proteomics of primary mesenchymal stem cells *Regen Med* **1**, 511–7.
2. Tian, Q., Stepaniants, S. B., Mao, M., Weng, L., Feetham, M. C., Doyle, M. J., Yi, E. C., Dai, H., Thorsson, V., Eng, J., Goodlett, D., Berger, J. P., Gunter, B., Linseley, P. S., Stoughton, R. B., Aebersold, R., Collins, S. J., Hanlon, W. A., and Hood, L. E. (2004) Integrated genomic and proteomic analyses of gene expression in Mammalian cells *Mol Cell Proteomics* **3**, 960–9.
3. Roche, S., Delorme, B., Oostendorp, R. A., Barbet, R., Caton, D., Noel, D., Boumediene, K., Papadaki, H. A., Cousin, B., Crozet, C., Milhavet, O., Casteilla, L., Hatzfeld, J., Jorgensen, C., Charbord, P., and Lehmann, S. (2009) Comparative proteomic analysis of human mesenchymal and embryonic stem cells: towards the definition of a mesenchymal stem cell proteomic signature *Proteomics* **9**, 223–32.
4. Noel, D., Caton, D., Roche, S., Bony, C., Lehmann, S., Casteilla, L., Jorgensen, C., and Cousin, B. (2008) Cell specific differences between human adipose-derived and mesenchymal-stromal cells despite similar differentiation potentials *Exp Cell Res* **314**, 1575–84.
5. Kastrinaki, M. C., Sidiropoulos, P., Roche, S., Ringe, J., Lehmann, S., Kritikos, H., Vlahava, V. M., Delorme, B., Eliopoulos, G. D., Jorgensen, C., Charbord, P., Haupl, T., Boumpas, D. T., and Papadaki, H. A. (2008) Functional, molecular and proteomic characterisation of bone marrow mesenchymal stem cells in rheumatoid arthritis *Ann Rheum Dis* **67**, 741–9.
6. Rabilloud, T. (1998) Use of thiourea to increase the solubility of membrane proteins in two-dimensional electrophoresis *Electrophoresis* **19**, 758–60.
7. Rabilloud, T. (2000) Proteome Research: Two-Dimensional Gel Electrophoresis and Identification Methods, Springer, Heidelberg.
8. Seo, J., Bakay, M., Chen, Y. W., Hilmer, S., Shneiderman, B., and Hoffman, E. P. (2004) Interactively optimizing signal-to-noise ratios in expression profiling: project-specific algorithm selection and detection p-value weighting in Affymetrix microarrays *Bioinformatics* **20**, 2534–44.
9. Shevchenko, A., Wilm, M., Vorm, O., and Mann, M. (1996) Mass spectrometric sequencing of proteins silver-stained polyacrylamide gels *Anal Chem* **68**, 850–8.

Chapter 32

Proteomic Analysis of Human Mesenchymal Stem Cells

Guo Li, Chu-yan Chan, Hua Wang, and Hsiang-fu Kung

Abstract

Umbilical cord (UC) and placenta (P) are generally believed to be potential alternatives to bone marrow (BM), as sources of mesenchymal stem cells (MSC) for cell therapy. They possess immunophenotypic and functional characteristics which are similar to that of BM-MSC, yet one of the crucial factors in determining the tissue regeneration process – the migration capacity – is still unclear. In our previous study, the migration capacity of BM- and P-MSC was found 5.9- and 3.2-fold higher than that of UC-MSC, respectively. By using 2-dimensional gel electrophoresis (2-DE) and combined MS and MS/MS analysis, six proteins were identified as differentially expressed among these MSC samples. Five out of the six proteins were known to be involved in cell migration as migration inhibiting or enhancing proteins.

Key words: Mesenchymal stem cells, Migration, 2-Dimensional gel electrophoresis, MS and MS/MS analysis

1. Introduction

Mesenchymal stem cells (MSC) take significant roles in tissue regeneration based on their abilities of self-renewal and multi-lineage differentiation. Bone marrow (BM), the major source of MSC, has been well recognized in clinical studies. However, the clinical application of BM-derived cells is greatly limited due to the relatively invasive procedure of sample collection as well as markedly reduction in cell number, proliferation and differentiation capacity with age (1). Therefore, any alternative source of MSC would be beneficial and invaluable for cell therapy in the near future.

Placenta (P) and umbilical cord (UC) have been postulated as new sources of MSC. They bear the advantage of relative ease of collection and more importantly, the much younger cellular age when compared to those from adult origins. Previous reports

Mohan C. Vemuri et al. (eds.), *Mesenchymal Stem Cell Assays and Applications*, Methods in Molecular Biology, vol. 698, DOI 10.1007/978-1-60761-999-4_32, © Springer Science+Business Media, LLC 2011

have demonstrated that the UC- and P-MSC share similar features with that of BM-MSC and that include the expression of cell surface antigens and multi-lineage differentiation potential (2–4). In fact, one of our previous studies had shown a higher migration capacity of BM- and P-MSC than that of UC-MSC (5).

Proteomic techniques have been widely applied to different disciplines of cancer metastasis as well as stem cell differentiation or in deriving related molecular mechanisms (6, 7). In this study, candidate proteins which are responsible for the differential migration capacity were deduced from protein expression profiles between BM-, P-, and UC-MSC. By the use of 2-dimensional gel electrophoresis (2-DE) and combined MS and MS/MS analysis, six differential expressed proteins were identified. The translational change of two of the candidate proteins, plasminogen activator inhibitor-1 (PAI-1), and cathepsin B (CTSB) were confirmed by western blotting. In fact, five of the six identified proteins have been reported to be involved in cell migration as migration inhibiting or enhancing proteins (8–24). Consistent with the reported migration capacity, the expression level of migration inhibiting proteins, such as PAI-1 and manganese superoxide dismutase, was much higher in the UC-MSC whilst those for migration enhancing proteins including CTSB, cathepsin D, and prohibitin were significantly lower (5).

2. Materials

2.1. Tissue Dissociation and Cell Culture

1. Culture medium: Minimum essential medium, alpha (α-MEM) supplemented with 10% fetal bovine serum (FBS), 2 mM L-Glutamine, 100 U/mL penicillin, and 100 μg/mL streptomycin (all reagents were from Invitrogen, Carlsbad, CA).
2. Collagenase type II (Invitrogen).
3. Solution of trypsin (0.05%) and ethylenediamine tetraacetic acid (EDTA) (1 mM) (Invitrogen).

2.2. MSC Cell Surface Markers Analysis

1. 0.05% Trypsin/EDTA (Invitrogen).
2. CD3-PE, CD16-FITC, CD19-FITC, CD33-FITC, CD34-PE, CD38-FITC, CD45-FITC (Miltenyi Biotec GmbH, Bergisch Gladbach, Germany), HLA-DR-FITC, CD29-Rhodamin, CD44-FITC, CD90-PE, CD105-PE and CD166-PE (AbD Serotec, Oxford, UK). Phycoerythin (PE) mouse IgG1 and FITC mouse IgG1 (BD Biosciences, San Jose, CA).
3. Coulter Epic XL MCL flow cytometry (Beckman Coulter).

2.3. Osteogenic and Adipogenic Differentiation

1. Adipogenesis induction medium: DMEM-low glucose medium (DMEM-LG) (Invitrogen) supplemented with 10% FBS (Invitrogen), 10^{-7} dexamethasone (Sigma-Aldrich),

50 μg/mL ascorbic acid-2 phosphate (Sigma-Aldrich), 50 μM indomethacin (Sigma-Aldrich), 10 μg/mL insulin (Sigma-Aldrich), and 0.45 mM 3-isobutyl-1-methyl-xanthinel (Sigma-Aldrich).

2. Osteogenesis induction medium: DMEM-LG (Invitrogen) supplemented with 10% FBS (Invitrogen), 10^{-8} M dexamethasone (Sigma-Aldrich), 50 μg/mL ascorbic acid-2 phosphate (Sigma-Aldrich), and 10 mM glycerolphosphate (Sigma-Aldrich).
3. Alizarin red S (2%; pH 4.1) diluted with ammonium hydroxide (Sigma-Aldrich).
4. Saturated (1/3) oil red O: oil red O was reconstituted in triethyl phosphate (TEP) and diluted with 60% TEP (see Note 1).

2.4. Protein Extraction for 2-DE

1. Cell scrapers (Invitrogen).
2. Cell lysis buffer: 8 M urea, 2 M thiourea, 2% CHAPS, 1% NP-40, 2 mM TBP, 1× Protease Inhibitor Mix, 1× Nuclease Mix, 1 mM PMSF, and 2% IPG buffer (GE Healthcare, Piscataway, NJ).
3. Bradford protein assay kit (Bio-Rad, USA).

2.5. Two-Dimensional Gel Electrophoresis

1. IPGphor II apparatus (GE Healthcare).
2. IPG strips (13 cm, pH 3–10, NL) (GE Healthcare).
3. Rehydration solution: 8 M Urea, 2% CHAPS, 0.4% DTT, 0.5% IPG buffer, 0.002% bromophenol blue.
4. Equilibration buffer: 6 M urea, 2% SDS, 30% glycerol, 0.002% bromophenol blue, 50 mM Tris–HCl, pH 6.8, 1% DTT.
5. 2.5% Uniform polyacrylamide gel slabs (150 × 158 × 1.5 mm).
6. Second dimensional separation Tris–glycine buffer: 25 mM Tris, 0.2 M glycine, 0.1% SDS.

2.6. Gel Staining and Image Analysis

1. Gel fixing buffer: 50% methanol containing 12% acetic acid and 0.0185% formaldehyde.
2. 50% Ethanol.
3. 0.02% Sodium thiosulfate.
4. 0.15% Silver nitrate (Sigma-Aldrich) containing 0.0278% formaldehyde.
5. Color development solution: 3% sodium carbonate containing 0.037% formaldehyde.
6. 5% Acetic acid.
7. Scanner (GS-800 calibrated densitometer, Bio-Rad, Hercules, CA).
8. Quantity One software (Bio-Rad).
9. Software PDQuest (version 8.0, Bio-Rad).

2.7. Tryptic In-Gel Digestion

1. Equal volume mixture of 30 mM potassium ferricyanide and 100 mM sodium thiosulfate solutions (Sigma-Aldrich).
2. 50 mM Ammonium bicarbonate.
3. 25 mM Ammonium bicarbonate and 50% acetonitrile (ACN).
4. 100% ACN (Merck, Darmstadt, Germany).
5. 10 μg/mL of trypsin (Promega Corporation, Madison, WI).
6. 25 mM Ammonium bicarbonate (pH 8.0).

2.8. MALDI-TOF/TOF Mass Spectrometric Analysis

1. MALDI-TOF/TOF tandem mass spectrometer ABI 4700 proteomics analyzer (Applied Biosystems, Foster City, CA).
2. Matrix solution: 4 mg/mL α-cyano-4-hydroxycinnamic acid in 35% ACN and 1% TFA.
3. 4000 Series Explorer™ Software v3.0 (Applied Biosystems).
4. Glu-Fibrinopeptide B from 4700 Calibration Mixture (Applied Biosystems).

2.9. Database Search

1. GPS Explorer™ software v3.5 (Applied Biosystems).
2. NCBInr 5825255 sequences (released on Jan 10, 2008).

2.10. SDS-Polyacrylamide Gel Electrophoresis (SDS-PAGE)

1. RIPA cell lysis buffer: 50 mM Tris–HCl pH 7.4; 150 mM NaCl; 1 mM PMSF; 1 mM EDTA; 5 μg/mL Aprotinin; 5 μg/mL Leupeptin; 1% Triton x-100; 1% Sodium deoxycholate; 0.1% SDS.
2. 2× SDS/Sample buffer: 100 mM Tris; 25% glycerol; 2% SDS; 0.01% bromophenol blue; 10% 2-Mercaptoethanol; pH 6.8 (see Note 2).
3. Bradford protein assay kit (Bio-Rad).
4. 30% Acrylamide/bis solution (Amersham) and *N,N,N,N′*-Tetramethyl-ethylenediamine (TEM-ED) (Bio-Rad) (see Note 3).
5. 1.5 M Tris–HCl pH 8.8.
6. 10% Aps (see Note 4).
7. 10% SDS.
8. 5× SDS running buffer: 15.1 g Tris; 94 g Glycine; 5g SDS; add ddH_2O until 1 L.
9. Prestained molecular weight markers (Invitrogen).

2.11. Western Blotting

1. 1× transfer buffer: 3.03 g Tris; 14.4 g Glycine; 200 mL Methanol; add ddH_2O until 1 L.
2. Tris-buffered saline with Tween (TBS/T): 10 mM Tris–HCl pH 7.4; 150 mM NaCl; 0.1% Tween 20 (see Note 5).

3. Blocking buffer: 5% non-fat milk in TBS/T.
4. Primary/secondary antibody dilution buffer: TBS/T with 3% BSA.
5. PVDF membrane (Bio-Rad).
6. 3MM chromatography paper (Whatman, Maidstone, UK).
7. Washing buffer: TBS/T.
8. Primary antibody: CTSB (sc-6493, goat polyclonal IgG), PAI-1 (sc-8979, rabbit polyclonal IgG) (Santa Cruz Biotechnology, Santa Cruz, CA).
9. Secondary antibody: donkey anti-goat IgG-HRP (sc-2020), goat anti-rabbit IgG-HRP (sc-2004) (Santa Cruz Biotechnology).
10. Enhanced chemiluminescent (ECL) reagents (GE Healthcare).
11. X-ray film (Fujifilm, Tokyo, Japan).

2.12. Stripping and Reprobing Blots

1. Stripping buffer: 62.5 mM Tris–HCl, pH 6.8, 2% (w/v) SDS. Store at room temperature. Warm to working temperature (37°C) and add 100 mM mercaptoethanol before use.
2. Washing buffer: TBS/T.
3. Primary antibody: Anti-actin (sc-1616, Santa Cruz Biotechnology).

3. Methods

In this protocol, MSCs are isolated from human bone marrow (BM), umbilical cord (UC), and placenta (P). They are enriched based on their adherent property on plastic surface as well as their propagation potential in culture medium containing 10% FBS. In agreement with previous reports (2–4), all MSC samples from the three sources were tested for and should express standard MSC cell surface antigens (i.e., positive for CD29, CD44, CD90, CD105, and CD166 and negative for CD45, HLA-DR, T-lineage CD3, NK-lineage CD16, B-lineage CD19, myeloid CD33, and CD38 and stem/precursor CD34 and CD133). In addition, the cultured MSCs should be able to be induced into adipocytes and osteocytes, demonstrating their multi-lineage differentiation potential.

After cultured cells have been confirmed as MSCs, 2-DE, and combined MS and MS/MS analysis is carried out to compare the protein expression profiles among the three different sources. As shown as an example below, two identified proteins were also further evaluated using western blotting.

3.1. Human MSC Isolation and Culture (see Note 6)

3.1.1. Umbilical Cord Digestion

1. Wash UC Wharton's jelly samples thoroughly with PBS to remove contamination.
2. Cautiously pull vessels away from the umbilical cord and tie each dissected vessel together at the ends.
3. Digest vessels overnight at 37°C in a 50 mL sterilized tube containing 25 mL of 0.2 mg/mL collagenase type II in serum-free DMEM-LG medium.
4. Remove the vessels from the mixture and dilute the suspension with 20 mL PBS to lower the viscosity.
5. Centrifuge at 1,500 rpm for 10 min and remove the supernatant.
6. Wash the cell pellet with 20 mL PBS and centrifuge at 1,000 rpm for 10 min.
7. Remove the supernatant and repeat step 6.
8. Remove supernatant and re-suspend the cell pellet in culture medium at 10^4 cells/mL.
9. Seed cells at 1 mL per T75 flask in 10 mL total culture medium.
10. Place cultures in a humidified atmosphere of 5% CO_2 at 37°C.
11. Proceed with Subheading 3.1.4.

3.1.2. Placenta Digestion

1. For placenta digestion, cut the whole tissue into 1–2 mm pieces and incubate in a 50 mL conical tube with 30 mL 0.2 mg/mL collagenase in serum-free DMEM-LG medium for 90 min.
2. Pass digested tissue through a 100 μm cell strainer to get rid of cell aggregates.
3. Wash sieved cell suspension in PBS and centrifuge at 1,000 rpm for 10 min.
4. Remove supernatant and repeat step 3.
5. Remove supernatant and re-suspend cell pellet in culture medium at 10^4 cells/mL.
6. Seed cells at 1 mL per T75 flask in 10 mL culture medium.
7. Place cultures in a humidified atmosphere of 5% CO_2 at 37°C.
8. Proceed with Subheading 3.1.4.

3.1.3. Bone Marrow Processing

1. Obtain bone marrow aspirate from healthy donors (see Note 7).
2. Enrich mononuclear cells by using density gradient centrifugation in Ficoll-Hypaque with specific gravity of 1.077 g/mL.
3. Harvest the mononuclear fraction and wash with PBS and centrifuge at 1,000 rpm for 10 min.

4. Remove supernatant and repeat step 3.
5. Remove supernatant and re-suspended cells at a concentration of 1×10^6 cells/mL in culture medium.
6. Seed cells at 1 mL per T75 flask in 10 mL culture medium.
7. Place cultures in a humidified atmosphere of 5% CO_2 at 37°C.
8. Proceed with Subheading 3.1.4.
9. Process all samples within 24 h after collection (see Note 8).

3.1.4. MSC Culture Maintenance

1. On day 2 after seeding cultures, aspirate the spent medium and replace with fresh culture medium to remove non-adherent cells (see Note 9).
2. Thereafter, replace culture medium twice per week and subculture or cryopreserved cells upon reaching ~90% confluence.
3. MSC at the seventh to ninth passages were used to generate the example data provided. For this study, individual variation of donors was taken into account and therefore two donors were employed for each MSC collection with the corresponding cells labeled as BM1-MSC, BM2-MSC, UC1-MSC, UC2-MSC, P1-MSC, and P2-MSC, respectively. The same passage of MSC was used for all experiments, while each experiment was repeated thrice (see Note 10).

3.2. MSC Surface Markers Analysis

1. Wash cultures twice with PBS.
2. Enzymatically detach cells using 0.05% trypsin/EDTA for 5 min at 37°C.
3. Stop enzymatic reaction by adding an equal volume of culture medium and collect the entire solution in a 50-mL conical tube.
4. Wash flask with PBS to collect any residual cells.
5. Centrifuge the tube at 1,000 rpm for 5 min and remove supernatant.
6. Re-suspend cells in a minimal volume of PBS supplemented with 1% BSA.
7. Count cells using preferred method and adjust concentration to $<1 \times 10^7$/mL by adding additional PBS supplemented with 1% BSA.
8. Stain 100 μL aliquots of cell suspensions with 20 μL (except where noted; see Note 11) fluorescein isothiocyanate (FITC)- and phycoerythrin (PE)-conjugated mAbs for 30 min at room temperature in the dark.
9. To analyze sample on the flow cytometer, evaluate nonspecific background by parallel staining 100 μL aliquots of cell suspension with isotype-matched IgG1-FITC and IgG1-PE.

10. Process cells using the ImmunoPrep reagent kit in the Q-Prep Immunology Workstation.
11. Analyze cell samples with the Coulter Epic XL MCL flow cytometer (see Note 12).

3.3. Osteogenic and Adipogenic Differentiation

1. To induce adipogenic and osteogenic differentiations of MSCs, the corresponding induction medium (as described in the Subheading 2) should be added. DMEM-LG supplemented with 10% FBS and the corresponding concentration of chemical reconstitution vehicles should be used as negative differentiation controls.
2. Change induction medium twice a week for 2–3 weeks.
3. At the end of induction, was the cell layer once with PBS, then fix cells with 4% paraformaldehyde for 10 min and wash twice with PBS.
4. The cells were washed with 60% Tep (V/V, water) for 5 min (skip this step for Alizarin red S staining) and then submerged with staining solution (2% Alizarin red S for osteogenic differentiation or with 1/3 saturated oil red O for adipogenic differentiation) for 30 min.
5. The whole staining process was carefully monitored using a microscope.
6. The layer was then washed again with ddH_2O for at least three times.
7. Observe results and create photo-documentation using a phase contrast microscope with a camera.

3.4. Protein Extraction for 2-DE

1. Source cultured MSCs from UC, P, and BM at 90% confluence and wash twice with PBS (see Note 13).
2. Enzymatically detach cells using 0.05% trypsin/EDTA for 5 min at 37°C.
3. Stop enzymatic reaction by adding an equal volume of culture medium and collect the entire solution in a 50 mL conical tube.
4. Wash flask with PBS to collect any residual cells.
5. Centrifuge the tube at 1,000 rpm for 5 min and remove supernatant.
6. Re-suspend cell pellets in PBS.
7. Repeat steps 8 and 9 three times.
8. Re-suspend cell pellets in lysis buffer, 10^5 cells/10 mL, and place samples on ice for 45 min.
9. Centrifuge samples at 14,000 rpm for 15 min at 4°C.
10. Collect the supernatant and store at –80°C.
11. Determine the protein concentration of the samples using the Bradford protein assay.

3.5. Two-Dimensional Gel Electrophoresis

1. Carry out isoelectric focusing (IEF) using the IPGphor II apparatus. Use IPG strips according to manufacturer's instructions.
2. Dilute samples such that 150 μg of protein is diluted to 250 μL in rehydration solution.
3. Carry out the rehydration step at low voltage (i.e., 30 V) for 10 h.
4. Perform IEF following a step-wise voltage incremental manner: 500 and 1,000 V for 1 h each, and 8,000 V afterwards until 64 kV/h.
5. After IEF, place IPG gel strips in an equilibration buffer.
6. Transfer the strips to equilibration solution containing 10 mg/mL DTT and gently agitated for 15 min.
7. Transfer the strips to equilibration solution containing 25 mg/mL iodoacetamide and leave under gentle agitation for 15 min before submission onto the 12.5% uniform polyacrylamide gel slabs.
8. Perform second dimensional separation in Tris–glycine buffer at a constant current setting of 15 mA/gel for 30 min and 30 mA/gel thereafter.
9. Terminate SDS-PAGE when the bromophenol blue dye front migrates to the lower end of the gels.

3.6. Gel Staining and Image Analysis

1. Stain gels using a modified silver-staining method which is compatible with the MS analysis. Carry out the entire staining procedure at room temperature with gentle agitation.
2. Fix gels in 50% methanol containing 12% acetic acid and 0.0185% formaldehyde for 1 h or overnight.
3. Wash gels in Milli-Q H_2O thrice (15 min per wash) followed by equilibration in 50% ethanol twice (20 min each).
4. Perform sensitization by immersing the gels in 0.02% sodium thiosulfate for 10 min.
5. Wash gels in H_2O thrice (5 min per wash) and then incubated in 0.15% silver nitrate containing 0.0278% formaldehyde for 1 h.
6. Carry out color development in 3% sodium carbonate containing 0.037% formaldehyde until protein spots can be clearly seen against the background.
7. Stop staining by transferring the gels into 5% acetic acid for at least 10 min.
8. Digitize gel images using a scanner and the Quantity One software.

9. Further analyze all images using the software PDQuest according to the manufacturer's user guide.
10. Set the spot detection parameters as follows: Ruby speckles, 50; smoothing, medium 7×7; background removal, floating ball 61; removal vertical streaks, 93; removal horizontal streaks, 93; sensitivity, 16.51; size scale, 5; and min peak, 868.
11. After manual editing, match spots on different gels and analyze.
12. Perform Statistical analysis using the independent samples *t*-test.
13. An example result is shown in Fig. 1.

3.7. Tryptic In-Gel Digestion

1. Manually excise spots of interest from the gels and place in a tube.
2. Soak gel chips in an equal volume mixture of 30 mM potassium ferricyanide and 100 mM sodium thiosulfate for 5 min.
3. After washing twice in H_2O, equilibrate gel chips in 50 mM ammonium bicarbonate for 20 min, then in 25 mM ammonium bicarbonate for 15 min followed by 50% ACN for 5 min.
4. Soak chips in 100% ACN until the gel turns opaque.
5. Rehydrate vacuum-dried gel chips with 10 μg/mL trypsin in 25 mM ammonium bicarbonate (pH 8.0) and digest for 16–18 h at 30°C.
6. After digestion, extract the liquid containing the tryptic peptides from the gel chips and transfer into a new tube.

3.8. MALDI-TOF/TOF Mass Spectrometric Analysis

1. Carry out mass spectrometry using the ABI 4700 Proteomics Analyzer, a MALDI-TOF/TOF tandem mass spectrometer analyzer.
2. For acquisition of mass spectra, spot 0.5 μL samples onto a MALDI plate, followed by 0.5 μL matrix solution.
3. Pilot mass data acquisitions with the 4000 Series Explorer™ Software v3.0 using batched processing and automatic switching between MS and MS/MS modes.
4. Acquire MS survey scan over the mass range *m/z* 800–3,500 in the reflectron positive ion mode and accumulate from 2,000 laser shots with acceleration of 20 kV.
5. Internally calibrate MS spectra using porcine trypsin autolytic products (*m/z* 842.509, *m/z* 1045.564, *m/z* 1940.935, and *m/z* 2211.104), which should result in mass errors of less than 30 ppm.

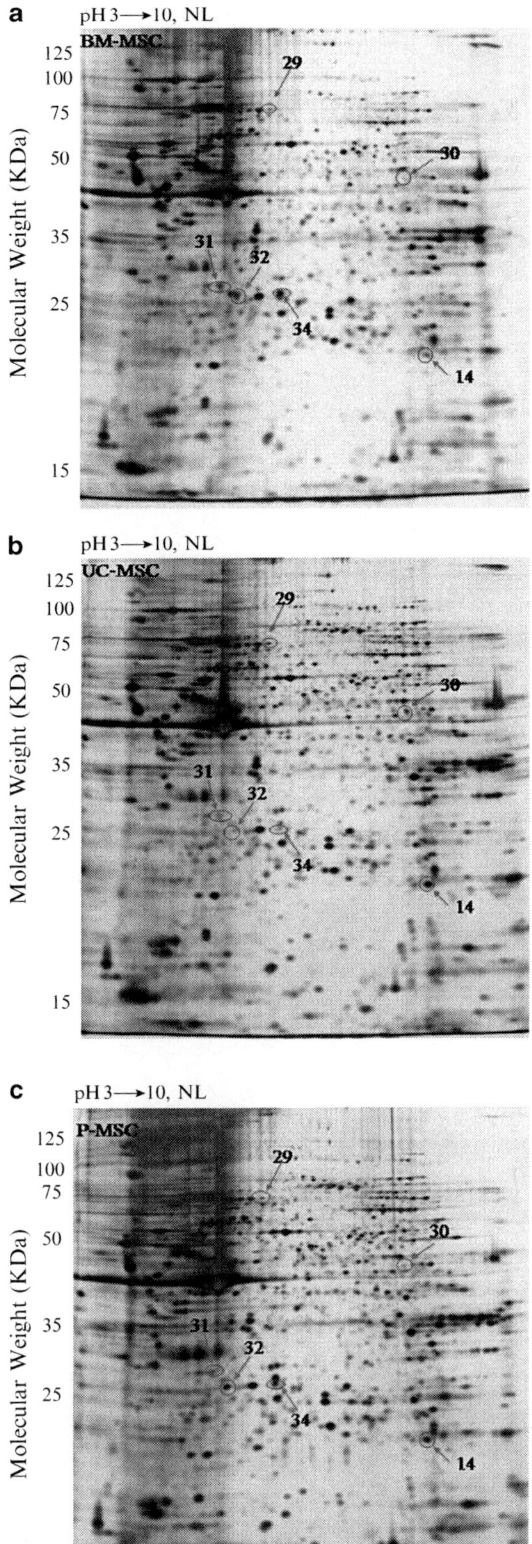

Fig. 1. 2-DE analysis of three types of MSCs. Whole cell extracts (150 μg) from these MSCs. Protein changes (indicated by *red circles*) were detected using gel analysis software after silver staining. (**a**) BM-MSC; (**b**) UC-MSC; (**c**) P-MSC. (Reproduced from Li et al. (5) Copyright Wiley-VCH Verlag GmbH & Co. KGaA, with permission).

6. Detect the MS peaks (MH^+) on minimum S/N ratio ≥20 and cluster area S/N threshold ≥25 without smoothing and raw spectrum filtering.
7. Exclude peptide precursor ions corresponding to contaminants including keratin and the trypsin autolytic products in a mass tolerance of ±0.2 Da.
8. Select the filtered precursor ions with a user-defined threshold (S/N ratio ≥50) for the MS/MS scan.
9. Perform fragmentation of precursor ions using MS/MS 1 kV positive mode with CID on and argon as the collision gas.
10. Accumulate MS/MS spectra from 3,000 laser shots using default calibration with Glu-Fibrinopeptide B from 4700 Calibration Mixture.
11. Detect the MS/MS peaks on minimum S/N ratio ≥3 and cluster area S/N threshold ≥15 with smoothing.

3.9. Database Search

1. Load the MS and MS/MS data into the GPS Explorer™ software v3.5 and search against NCBInr 5825255 sequences (released on Jan 10, 2008) by Mascot search engine v1.9.05 using combined MS (peptide-mass-fingerprint approach) with MS/MS (DeNovo sequencing approach) analysis for protein identification.
2. Use the following search parameters: monoisotopic peptide mass (MH^+); 800–3,500 Da; one missed cleavage per peptide; enzyme, trypsin; taxonomy, Homo sapiens; pI, 0–14; precursor-ion mass tolerance, 50 ppm; MS/MS fragment-ion mass tolerance, 0.1 Da; variable modifications, allow carbamidomethylation for cysteine, and oxidation for methionine.
3. Exclude known contaminant ions corresponding to keratin and trypsin from the peak lists before database searching. Report the top ten hits for each protein search.
4. In an MS (PMF) database search, determine a protein score for each protein matched from the MS peak lists.
5. In an MS/MS database search, determine an ion score for each peptide matched from the MS/MS peak list.
6. Determine the CI% calculation rates the confidence level of the score. If the score returned from the search is equal to the Mascot Significance Level for the search, the score is given a 95% CI (see Notes 14 and 15).
7. Identify accepted proteins to report when their protein score CI%, sequence coverage is greater than 99 and 15%, respectively, as well as at least four matched peaks. If a protein spot matched multiple members of a protein family, select the protein candidate with pI value nearest to observed one to report.

3.10. Western Blotting

1. Resolve protein samples (10 μg) by 12.5% SDS-PAGE and transfer to a PVDF membrane.
2. Block membranes with 5% non-fat dry milk for 1 h and allow to react with antibodies (i.e., anti-human CTSB, anti-human PAI-1, or anti-human actin) overnight.
3. Wash membranes thrice with blocking solution and incubate with corresponding secondary antibodies conjugated with horseradish peroxidase.
4. Visualize signals using the ECL kit.
5. After signal exposure, strip the membrane, and reprobe with an anti-actin antibody.

4. Notes

1. Any precipitation seen before use should be filtered off one or more times with Whatman filter paper to guarantee clarity.
2. When preparing 2× SDS/Sample buffer, adjust the pH to 6.8 before adding 2-mercaptoethanol.
3. Acrylamide/bis solution is a neurotoxic solution in liquid form. Extra caution must be taken to avoid direct contact or excessive exposure.
4. Dissolve Aps in H_2O, aliquot (200 μL) and immediately freeze at –20°C.
5. Due to the high viscosity of Tween 20, the end of 1-mL pipette tip can be cut to facilitate pipetting.
6. Human UC and P were donated by two pregnant women, each with formal consent.
7. For the described example results, aspirates were obtained from a 38-year-old male (BM1) and a 17-year-old female (BM2), each with formal consent.
8. Experiments were carried out according to the protocols approved by The Chinese University of Hong Kong – New Territories East Cluster Clinical Research Ethics Committee.
9. MSC has relatively strong adherent property to plastic surface; therefore, it is important to replace the primary culture medium every 2 days to avoid unnecessary attachment or cell growth.
10. MSC at the seventh to ninth passages were used to generate the example data provided. For this study, individual variation of donors was taken into account and therefore two donors were employed for each MSC collection with the corresponding cells labeled as BM1-MSC, BM2-MSC, UC1-MSC, UC2-MSC, P1-MSC, and P2-MSC, respectively. The same passage

of MSC was used for all experiments, while each experiment was repeated thrice. Cultured MSC was cryopreserved in a medium of DMEM-LG containing 20% FBS, 10% DMSO.

11. Stain 100 μL aliquots of cell suspensions with 20 μL (except where noted) fluorescein isothiocyanate (FITC)- and phycoerythrin (PE)-conjugated mAbs for 30 min at room temperature in the dark. The suggested markers for analysis include CD3-PE, CD16-FITC, CD19-FITC, CD33-FITC, CD34-PE, CD38-FITC, CD45-FITC, CD133-PE (10 μL used for staining), HLA-DR-FITC, CD29-Rhodamin (10 μL used for staining), CD44-FITC, CD90-PE, CD105-PE, and CD166-PE.
12. The stability of the settings of the Coulter Epic XL MCL flow cytometer was calibrated before each sample analysis using the control fluorescent microspheres of Flow Check and Flow Set (Beckman Coulter). The adjustment of settings for fluorescent color compensation was accomplished using the EXPO32 ADC analysis software (Applied Cytometry Systems, Sacramento, CA, USA) by discriminating the spectral overlap of leukocytes stained with CD45-FITC and CD45-PE. There are at least 7,000 events acquired using the System II software (Beckman Coulter). In this study, sequential gating of forward scatter versus side scatter followed by FITC and PE staining was used. Simultaneous interference from debris, cell aggregates and non-specific antibody bindings was eliminated. The quantity of cells, which exhibited low forward scatter and low side scatter and showed fluorescence of FITC and PE, was expressed in percentage of nucleated cells following the acquisition of at least 10^5 events. Background fluorescence was kept at <2% using isotype controls.
13. Cell number was adjusted for each passage, thus to make sure the cells used in each experimental group were around 90% of confluence at the same cultivation period.
14. The Mascot Significance Level is the values above which protein scores (MS data) and ion scores (MS/MS data) are considered to be statistically non-random at the 95% CI.
15. The closer the CI% value is to 100%, the more likely the protein is correctly identified.

Acknowledgements

We would like to thank the donors of bone marrow, umbilical cord, and placenta. This work was supported by the Li Ka Shing Institute of Health Sciences Grant and RGC project: CUHK 7422_03M.

References

1. Panepucci, R.A., Siufi, J.L., Silva, W.A. Jr., Proto-Siquiera, R. Neder, L., Orellana, M., et al. (2004) Comparison of gene expression of umbilical cord vein and bone marrow-derived mesenchymal stem cells *Stem Cells* **22,** 1263–1278.
2. Musina, R.A., Bekchanova, E.S., Belyavskii, A.V., Sukhikh, G.T. (2006) Differentiation potential of mesenchymal stem cells of different origin *Bull. Exp. Biol. Med.* **141,** 147–151.
3. Miao, Z., Jin, J., Chen, L., Zhu, J., Huang, W., Zhao, J., et al. (2006) Isolation of mesenchymal stem cells from human placenta: comparison with human bone marrow mesenchymal stem cells *Cell Biol. Int.* **30,** 681–687.
4. Lu, L.L., Liu, Y.J., Yang, S.G., Zhao, Q.J., Wang, X., Gong, W., et al. (2006) Isolation and characterization of human umbilical cord mesenchymal stem cells with hematopoiesis-supportive function and other potentials *Haematologica* **91,** 1017–1026.
5. Li, G., Zhang, X.A., Wang, H., Wang, X., Meng, C.L., Chan, C.Y., et al. (2009) Comparative proteomic analysis of mesenchymal stem cells derived from human bone marrow, umbilical cord, and placenta: implication in the migration *Proteomics* **9,** 20–30.
6. Ding, S.J., Li, Y., Tan, Y.X., Jiang, M.R.,Tian, B., Liu, Y.K., et al. (2004) From proteomic analysis to clinical significance: overexpression of cytokeratin 19 correlates with hepatocellular carcinoma metastasis *Mol. Cell. Proteomics* **3,** 73–81.
7. Lee, H.K., Lee, B.H., Park, S.A., Kim, C.W. (2006) The proteomic analysis of an adipocyte differentiated from human mesenchymal stem cells using two-dimensional gel electrophoresis *Proteomics* **6,** 1223–1229.
8. Halliwell, B., Gutteridge, J. M. C. (1986) *Free Radicals in Biology and Medicine* Oxford University Press, New York, 86–106.
9. Oberley, L. W., Buettner. G. R. (1979) Role of superoxide dismutase in cancer: a review *Cancer Res.* **39,** 1141–1149.
10. Safford, S.E., Oberley, T.D., Urano, M., St Clair, D.K. (1994) Suppression of fibrosarcoma metastasis by elevated expression of manganese superoxide dismutase *Cancer Res.* **54,** 4261–4265.
11. McClung, J. K., Jupe, E. R., Liu, X. T., Dell'Orco, R. T. (1995) Prohibitin: potential role in senescence, development, and tumor suppression *Exp. Gerontol.* **30**, 99–124.
12. Rajalingam, K., Wunder, C., Brinkmann, V., Churin, Y., Hekman, M., Sievers, C., et al. (2005) Prohibitin is required for Ras-induced Raf-MEK-ERK activation and epithelial cell migration *Nat. Cell Biol.* **7,** 837–843.
13. Aznavoorian, S., Murphy, A.N., Stetler-Stevenson, W.G., Liotta, L.A. (1993) Molecular aspects of tumor cell invasion and metastasis *Cancer* **71,** 1368–1383.
14. Del Rosso, M., Fibbi, G., Pucci, M., D'Alessio, S., Del, Rosso, A., Magnelli, L., et al. (2002) Multiple pathways of cell invasion are regulated by multiple families of serine proteases *Clin. Exp. Metastasis* **19,** 193–207.
15. Egeblad, M., Werb, Z. (2002) New functions for the matrix metalloproteinases in cancer progression. *Nat. Rev. Cancer* **2,** 161–174.
16. Turk, V., Kos, J., Turk, B. (2004) Cysteine cathepsins (proteases) – on the main stage of cancer? *Cancer Cell* **5,** 409–410.
17. Podgorski, I., Sloane, B.F. (2003) Cathepsin B and its role(s) in cancer progression *Biochem. Soc. Symp.* **70,** 263–276.
18. Yan, S., Sloane, B.F. (2003) Molecular regulation of human cathepsin B: implication in pathologies *Biol. Chem.* **384,** 845–854.
19. Vasiljeva, O., Papazoglou, A., Kruger, A., Brodoefel, H., Korovin, M., Deussing, J., et al. (2006) Tumor cell-derived and macrophage-derived cathepsin B promotes progression and lung metastasis of mammary cancer *Cancer Res.* **66,** 5242–5250.
20. Wickramasinghe, N.S., Nagaraj, N.S., Vigneswaran, N., Zacharias, W. (2005) Cathepsin B promotes both motility and invasiveness of oral carcinoma cells *Arch. Biochem. Biophys.* **436,** 187–195.
21. Oh-e, H., Tanaka, S., Kitadai, Y., Shimamoto, F., Yoshihara, M., Haruma, K. (2001) Cathepsin D expression as a possible predictor of lymph node metastasis in submucosal colorectal cancer *Eur. J. Cancer* **37,** 180–188.
22. Ikeguchi, M., Fukuda, K., Oka, S., Hisamitsu, K., Katano, K., Tsujitani, S., Kaibara, N. (2001) Micro-lymph node metastasis and its correlation with cathepsin D expression in early gastric cancer *J. Surg. Oncol.* **77,** 188–194.
23. Glondu, M., Liaudet-Coopman, E., Derocq, D., Platet, N., Rochefort, H., Garcia, M. (2002) Down-regulation of cathepsin-D expression by antisense gene transfer inhibits tumor growth and experimental lung metastasis of human breast cancer cells *Oncogene* **21,** 5127–5134.
24. Praus, M., Collen, D., Gerard, R.D. (2002) Both u-PA inhibition and vitronectin binding by plasminogen activator inhibitor 1 regulate HT1080 fibrosarcoma cell metastasis *Int. J. Cancer* **102,** 584–591.

Chapter 33

Metabolic Labeling and Click Chemistry Detection of Glycoprotein Markers of Mesenchymal Stem Cell Differentiation

Courtenay Hart, Lucas G. Chase, Mahbod Hajivandi, and Brian Agnew

Abstract

Human mesenchymal stem cells (hMSCs) are multipotent stem cells that can differentiate into a variety of cell types in vitro including osteoblasts, adipocytes, and chondrocytes. Here we apply a metabolic labeling approach to characterize changes in cellular glycoprotein expression during hMSC differentiation and to identify glycoprotein markers unique to differentiated cell types. The two-step labeling method involves the metabolic incorporation of unnatural azido-modified sugars into protein glycans and subsequent ligation with fluorescent azide-reactive detection probes utilizing the copper (I)-catalyzed cycloaddition reaction between azides and alkynes, or "click" chemistry. Metabolic labeling of cell surface *O*-linked or sialic acid-containing glycoproteins, or intracellular *O*-GlcNAc-modified proteins was accomplished by feeding cells the tetraacetylated azide-modified sugar precursors, GalNAz, ManNAz, or GlcNAz, respectively, for 48–72 h prior to harvesting the cells. The cells were then lysed, and protein extracts were reacted with a fluorescent alkyne detection probe. Labeled glycoproteins were analyzed by 1D and 2D gel electrophoresis and detected by fluorescence imaging. Our results demonstrate highly sensitive labeling of *O*-linked, sialic acid-containing, and *O*-GlcNAc modified proteins in all cell types without affecting cell growth or morphology. Selective labeling of sialic acid-containing glycoproteins by ManNAz was validated by loss of labeling following digestion with sialidase A. Significant changes in cellular glycoprotein profiles were seen upon differentiation into different cell types, and several putative glycoprotein markers were identified by MALDI peptide fingerprinting. One of these identified proteins, Galectin 1, is validated and shown for the first time to be posttranslationally modified by *O*-glycosylation, most likely by *O*-linked *N*-acetylglucosamine (*O*-GlcNAc).

Key words: MSC, Mesenchymal stem cells, Glycoproteins, Cell surface markers, Metabolic labeling, Click chemistry, Azide, Alkyne

Mohan C. Vemuri et al. (eds.), *Mesenchymal Stem Cell Assays and Applications*, Methods in Molecular Biology, vol. 698, DOI 10.1007/978-1-60761-999-4_33, © Springer Science+Business Media, LLC 2011

1. Introduction

Human mesenchymal stem cells (hMSCs) are multipotent cells that can differentiate into a variety of cell types including osteoblasts, adipocytes, and chondrocytes. Since first described by Friedenstein et al. (1), the ability to expand and differentiate hMSCs in vitro has been greatly defined (2, 3). Despite these advancements, many of the molecular mechanisms behind these processes are not completely understood. Although numerous efforts have been made to further expand our knowledge of the gene expression (4–7) and proteomic (8, 9) characteristics of MSCs and their differentiated progeny, future work is necessary to better understand the underlying biology behind these cells.

The term "click chemistry" was first coined by Barry Sharpless in 2002 as a set of chemical reactions meeting an established set of criteria for the building of chemical compound libraries (10). One of these reactions in particular, the copper(I) catalyzed 1,3-dipolar cycloaddition reaction, described by the Meldal and Sharpless groups independently in 2002, proved to be an exceptionally powerful bioconjugation tool for the labeling and detection of biological macromolecules and is now generally termed as the "click reaction" (Fig. 1) (11, 12). The click reaction is often described as bioorthogonal, in that the reaction components, azides, and alkynes are not normally found in biological systems and will react selectively even within a highly complex background of biological molecules. This results in low background and high sensitivity when used for the detection of biomolecules. Additionally, the small size of the reaction components enables their use in in vivo metabolic labeling schemes as incorporation of unnatural alkyne or azide-tagged substrate molecules are often permissible to the cellular enzymatic machinery.

Here we demonstrate the coupling of metabolic labeling with click chemistry detection in the labeling of glycoproteins from hMSCs and their differentiated progeny. The first labeling step employs the use of metabolic labeling technology first demonstrated

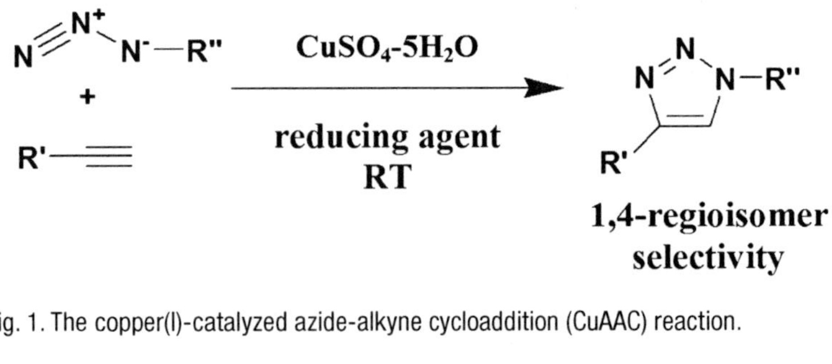

Fig. 1. The copper(I)-catalyzed azide-alkyne cycloaddition (CuAAC) reaction.

by Bertozzi and coworkers, whereby tetraacetylated forms of the unnatural azido-modified sugars, *N*-azidoacetylmannosamine (ManNAz), *N*-azidoacetylgalactosamine (GalNAz), and *N*-azidoacetylglucosamine (GlcNAz), are fed to cultured cells for 48–72 h (13, 14). After entry into the cells, the acetyl groups are cleaved by intracellular esterases and the sugars enter specific synthetic metabolic pathways. ManNAz is metabolized by cultured cells and converted to *N*-azidoacetylsialic acid on *N*- and *O*-linked cell surface and secreted glycoproteins. GalNAz replaces GalNAc residues in core cell surface and secreted *O*-linked glycoproteins and is also incorporated into *O*-GlcNAc-modified proteins inside the cell (unpublished data). Finally, GlcNAz is specifically incorporated into *O*-GlcNAc-modified proteins on the cell interior. After feeding, cells are lysed and the azide-labeled glycoproteins are reacted with an alkyne detection probe in the presence of copper(I). The labeled glycoproteins are then separated by 1D or 2D gel electrophoresis and detected by fluorescence gel scanning and image collection.

2. Materials

2.1. Human MSC Expansion and Differentiation

1. Cryopreserved human MSCs (Passage 4, 4-donor pool; 5×10^5 cells/vial).
2. Serum-containing medium (SCM): Composed of low glucose Dulbecco's Modified Eagle's Medium supplemented with 10% MSC-Qualified fetal bovine serum, 2 mM L-Glutamine, and 5 μg/ml Gentamicin (all from Invitrogen, Carlsbad, CA).
3. TrypLE™ Express stable trypsin replacement enzyme (Invitrogen).
4. StemPro® Adipogenesis, Osteogenesis, and Chondrogenesis Differentiation reagents (Invitrogen).
5. Dulbecco's Phosphate-Buffered Saline without calcium and magnesium (DPBS; Invitrogen).
6. T225 tissue culture flasks.
7. 24-well tissue culture plates.
8. 50 ml conical tubes.
9. 70% Isopropanol.

2.2. Characterization of Differentiated MSCs

1. Alizarin Red S solution: 2% Alizarin Red S (Sigma, St. Louis, MO) in distilled water; pH adjusted to 4.1–4.3 with ammonium hydroxide; 0.22 μm filtered.
2. Oil Red O: 0.5% Oil Red O (Sigma) in isopropanol; 0.22 μm filtered.

3. Alcian Blue solution: 1% Alcian Blue (Sigma) in 0.1 N HCl; 0.22 μm filtered.
4. Dulbecco's Phosphate-Buffered Saline with calcium and magnesium (DPBS; Invitrogen).
5. Hanks' Balanced Salt Solution with calcium and magnesium (HBSS; Invitrogen).
6. Distilled Water (Invitrogen).
7. 4% Formaldehyde: Prepare 4% formaldehyde by diluting stock 16% formaldehyde (Electron Microscopy Sciences, Hatfield, PA) in DPBS (Invitrogen). Store aliquots at –20°C and thaw fresh before use.
8. 0.1 N HCl solution: Dilute 10 N stock HCl in distilled water (Invitrogen).

2.3. Azido Sugar Labeling

1. Click-iT® GalNAz metabolic glycoprotein labeling reagent for *O*-linked glycoproteins (Invitrogen). Dissolve at 17.2 mg/ml in DMSO to make a 40 mM (1,000×) stock solution. Store in aliquots at –20°C.
2. Click-iT® ManNAz metabolic glycoprotein labeling reagent for sialic acid glycoproteins (Invitrogen). Dissolve at 17.2 mg/ml in DMSO to make a 40 mM stock solution. Store in aliquots at –20°C.
3. Click-iT® GlcNAz metabolic glycoprotein labeling reagent for *O*-GlcNAc-modified proteins (Invitrogen). Dissolve at 17.2 mg/ml in DMSO to make a 40 mM stock solution. Store in aliquots at –20°C.
4. 6-well and 12-well culture plates.
5. Cell Dissociation Buffer, enzyme free, PBS based (Invitrogen).
6. Protease inhibitor cocktail (Roche Applied Sciences, Indianapolis, IN). Store in aliquots at –20°C.
7. Cell lysis buffer: 100 mM Tris–HCl, pH 8.0 containing 1% (w/v) SDS. Store at room temperature. Immediately before use add protease inhibitor cocktail to 1% (w/v).
8. Vibra-Cell™ Sonicator (Sonics & Materials Inc., Newtown, CT) or Benzonase® endonuclease (Sigma).
9. EZQ® protein quantitation kit (Invitrogen) or equivalent.
10. Chloroform.
11. Methanol.

2.4. Endoglycosidase Digestion

1. Protein solubilization buffer: 100 mM Tris–HCl, pH 8.0 containing 1% (w/v) SDS. Store at room temperature.
2. Sialidase A and 5× Sialidase A reaction buffer: 250 mM Sodium phosphate pH 6.0 (Prozyme, Hayward, CA).

3. Water bath set at 37°C.
4. Chloroform.
5. Methanol.

2.5. Click-iT® Protein Labeling

1. Protein solubilization buffer: 100 mM Tris–HCl, pH 8.0 containing 1% (w/v) SDS. Store at room temperature.
2. Click-iT® Tetramethylrhodamine (TAMRA) Protein Analysis Detection Kit (Invitrogen).
3. Click-iT® Biotin Protein Analysis Detection Kit (Invitrogen).
4. Vortexer equipped with a microcentrifuge tube holder or an end-over-end rotator capable of holding microcentrifuge tubes.
5. Chloroform.
6. Methanol.

2.6. 1D SDS-Polyacrylamide Gel Electrophoresis (SDS-PAGE)

1. XCell SureLock® mini-cell electrophoresis system (Invitrogen).
2. NuPAGE® 4–12% Bis-Tris Mini gels, 1.0 mm (Invitrogen).
3. MOPS SDS Running Buffer (Invitrogen).
4. NuPAGE® LDS sample buffer (Invitrogen).
5. 2 M DTT.
6. Mark 12™ unstained molecular weight markers (Invitrogen) or equivalent for gels that will be stained with SYPRO® Ruby protein gel stain.
7. SeeBlue® Plus2 prestained molecular weight markers (Invitrogen) or equivalent for gels that will be western blotted.

2.7. 2D SDS-Polyacrylamide Gel Electrophoresis (SDS-PAGE)

1. SDS solubilization buffer: 100 mM Tris–HCl, pH 8.0 containing 1% (w/v) SDS. Store at room temperature.
2. 10% SDS.
3. 100 mM Tris base, pH 9.1.
4. *N,N*-Dimethylacrylamide (DMA) (Sigma).
5. 200 mM Tributylphosphine (TBP) (Sigma).
6. 2D sample buffer: 7 M urea, 2 M thiourea, 2% CHAPS, 2% ASB-14. Incubate at 30°C and shake periodically to dissolve the urea; 0.22 μm filtered. Store in aliquots at –20°C.
7. 2 M DTT.
8. pH 4–7 carrier ampholytes (Invitrogen).
9. Bromophenol blue.
10. ZOOM® IPGRunner™ Mini-Cell (Invitrogen).

11. ZOOM® pH 4–7 IEF strips (Invitrogen).
12. ZOOM® IPGRunner™ cassettes (Invitrogen).
13. ZOOM® equilibration trays (Invitrogen) or equivalent.
14. XCell SureLock® mini-cell electrophoresis system (Invitrogen).
15. NuPAGE® 4–12% Bis-Tris Mini gels, 1.0 mm, IPG well (Invitrogen).
16. MOPS SDS Running Buffer (Invitrogen).
17. NuPAGE® LDS sample buffer (Invitrogen).
18. Mark12™ unstained molecular weight markers (Invitrogen) or equivalent for gels that will be stained with SYPRO® Ruby protein gel stain.
19. SeeBlue® Plus2 prestained molecular weight markers (Invitrogen) or equivalent for gels that will be western blotted.
20. Orbital shaker or rocking platform.

2.8. IEF Fractionation

1. ZOOM® IEF Fractionator (Invitrogen).
2. 2D sample buffer: 7 M urea, 2 M thiourea, 2% CHAPS, 2% ASB-14, trace bromophenol blue. Incubate at 30°C and shake periodically to dissolve the urea; 0.22 μm filtered. Store in aliquots at −20°C.
3. 2 M DTT.
4. ZOOM® Focusing Buffer pH 3–7 (Invitrogen).
5. ZOOM® Focusing Buffer pH 7–12 (Invitrogen).
6. ZOOM® Disks (pH 3.0, 4.6, 5.4, 6.2, 7.0, and 10) (Invitrogen).
7. Anode buffer: 7 M urea, 2 M thiourea, 8.25× Novex® IEF Anode Buffer (50× stock), Invitrogen). Make fresh before use. Incubate at 30°C and shake periodically to dissolve the urea.
8. Cathode buffer: 7 M urea, 2 M thiourea, 1× Novex® IEF Cathode buffer (10× stock, Invitrogen). Make fresh before use. Incubate at 30°C and shake periodically to dissolve the urea.
9. Orbital shaker or rocking platform.

2.9. Imaging and Staining for Total Protein

1. SYPRO® Ruby protein gel stain (Invitrogen).
2. Gel fixation solution: 50% (v/v) methanol and 7% (v/v) acetic acid in distilled water. Store at room temperature.
3. Gel destain solution: 10% (v/v) methanol and 7% (v/v) acetic acid in distilled water. Store at room temperature.
4. Orbital shaker or rocking platform.
5. Microwave (optional).

6. Fluorescence imaging system with 532 nm illumination.
7. Fluorescence imaging system with UV transillumination or 473–488 nm illumination.
8. Image analysis software.

2.10. Multiplexed Western Blotting

1. iBlot® Gel Transfer Device (Invitrogen).
2. iBlot® Transfer Stack, Nitrocellulose (Invitrogen).
3. Transfer Buffer: 2× NuPAGE® Transfer Buffer (20× stock, Invitrogen), 10% methanol.
4. Wash buffer (10×): 0.5 M Tris–HCl pH 7.4, 1.5 M NaCl, 0.5% Tween 20.
5. Blocking buffer: 0.5× SeaBlock™/EIA/WB (East Coast Bio, North Berwick, ME), 25 mM Tris–HCl, pH 7.4, 75 mM NaCl, 0.05% Tween 20, 2 mM sodium azide.
6. Qdot® 625 streptavidin conjugate (Invitrogen).
7. Rabbit anti-galectin-1 polyclonal antibody (Abcam).
8. Rabbit anti-CTD105 for detecting hMSCs (Abcam).
9. Rabbit anti-fatty acid binding protein 4 for detecting adipocytes (Abcam).
10. Mouse anti-bone sialoprotein, for detecting osteoblasts (Calbiochem).
11. Goat anti-rabbit or goat anti-mouse horseradish peroxidase (GAR-HRP) secondary antibody (Invitrogen).
12. Novex® ECL (Invitrogen) or equivalent.
13. Orbital shaker or rocking platform.
14. Fluorescence imaging system with UV epi or transillumination.
15. Chemiluminescent imager or X-ray film.

2.11. Mass Spectrometry

1. 25 mM ammonium bicarbonate.
2. In gel digestion trypsin solution: 10 ng/μl of trypsin (Promega, Madison, WI) prepared in 25 mM ammonium bicarbonate, pH 8.2.
3. Destaining buffer: 50% acetonitrile (ACN) in 25 mM ammonium bicarbonate, pH 8.2.
4. Dehydration buffer: 100% ACN.
5. Reconstitution sample buffer: 50% ACN in 0.1% triflouro acetic acid.
6. Matrix solution: super saturated α-cyano-4-hydroxycinnamic acid (CHCA) dissolved in 50% ACN in 0.1% triflouro acetic acid.
7. Speed vac (Savant).

8. 4700 Protein Analyzer with MALDI/TOF analysis (Applied Biosystems, Foster City, CA).
9. 4700 protein analyzer target plate (Applied Biosystems).
10. 4700 peptide standard (Applied Biosystems).
11. GPS explorer (Version 2.0, Applied Biosystems).
12. Mascot search algorithm (Version 2.0).

3. Methods

3.1. Human MSC Expansion and Differentiation

1. Quickly thaw cryopreserved human MSCs in a 37°C water batch (<2 min) until a small amount of ice remains. Spray vial with 70% isopropanol and transfer to a laminar flow hood.
2. Gently pipette cells up and down to mix and transfer into a 50 ml conical tube.
3. Add 9 ml prewarmed (37°C) SCM dropwise (~3 drops every 5 s) to the tube.
4. Add 30 ml prewarmed SCM to the tube and place entire contents in a T225 culture flask.
5. After 24 h make a complete medium change and thereafter every 2–3 days (passage 1).
6. Upon reaching 70–90% confluence, wash cells 1× with DPBS and detach cells from the flask by adding 5 ml prewarmed (37°C) TrypLE Express. Incubate flask in the incubator for 5 min.
7. After incubation, gently tap flask to facilitate detachment and add 5 ml prewarmed SCM. Collect cell solution in a conical tube. Wash flask with prewarmed SCM and collect.
8. Centrifuge tube at $290 \times g$ for 5 min. Reconstitute cell pellet in prewarmed SCM and count cells using preferred method.
9. Re-seed harvested MSCs at 3.0×10^3 cells/cm^2 in an appropriate number of T225 culture flasks in 40 ml SCM per flask (passage 2).
10. For osteoblast differentiation, after 24 h (postseed) replace passage 2 SCM with StemPro® Osteogenesis Differentiation Medium.
11. For adipocyte differentiation, upon reaching absolute (>100%) cell confluence, replace SCM during passage 2 with StemPro® Adipogenesis Differentiation Medium.
12. For chondrocyte differentiation, harvest passage 1 MSCs as described above and resuspend harvested cells at 8×10^6 cells/ml in SCM. Place 10 μl drops of cell solution into the middle of each well of a 24-well plate. Incubate plates in a humidified

chamber inside an incubator for 2 h. After incubation, add StemPro® Chondrogenesis Differentiation Medium to each well.

13. Make complete osteogenesis, adipogenesis, and chondrogenesis differentiation medium changes every 3–4 days until desired level of differentiation is achieved.
14. To monitor cell differentiation, culture cells similarly in 24-well plates and analyze after desired time (14–21 days) with Alizarin Red S for osteoblasts, Oil Red O for adipocytes, or Alcian Blue for chondrocytes (see Subheading 3.2 and Fig. 2).

3.2. Characterization of Differentiated MSCs

3.2.1. Oil Red O Staining of Differentiated Adipocytes

1. After desired period of differentiation, remove medium and add 0.5% Oil Red O solution for 30 min at room temperature.
2. After incubation, remove staining solution and rinse cells twice with distilled water and visualize under a microscope for the presence of stained lipid vesicles.

3.2.2. Alizarin Red S Staining of Differentiated Osteoblasts

1. After desired period of differentiation, remove medium and rinse cultures twice with HBSS (with calcium and magnesium).
2. Fix cells with 4% formaldehyde solution for 30 min.
3. After fixation, rinse cells twice with distilled water and incubate with 2% Alizarin Red S solution for 2 min.

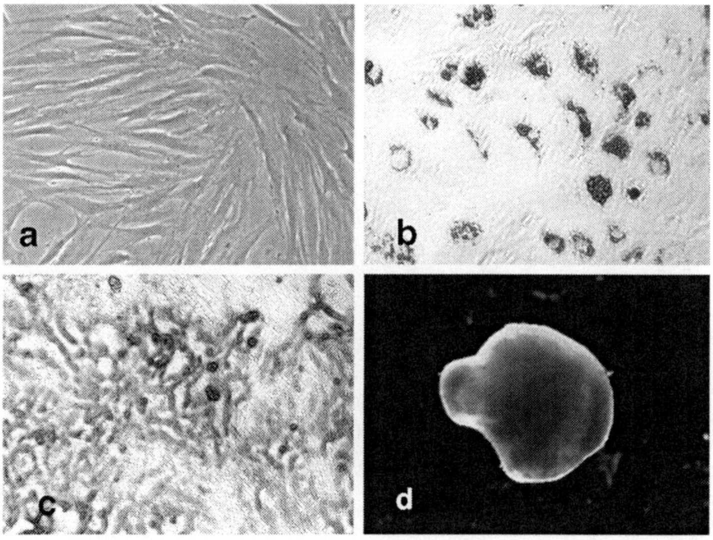

Fig. 2. Differentiation of human mesenchymal stem cells into adipocytes, osteoblasts, and chondrocytes. Passage 2 human MSCs (**a**) differentiated for 18 days and characterized using (**b**) Oil Red O (adipocyte), (**c**) Alizarin Red S (osteoblast), and (**d**) Alcian Blue (chondrocyte).

4. After incubation, remove staining solution and rinse cultures three times with distilled water and visualize under a microscope for the presence of red stained Alizarin Red S-calcium complex.

3.2.3. Alcian Blue Staining of Differentiated Chondrocytes

1. After desired period of differentiation, remove medium and rinse cultures twice with HBSS (with calcium and magnesium).
2. Fix cultures with 4% formaldehyde solution for 30 min.
3. After fixation, rinse cultures twice with DPBS.
4. Incubate cultures with 1% Alcian Blue solution for 30 min at room temperature.
5. After incubation, remove staining solution and rinse cultures three times with 0.1 N HCl solution.
6. Remove acidic solution and replace with an equal volume of distilled water.
7. Visualize cultures under a microscope for the presence of blue stained cells within the chondrogenic pellet (see Note 1).

3.3. Azido Sugar Labeling

1. Label passage 2, Day 5 expansion cultures of hMSC cultures or passage 2, days 14–21 differentiated adipocyte, chondrocyte, and osteoblast cultures with 40 μM GalNAz, ManNAz, or GlcNAz tetraacetylated azido sugar by adding the appropriate volume of 1,000× stock solutions to the respective differentiation media (see Notes 2 and 3; Fig. 3). Mix gently and continue to incubate at 37°C, 5% O_2.
2. After 72 h post-treatment, remove medium by aspiration and wash cells three times with DPBS. At this point, detach the cells using nonenzymatic dissociation buffer or a cell scraper.
3. Collect the cell suspension in a 50 ml conical tube and centrifuge at 290 × *g* for 5 min.
4. Discard the supernatant and resuspend the cell pellet in a minimum volume of DPBS (see Note 4).
5. Transfer the cell solution to a 15 ml conical tube and spin at 290 × *g* for 5 min. Carefully remove the supernatant and discard. Flash freeze the cell pellet in liquid nitrogen and store at −80°C.
6. Prepare the lysis buffer by adding protease inhibitors to 1% (w/v) to 100 mM Tris (pH 8) containing 1% SDS (see Note 5).
7. Prepare 20 μl of lysis buffer per cm^2 culture flask surface area (e.g., cells isolated from a T225 culture flask (80–100% confluence) would be lysed in 4.5 ml of lysis buffer).
8. Briefly vortex the tube and incubate at room temperature for 15–30 min.
9. Sonicate the sample with a probe sonicator set at 50% power using 4 × 10 s bursts to cleave the DNA (see Notes 6 and 7).

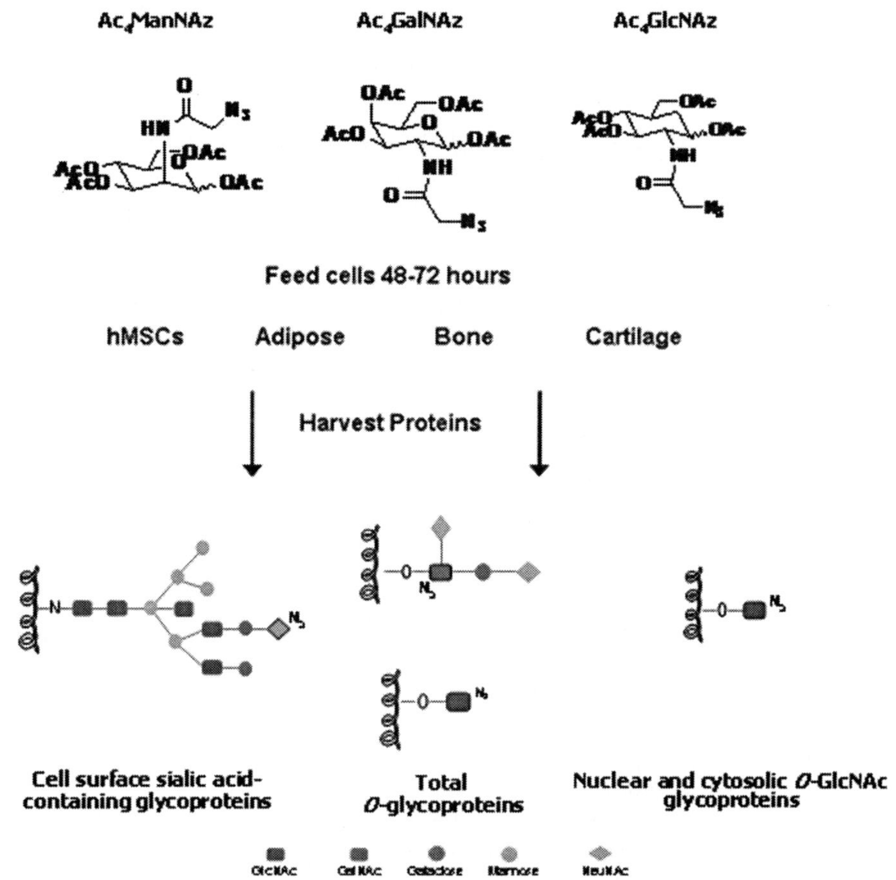

Fig. 3. Scheme for metabolic labeling of hMSCs and differentiated cell types with tetraacetylated azido sugars.

10. Vortex the lysate for 5 min.
11. Centrifuge the cell lysate at 20,000 × *g* at 4°C for 10 min to spin down insoluble debris (see Note 8).
12. Transfer the supernatant to a clean microcentrifuge tube and determine the protein concentration using the EZQ® protein quantitation Kit or alternate method (see Note 9).
13. Adjust lysate concentration to 2 mg/ml with lysis buffer (see Note 10).
14. To precipitate the proteins, transfer 200 μl aliquots of lysate into 1.5 ml microcentrifuge tubes.
15. Add 600 μl of methanol and vortex briefly. Add 150 μl of chloroform and vortex briefly. Add 400 μl of water and vortex briefly.
16. Centrifuge for 5 min at 13,000–18,000 × *g*.
17. Carefully remove and discard as much of the upper aqueous phase as possible while leaving the interface layer containing the protein precipitate intact.

18. Add 450 μl of methanol and vortex briefly.
19. Centrifuge for 5 min at 13,000–18,000 × *g* to pellet the protein.
20. Carefully discard the supernatant and cover the tube with a lint-free tissue and keep the tube cap open. Allow the pellet to air-dry for 15 min to overnight. Store dry pellets at −20°C (see Note 11).

3.4. Endoglycosidase Digestion

1. If the proteins were precipitated, resolubilize to 2 mg/ml in 100 mM Tris–HCl, pH 8.0 containing 1% SDS. Vortex for 10–15 min. If a significant pellet is visible, vortex for a longer time or repeat the probe sonication to solubilize more proteins.
2. Centrifuge for 5 min at 13,000–18,000 × *g* to remove any insoluble protein.
3. To 100 μl of 2 mg/ml protein lysate add 30 μl of 5× Sialidase reaction buffer (250 mM sodium phosphate pH 6.0) and 14 μl of water.
4. Add 6 μl Sialidase A (5 U/ml) to obtain a final concentration of 0.2 U/ml.
5. Cap tightly, vortex 5 s, and incubate in a water bath at 37°C for 2 h.
6. Precipitate the proteins using the chloroform/methanol precipitation method. Add 300 μl of methanol and vortex briefly. Add 75 μl of chloroform and vortex briefly. Add 200 μl of water and vortex briefly.
7. Centrifuge for 5 min at 13,000–18,000 × *g*.
8. Carefully remove and discard as much of the upper aqueous phase as possible while leaving the interface layer containing the protein precipitate intact.
9. Add 225 μl of methanol and vortex briefly.
10. Centrifuge for 5 min at 13,000–18,000 × *g* to pellet the protein.
11. Carefully discard the supernatant. Cover the tube with a lint-free tissue and keep the tube cap open. Allow the pellet to air-dry for 15 min to overnight. Store dry pellets at −20°C.

3.5. Click-iT® Protein Labeling

1. If the proteins were precipitated, resolubilize to 2 mg/ml in 100 mM Tris–HCl, pH 8.0 containing 1% SDS. Vortex for 10–15 min. If a significant pellet is visible, vortex for a longer time or repeat the probe sonication to solubilize more protein.
2. Centrifuge for 5 min at 13,000–18,000 × *g* to remove any insoluble protein.

3. Prepare Click-iT® detection reagent solutions as recommended by the manufacturer.
4. Click-iT® Reaction volumes can be scaled up or down as required. Generally, 50 μg is sufficient for 1D SDS-PAGE, 200 μg is sufficient for 2D SDS-PAGE, and 2 mg is sufficient for IEF fractionation followed by 1D or narrow pH range 2D SDS-PAGE.
5. To label 100 μg of lysate: To 50 μl of 2 mg/ml protein lysate add 100 μl of 2× Click-iT® reaction Buffer containing TAMRA or biotin alkyne detection reagent and 10 μl water. Cap the tube and vortex for 5 s.
6. Add 10 μl of $CuSO_4$, cap the tube, and vortex for 5 s.
7. Add 10 μl of Click-iT® Reaction Buffer Additive 1, cap the tube, and vortex for 5 s.
8. Wait for 2–5 min, then add 20 μl of Click-iT® Reaction Buffer Additive 2, cap the tube, and vortex for 5 s (see Note 12). The solution will turn bright orange.
9. Vortex continuously or rotate end-over-end for 20–60 min. If using TAMRA alkyne detection reagent, cover the tube with foil to minimize light exposure (see Note 13).
10. Precipitate the proteins using the chloroform/methanol precipitation method. Add 600 μl of methanol and vortex briefly.
11. Add 150 μl of chloroform and vortex briefly.
12. Add 400 μl of water and vortex briefly.
13. Centrifuge for 5 min at 13,000–18,000 × *g*.
14. Carefully remove and discard as much of the upper aqueous phase as possible while leaving the interface layer containing the protein precipitate intact. The upper aqueous phase will be bright orange. The lower phase will be pink for the TAMRA reagent and colorless for the biotin reagent.
15. Add 450 μl of methanol and vortex briefly.
16. Centrifuge for 5 min at 13,000–18,000 × *g* to pellet the protein. Carefully discard the supernatant.
17. Add an additional 450 μl of methanol and vortex briefly.
18. Centrifuge for 5 min at 13,000–18,000 × *g* and discard the supernatant. This second methanol wash step serves to remove residual reaction components.
19. Cover the tube with a lint-free tissue and keep the tube cap open. Allow the pellet to air-dry for 15 min to overnight. Store dry pellets at −20°C.

3.6. 1D SDS-PAGE

1. Assemble the XCell SureLock® mini-cell unit with one or two NuPAGE® 4–12% Bis-Tris gels and fill the buffer chamber with 700 ml of 1× MOPS SDS running buffer.
2. Resolubilize the precipitated Click-iT® labeled protein extract to approximately 2 mg/ml in 1× NuPAGE® LDS sample buffer containing 50 mM DTT. Vortex for 10–15 min.
3. At this point, the EZQ® protein quantitation should be repeated to obtain accurate protein concentrations (see Note 9).
4. Heat samples for 10 min in a heat block set at 70°C.
5. Briefly spin the sample before loading to remove any insoluble material.
6. Suggested loading amounts are 10–20 µg lysate per sample. For gels that will be stained for total protein include at least one lane of Mark12™ molecular weight markers. For gels that will be western blotted include at least one lane of SeeBlue® Plus2 prestained molecular weight markers. A 1/4× dilution of Mark12™ markers or SeeBlue® Plus2 markers provides an optimal signal when stained with SYPRO® Ruby total protein stain or when imaging Qdot® 625 for western blot detection, respectively.
7. Perform electrophoresis for 10 min at 100 V to allow the proteins to enter the gel and then increase the voltage to 150–200 V until the blue LDS sample buffer dye has entered the foot of the gel. Cover the electrophoresis box with a loose piece of aluminum foil during electrophoresis to reduce exposure of TAMRA Click-iT® dye to light.
8. For biotin Click-iT® labeled samples proceed to Subheading 3.10. Multiplexed Western Blotting. For TAMRA Click-iT® labeled samples proceed to Subheading 3.9. Imaging and Staining for Total Protein (see Fig. 4).

3.7. 2D SDS-PAGE (see Note 14)

1. Resolubilize the precipitated Click-iT® labeled protein extract to approximately 2 mg/ml in 100 mM Tris pH 8.0 containing 1% (w/v) SDS. Vortex for 10–15 min.
2. Reduce and alkylate proteins: For 100 µl of 2 mg/ml lysate add 49 µl water, 5 µl of 10% SDS, and 40 µl of 100 mM Tris base, pH 9.1. Cap the tube and vortex for 5 s.
3. Add 5 µl of 200 mM TBP. Cap the tube and vortex for 5 s.
4. Add 1 µl of dimethyl acrylamide. Cap the tube and vortex for 5 s.
5. Incubate in a heat block at 65°C for 10 min.
6. Vortex or rotate end-over-end at room temperature for 20 min.
7. Precipitate the proteins using the chloroform/methanol precipitation method. Add 600 µl of methanol and vortex briefly.

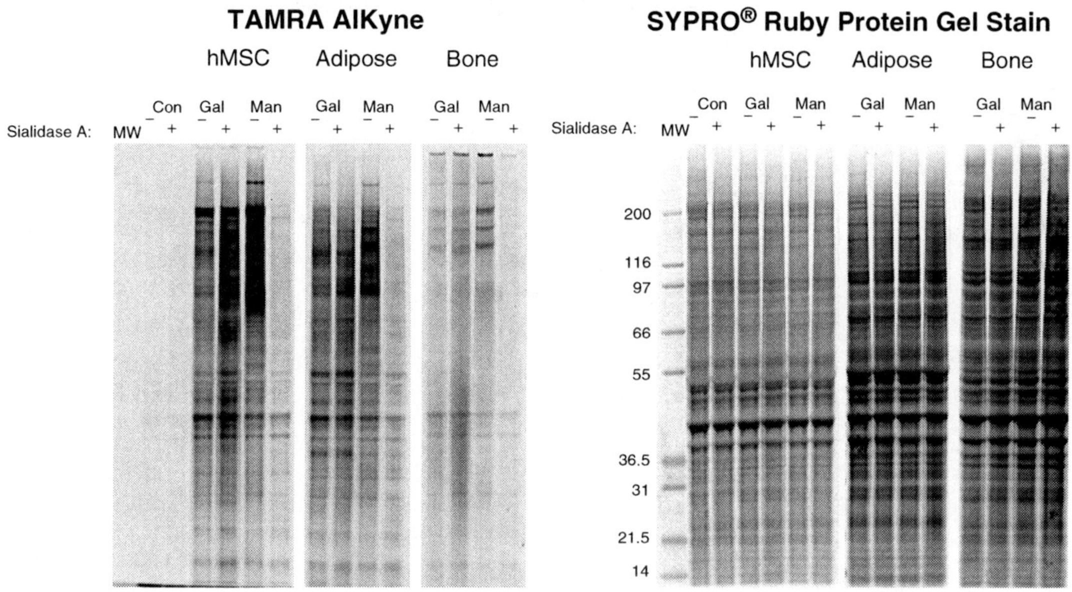

Fig. 4. Specificity of metabolic azido sugar labeling: hMSC, adipose, and bone cells were labeled with GalNAz (Gal) and ManNAz (Man) and treated with (+) or without (–) sialidase. The proteins were then reacted with TAMRA-alkyne dye and separated by 1D gel electrophoresis. The *left panel* shows TAMRA labeling and the *right panel* shows poststaining with SYPRO® Ruby total protein gel stain. The results show that the signals for ManNAz-labeled samples (sialic acid labeling) treated with sialidase are significantly decreased while the GalNAz-labeled samples are unaffected.

Add 150 μl of chloroform and vortex briefly. Add 400 μl of water and vortex briefly.

8. Centrifuge for 5 min at 13,000–18,000 × *g*. Carefully remove and discard as much of the upper aqueous phase as possible while leaving the interface layer containing the protein precipitate intact.
9. Add 450 μl of methanol and vortex briefly.
10. Centrifuge for 5 min at 13,000–18,000 × *g* to pellet the protein. Carefully discard the supernatant.
11. Cover the tube with a lint-free tissue and keep the tube cap open. Allow the pellet to air-dry for 15 min to overnight. Store dry pellets at –20°C.
12. Resolubilize the reduced/alkylated and precipitated Click-iT® labeled protein extract to approximately 2 mg/ml in 2D sample buffer. Vortex for 10–15 min.
13. At this point the EZQ protein quantitation should be repeated to obtain accurate protein concentrations (see Note 9).
14. For two 2D mini-gels, dilute 106 μg lysate in additional 2D sample buffer to bring the volume to 349 μl. Add 9 μl of 2 M DTT, 2 μl of pH 4–7 ampholytes, and a trace amount of bromophenol blue dye. Cap the tube and vortex 5 s.

15. Centrifuge for 5 min at 13,000–18,000 × g to remove any insoluble protein.
16. Load 170 μl (50 μg lysate) each into two wells of a ZOOM® IPGRunner™ cassette and add a ZOOM® pH 4–7 IEF strip being careful not to introduce any air bubbles. Cover the cassette openings with the provided tape and rehydrate the strips for 1–2 h.
17. Remove the tape and plastic well forms from the IPGRunner™ cassette. Add a drop of 2D sample buffer to each of the IEF strip ends to help reduce the formation of air bubbles and then cover the IEF strip ends with the provided piece of filter paper, which has been wet in deionized water and blotted to remove excess water.
18. Assemble the ZOOM® IPGRunner™ mini-cell apparatus according to the manufacturer's directions and focus the samples for 20 min at 200 V, 25 min at 450 V, 20 min at 700 V, and 55 min at 2,000 V.
19. Disassemble the ZOOM® IPGRunner™ apparatus and remove the focused IPG strips. Place the strips in an equilibration tray and incubate 10 min in 2 ml of 1× NuPAGE® LDS buffer containing 50 mM DTT on an orbital shaker at 40 rpm.
20. Prepare a NuPAGE® 4–12% Bis-Tris gel with an IPG well: remove the comb and tape, stand the gel up vertically in a rack on the benchtop, and fill the well with 1× MOPS running buffer.
21. Using tweezers, insert the equilibrated IPG strip into the well, cutting off the plastic tabs on the strip. Avoid trapping air bubbles under the strip. Push the strip down into the bottom of the well so that the entire strip makes contact with the top of the gel.
22. Assemble the XCell SureLock® mini-cell unit with one or two loaded NuPAGE® gels and fill the buffer chamber with 700 ml of 1× MOPS SDS running buffer. For gels that will be stained for total protein load Mark12™ molecular weight markers into the provided marker lane. For gels that will be western blotted load SeeBlue® Plus2 prestained molecular weight markers. A 1/4× dilution of Mark12™ markers or SeeBlue® Plus2 markers provides an optimal signal when stained with SYPRO® Ruby total protein stain or when imaging Qdot® 625 for western blot detection, respectively.
23. Perform electrophoresis for 10 min at 100 V to allow the proteins to enter the gel and then increase the voltage to 150–200 V until the blue LDS sample buffer dye has entered the foot of the gel. Cover the electrophoresis box with a loose piece of aluminum foil during electrophoresis to reduce exposure of TAMRA Click-iT® dye to light.

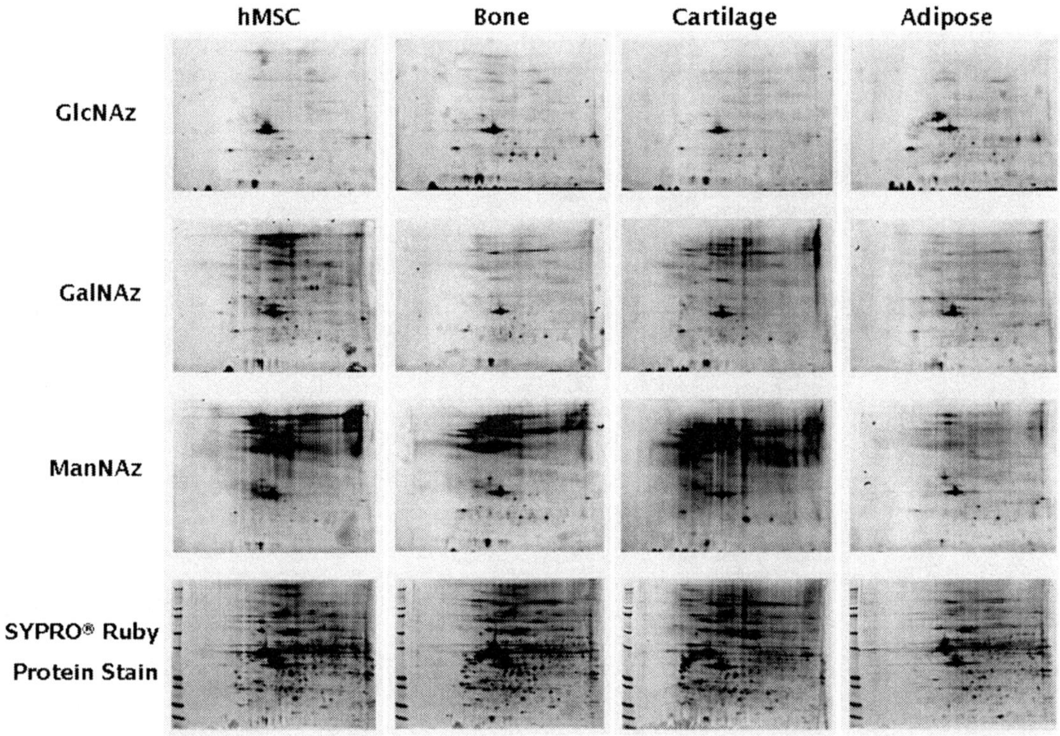

Fig. 5. 2D gel analysis of hMSC, bone, cartilage, and adipose glycoproteins azido sugar-labeled proteins reacted with TAMRA alkyne dye were separated by 2D gel electrophoresis and imaged as described in the text. Adipose tissue shows markedly less sialic acid-labeled proteins while *O*-linked glycoproteins are decreased in bone and adipose tissue.

24. Following electrophoresis, immediately image the TAMRA Click-iT® signal and poststain the gel with SYPRO® Ruby gel stain as detailed in Subheading 3.9 (see Figs. 5 and 6).

3.8. IEF Fractionation (see Note 14)

1. Resolubilize approximately 2 mg of precipitated Click-iT® labeled protein extract in 3.33 μl of 2D sample buffer. Vortex for 10–15 min.
2. Add 100 μl of 2 M DTT, 35 μl of ZOOM® Focusing Buffer pH 3–7 and 35 μl of ZOOM® Focusing Buffer pH 7–12. Cap the tube and vortex 5 s.
3. Centrifuge for 5 min at 3,000 × *g* to remove any insoluble protein.
4. Assemble the ZOOM® IEF Fractionator device according to the manufacturer's instructions, inserting pH 3.0, 4.6, 5.4, 6.2, 7.0, and 10.0 ZOOM® Disks into the core chamber.
5. Fill the anode buffer chamber with 17.5 ml of freshly prepared anode buffer.
6. Fill the cathode buffer chamber with 17.5 ml of freshly prepared cathode buffer.

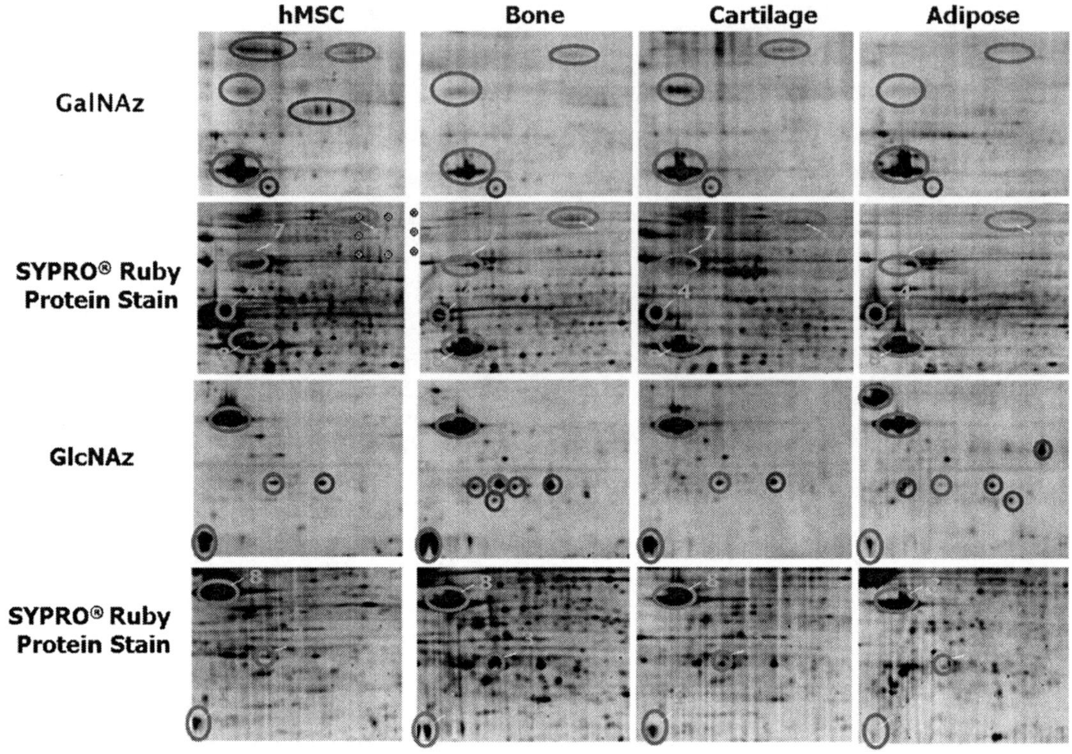

Fig. 6. Differentially labeled protein identification: Expanded regions of TAMRA 2D gels shown in Fig. 5 paired with the same gel after poststaining with SYPRO® Ruby protein gel stain. Circled spots represent differentially modified or expressed proteins of interest, several of which were identified by mass spectrometry (see Table 1).

7. Fill the five central sample chambers with 650 μl of prepared sample and insert the chamber plugs.
8. Focus at 100 V for 20 min, 200 V for 80 min, and 600 V for 80 min. The bromophenol blue dye will migrate toward the anode reservoir during the run turning the anode buffer yellow.
9. Remove the chamber plugs and transfer the focused samples into microcentrifuge tubes. To increase the focused sample recovery, add an additional 200 μl of 2D sample buffer to each chamber and rock on an orbital shaker for 10 min.
10. Add this sample to the respective tubes of focused samples. The final volume of each sample will now be approximately 850 μl, but will vary slightly depending on the protein concentration of each individual pH range.
11. Precipitate the proteins using the chloroform/methanol precipitation method. The sample can be precipitated all together, or divided into aliquots of approximately 200 μl. For each

200 μl aliquot add 600 μl of methanol and vortex briefly. Add 150 μl of chloroform and vortex briefly. Add 400 μl of water and vortex briefly.

12. Centrifuge for 5 min at 13,000–18,000 × *g*. Carefully remove and discard as much of the upper aqueous phase as possible while leaving the interface layer containing the protein precipitate intact.
13. Add 450 μl of methanol and vortex briefly.
14. Centrifuge for 5 min at 13,000–18,000 × *g* to pellet the protein. Carefully discard the supernatant. Cover the tubes with a lint-free tissue and keep the tube caps open.
15. Allow the pellets to air-dry for 15 min to overnight. Store dry pellets at −20°C (see Fig. 7).

3.9. Imaging and Staining for Total Protein

1. Immediately image the TAMRA signal. Remove the top cassette plate. Use water to rinse off some of the SDS on the side of the gel that will be in contact with the imager tray. Add a

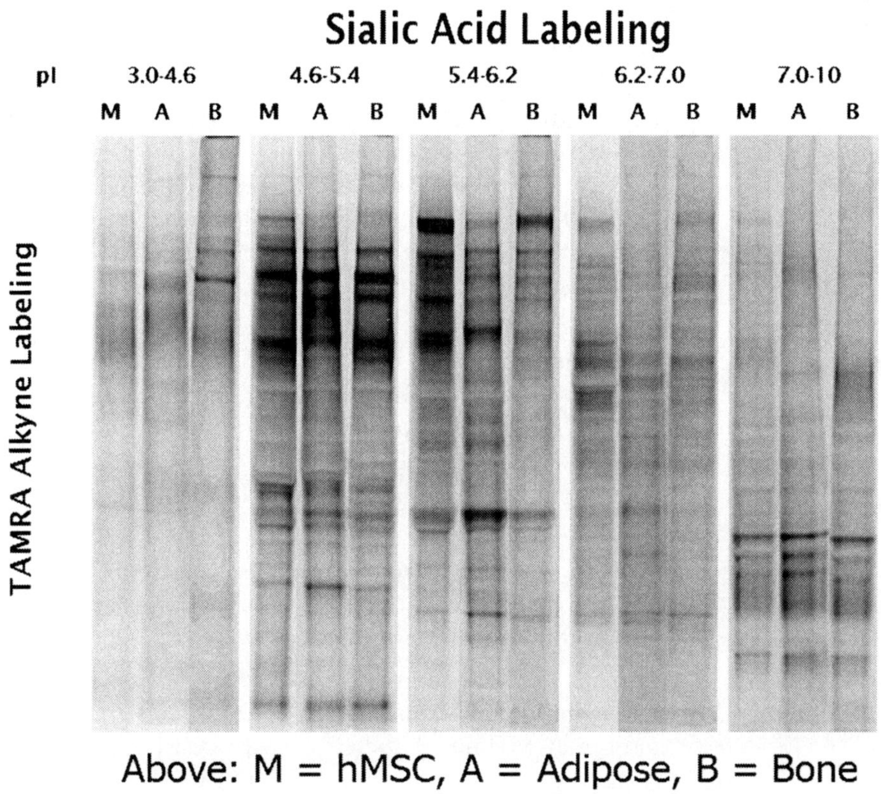

Fig. 7. ZOOM® IEF fractionation of hMSC, bone, and adipose proteins: TAMRA-labeled glycoproteins were separated by liquid fractionation and then, in the second dimension, by 1D gel electrophoresis. The approach allows for better separation of high molecular weight (HMW) glycoproteins than the standard 2D approach using isoelectric focusing strips allowing for precise excision of HMW bands for mass spectrometry determinations.

little water to the imager tray to prevent the gel from sticking, and then lay the gel down while separating it from the bottom cassette plate. Gently press out any air bubbles that formed between the gel and the imager tray.

2. Image the TAMRA Click-iT® signal: The TAMRA Click-iT® dye has an excitation maximum at 545 nm and an emission maximum at 580 nm. Labeled gels are best imaged using a laser-based gel scanner with an excitation source of 532 nm. Emission is best captured using a 555 or 580 nm long-pass filter or band-pass filter that can capture 580 nm light. A 300 nm UV illumination-based gel imager equipped with an emission filter that can capture 580 nm light can also be used, but the TAMRA Click-iT® signal intensity and sensitivity will be reduced.
3. Fix the protein in the gel and remove SDS: Place the gel into a suitable container (e.g., polypropylene or polycarbonate dishes such as Servin' Saver or Stain Shield containers from Rubbermaid).
4. Incubate the gel two times for 30 min each in 100 ml gel fixation solution on an orbital shaker with continuous agitation at 50 rpm (see Note 15).
5. Stain for total proteins: Incubate the gel in 50 ml of SYPRO® Ruby protein gel stain overnight at room temperature with continuous agitation on an orbital shaker at 50 rpm or for 30–60 min combined with microwave heating for three 20–30 s bursts that heats the stain solution to near boiling alternating with agitation on an orbital shaker at 50 rpm (see Note 16).
6. Destain: To reduce formation of dye speckles, briefly remove the gel and wipe down the container with ethanol or transfer the gel to a clean container before adding destain solution. Incubate the gel one time for 30 min in 100 ml of gel destain solution with continuous agitation on an orbital shaker at 50 rpm.
7. Wash: In order to reduce possible corrosion on imagers due to the destain solution; it is advisable to wash the gels in water for a minimum of 5 min.
8. Image total proteins: SYPRO® Ruby dye has an excitation maximum at 280 nm and at 450 nm and an emission maximum at 610 nm. Labeled gels are best imaged using a laser-based gel scanner with an excitation source of 473–488 nm or a 300 nm UV or a blue-light transilluminator. Emission is best captured using a 555 or 580 nm long-pass filter or band-pass filter that can capture 610 nm light.

3.10. Multiplexed Western Blotting

Perform all western blot incubations on an orbital shaker or rocking platform at room temperature.

1. Electrophoretically transfer biotin Click-iT® labeled samples that were separated by 1D or 2D SDS-PAGE to nitrocellulose membranes using the iBlot® gel transfer device (see Note 17).
2. Soak the gels 10 min in 2× NuPAGE® transfer buffer containing 10% methanol before placing on top of the bottom iBlot® transfer stack.
3. Cover the gel with a piece of wet filter paper and the top iBlot® transfer stack. After the device has completed the 7 min transfer, remove the membrane from the transfer stack and allow to air-dry completely. The colored molecular weight markers should be clearly visible.
4. Wash the membrane two times for 5 min each with 20 ml of ultra pure water to remove gel and transfer buffer components.
5. Block the membrane for 1 h in 10 ml of blocking buffer.
6. Incubate the membrane for 1 h up to overnight (as recommended by the manufacturer) in the manufacturer's recommended dilution of primary antibody in 10 ml of blocking buffer.
7. Wash the membrane 3 times for 5 min each with 20 ml 1× wash buffer.
8. Co-incubate the membrane for 1 h in a 1:2,000 dilution of GAR or goat anti-mouse HRP secondary antibody and with a 1:2,000 dilution of Qdot® 625 streptavidin conjugate (0.5 nM) in 10 ml of blocking buffer (see Note 18).
9. Wash the membrane three times for 5 min each with 20 ml 1× wash buffer.
10. Wash the membrane one time for 2 min with 20 ml ultra pure water.
11. Image the Qdot® 625 signal: Qdot® 625 excites best with UV epi or transillumination. Laser-based scanners with an excitation source of 473–488 nm are also acceptable. Qdot® 625 has an emission maximum at 625 nm, so emission is best captured using a 555 or 580 nm long-pass filter or band-pass filter that can capture 625 nm light. Exposure times vary with the intensity of the illumination source and concentration of the labeled proteins from 2 s to 2 min. Do not let the membrane dry during imaging as that will reduce the activity of the HRP enzyme for the subsequent chemiluminescent detection step. Simply squirt a little water onto the surface of the imager and place the membrane face up over the water. Do not wrap the membrane in plastic wrap as it will contribute to background fluorescence, interfering with detection sensitivity (see Notes 19 and 20).

12. If the same imager is to be used for imaging of the ECL chemiluminescent signal, then simply prepare the ECL chemiluminescent substrate as directed by the manufacturer and pipette approximately 2 ml of substrate over the surface of the membrane, completely covering the surface. Try not to move the membrane from its position on the imager tray. Allow the reaction to develop for 1–5 min and then expose for 1 min to several min depending on the signal produced and the imaged desired. If longer development time or exposures are needed, then overlay a clean piece of plastic, such as copier transparency film, over the membrane to prevent drying. The chemiluminescent and Qdot® 625 images can now be directly overlayed without any manipulation of the orientation of the images.
13. If it is necessary to move the membrane to another imager, then briefly return the membrane to the western staining dish and immerse in 20 ml water to rewet.
14. Lay the membrane antigen side up on a clean piece of plastic sheet, such as copier transparency film, and pipette approximately 2 ml of prepared ECL chemiluminescent substrate over the surface of the membrane, completely covering the surface of the blot.
15. Allow the reaction to develop for 1–5 min. Tilt the membrane to drain off excess substrate and then overlay another clean piece of plastic over the membrane to prevent drying during imaging.
16. Place the protected membrane in an imaging system capable of detecting chemiluminescence or a film cassette and expose for 1 min to several min depending on the signal produced and the imaged desired (see Fig. 8)

3.11. Mass Spectrometry

1. Excise gel spots and mince in 1 mm cube pieces and transferred to a clean microcentrifuge tube.
2. Destain minced gel pieces in 500 µl of 50% ACN in 25 mM ammonium bicarbonate buffer until clear.
3. After removing the destaining solution, dehydrate the gel pieces in 200 µl of ACN and dry for 2 min in a speedvac.
4. Rehydrate gel plugs for 1 h on ice with a minimal volume of trypsin solution (10 ng/µl in 25 mM ammonium bicarbonate buffer) and incubate overnight at 37°C.
5. Extract the digested peptides with 50 µl of 5% TFA, followed by extraction with 50 µl of 5% TFA containing 50% ACN.
6. Combine the peptide extracts and dry in a speed vac.
7. Resuspend the dried peptides in 10 µl of 0.1% TFA containing 5% ACN.

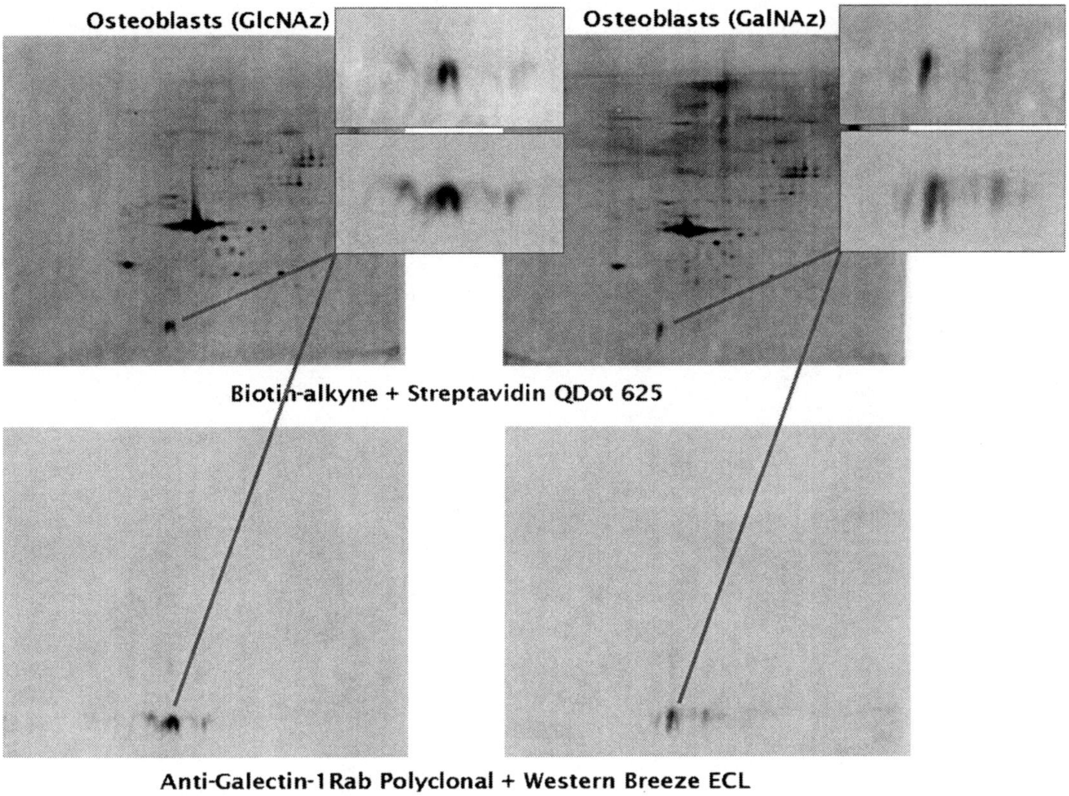

Fig. 8. Multiplex Western blot analysis of galectin 1: GalNAz- and GlcNAz-labeled bone extracts were labeled with biotin-alkyne then blotted and detected with streptavidin Qdot 625 conjugate as described in the text. The *blots* were secondarily incubated with anti-galectin 1 antibody and detected with Western Breeze ECL. The exact overlay of both GalNAz and GlcNAz *O*-linked sugar staining with the anti-galectin 1 antibody strongly suggests galectin 1 is modified by *O*-glycosylation, and most likely by *O*-*N*-acetylglucosamine, or *O*-GlcNAc.

8. Mix 1 μl of resuspended peptide mixture with 1 μl of matrix solution.
9. Deposit 0.5 μl of the peptide/matrix mixture on the probe surface.
10. Analyze the samples with the 4700 Protein Analyzer, MALDI-TOF/TOF Instrument.
11. Process raw data without smoothing or baseline correction with the GPS Explorer Software.
12. Perform database searches against the NCBI nr database using Mascot. The mass tolerance for MS and MS/MS fragment ions were set at 200 ppm and 0.5 Da, respectively, with one missed cleavage allowed. Variable oxidation of methionine reside was considered (see Table 1).

Table 1
Identification of GalNAz- or GlcNAz-labeled protein excised from 2D gels

Gel spot	Protein ID	SWISS PROT Acc. No.	Mass	pI	Seq cov (%)
1.	Galectin-1	P09382	14,584	5.34	42
2.	Adipocyte fatty acid binding protein	P15090	14,588	6.80	60
3.	NADH-ubiquinone reductase	O75489	26,415	5.48	36
3.	Heat shock 27 kDa protein	P04792	22,783	5.98	31
4.	Vimentin	P08670	53,521	5.06	26
5–6.	Alpha 2 type VI collagen	P12110	106,510	5.85	16
5–6.	Vinculin	P18206	123,668	5.51	21
7.	Heat shock 70 kDa protein 8	P11142	70,898	5.37	31
8.	Actin isoforms	P63261	41,662	5.31	43

4. Notes

1. For more extensive characterization of chondrogenesis, it is recommended to analyze chondrogenic pellets through a more detailed histochemical analysis as detailed elsewhere within this text.
2. Azido sugar labeling concentration range (25–50 μM) and labeling times can vary from 24 to 72 h. These parameters should be titrated for each specific cell type.
3. Mouse animal models can be metabolically labeled with azide sugars by IP administration of azide sugars once per day for 7 days at a dose of 200 mg/kg (13).
4. Azido sugar labeled cells can also be fixed, click labeled, and analyzed by fluorescence microscopy or flow cytometry (see product references for cell imaging reagents).
5. Avoid adding DTT or other reducing agents to the cell lysis buffer or other buffers before performing the Click-iT® labeling procedure. These reagents will reduce the azide and make it unavailable for the click reaction.
6. Additional 10 s bursts can be performed until the viscosity decreases to that of the lysis buffer. In order to keep samples from overheating, the sample tubes should be stored on ice during sonication.

7. If a probe sonicator is not available, add 250 U of Benzonase® (Merck KGaA) endonuclease per ml of lysis buffer to digest DNA.
8. Generally, hMSC lysate will have a small pellet, adipocyte lysate will have a medium pellet with some floating material, and chondrocyte and osteoblast lysates will have larger pellets.
9. The EZQ® Protein Quantitation Kit is a convenient protein quantitation method, as it is compatible with SDS, LDS buffer, and urea and detergents in the 2D sample buffers and only requires 1 μl of sample.
10. Samples can be stored in lysis buffer at –20°C up to 1 week. Longer storage is not recommended, as proteins tend to become insoluble in 1% SDS over time.
11. Chloroform/Methanol precipitation can be scaled up or down using the same reagent ratios.
12. It is important to add the Click-iT® Reaction Buffer Additive 2 after waiting for 2–5 min after addition of Click-iT® Reaction Buffer Additive 1. A shorter wait period may result in a decrease in the labeled signal.
13. Once the samples have been labeled with TAMRA Click-iT® dye, reduce overall light exposure to the samples during subsequent steps by covering the tubes, gel electrophoresis apparatus, IPGRunner™ cassettes, etc. with foil to prevent photobleaching.
14. 2D gel separation is best for lower molecular weight glycoproteins while the ZOOM® liquid fractionation with 1D gel separation is better for high MW glycoproteins or very low abundant glycoproteins.
15. Gels may be left in fix solution longer up to overnight, in which case a single incubation in fix solution is adequate.
16. Overnight room temperature staining with SYPRO® Ruby gel stain is recommended when saturation staining of highly abundant or heavily loaded proteins is desired, such as when staining 1D gels of total cell lysate proteins. Microwave staining with SYPRO® Ruby gel stain is recommended for 2D mini-gels or 1D gels in which low background and sensitive detection of low abundant proteins is desired.
17. Multiplexed western detection can also be performed using PVDF membranes, but wet UV transillumination results in lower sensitivity and high background. For best results with PVDF membranes, image with epi-UV illumination.
18. Multiplexed western detection with Qdot® 625 streptavidin conjugate is also compatible with alkaline phosphatase secondary conjugates and colorimetric detection, such as with BCIP/NBT.

19. ECL signal can be detected with UV excitation and emission filters used for Qdot® 625 imaging, so the Qdot® 625 signal must be imaged before adding the ECL reagents. Subsequent to adding the ECL reagents, the membrane will show a composite of both signals when imaged wet and dried using the optimal Qdot® 625 settings.
20. WesternBreeze® chemiluminescent detection kits (Invitrogen) contain an alkaline phosphatase reactive substrate (CDP-Star®) that is not detected with UV excitation and emission filters used for Qdot® 625 imaging. Membranes can be imaged wet and dried for Qdot® 625 subsequent to adding WesternBreeze® CDP-Star® reagent.

References

1. Friedenstein A.J., Gorskaja J.F., Kulagina N.N. "Fibroblast precursors in normal and irradiated mouse hematopoietic organs." Exp. Hematol. 4(5), 267–274 (1976).
2. Pittenger M.F., Mackay A.M., Beck S.C., Jaiswal R.K., Douglas R., Mosca J.D. "Multilineage potential of adult human mesenchymal stem cells." Science 284(5411), 143–147 (1999).
3. Chamberlain G., Fox J., Ashton B., Middleton J. "Concise review: mesenchymal stem cells: their phenotype, differentiation capacity, immunological features, and potential for homing." Stem Cells 25(11), 2739–2749 (2007).
4. Ng F., Boucher S., Koh S., Sastry K.S., Chase L., Lakshmipathy U. "PDGF, TGF-beta, and FGF signaling is important for differentiation and growth of mesenchymal stem cells (MSCs): transcriptional profiling can identify markers and signaling pathways important in differentiation of MSCs into adipogenic, chondrogenic, and osteogenic lineages." Blood 112(2), 295–307 (2008).
5. Tormin A., Brune J.C., Olsson E., Valcich J., Neuman U., Olofsson T. "Characterization of bone marrow-derived mesenchymal stromal cells (MSC) based on gene expression profiling of functionally defined MSC subsets." Cytotherapy 11(2), 114–128 (2009).
6. Solchaga L.A., Penick K., Goldberg V.M., Caplan A.I., Welter J.F. "Fibroblast growth factor-2 enhances proliferation and delays loss of chondrogenic potential in human adult bone marrow-derived mesenchymal stem cells." Tissue Eng. Part A 16(3), 1009–1019 (2010).
7. Piek E., Sleumer L.S., van Someren E.P., Heuver L., de Haan J.R., de Grijs I. "Osteo-transcriptomics of human mesenchymal stem cells: accelerated gene expression and osteoblast differentiation induced by vitamin D reveals c-MYC as an enhancer of BMP2-induced osteogenesis." Bone 46(3), 613–627 (2010).
8. Roche S., Delorme B., Oostendorp R.A., Barbet R., Caton D., Noel D. "Comparative proteomic analysis of human mesenchymal and embryonic stem cells: towards the definition of a mesenchymal stem cell proteomic signature." Proteomics 9(2), 223–232 (2009).
9. Li G., Zhang X.A., Wang H., Wang X., Meng C.L., Chan C.Y. "Comparative proteomic analysis of mesenchymal stem cells derived from human bone marrow, umbilical cord, and placenta: implication in the migration." Proteomics 9(1), 20–30 (2009).
10. Kolb H.C., Finn M.G., Sharpless K.B. "Click chemistry: diverse chemical function from a few good reactions." Angew. Chem. Int. Ed. Engl. 40(11), 2004–2021 (2001).
11. Rostovtsev V.V., Green G.L., Fokin V.V., Sharpless K.B. "A stepwise Huisgen cycloaddition process: copper(I)-catalyzed regioselective 'ligation' of azides and terminal alkynes." Angew. Chem. Int. Ed. 41, 2596–2599 (2002).
12. Tornøe C.W., Christensen C., Meldal M. "Peptidotriazoles on solid phase: [1,2,3]-triazoles by regiospecific copper(1)-catalyzed 1,3-dipolar cycloadditions to terminal alkynes to azides." J. Org. Chem. 76, 3057–3064 (2002).
13. Prescher J.A., Dube D.H., Bertozzi C.R. "Chemical remodeling of cell surfaces in living animals." Nature 430, 873–877 (2002).
14. Luchansky S.J., Hang H.C., Saxon E., Grunwell J.R., Yu C., Dube D.H., Bertozzi C.R. "Constructing azide-labeled cell surfaces using polysaccharide biosynthetic pathways." Methods Enzymol. 362, 249–272 (2003).

Chapter 34

BacMam-Mediated Gene Delivery into Multipotent Mesenchymal Stromal Cells

Michael O'Grady, Robert H. Batchelor, Kelly Scheyhing, Christopher W. Kemp, George T. Hanson, and Uma Lakshmipathy

Abstract

Baculoviruses have been used over the last several decades for high-level protein production in insect cells. Recently, modified baculovirus containing a mammalian promoter, known as BacMam virus, has been shown to give high transduction efficiencies across several cell types with minimal cytopathic effects. Cell types amenable to BacMam transduction include primary and adult stem cells. The shuttle vectors used in the construction of BacMam viruses can hold gene fragments up to 38 kb in size, and multiple BacMam viruses can be used in a single transduction for the delivery of more than one gene. BacMam technology has been used in the delivery and expression of targeted fluorescent protein cellular markers, small interfering RNAi, and extensively in the development of cell-based assays. BacMam offers an ideal method for the delivery and expression of large genes in hard-to-transfect cells such as primary and adult stem cells. In this chapter, we describe methods of generating high titer stocks of BacMam for transducing MSC and their derivatives.

Key words: Baculovirus, BacMam, Viral delivery, Transient, Transduction, Multipotent mesenchymal stromal cells

1. Introduction

Multipotent mesenchymal stromal cells or mesenchymal stem cells (MSCs) have been isolated from multiple tissue sources such as adult bone marrow, adipose tissue, skeletal muscle, lung, deciduous teeth, and human umbilical cord (1–4). MSCs are defined as cells that attach to plastic, express the surface markers, CD105, CD90, and CD73, and lack CD34 and CD45, while possessing the ability to differentiate into mesenchymal lineages such as adipocytes, chondroblasts, osteoblasts, tendonocytes, myocytes, and hematopoietic-supporting stroma (5–12). The ability of MSCs to differentiate into

Mohan C. Vemuri et al. (eds.), *Mesenchymal Stem Cell Assays and Applications*, Methods in Molecular Biology, vol. 698, DOI 10.1007/978-1-60761-999-4_34, © Springer Science+Business Media, LLC 2011

several connective tissue cell types makes them appealing for cell therapy (13–15). Genetic engineering of MSCs would extend such applications to a variety of gene disorders and also provide a powerful tool to dissect the basic biology of these cells (16). Several methods have been reported for the gene transfer and expression of transgenes in these cells (17–21). In particular, baculoviruses have been shown to transduce MSCs efficiently without compromising their proliferation or differentiation potential (22).

Baculoviruses are a diverse group of viruses that contain double-stranded, circular supercoiled DNA. The *Autographa californica multiple nucleopolyhedrovirus* (AcMNPV) is the most widely used in biotechnology and contains a circular double-stranded DNA genome that is packed in a rod-shaped nucleocapsid. The baculovirus/insect cell system has been utilized for the production of several recombinant proteins in insect cells (23, 24). BacMam versions are modified AcMNPVs that carry a mammalian expression cassette for expression of transgenes in mammalian cells (23). The AcMNPV genome can accommodate a DNA insert of up to 38 kb, and the expression of the gene is transient. BacMam gene expression typically lasts for 4–5 days, unless drug selection is utilized to select for the rare cells that have undergone random genomic integration (25–28). Since the commonly used BacMam systems are unable to replicate in mammalian cells, they have minimal cytopathic effects and low safety requirements for handling the virus (29). These features have enabled the use of BacMam to achieve efficient gene delivery and expression in several hard-to-transfect mammalian cell types (27, 30–33).

Several methods have been reported for the generation of BacMam viruses (34–38). Here, we describe the Bac-to-Bac system for generation of recombinant baculoviruses for mammalian cell expression. The various steps involved in the generation of virus are described along with a detailed protocol for transducing MSCs.

2. Materials

2.1. Generation of Bacmid DNA

1. Max Efficiency DH10Bac transformation competent cells, (Life Technologies).
2. LB Agar Plates with 50 μg/ml kanamycin, 7 μg/ml gentamicin, 10 μg/ml tetracycline, 100 μg/ml Bluo-gal, and 40 μg/ml IPTG, PureLink HiPure Plasmid Midiprep Kit (Life Technologies) Platinum PCR Supermix from Life Technologies (Life Technologies).
3. Bacmid M13 Forward Primer 5′ccagtcacgacgttgtaaaacg3′ (see Note 1).
4. Bacmid M13 Reverse Primer 5′agcggataacaatttcacacagg3′.

5. Gene-Specific Primers.
6. 0.8% egels (Life Technologies).

2.2. Generation of Virus

1. Sf9 Cells (SFM Adapted) or Sf21 Cells (SFM Adapted) (Life Technologies).
2. SFM Media: Sf900 II SFM Medium supplemented with 0.5× Antibiotic-Antimycotic (Life Technologies).
3. Grace's unsupplemented medium (Life Technologies).
4. Cellfectin II Reagent (Life Technologies).
5. Heat-Inactivated Fetal Bovine Serum (Life Technologies).
6. Sterile 50 ml amber HDPE vial (VWR).
7. Sterile 50 ml amber 4 ml tubes (VWR).
8. Water bath equilibrated to 42°C.
9. Leveling table acrylic (VWR).
10. 10 cm Tissue culture-grade petri dishes.
11. 5 ml Polystyrene tubes.
12. 4% Agarose gel (Life Technologies).
13. Neutral Red (Life Technologies).
14. 0.5 M NaOH.
15. 5 M ammonium acetate.
16. 0.1× TE (Life Technologies).
17. 0.1 mM EDTA (Life Technologies).
18. Platinum PCR Supermix from Life Technologies (Life Technologies).
19. Gene of interest Forward Primer: 5′tgctaccatggtgagcaagggc-gag3′ (GFP is example given here).
20. Gene of interest Reverse Primer: 5′ctcactattacttgtacagctcgt 3′.
21. 500 kDa MWCO hollow-fiber ultrafilter (Cat.# X2-500S-300-02N; Spectrum Laboratories, Inc)
22. 500 kDa MWCO hollow-fiber ultrafilter (Spectrum Laboratories, Inc.).
23. Masterflex tubing (Cole-Parmer).
24. C-Flex tubing (Cole-Parmer).
25. Reservoir Cap, 38 mm (FiberCell Systems, Inc.).

2.3. MSC Transduction

1. StemPro® Adipose-derived mesenchymal stem cells (AdSC) (Life Technologies).
2. MesenPRO RS™ (Life Technologies).
3. Premo™ Cameleon (BacMam control virus for efficient transduction) (Life Technologies).

4. DPBS (PBS without calcium chloride, without magnesium chloride) (Life Technologies).
5. TrypLE Select (Life Technologies).

3. Methods

3.1. Generation of Bacmid DNA

In the Bac-to-Bac method, the gene of interest is cloned into a transfer vector such as pFastBac-CMV or pDEST8-CMV that carries the gentamicin resistance gene and the mammalian promoter-gene flanked by the Tn7 sites. An expression bacmid is created by transforming the transfer vector containing the gene of interest into DH10Bac, an *Escherichia coli* bacterial strain that carries the parent bacmid (baculovirus genome) comprising the mini-F replicon, kanamycin resistance gene, att Tn7 site, LacZα complementation factor, and the helper plasmid that provides the transposition proteins required for the insertion of the mini-Tn7 from the donor plasmid into its target site on the parent bacmid. The site-specific transposition of the transfer vector into the parent bacmid results in disruption of the LacZ gene and subsequent white recombinant colonies (Fig. 1). The large viral genome termed a Bacmid is isolated from these bacterial colonies and molecular characterization is carried out before proceeding to virus generation in insect cells.

3.1.1. Generation of Recombinant Bacmids

1. Thaw a vial of MAX Efficiency® DH10Bac™ competent cells on ice for each transformation.
2. Add 100 and 500 pg of expression clone DNA to two separate vials and mix gently by flicking and incubate cells on ice for 30 min (see Note 2).
3. Heat-shock the cells for 45 s at 42°C without shaking and immediately transfer the tubes to ice and chill for 2 min.
4. Add 900 μl of room temperature S.O.C. Medium and Shake tubes at 37°C at 225 rpm for 4 h.
5. Plate 100 μl of the above mixture onto fresh LB agar plate containing 50 μg/ml kanamycin, 7 μg/ml gentamicin, 10 μg/ml tetracycline, 100 μg/ml Bluo-gal, and 40 μg/ml IPTG. Incubate plates for 48–72 h at 37°C.
6. Pick ten white, well-separated colonies from each plate and streak onto a fresh LB agar plate containing 50 μg/ml kanamycin, 7 μg/ml gentamicin, 10 μg/ml tetracycline, 100 μg/ml Bluo-gal, and 40 μg/ml IPTG. Incubate plates for 48–72 h at 37°C (see Note 3).

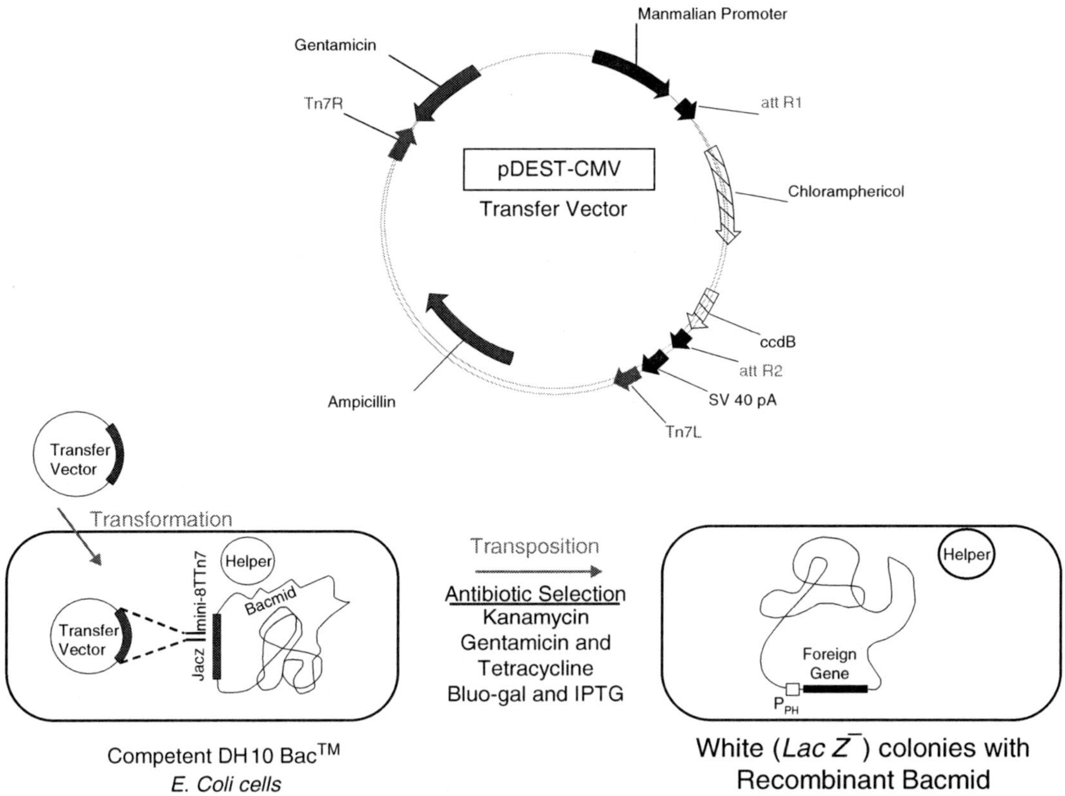

Fig. 1. Gene of interest is cloned into the transfer vector using Multisite Gateway Cloning. The transfer vector is transformed into competent DH10Bac *E. coli* cells for the generation of Bacmids, a process that involves site-specific transposition of the mammalian expression elements on the transfer vector to the parent Bacmid genome contained in the Dh10Bac cells.

3.1.2. Bacmid DNA Purification (see Note 4)

1. Pick five true white colonies each from 100 to 500 ng transformation plates and inoculate into 6 ml of LB containing 50 μg/ml kanamycin, 7 μg/ml gentamicin, 10 μg/ml tetracycline. Shake 37°C for at least 72 h to ensure good growth.
2. Generate 3 ml Glycerol stocks by adding 10% glycerol and aliquoting 1 ml in cryovials. Store at −80°C as parent stocks. Use the remaining 3 ml for Bacmid DNA extraction.
3. Spin down DH10Bac *E. coli* cells at 3,000×*g* for 5 min and use reagents from Purelink MidPrep kit for subsequent steps. Resuspend cell pellets in 300 ml P1 buffer.
4. Add 300 ml P2 buffer, invert gently several times, and incubate for 5 min at room temperature.
5. Add 300 ml P3 buffer, invert gently several times, and incubate 15 min on ice.
6. Spin for 10 min at max speed.
7. Transfer supernatant to new tube and add 800 ml of chilled Isopropanol.

8. Centrifuge 20 min at max speed and discard supernatant.
9. Wash pellet with 500 µl 70% ethanol and centrifuge for 20 min at max speed.
10. Remove ethanol, dry pellet in vacuum dryer.
11. Resuspend pellet in 50 µl 10 mM Tris, pH 8.0 or TE (see Note 5).
12. Determine DNA concentration/quality and store at 4°C.

3.1.3. Bacmid PCR Amplification (see Note 6)

1. Dilute each primer to 100 µM to prepare working stocks of 10 µM.
2. PCR Master Mix: for each reaction use: 1 µl of template + 1 µl (10 µM) forward primer + 1 µl (10 µM) reverse primer and 16 µl Supermix.
3. PCR amplification appropriate for gene-specific primers and M13 primers.
4. Load on 0.8% egel along with DNA ladder.
5. Confirm presence of gene-specific amplicon and absence of ~350 bp wild-type amplicon generated with M13 primers.

3.2. Generation of Virus

The Bacmid generated is transfected into insect cells for the generation of P0 virus. This virus can be further expanded by expansion to generate Bulk HTS virus stocks or subjected to plaque purification to isolate clonal virus for further amplification and generation of high titer stocks (Fig. 2).

3.2.1. Insect Cell Expansion

1. Obtain Sf9 or Sf21 insect cell stocks (see Note 7).
2. Store cells in liq N_2 until ready for use.
3. Rapidly thaw frozen vial in 37°C water bath and transfer the contents of the vial into a 125 ml culture flask with 27 ml of prewarmed Sf-900 II SFM media (see Note 8). Incubate at

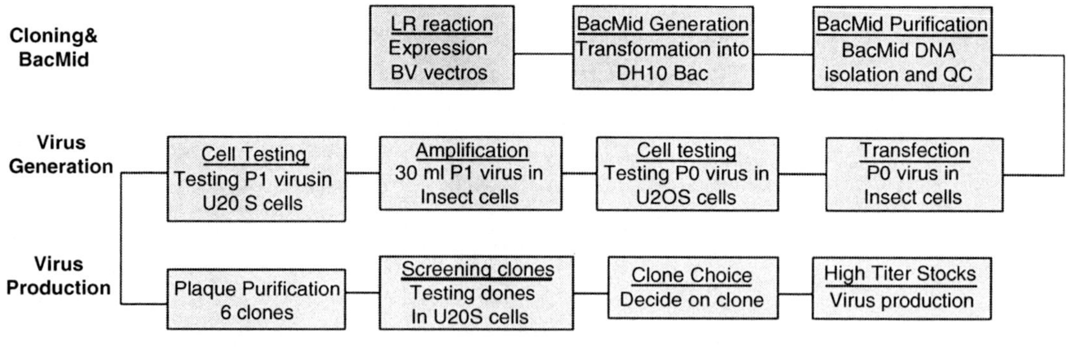

Fig. 2. Detailed workflow of virus generation and production of high titer virus stocks in the range of 0.5–2.5 × 10^8 pfu/ml. U-2 OS cells, an osteosarcoma cell line, are commonly used as the cell line for testing early virus stocks due to their efficient uptake of BacMam particles.

28 + 0.5°C nonhumidified air-regulated incubator on an orbital shaker platform rotating at 125–150 rpm. Loosen caps to allow aeration or use vented caps.

4. Once the culture has reached $>2 \times 10^6$ viable cells/ml, determine total and viable cell counts.
5. Expand cells by seeding shake culture flasks at 1×10^6 viable cells/ml by appropriate diluting in medium (see Notes 9 and 10).

3.2.2. Insect Cell Cryopreservation

1. Grow cells to desired density in culture flasks and harvest cells in midlog exponential growth at a viability of >90%.
2. Determine viable and total cell counts and calculate the volume of freezing medium required to yield a final cell density of $>1 \times 10^7$ viable cells/ml.
3. Freezing media: freshly made with half growth medium + half conditioned growth media (day 2–4 cell conditioned media collected from insect cell cultures during subculture procedure) + 7.5% DMSO.
4. Centrifuge cells from cell suspension at $100 \times g$ for 5–10 min, discard supernatant, and resuspend cell pellet in chilled freezing medium.
5. Dispense aliquots into 1.5–2.0 ml cryovials and freeze under controlled rate.

3.2.3. Virus Generation

1. Generate virus from at least three Bacmid clones.
2. Seed one well of a 6-well dish for each transformation with $8–10 \times 10^5$ cells/well in SFM media.
3. Include one negative control well (no virus) as a visual aid to compare and analyze cytopathic effect in the transfected wells.
4. Incubate at 27°C for at least 1 h to allow cells to attach.
5. Transfection reaction:
   (a) Use Cellfectin II protocol: Mix Cellfectin® II before use, and dilute 8 µl in 100 µl Grace's Medium, Unsupplemented (without antibiotics and serum). Vortex briefly to mix (see Note 11).
   (b) Dilute 1 µg baculovirus DNA in 100 µl Grace's Insect Medium, Unsupplemented (without antibiotics and serum). Mix gently.
   (c) Combine the diluted DNA with diluted Cellfectin® II (total volume ~210 µl). Mix gently and incubate for 15–30 min at room temperature.
   (d) Add ~210 µl DNA–lipid mixture or transfection mixture (Step 5c) dropwise onto the cells from Step 5c.

(e) Incubate cells at 27°C for 3–5 h.

(f) Remove the transfection mixture and replace with 2 ml of complete growth medium (add 2 ml/well Sf-900 II SFM media containing 3% (v/v) heat-inactivated FBS at or 1% (w/v) BSA to each well) (see Note 12).

(g) Incubate at 27°C, for 4–6 days until you see a cytopathic effect.

6. Collecting P0 virus and proceed to cell testing:

(a) Collect media from each well into a 15-ml conical tube.

(b) Centrifuge at 800 × *g* for 5 min at room temperature and transfer supernatant to fresh 2-ml cryo tube.

(c) Store labeled tubes of P0 virus at 4°C in the dark (see Note 13).

3.2.4. Bulk Amplification (see Note 14)

1. Harvest P0 at 72 h posttransfection.
2. Set up insect cell culture the night before so that they are at ~2×10^6/ml on day of infection (see Note 15).
3. At the time of infection, add heat-inactivated FBS at 3% (v/v) or 1% BSA (w/v) to the culture.
4. Infect cultures at a cell concentration of 2×10^6 cell/ml and at an MOI of 0.1–0.08 pfu/cell (assuming ~1×10^7 pfu/ml of the P0 virus).
5. Count culture the following day and, if required, dilute accordingly to maintain the cell concentration at $<5 \times 10^6$ cells/ml.
6. Monitor cell concentration daily (maintain cell density at $<5 \times 10^6$/ml) until cell growth has stopped.
7. Continue to monitor culture and harvest virus 3 days post growth cessation or when culture viability has reached approximately 50–70%, whichever is sooner.
8. Harvest by centrifugation at 1,000 × *g* for 15 min at 4°C.
9. Filter virus stocks for clarity using a 0.22 μm rated hydrophilic polyethersulfone membrane.
10. Pour virus into sterile Amber bottles and store the virus at 4°C in the dark in light-proof bags.

3.2.5. Plaque Purification (see Notes 16 and 17)

1. Of the three Bacmids used for P0 virus generation, choose the best performing P0 virus from BacMam transduction on U-2 OS cells.
2. For each virus to be plaque-purified, prepare four petri dishes.
3. Fill each petri dish with 10 ml SFM media to uniformly coat the dish surface.

4. Harvest Sf21 insect cells in their log phase, determine the cell density, and seed exactly 5×10^6 cells per 10 cm petri dish (see Note 18).
5. After addition of the cells, rock the dishes gently to distribute the cell evenly for 10 min (do not swirl) and allow the cells to attach at room temperature for 1 h on a Plexiglas-leveled table.
6. Perform serial tenfold dilutions of virus in SFM media in polystyrene tubes by diluting 100 μl of the 48-h supernatant virus in the first tube to 3 ml SFM media and dilute to 10, 1, and 0.1 μl of virus with a volume of 3 ml for each tube.
7. Gently remove media from petri dishes and add 3 ml of the diluted virus to the dish. Allow infection for 1 h at room temperature on a leveled surface with gentle rocking every 15 min.
8. Meanwhile, melt 2.5% agarose in a 70°C water bath for 10 min and move to 42°C water bath. Pipette 25 ml 4% agarose in a 50 ml polypropylene tube and place the tube back in the 42°C water bath. Pipette 25 ml of SFM media into a 50 ml tube and equilibrate at 42°C.
9. After 1 h of infection, aspirate the media from each petri dish and carefully remove all liquid.
10. Add the 25 ml 2.5% agarose to 25 ml of the SFM media both maintained at 42°C and overlay each dish with 10 ml agarose/medium mixture.
11. Set the mixture aside in the culture hood with the lid cracked open and leave the dishes undisturbed for 20 min to allow the agarose to solidify (see Note 19).
12. Put the plates in a humidified box or ziplock bag, prepared by placing wet tissue towels (with a few drops of 5 mM EDTA to prevent growth of bacteria and mold) in a 27°C incubator for 6–7 days.
13. Plaques become visible as holes in the monolayer after 6–7 days, but may take up to 2 weeks. Under a microscope, plaques have a central area of cell death with surrounding areas of swollen cells.
14. Mark each plaque and pick colonies that are well-separated (typically from higher dilutions that have fewer plaques).
15. Plaques can be picked by aspiration of the agarose overlaying the marked plaque using a sterile Pasteur pipette and depositing the plug in 1 ml SFM media + 3% FBS in a cryovial.
16. Allow the virus to diffuse from the agarose for 1 h or longer before proceeding to amplification.
17. Typically, ten plaques are picked and PCR screened for gene of interest (see above).

18. Allow virus to diffuse from the agarose plug for at least 1 h.
19. Seed 96-well dish with 1×10^4 insect cells/well (use multichannel pipettor to pipette 100 μl of 1×10^5 cells/ml solution). Allow cells to attach for at least 1 h.
20. Carefully remove the media from insect cells and add 100 μl of primary plaque eluate. Include negative (no virus) control wells. Incubate the cells with virus for 1–2 h.
21. Remove virus from the wells and refeed insect cells with 100 μl SFM media and allow infection to proceed for 48–72 h.
22. After 72 h, aspirate the media and lyse the cells by addition of 200 μl 0.5 M NaOH. Allow lysis to occur for 1 min and neutralize by addition of 40 μl mM ammonium acetate.
23. The lysate can be stored at −20°C or immediately used for PCR.
24. Dilute 5 μl of lysate in 95 μl 0.1× TE in 96-well plates.
25. Set up PCR reaction in total volume of 20 μl by adding 1 μl each of forward and reverse primers and 13 μl of Platinum PCR Supermix.
26. PCR according to amplification condition for gene-specific primer and visualize amplicons on 0.8% egel.

3.2.6. Virus Amplification (see Note 20)

1. Harvest midlog phase insect cells with over 95% viability and seed $1.5–2.0 \times 10^6$ cells in 5 ml of SFM media + 3% FBS.
2. Add 100 μl of viral eluate from plaque purification.
3. Incubate at 27°C and harvest P1 virus when widespread cell death occurs, usually 7 days.
4. Harvest virus by spinning down cells at $800 \times g$ for 5 min at room temperature.
5. Freeze down 1 vial each of 1 ml P1 virus as Master stock at −80 and 4°C.
6. Store labeled tubes of P1 virus at 4°C in the dark or proceed with titering/testing/amplification of the remaining 3 ml of P1 virus.

3.2.7. Virus Titering

1. For each virus to be plaque-purified, prepare 6–8 petri dishes per dilution in triplicates
2. Fill each petri dish with 10 ml SFM media to uniformly coat the dish surface.
3. Determine the cell density of Sf21 insect cells and seed exactly 4×10^6 cells per 10 cm petri dish (see Note 18).
4. After addition of the cells, rock the dishes gently to distribute the cells evenly (do not swirl) and allow the cells to attach at room temperature for 1 h on a Plexiglas-leveled table.

5. Perform serial tenfold dilutions of virus in SFM media in polystyrene tubes by diluting 100 μl of the 48-h supernatant virus in the first tube to 3 ml SFM media and dilute to 10, 1, 0.1, 0.01, and 0.001 μl of virus with a volume of 3 ml for each tube.
6. Gently remove media from petri dishes and add 3 ml of the diluted virus to the dish. Allow infection for 1 h at room temperature on a leveled surface with gentle rocking every 15 min.
7. Meanwhile, melt 4% agarose in a 70°C water bath for 10 min and move to 42°C water bath. Pipette 25 ml agarose in a 50 ml polypropylene tube and place the tube in a water bath equilibrated to 42°C. Pipette 25 ml of SFM media into a 50 ml tube and equilibrate at 42°C.
8. After 1 h of infection, aspirate the media from each petri dish and carefully remove all liquid.
9. Add the 25 ml 4% agarose to 25 ml of the SFM media both maintained at 42°C and overlay each dish with 12.5 ml agarose/medium mixture.
10. Set the mixture aside in the culture hood with the lid cracked open and leave the dishes undisturbed for 20 min to allow the agarose to solidify (see Note 19).
11. Wrap plates tightly with Parafilm and invert so that the agarose layer is on top. Put the plates in a humidified box, prepared by placing wet tissue towels (with a few drops of water bath antifungal solution to prevent growth of mold) in a plastic box tightly sealed with Parafilm at 27°C incubator for 4–5 days.
12. Plaques become visible as holes in the monolayer after 6–7 days, but may take up to 2 weeks. Under a microscope, plaques have a central area of cell death with surrounding areas of swollen cells. Plaques can be stained with MTT or neutral red for easy visualization and counting.
13. *MTT staining*: Overlay a second layer of agarose layer with 4 ml of 4% agarose + 4 ml of SFM with 0.4 ml MTT (10 mg/ml in water) per plate, prewarmed to 42°C. Incubate for 1 h at room temperature to allow agarose to solidify, and then return plates to incubator. The plaques are visible as clear holes in the purple MTT-stained cellular monolayer.
14. *Neutral red staining*: Dilute 1 ml of neutral red in 30 ml of PBS. Add 1 ml of the neutral red solution to the plate and incubate the plates in the dark for 1–2 h. Monolayer appears as rust color with clear plaques.
15. Disregard plates with too many plaques to accurately count. Average the number of plaques in the remaining plates to calculate the titer using the formula:

$$\text{pfu/ml} = \frac{\text{number of plaques (pfu)}}{\text{(dilution factor)(ml of inoculum)}}.$$

3.2.8. P2 Virus Amplification

1. Grow 50–1,000 ml of insect cells at a density of 1×10^6 cells/ml in SFM containing 3–5% serum in an appropriately sized flask.
2. Add P1 virus stock at an MOI of 0.01–0.1 from chosen clone and allow infection to proceed for 7 days (see Note 21).
3. Harvest virus supernatant by centrifugation and store virus in sterile amber 50 ml tubes.
4. Aliquot at least ten cryovials of P2 virus and store at –80°C as working stocks (see Note 22).
5. Aliquot remaining P2 Virus in amber eppendorfs or store in 50 ml amber tubes at 4°C as working stocks for downstream testing (see Note 23).

3.2.9. Virus Concentration and Purification (Optional)

1. For volumes between 200 ml and 3 l, use Spectrum Laboratories, Inc. 500 kDa MWCO hollow-fiber ultrafilter (Cat.# X2-500S-300-02N). For volumes between 3 and 10 l, use Spectrum Laboratories, Inc. 500 kDa MWCO ultrafilter (Cat.# 3244229-100-01P).
2. The filters require a recirculation loop in order to function properly. To create the loop, attach a 36 in. (or best length) section of Masterflex-25 tubing to the inlet side of the hollow-fiber filter. Attach a 36 in. (or best length) section of C-Flex tubing to the effluent side of the hollow-fiber and secure both sections with nylon cable ties. Attach the two ends of the tubing attached to the inlet and outlet of the filter to a Reservoir Cap. This Reservoir Cap is designed to attach to standard Nalge square media bottles. Using the Luer connectors supplied with the ultrafilters, attach a 3–6 in. section of the C-Flex tubing to the two outlet ports on the hollow-fiber.
3. Connect the Reservoir cap to a clean Nalge square 1-l media bottle and place the length of Masterflex-25 tubing in an appropriate peristaltic pump. Flush the filter with water for approximately 10–15 min with the two outlet ports on the ultrafilter clamped off. Replace the water in the bottle and restart the pump. Open the ports on the ultrafilter and allow water to flow through the ports for 5–10 min. Disconnect the Reservoir Cap from the bottle and drain the system. The system is now ready for use.
4. Transfer the virus stock to be concentrated/diafiltered to a clean, sterile bottle with a 38 mm finish. Attach the Reservoir Cap assembly and close the outlet ports.

5. Start the recirculation using the peristaltic pump and adjust the flow to between 100 ml and 1-l/min.
6. When the flow has stabilized, open one of the outlet ports and collect the permeate fluid into a clean, sterile bottle.
7. Continue recirculation until the desired level of concentration has been achieved. If diafiltration is required, proceed to step 8. Otherwise, proceed to step 9.
8. For diafiltration, continue the recirculation until the volume of the virus stock has been reduced by a factor of tenfold. Close the outlet port and stop the pump. Resuspend the concentrated virus to the original volume with 25 mM Hepes buffer with 300 mM NaCl and 1% w/v BSA or 3% heat inactivated FBS pH 7.0. Restart the recirculation and open the outlet port on the filter. When the volume has been reduced by a factor of 10, repeat the resuspension step using the same buffer. When the volume has been reduced to the desired level, close the outlet ports, stop the pump, and drain the system.
9. Filter sterilize the concentrated virus through a standard 0.22 μm filter unit and store at 2–8°C protected from light (see Notes 24 and 25).

3.3. MSC Culture

3.3.1. MSC Thawing and Maintenance

1. Warm MSC media in 37°C water bath or incubator.
2. Quickly thaw vial of frozen MSC in 37°C water bath and transfer content to 50 ml centrifuge tube.
3. Slowly pipette 10 ml prewarmed MB media into tube.
4. Transfer cell solution into T-175 flask containing 20 ml prewarmed MesenPro RS medium.
5. Once flasks are seeded, moderately swirl the flasks on the incubator shelf in figure 8 pattern to homogenously distribute the cells across the culture surface.
6. Incubate vessels at 37°C in a 5% CO_2 humidified incubator.
7. Check flasks every 2–3 days to monitor health of cells. Change media if flask is becoming acidic.
8. Completely replace old MSC media with fresh MSC media.
9. Refeed cultures every 2–3 days.
10. Follow refeeding regimen until culture vessels reach ~80% confluency. At this stage, the cultures are ready to be passaged.

3.3.2. MSC Propagation

1. After reaching near confluency, remove media and wash cells with DPBS.
2. Add 5 ml TrypLE Select per T-175 flask. Incubate for 10 min in 37°C incubator.

3. With moderate force, tap the sides of the flask against the palm of your hand to detach the cells from the culture surface.
4. Triturate cell solution with serological pipette to break up loose clumps.
5. Aseptically remove 100 μl from triturated cell solution and place in a tube (12 × 75 mm glass tubes are convenient for this method) for hemacytometer counts (see below for counting method).
6. Transfer cell suspension to 15 ml centrifuge tube and spin for 5–7 min at 100 × *g*, RT. Perform cell count while tubes are spinning in centrifuge.
7. After centrifuging, aspirate supernatant, resuspend pellet in 3 ml prewarmed MesenPro RS medium, and calculate cell concentration based on hemacytometer counts.
8. Plate culture vessels at a seeding density of 3,000–5,000 cells/cm^2 in MesenPro RS medium.
9. Incubate culture vessels at 37°C in 5% CO_2 humidified incubator.
10. Passage expansion flasks when they reach ~80% confluency.

3.4. MSC Transduction

MSC can be transduced with BacMam either in suspension or as attached cells. Cells transduced in suspension are recommended due to more control over cell density and viral multiplicity of infection (Fig. 3a). MSC can also be cotransduced with more than one BacMam virus to achieve parallel delivery of more than one gene of interest (Fig. 3b).

3.4.1. Suspension Transduction

1. Wash MSC with PBS (–Ca^{2+}/–Mg) and add 2–3 ml of TrypLE.
2. Harvest cells and transfer to 15-ml conical tube, add medium up to 10 ml.
3. Spin cells at 1,000 rpm for 2 min at room temperature.
4. Add 5.5 ml of BacMam transduction mixture (~100–200 MOI [pfu/cell] in PBS w/o Ca^{2+}/Mg) to 4–6 × 10^6 cells and incubate at room temperature for 20 min with gentle agitation.
5. After incubation, spin cells as in step 3 above, remove transduction mix, and seed cells in appropriate media.
6. Incubate cells at 37°C and 5% CO_2 for 24–48 h. If expression is low, more viruses can be added the next day. Due to the low cytopathic effect of BacMam, serial addition is possible over 3–4 days.

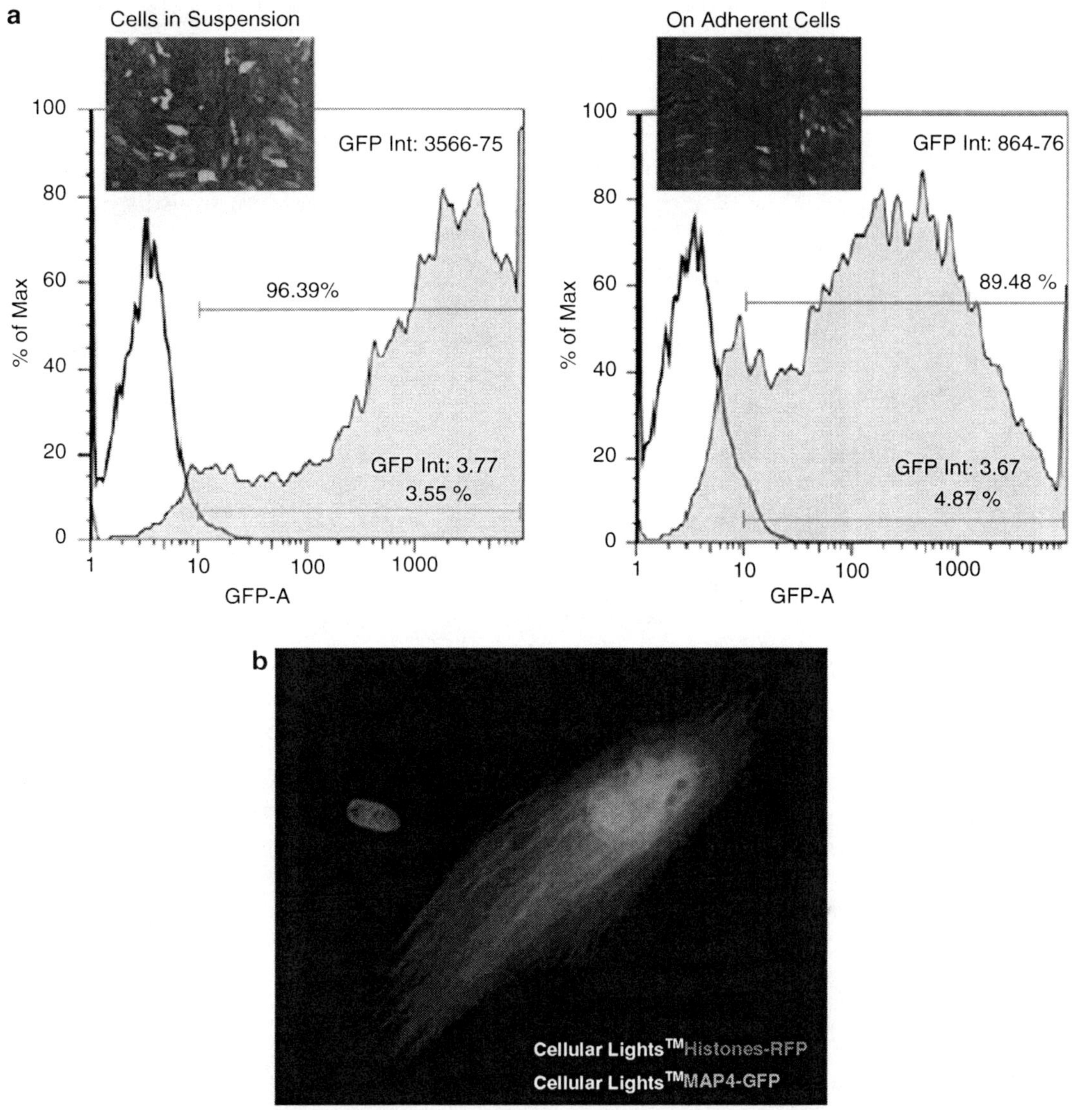

Fig. 3. BacMam transduction in adipose-derived mesenchymal stem cells (AdSC). (**a**) AdSC transduced with Premo™ Cameleon either in suspension or on adherent cells. One ml of the reagent was added to 1×10^6 cells and incubated at room temperature for 1 h. Cells were allowed to recover overnight and analysis carried out 24 h posttransduction. Images were collected using Ziess Axiovert Fluorescence microscope at 10× magnification. FACS analysis was carried out on the FACS Canto flow cytometer and data analyzed using FloJo software to determine percent and intensity of GFP-expressing AdSC. (**b**) Cotransduction of AdSC with Cellular Lights™ Histones-RFP (Nuclear Staining) and MAP4-GFP (Cytoplasmic) to label cell nucleus and microtubules, respectively, in live cells.

3.4.2. Adherent Transduction

1. Wash ~90% confluent MSC with PBS (−/−) and add 1 ml of TrypLE.
2. Transfer cells to 15 ml conical tube and add medium up to 5 ml.
3. Spin cells at 1,000 rpm for 2 min at room temperature.

4. Aspirate medium and replate cells at 1:2 split ratio.
5. Upon reaching ~80% confluent MSC culture, aspirate cell culture media and wash once with PBS.
6. Add 5.5 ml of the BacMam transduction mixture (~100–200 MOI (pfu/cell] in PBS w/o Mg/Ca^{2+}) to the plate and incubate cells 20 min at RT, aspirate BacMam transduction mix, and add regular growth media (see Note 26).
7. Incubate cells at 37°C and 5% CO_2 for 24–48 h for expression. If expression is low, more viruses can be added the next day. Due to the low cytopathic effect of BacMam, serial addition is possible over 3–4 days (see Notes 26a–g).

4. Notes

1. Bacmid forward primer binds complimentary to sequence between gene of interest and SV40-pA when present.
2. As low as 1 pg of Baculovirus DNA is recommended for transformation into DH10Bac for the generation of Bacmids, but may require as high as 0.5–1 μg for some constructs.
3. Bacmids are over 150 kb, be careful not to shear the DNA.
4. During Bacmid generation, a second round of patching of the white recombinant colonies on fresh LB agar plate containing 50 μg/ml kanamycin, 7 μg/ml gentamicin, 10 μg/ml tetracycline, 100 μg/ml Bluo-gal, and 40 μg/ml IPTG is recommended to ensure that the colonies are truly white and not a mixture of white/blue.
5. During Bacmid DNA Purification, use of wide-bore tips is recommended so that the large Bacmid DNA does not shear.
6. For Bacmid sequencing, M13 Forward Primer sequences the Tn7 Left site and M13 Reverse Primer sequences the Tn7R.
7. Sf9 is a subclone of Sf21 cells and is more homogeneous, but grows slower than Sf21.
8. Insect cells have similar growth rates in Sf900 II media or Sf900 III media, either of the media can be used to grow insect cells.
9. Insect cells need to be subcultured a minimum of three passages before use in downstream applications.
10. To reduce accumulation of cell debris and metabolic waste by-products, it is recommended that insect cells are centrifuged at $100 \times g$ for 5–10 min and cell pellet resuspended in fresh Sf900 II SFM media once every 2–3 weeks.

11. During transfection of Bacmid into insect cells, diluted Cellfectin II mixture can be kept at room temperature for up to 30 min.
12. Addition of 3–10% FBS into virus is recommended for stability of the virus with storage. BSA cannot replace FBS since oxidation is not the issue, but rather proteolysis.
13. It is best to proceed with cell testing and further virus plaque purification and amplification without long-term storage of P0 virus.
14. More than one round of bulk amplification is NOT recommended since probability of contamination with noninfectious particles is high.
15. It is important to mix the insect cell cultures thoroughly before dilution and count the number of insect cells after dilution.
16. One round of plaque purification is essential to prevent amplification of defective virus particles and for clonal selection and amplification of desired functional virus.
17. Two rounds of plaque purification is recommended for BioPharmaceutical products.
18. It is critical to have a uniform monolayer of insect cells to visualize plaques. Too many cells result in small plaques and too little will result in gaps in the monolayer which can be falsely read out as plaques.
19. During plaque assay, it is recommended that the plates should be maintained moist and not allowed to dry out too much
20. Successive amplifications of virus need to be carried out using a low (<0.1) MOI to prevent propagation of defective viruses. Higher MOI leads to mutation of the virus resulting in deleted genomes with smaller genome size and hence preferential amplification of the defective particles.
21. Dilute virus as necessary in SFM + 3–5% FBS.
22. Virus tends to clump with freezing, and in general, viral titer decreases by a fold with every freeze/thaw. Freezing is recommended only for long-term storage.
23. Viral stocks are stable at 4°C for 6 months –1 year. For longer storage, freeze down aliquots of virus at –80°C.
24. Filtering of virus can cause loss of titer.
25. Concentrated viruses tend to clump, but can be minimized by resuspension of the virus in 2× PBS.
26. Efficiency is increased if virus is allowed to attach and enter cells at 25°C. Helpful tips for handling BacMam are listed below:

- Baculoviruses are BSL1-classified agents and handling precautions are similar to that for immortalized, primary and stem cell lines.
- Viruses tend to clump and need to be mixed thoroughly by pipetting up and down before use.
- Baculovirus should be stored at 4°C in the dark. Under these conditions, viruses are stable for nearly 1 year. Longer periods of storage can be achieved by storage at –80°C, but the viral titers generally decline by 1–2 logs.
- Transduction in suspension works best probably due to more exposed surface area of the cells for virus binding.
- Presence of bicarbonate in media has been shown to be inhibitory during initial period of viral transduction.
- Incubation of the virus with cells needs to be carried out at room temperature since 37°C inactivates the baculovirus rapidly.
- For harder to transfect cell types, it may be best to incubate the cell pellet with virus for 20–30 min at room temperature and continue incubation with virus overnight in regular growth media.

References

1. Caplan AI. (1991). Mesenchymal stem cells. Journal of Orthopaedic Research 9:641–650.
2. Noth U, AM Osyczka, R Tuli, NJ Hickok, KG Danielson and RS Tuan. (2002). Multilineage mesenchymal differentiation potential of human trabecular bone-derived cells. Journal of Orthopaedic Research 20:1060–1069.
3. Sarugaser R, D Lickorish, D Baksh, MM Hosseini and JE Davies. (2005). Human umbilical cord perivascular (HUCPV) cells: a source of mesenchymal progenitors. Stem Cells 23:220–229.
4. Tuan RS, G Boland and R Tuli. (2003). Adult mesenchymal stem cells and cell-based tissue engineering. Arthritis Research & Therapy 5:32–45.
5. Awad HA, DL Butler, GP Boivin, FN Smith, P Malaviya, B Huibregtse and AI Caplan. (1999). Autologous mesenchymal stem cell-mediated repair of tendon. Tissue Engineering 5:267–277.
6. Bruder SP, N Jaiswal and SE Haynesworth. (1997). Growth kinetics, self-renewal, and the osteogenic potential of purified human mesenchymal stem cells during extensive subcultivation and following cryopreservation. Journal of Cellular Biochemistry 64:278–294.
7. Bruder SP, AA Kurth, M Shea, WC Hayes, N Jaiswal and S Kadiyala. (1998). Bone regeneration by implantation of purified, culture-expanded human mesenchymal stem cells. Journal of Orthopaedic Research 16:155–162.
8. Dennis JE, A Merriam, A Awadallah, JU Yoo, B Johnstone and AI Caplan. (1999). A quadripotential mesenchymal progenitor cell isolated from the marrow of an adult mouse. Journal of Bone and Mineral Research 14:700–709.
9. Ferrari G, G Cusella-De Angelis, M Coletta, E Paolucci, A Stornaiuolo, G Cossu and F Mavilio. (1998). Muscle regeneration by bone marrow-derived myogenic progenitors. Science 279:1528–1530.
10. Galmiche MC, VE Koteliansky, J Briere, P Herve and P Charbord. (1993). Stromal cells from human long-term marrow cultures are mesenchymal cells that differentiate following a vascular smooth muscle differentiation pathway. Blood 82:66–76.
11. Pittenger MF, AM Mackay, SC Beck, RK Jaiswal, R Douglas, JD Mosca, MA Moorman, DW Simonetti, S Craig and DR Marshak. (1999). Multilineage potential of adult human mesenchymal stem cells. Science 284:143–147.

12. Young RG, DL Butler, W Weber, AI Caplan, SL Gordon and DJ Fink. (1998). Use of mesenchymal stem cells in a collagen matrix for Achilles tendon repair. Journal of Orthopaedic Research 16:406–413.

13. Horwitz EM, PL Gordon, WK Koo, JC Marx, MD Neel, RY McNall, L Muul and T Hofmann. (2002). Isolated allogeneic bone marrow-derived mesenchymal cells engraft and stimulate growth in children with osteogenesis imperfecta: implications for cell therapy of bone. Proceedings of the National Academy of Sciences of the United States of America 99:8932–8937.

14. Koc ON, SL Gerson, BW Cooper, SM Dyhouse, SE Haynesworth, AI Caplan and HM Lazarus. (2000). Rapid hematopoietic recovery after coinfusion of autologous-blood stem cells and culture-expanded marrow mesenchymal stem cells in advanced breast cancer patients receiving high-dose chemotherapy. Journal of Clinical Oncology 18:307–316.

15. Petite H, V Viateau, W Bensaid, A Meunier, C de Pollak, M Bourguignon, K Oudina, L Sedel and G Guillemin. (2000). Tissue-engineered bone regeneration. Nature Biotechnology 18:959–963.

16. Hamada H, M Kobune, K Nakamura, Y Kawano, K Kato, O Honmou, K Houkin, T Matsunaga and Y Niitsu. (2005). Mesenchymal stem cells (MSC) as therapeutic cytoreagents for gene therapy. Cancer Science 96:149–156.

17. Ferreira E, E Potier, D Logeart-Avramoglou, S Salomskaite-Davalgiene, LM Mir and H Petite. (2008). Optimization of a gene electrotransfer method for mesenchymal stem cell transfection. Gene Therapy 15:537–544.

18. Bosch P and SL Stice. (2007). Adenoviral transduction of mesenchymal stem cells. Methods in Molecular Biology 407:265–274.

19. Kallifatidis G, BM Beckermann, A Groth, M Schubert, A Apel, A Khamidjanov, E Ryschich, T Wenger, W Wagner, A Diehlmann, R Saffrich, U Krause, V Eckstein, J Mattern, M Chai, G Schutz, AD Ho, MM Gebhard, MW Buchler, H Friess, P Buchler and I Herr. (2008). Improved lentiviral transduction of human mesenchymal stem cells for therapeutic intervention in pancreatic cancer. Cancer Gene Therapy 15:231–240.

20. Zhang XY, VF La Russa and J Reiser. (2004). Transduction of bone-marrow-derived mesenchymal stem cells by using lentivirus vectors pseudotyped with modified RD114 envelope glycoproteins. Journal of virology 78:1219–1229.

21. Uchibori R, T Okada, T Ito, M Urabe, H Mizukami, A Kume and K Ozawa. (2009). Retroviral vector-producing mesenchymal stem cells for targeted suicide cancer gene therapy. The Journal of Gene Medicine 11:373–381.

22. Lee HP, YC Ho, SM Hwang, LY Sung, HC Shen, HJ Liu and YC Hu. (2007). Variation of baculovirus-harbored transgene transcription among mesenchymal stem cell-derived progenitors leads to varied expression. Biotechnology and Bioengineering 97:649–655.

23. Kost TA and JP Condreay. (1999). Recombinant baculoviruses as expression vectors for insect and mammalian cells. Current Opinion in Biotechnology 10:428–433.

24. Summers MD. (2006). Milestones leading to the genetic engineering of baculoviruses as expression vector systems and viral pesticides. Advances in Virus Research 68:3–73.

25. Cheshenko N, N Krougliak, RC Eisensmith and VA Krougliak. (2001). A novel system for the production of fully deleted adenovirus vectors that does not require helper adenovirus. Gene Therapy 8:846–854.

26. Condreay JP, SM Witherspoon, WC Clay and TA Kost. (1999). Transient and stable gene expression in mammalian cells transduced with a recombinant baculovirus vector. Proceedings of the National Academy of Sciences of the United States of America 96:127–132.

27. Huser A and C Hofmann. (2003). Baculovirus vectors: novel mammalian cell gene-delivery vehicles and their applications. American Journal of Pharmacogenomics 3:53–63.

28. Merrihew RV, WC Clay, JP Condreay, SM Witherspoon, WS Dallas and TA Kost. (2001). Chromosomal integration of transduced recombinant baculovirus DNA in mammalian cells. Journal of Virology 75:903–909.

29. Kost TA and JP Condreay. (2002). Innovations-biotechnology: baculovirus vectors as gene transfer vectors for mammalian cells: biosafety considerations. Applied Biosafety 7:167–169.

30. Kost TA and JP Condreay. (2002). Recombinant baculoviruses as mammalian cell gene-delivery vectors. Trends in Biotechnology 20:173–180.

31. Kost TA, JP Condreay and DL Jarvis. (2005). Baculovirus as versatile vectors for protein expression in insect and mammalian cells. Nature Biotechnology 23:567–575.

32. Hu YC. (2008). Baculoviral vectors for gene delivery: a review. Current Gene Therapy 8:54–65.

33. Boyce FM and NL Bucher. (1996). Baculovirus-mediated gene transfer into mammalian cells. Proceedings of the National Academy of Sciences of the United States of America 93:2348–2352.

34. Ernst WJ, RM Grabherr and HW Katinger. (1994). Direct cloning into the Autographa californica nuclear polyhedrosis virus for generation of recombinant baculoviruses. Nucleic Acids Research 22:2855–2856.

35. Hartig PC, MC Cardon and CY Kawanishi. (1991). Generation of recombinant baculovirus via liposome-mediated transfection. BioTechniques 11:310, 312–313.

36. Murphy CI, H Piwnica-Worms, S Grunwald, WG Romanow, N Francis and HY Fan. (1997). Maintenance of insect cell cultures and generation of recombinant baculoviruses. Current Protocols in Molecular Biology (Ausubel FM., Brent R., Kingston RE., et al., Eds.) John Wiley & Sons, New York, pp. 16.10.1–16.10.17.

37. Karkkainen HR, HP Lesch, AI Maatta, PI Toivanen, AJ Mahonen, MM Roschier, KJ Airenne, OH Laitinen and S Yla-Herttuala. (2009). A 96-well format for a high-throughput baculovirus generation, fast titering and recombinant protein production in insect and mammalian cells. BMC Research Notes 2:63.

38. Hu YC. (2005). Baculovirus as a highly efficient expression vector in insect and mammalian cells. Acta Pharmacologica Sinica 26:405–416.

Chapter 35

Cell Surface Engineering of Mesenchymal Stem Cells

Debanjan Sarkar, Weian Zhao, Ashish Gupta, Wei Li Loh, Rohit Karnik, and Jeffrey M. Karp

Abstract

By leveraging the capacity to promote regeneration, stem cell therapies offer enormous hope for solving some of the most tragic illnesses, diseases, and tissue defects world-wide. However, a significant barrier to the effective implementation of cell therapies is the inability to target a large quantity of viable cells with high efficiency to tissues of interest. Systemic infusion is desired as it minimizes the invasiveness of cell therapy, and maximizes practical aspects of repeated doses. However, cell types such as mesenchymal stem cells exhibit a poor homing capability or lose their capacity to home following culture expansion (i.e. FASEB J 21:3197–3207, 2007; Circulation 108:863–868, 2003; Stroke: A Journal of Cerebral Circulation 32:1005–1011; Blood 104:3581–3587, 2004). To address this challenge, we have developed a simple platform technology to chemically attach cell adhesion molecules to the cell surface to improve the homing efficiency to specific tissues. This chemical approach involves a stepwise process including (1) treatment of cells with sulfonated biotinyl-*N*-hydroxy-succinimide to introduce biotin groups on the cell surface, (2) addition of streptavidin that binds to the biotin on the cell surface and presents unoccupied binding sites, and (3) attachment of biotinylated targeting ligands that promote adhesive interactions with vascular endothelium. Specifically, in our model system, a biotinylated cell rolling ligand, sialyl Lewis^x (SLeX), found on the surface of leukocytes (i.e., the active site of the P-selectin glycoprotein ligand (PSGL-1)), is conjugated on MSC surface. The SLeX engineered MSCs exhibit a rolling response on a P-selectin coated substrate under shear stress conditions. This indicates that this approach can be used to potentially target P-selectin expressing endothelium in the more marrow or at sites of inflammation. Importantly, the surface modification has no adverse impact on MSCs' native phenotype including their multilineage differentiation capacity, viability, proliferation, and adhesion kinetics. We anticipate that the present approach to covalently modify the cell surface and immobilize required ligands is not limited to MSCs or the SLeX ligand. Therefore, this technology should have broad implications on cell therapies that utilize systemic administration and require targeting of cells to specific tissues. The approach may also be useful to promote specific cell–cell interactions. In this protocol, we describe the conjugation of SLeX on MSC surface and methods to study cell rolling behaviors of SLeX-modified MSCs on a P-selectin coated substrate using an in vitro flow chamber assay. We also provide a brief description of cell characterization assays that can be used to examine the impact of the chemical modification regimen.

Key words: Mesenchymal stem cell, Homing, Trafficking, Cell rolling, Sialyl Lewis^x, Covalent modification

Mohan C. Vemuri et al. (eds.), *Mesenchymal Stem Cell Assays and Applications*, Methods in Molecular Biology, vol. 698, DOI 10.1007/978-1-60761-999-4_35, © Springer Science+Business Media, LLC 2011

1. Introduction

Human mesenchymal stem cells (MSCs), also referred to as multipotent stromal cells, have generated great interest as therapeutic agents to treat disease and promote tissue regeneration due to their multilineage differentiation capacity, their role in angiogenesis, and their profound immunomodulatory functions (1–6). MSCs are currently delivered to treat cardiovascular disease and heart attacks, brain and spinal cord injury, cartilage and bone injury, Crohn's disease, and graft versus host disease during bone marrow transplantation. However, a major challenge is to deliver a large quantity of viable stem cells with high engraftment efficiency. Systemic delivery of the cells is preferable over local injection because it ensures that the cells remain in close proximity to oxygen and nutrient rich blood vessels, and particularly useful in situations where the site of injury is difficult to reach or not suitable for direct injection. Unfortunately, the homing efficiency of systemically infused MSCs to the targeted site is low (typically less than 1%) and unpredictable (7). The limited homing efficiency of MSCs is primarily due to lack of relevant surface adhesion molecules and cytokine receptors. In particular, during culture expansion, a crucial step to achieve sufficient cell numbers for systemic infusion, MSCs lose the expression of certain cell surface adhesion receptors/ligands which are critical for the homing of these cells to the targeted sites (8–10). MSCs have previously been enzymatically (10) and genetically (11) modified to improve their homing efficiency. However, enzymatic modification is limited by the potential surface receptors that can be converted to homing receptors, and gene therapy may not be desired due to unforeseen long-term safety concerns. We recently introduced a simple alternative strategy to covalently couple sialyl Lewis^{x} (SLeX) moiety, a cell rolling ligand, onto the surface of MSCs through biotin-streptavidin chemical modifications (Fig. 1). Without compromising MSC viability, adhesion, proliferation, and multidifferentiation potential, the SLeX engineered MSCs exhibit a rolling response in vitro on substrates coated with P-selectin (12). Specifically, the free amine groups present on the surface of cells react with the *N*-hydroxy-succinimide group of biotinyl-*N*-hydroxy-succinimide to biotinylate the cell surface (12, 13). The biotinylation efficiency is improved with sulfonated biotinyl-*N*-hydroxy-succinimide (sulfo-NHS-Biotin, BNHS) because BNHS is hydrophilic which limits its transport through the cell membrane and thus maximizes interaction of the NHS group with amines on the cell membrane. This simple coupling scheme has minimal impact on cell phenotype which makes this approach a generic method to conjugate any biotinylated ligands on the cell surface and to potentially target any cell type to specific

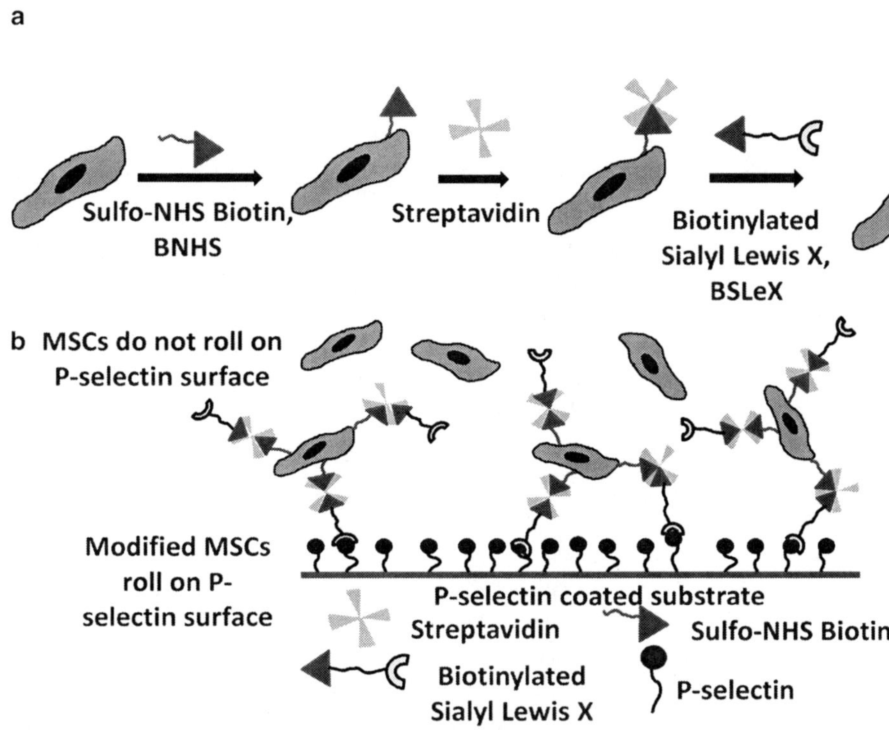

Fig. 1. (**a**) Modification of MSCs by SLeX using biotin–streptavidin chemistry. (**b**) Schematic presentation for rolling of SLeX-modified MSCs on a P-selectin coated surface.

tissues via systemic cell infusion. This chapter describes detailed protocols for chemically modifying cells to enhance their homing response, characterizing cell phenotype following chemical modification, and studying the rolling response of the engineered cells on P-selectin coated substrates.

2. Materials

2.1. MSC Culture and Characterization

1. Complete culture medium (CCM): Supplement 500 mL (αMEM) with L-glutamine, without deoxyribonucleosides and ribonucleosides (Invitrogen, Carlsbad, CA) with 90 mL (15% v/v) Fetal bovine serum (FBS), premium select (Atlanta Biologicals, Lawrenceville, GA), 6 mL L-glutamine (Invitrogen) and 6 mL penicillin/streptomycin (Invitrogen). Store at 4°C in dark for up to 2 weeks. Before experimentation, warm the required amount of medium to 37°C αMEM = α Minimum Essential Medium.
2. Phosphate buffer saline (PBS) without Ca^{2+} and Mg^{2+}, sterile filtered, cell culture tested (Sigma Aldrich, St. Louis, MO).

3. Trypsin-EDTA solution (5.0 g/L porcine trypsin and 2.0 g/L EDTA•4Na^+ in 0.9% sodium chloride), 10× (Sigma-Aldrich); diluted to 1× with PBS.
4. Nonenzymatic Cell Dissociation Solution 1× (prepared in Hanks' Balanced Salt Solution without Ca^{2+} and Mg^{2+}, sterile filtered) (Sigma Aldrich).
5. FITC Mouse Anti-Human CD90 (BD Biosciences, San Jose, CA), PE-Cy™5 Mouse Anti-Human CD29 (BD Biosciences), FITC Mouse Anti-Human CD34 (BD Biosciences), FITC Mouse Anti-Human CD45 (BD Biosciences), FITC Mouse Ig Isotype Control (BD Biosciences), PE Mouse Ig Isotype Control (BD Biosciences).
6. FACS buffer solution: Prepare 1% FACS buffer solution by adding 1 mL FBS to 99 mL of PBS and store at 4°C for no more than 15 days.

2.2. Generating Surface Modified Cells

1. Biotinylation solution: Prepare 1 mM of Sulfonated biotinyl-*N*-hydroxy-succinimide (EZ-Link Sulfo-NHS-LC-Biotin, BNHS); (Thermo Fisher Scientific, Rockford, IL) solution by dissolving 1.15 mg of BNHS in 2 mL of PBS at room temperature (see Note 1).
2. Biotin (Sigma Aldrich) solution: Prepare 1 mM of Biotin (Sigma Aldrich)solution by dissolving 1.08 mg of biotin in 5 mL of PBS at room temperature.
3. Thermo Scientific Pierce Purified Streptavidin (Thermo Fisher Scientific) solution: Prepare a stock solution of streptavidin in PBS with a concentration of 1 mg/mL in PBS. Store at 4°C in for no more than 7 days. Before experimentation, dilute the stock solution to 50 μg/mL with PBS.
4. Biotinylated Sialyl Lewis^x (Sialyl Le^x-PAA-biotin, BSLeX) (Glycotech, Gaithersburg, MD) solution: Prepare a stock solution by dissolving 0.5 mg BSLeX with 0.5 mL of PBS to prepare a stock solution of 1 mg/mL. Aliquot the stock solution into eppendorf tubes with 50 μL in each and freeze it at −20°C. Before experimentation, thaw the frozen stock at room temperature and dilute the samples to 5 μg/mL with PBS (see Note 2).

2.3. Visualization of Cell Surface Biotinylation

1. Streptavidin, Rhodamine (TRITC) Conjugated (Thermo Fisher Scientific) solution: Prepare a stock solution of TRITC–streptavidin in PBS with a concentration of 1 mg/mL in PBS. Store at 4°C in dark for no more than 7 days. Before experimentation, dilute the stock solution to 50 μg/mL with PBS.
2. Immunofluorescence inverted (phase) microscope (with image processing software, e.g., ImageJ) with a rhodamine filter (excitation: 540 nm; emission: 625 nm).

2.4. Characterization Assays for Surface Modified Cells

1. Trypan blue (0.4%) in 0.81% sodium chloride and 0.06% potassium phosphate dibasic (Sigma Aldrich).
2. hMSC Osteogenic BulletKit (Lonza, Walkersville, MD).
3. Neutral Formalin Buffer: Add 4 g of Sodium phosphate dibasic (Sigma Aldrich), 1 g of Sodium phosphate monobasic (Sigma Aldrich) and 25 mL of 37 wt. % of Formaldehyde solution (Sigma Aldrich) to 225 mL of distilled water.
4. Alkaline Phosphatase staining solution: Dissolve 5 mg of Naphthol AS-MX phosphate disodium salt (Sigma Aldrich) in 200 μL of dimethyl formamide (Sigma Aldrich). Mix 25 mL of 0.2 M Tris–HCl buffer (pH 8.3) and 25 mL of distilled water in a graduated cylinder. To this add 200 μL of Naphthol AS-MX phosphate solution (in dimethyl formamide). To this add 30 mg of Fast Red Violet LB salt (Sigma Aldrich) and mix thoroughly. Filter this solution through Whatman's No. 1 filter paper and use immediately for the experiment. Discard any unused solution.
5. Von Kossa staining solution: Prepare 2.5% silver nitrate ($AgNO_3$) solution by dissolving 2.5 g of silver nitrate (Sigma Aldrich) in 100 mL of distilled water (see Note 5).
6. hMSC Adipogenic BulletKit (Lonza).
7. Hemotoxylin Solution (Sigma Aldrich).
8. Adipogenic staining solution (Oil Red O staining): Prepare Oil Red O staining stock solution by dissolving 0.5 g of Oil Red O (Sigma Aldrich) in 100 mL isopropyl alcohol. Dissolve completely and store in a tightly capped bottle at room temperature in the dark for up to 3 months. Before experimentation, prepare Oil Red O working solution by mixing three parts (by volume) of stock solution with two parts (by volume) of PBS. Mix thoroughly and wait for 10 min. Filter the solution through a 0.2-μm filter and wait 10 min before use. Discard any unused solution.

2.5. Flow Chamber Assay to Assess Modified MSC Rolling

1. Immunofluorescence inverted (phase) microscope with image processing software (e.g., ImageJ).
2. Syringe pump (Harvard Apparatus, MA).
3. Parallel Plate Circular Flow Chamber Kit (Glycotech) including gasket size of flow width = 0.25 cm, thickness = 0.005 in.
4. Parafilm (VWR, West Chester, PA).
5. P-selectin solution: prepare a stock solution of recombinant human P-selectin (Fc chimera) (R&D system, Minneapolis, MN) in PBS at concentration of 100 μg/mL. Aliquot and store the stock solution at –20°C for up to 3 months. Before experimentation, thaw the frozen stock at room temperature

and dilute to 5 μg/mL with PBS. Discard any unused solution. Avoid repeated freezing and thawing of the stock solutions.

2.6. Equipments

1. Biological safety Cabinet, Class II.
2. Incubator humidified with 5% CO_2, maintained at 37°C.
3. Hemocytometer with coverslip.
4. Flow cytometer with Cell Quest software.

3. Methods

The cell surface modification protocol described below is optimized to obtain reliable and reproducible results under the given set of conditions. It is important to confirm that biotinylation and subsequent conjugation of streptavidin functionalizes the cell surface effectively. This can be examined by fluorescent microscopic analysis. The effect of cell surface modification on the native phenotype must be extensively characterized. The protocol below is not exhaustive and thus the reader is encouraged to familiarize with other protocols described elsewhere within this text for characterizing the phenotype of MSCs.

3.1 MSC Culture and Characterization

3.1.1. Cell Culture

1. Seed MSCs in a T175 cm² flask at a seeding density of 1×10^3 cells/cm² with 30 mL of CCM. Place cells in an incubator at 37°C with 5% CO_2.
2. Change the medium every 2 days with fresh and prewarmed (37°C) CCM.
3. Monitor the cells everyday using an inverted phase microscope (brightfield) at 10× to observe the level of confluency.
4. At 80% confluence, wash the cells with PBS, harvest the cells from the flasks with 7 mL of trypsin-EDTA (1×) at 37°C for 3 min, followed by centrifugation at 151 ×*g* for 3 min at room temperature.
5. Remove the supernatant and resuspend the cells in 1 mL of CCM.
6. Count the cells using a hemocytometer and replate them in appropriate culture flasks/plates at a seeding density of 1×10^3 cells/cm² (see Note 9).

3.1.2. Flow Cytometry

Flow cytometry is a convenient method to determine the presence of surface molecules on MSCs. It is recommended to perform flow cytometric analysis to detect two positive (e.g., CD90 and CD29) and two negative (e.g., CD34 and CD45) markers on MSCs.

1. Harvest MSCs with trypsin-EDTA (see Subheading 3.1.1, step 4) and resuspend the cells in 100 μL PBS at a concentration of 1×10^6cells/mL.
2. Stain each cell aliquot with 10–20 μL of the appropriate conjugated antibody solution (i.e., FITC Mouse Anti-Human CD90, PE-Cy™5 Mouse Anti-Human CD29, FITC Mouse Anti-Human CD34, FITC Mouse Anti-Human CD45, FITC Mouse Ig Isotype Control, PE Mouse Ig Isotype Control) according to manufacturer's recommendations for 20 min at 4°C in the dark.
3. Wash the samples twice with PBS and resuspend in 100 μL FACS buffer at a concentration of 1×10^6cells/mL.
4. Examine the cells using a flow cytometer, collect 10,000 events for each sample and analyze with Cell Quest software (see Note 14).

3.2. Generating Surface Modified Cells

1. Seed 5×10^5 cells in a T25 cm^2 flask with 5 mL CCM and allow the cells to adhere overnight in an incubator at 37°C and 5% CO_2.
2. Remove the medium and wash twice with 1 mL of prewarmed PBS.
3. Add 1 mL of 1 mM BNHS solution and incubate for 10 min at room temperature.
4. Remove the BNHS solution and wash twice with 1 mL of PBS to remove the unreacted and adsorbed BNHS.
5. Add 1 mL of streptavidin (50 μg/mL) and incubate for 10 min at room temperature.
6. Remove the streptavidin solution and rinse the flask twice with 1 mL PBS to remove unreacted and adsorbed streptavidin.
7. Add 1 mL of BSLeX (5 μg/mL) and incubate for 10 min at room temperature.
8. Remove the BSLeX solution and rinse the flask twice with 1 mL PBS to remove the unreacted and adsorbed BSLeX.
9. Harvest the cells from the flask with 1 mL of nonenzymatic cell dissociation solution (1×) at 37°C for 3 min, followed by centrifugation at 1,500 rpm for 3 min at room temperature (see Note 10).
10. Remove the supernatant and resuspend the cells in 1 mL CCM. Count the cells using a hemocytometer before performing cell characterization studies.

3.3. Visualization of Cell Surface Biotinylation

To determine if cells have been modified following BNHS treatment, TRITC–streptavidin is added to biotinylated cells and the fluorescence intensity of the cells is measured. As biotin on the cell surface should bind strongly to streptavidin, this interaction can be visualized by using fluorescently labeled streptavidin. The negative controls used for this experiment include cells treated with only TRITC–streptavidin (to assess potential adsorption of the fluorescent streptavidin to the cell surface), and cells treated with biotin without NHS and then TRITC–streptavidin (the adsorbed biotin should not be as stable as the NHS-biotin and consequently the fluorescent signal will not be as strong).

1. Plate 5×10^3 cells per well in a 96-well plate with 100 μL of CCM per well. Incubate the plates at 37°C and 5% CO_2. Complete cell attachment typically takes 3–12 h.
2. Remove the CCM from the wells using a micropipette and gently wash the wells twice with 100 μL of room temperature PBS.
3. Add 100 μL of 1 mM BNHS solution and incubate for 10 min at room temperature. To serve as negative controls, incubate individual wells with cells with 100 μL of PBS or 100 μL of 1 mM biotin solution for 10 min at room temperature.
4. Remove the BNHS solution and wash the wells twice with 100 μL of PBS to remove the unreacted and nonspecifically adsorbed BNHS.
5. To characterize cell surface biotinylation, add 50 μL of TRITC–streptavidin (50 μg/mL) to each well and incubate for 10 min in the dark at room temperature.
6. Remove the TRITC–streptavidin solution from the wells and wash twice with 100 μL of PBS to remove the unreacted and adsorbed streptavidin.
7. Add 100 μL of CCM to each well and incubate at 37°C at 5% CO_2 for 5 min.
8. Examine cell surface biotinylation by detecting and measuring the fluorescence signal of rhodamine from the TRITC–streptavidin conjugated to the biotin on the cell surface using an immunofluorescence microscope with a rhodamine filter (excitation: 540 nm; emission: 625 nm) (Fig. 2).

3.4. Characterization Assays for Surface Modified Cells

The characterization of cell phenotype after the cell surface modification should be performed after complete surface modification of the cells, including all three conjugation steps described in Subheading 3.2.

3.4.1. Viability

1. Determine the viability of the modified cells (with BNHS followed by streptavidin and BSLeX) by thoroughly mixing 100 μL of cell suspension with 100 μL of 0.4% trypan blue solution.
2. Add 10 μL of the trypan-cell suspension into both sides of the hemocytometer chambers and count viable and nonviable cells. Nonviable cells will stain blue, whereas viable cells will

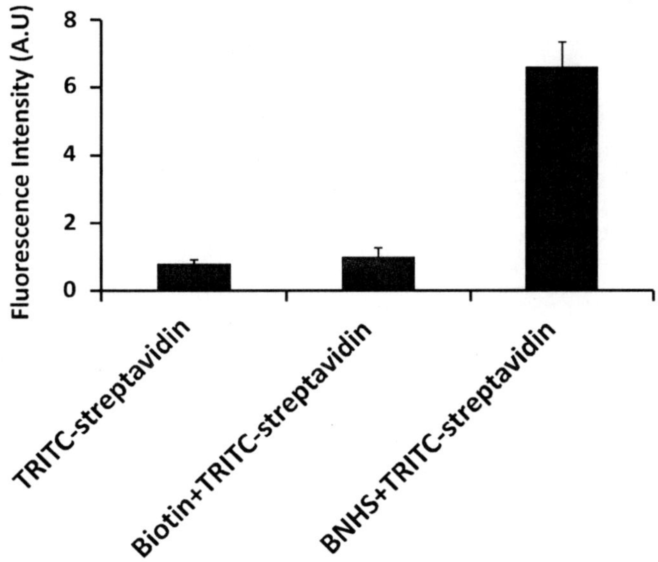

Fig. 2. Covalent modification of the MSCs by biotinylation with BNHS measured as a function of fluorescent signal of rhodamine from TRITC–streptavidin. The higher fluorescent signal from TRITC–streptavidin conjugated to BNHS modified MSCs, compared to MSCs treated by TRITC–streptavidin (without biotin) and MSCs treated by biotin (not the NHS form) and TRITC–streptavidin, shows that BNHS treatment of MSCs covalently biotinylates the cell surface (adapted from ref. (12) with permission American Chemical Society).

be remain unstained (see Note 6). Calculate the viability as follows:

$$\text{Viability}(\%) = \frac{\text{Total viable cells (unstained)}}{\text{Total cells (stained and unstained)}} \times 100.$$

3.4.2. Cell Proliferation

1. Plate 5×10^5 viable cells into a T25 cm^2 flask with 5 mL CCM and place the flask in the incubator at 37°C and 5% CO_2 (see Note 3).
2. After 24 h, count the number of adherent cells in ten randomly selected fields under the inverted phase microscope (brightfield) using the 10× objective and calculate the average number of cells per flask as follows (see Note 4):

$$\text{Total number of cells within T25 flask} \\ = \frac{\text{Average number of cell}}{\text{Area of the field of view of microscopeat } 10\times (\text{in cm}^2)} \times 25\text{cm}^2.$$

3. Repeat the visual count at different time points (i.e., day 2, 4, 6 and 8) and calculate the total number of cells at each given time point (see Note 11).

3.4.3. Adhesion Kinetics

1. Plate 5×10^3 viable cells (based on the viability determined from Subheading 3.4.1) into three separate wells of a 96 well plate in 100 μL CCM per well and incubate at 37°C and 5% CO_2.

2. Remove the media from each well, wash the wells thrice with PBS to remove the nonadherent cells and count the number of cells within three randomly selected fields under the inverted phase microscope (with 10× objective).
3. Continue to examine the number of adherent cells at different time points, specifically at 10, 30, and 90 min. Calculate the total number of adherent cells using the following formula (see Note 4):

$$\text{Total number of adherent cells per well} = \left\{ \frac{\text{Average number of cell in the field of view of microscope at } 10\times}{\text{Area of the field of view of microscope at } 10\times (\text{in cm}^2)} \right\} \times \text{Surface area of each well (in cm}^2\text{)}.$$

Calculate the percent adhesion as follows:

$$\text{Adherence}(\%) = \frac{\text{Total number of adherent cells per well}}{5{,}000} \times 100.$$

3.4.4. Osteogenic Differentiation and Staining

1. Prepare osteogenic media from the hMSC Osteogenic BulletKit by following the manufacturer's instructions. Store the media at 4°C in the dark for the duration of the osteogenic culture period.
2. Seed MSCs into a 24 well plate at a density of 2×10^4 viable cells per well in 500 μL CCM. Incubate the plate overnight at 37°C and 5% CO_2 to permit cell attachment.
3. Remove the CCM and add 500 μL of osteogenic medium and return the plate to the incubator.
4. Refresh complete osteogenic medium every 3 days for 21 days. Monitor the progress of differentiation under an inverted phase microscope by visual examination of the change in the cultures.
5. After 21 days of osteogenic culture, aspirate the medium, rinse with 500 μL of PBS and add 500 μL of 10% Neutral formalin buffer for 30 min at room temperature.
6. Discard the formalin buffer and rinse twice with PBS. Add 500 μL of alkaline phosphatase staining solution and incubate for 30–45 min at room temperature to develop the stain.
7. Discard the staining solution and wash thrice with PBS. A positive alkaline phosphatase stain results in cell membranes staining pink to red.
8. To examine colocalization of early and late markers for osteogenesis (alkaline phosphatase and calcium phosphate mineral, respectively), remove PBS from alkaline phosphatase stained wells and add 500 μL of von Kossa staining solution (2.5% $AgNO_3$) in the dark for 30 min.

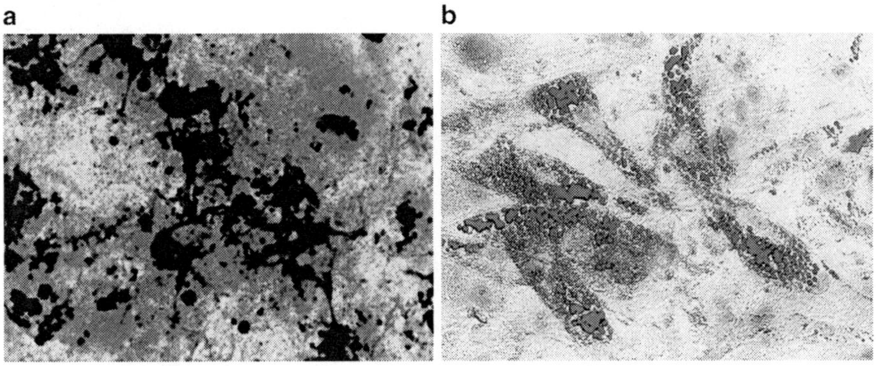

Fig. 3. (**a**) Osteogenic differentiation 23 days after induction shows positive alkaline phosphatase and von Kossa staining of the BNHS modified cells. Positive staining is indicated by black foci (von Kossa staining) surrounded by cells that are red/pink staining (alkaline phosphatase) (**b**) Adipogenic differentiation 23 days after induction shows positive Oil Red O staining for oil droplet formation together with hemotoxylin staining (for contrast) for the BNHS modified cells. Image taken with 20× magnification.

9. Discard the solution and rinse wells twice with double-distilled water.
10. Examine staining under an inverted phase microscope. Figure 3a shows the alkaline phosphatase and von Kossa stain for modified MSCs with phase inverted microscope with a 20× objective (see Note 12).

3.4.5. Adipogenic Differentiation and Staining

1. Prepare the adipogenic induction and maintenance media from hMSC Adipogenic BulletKit by following the manufacturer's instructions. Store the media at 4°C in dark for the duration of the adipogenic culture period.
2. Seed MSCs into 24 well plate at a density of 2×10^4 viable cells per well in 500 μL CCM. Incubate the plate overnight at 37°C at 5% CO_2 to allow cell adherence.
3. Aspirate the CCM and add 500 μL adipogenic induction media to each well and incubate at 37°C at 5% CO_2 for 3 days followed by incubation with 500 μL of adipogenic maintenance media for another 3 days. Repeat this alternating cycle of induction/maintenance medium for two additional cycles (12 additional days).
4. After 18 days of induction/maintenance cycling, refresh the cultures with adipogenic maintenance medium for 3 days. The total number of days for the adipogenic culture is 21. Monitor the progress of differentiation under an inverted phase microscope by visual examination of the change in the cultures.
5. After 21 days of adipogenic culture, aspirate the medium, rinse with 500 μL of PBS, and add 500 μL of 10% Neutral formalin buffer for 30 min at room temperature.

6. Discard the formalin buffer and rinse twice with PBS.
7. Add 500 µL of Oil Red O working stain solution for 50 min at room temperature.
8. Remove the staining solution, rinse thrice with 500 µL of PBS, and counterstain with 500 µL Hemotoxylin solution for 5 min.
9. Examine stained cultures under a phase inverted microscope. Figure 3b for an example of stained adipogenic cultures for modified MSCs with phase inverted microscope at 20× (see Note 13).

3.5. Flow Chamber Assay to Assess Modified MSC Rolling

Flow chambers emulate the dynamic shear conditions experienced by cells as they travel through the blood stream interacting with the endothelial surface within postcapillary venules.

3.5.1. P-Selectin Coating

1. Prepare fresh P-selectin solution at a concentration of 5 µg/mL in PBS.
2. Add 1 mL of P-selectin solution to a 35 mm tissue culture treated dish and incubate for 18 h at room temperature on a shaker at 50 rpm. Make sure to prevent any evaporation of the solution during the incubation by sealing the culture dish with parafilm.
3. Aspirate the liquid and wash twice with PBS. The dish is now ready for the flow chamber assay.

3.5.2. Flow Chamber Assembly

1. Assemble flow system apparatus by connecting the inlet, outlet, and vacuum lines to the flow chamber deck (Fig. 4) (14). Select the appropriate gasket size (gasket size determines the shear stress in the flow chamber) and attach securely to the bottom of the flow chamber deck. Fill the lines with cell culture medium and remove all air bubbles from system.
2. Attach a P-selectin coated dish to the flow chamber deck by holding the deck inverted. The vacuum should seal the dish on deck. Examine the dish to ensure that it is attached with no air bubbles in the flow path.
3. Place the assembled system on the microscope stage (Fig. 4).

3.5.3. Preparation of SLeX-Modified Cells

1. Seed 5×10^5 cells into a T25 cm^2 flask in 5 mL CCM. Allow the cells to adhere overnight in an incubator at 37°C and 5% CO_2.
2. After overnight incubation, remove the medium, and wash twice with 1 mL of prewarmed PBS (see Note 7).
3. Functionalize the MSCs with SLeX as described in Subheading 3.2.

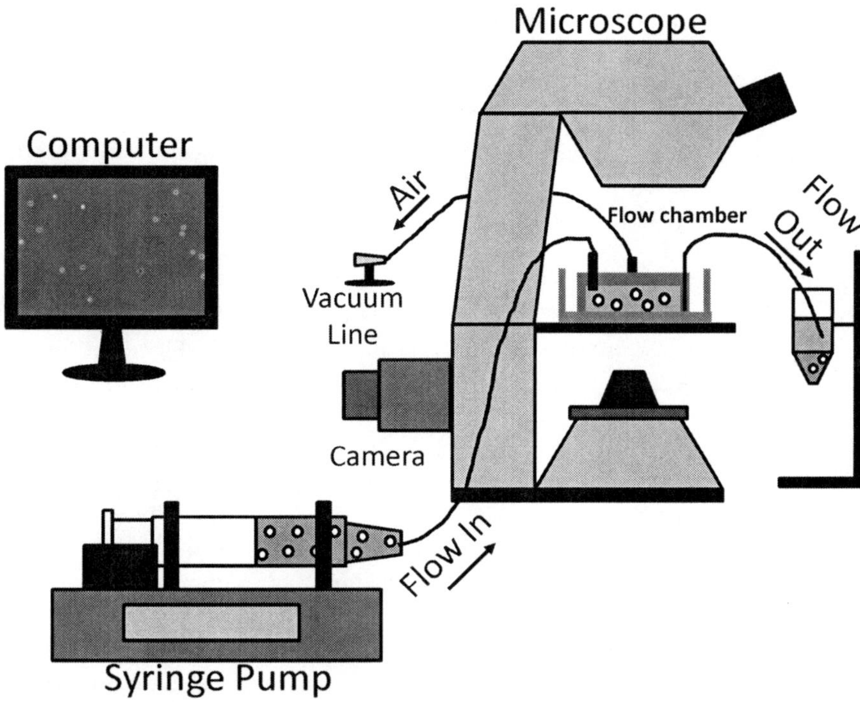

Fig. 4. Apparatus for studying cell adhesion under shear stress conditions including a flow chamber integrated with an imaging system and syringe pump (adapted from ref. (14) with permission from JoVE).

4. After the modification, harvest the cells from the flask with 1 mL of nonenzymatic cell dissociation solution (1×) at 37°C for 3 min, followed by centrifugation at 151 ×*g* for 3 min at room temperature.
5. Remove the supernatant and resuspend the cells in 1 mL CCM.
6. Count the cells using a hemocytometer and adjust the volume of the cell suspension to 1×10^5 cells/mL and proceed with Subheading 3.5.4.

3.5.4. Rolling Assay with SLeX-Modified MSCs

The multistep cell homing cascade comprises an initial rolling step, firm adhesion, and transendothelial migration of the cells. Cell rolling is critical to initiate the homing cascade through a reduction in cell velocity to slow down the cells from free stream. The rolling response is mediated by the adhesion bonds between selectin and their complementary ligands and these bonds dissociate readily under shear forces. The fast association kinetics of rolling ligands enables tethering of cells flowing in the blood stream and brings the cells into contact with the surface endothelium. As a consequence, there is a large enhancement of other receptor-ligand interactions between the cell and the endothelium (cytokine activation and firm adhesion). Consequently, the firm adhesion ligands can engage and firm adhesion is enhanced.

The firm adhesion of cells leads to the complete arrest of the cells on the endothelium and is essential to permit the transmigration of the cells through endothelium. Thus, it is critical to consider rolling and firm adhesion of the modified cells.

For flow assays under shear stress, the adhesive interactions of the cells are broadly classified either as rolling or firmly adherent (see Note 8). The criterion for these classifications are as follows: (1) Rolling cells: A cell is classified as rolling if it rolls for more than 10 s while remaining in the field of view and if it travels at an average velocity of less than 50% of the calculated free stream velocity of a noninteracting cell (15). (2) Firmly adherent cells: A cell is classified as firmly adherent if it remains in the field of view without any displacement for 10 s (16). The rolling characteristics of the cells are characterized by rolling flux (i.e., the number of cells rolling at a given time within a particular field of view) and the velocity of the rolling cells. The firm adhesion characteristics of the cells are characterized as firm binding flux (i.e., the number of cells remaining firmly adherent at a given time within a particular field of view). The adhesive interactions of the SLeX-modified MSCs on P-selectin are analyzed to detect the rolling response and/or firm adhesion of the cells on P-selectin and are compared to that of unmodified MSCs. The following protocol describes the flow chamber assay to examine the adhesive interactions of SLeX engineered MSCs on P-selectin.

1. Place the cell suspension in a sterile syringe and fit the syringe to the syringe pump. Set the flow rate to obtain the desired range of shear stress in the flow chamber. The selection of relevant shear stress should be corresponding to the physiological shear stresses experienced within animal models (17). The shear stress can be calculated from the following equation:

$$\tau_w = \mu\gamma = 6\mu Q / a^2 b,$$

   where τ_w = wall shear stress, dyn/cm^2; μ = apparent viscosity of the medium (P); a = channel height (i.e., gasket thickness, cm); b = channel width (i.e., gasket width, cm); Q = volumetric flow rate (mL/s)

2. Set all other parameters of the syringe pump according to manufacturer's instructions. Specifically set the flow (infuse mode), flow rate, and syringe diameter to achieve the desired shear stress within the flow chamber.

3. Initiate the flow at lower shear stress (0.5 dyn/cm^2) to flow in the cell suspension within the chamber for 5–10 min so that there are approximately 30 cells within the field of view.

4. Allow the cells to settle for 10 min to interact with the P-selectin coated substrate.

5. Turn on the syringe pump to initiate the flow (corresponding to particular shear stress) and record/capture the interaction

of the modified MSCs with P-selectin through the CCD camera connected to the microscope (in 10× and brightfield) for 1 min. Save the files for future analysis.

6. Initiate the flow at the next higher flow rate (i.e., wall shear stress) after a pause of a given time (e.g., 1–5 min). Note that the flow is continued here after a stop. This is to allow interaction of the cells with the surface under static condition to take place.
7. Analyze the results to examine the adhesive interactions of the cells within the flow chamber. Calculate the number of cells interacting with P-selectin by counting the number of cells that remain within the field of view of the microscope for a given period of time (e.g., 10 s). This calculation will provide the total number of interacting cells which includes both the rolling cells and firmly adherent cells. Classify this data as rolling flux (for rolling cells) and firmly binding flux (for firmly adherent cells). The analysis of the adhesive interactions should be performed over a range of shear stresses to examine the effect of shear stress on the rolling flux and firmly binding flux.
8. Refer to Fig. 5a for an example of the velocity of MSCs modified with SLeX compared to PBS treated MSCs (unmodified MSCs) on P-selectin coated substrate at shear 0.366 dyn/cm^2. In addition, see Fig. 5b as an example of the number of cells interacting with the P-selectin substrate for MSCs modified with SLeX and PBS treated MSCs at a shear stress of 0.36 dyn/cm^2 using the flow chamber assay.
9. See Fig. 5c for an example of the effect of shear stress on the rolling velocity of the SLeX engineered MSCs on the P-selectin surface. This data indicates that modification of MSCs with SLeX leads to specific interaction with P-selectin leading to an induced rolling interaction in a flow chamber assay.

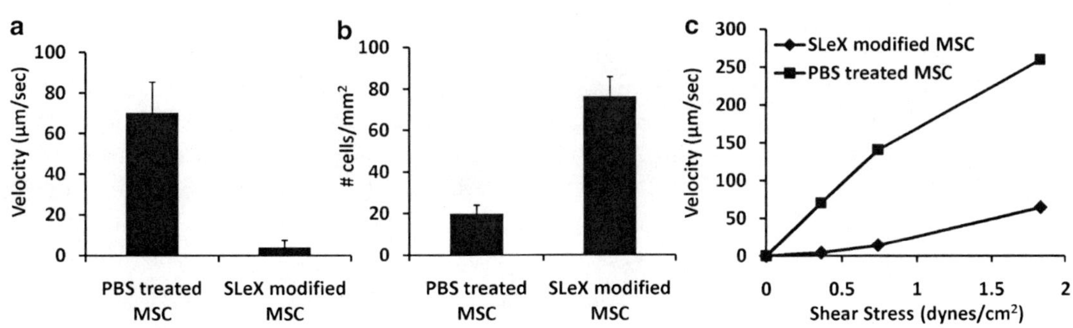

Fig. 5. (**a**) Velocity and (**b**) Flux of SLeX-modified MSCs and controls on P-selectin treated surfaces at the shear stress of 0.366 dyn/cm^2. (**c**) Effect of shear stress on the velocity of SLeX-modified MSCs and unmodified MSCs on P-selectin treated surface (adapted from ref. (12) with permission American Chemical Society).

4. Notes

1. Always prepare fresh BNHS solution before experimentation. Discard any unused solution and do not store it for later use.
2. Avoid repeated freezing and thawing of the stock solution to retain the functional activity of SLeX.
3. During the proliferation assay, change the media of the flask every 2 days with fresh prewarmed CCM.
4. The area for the field of view at a given magnification is different for specific microscopes and must be determined specifically for each microscope.
5. The alkaline phosphatase and von Kossa staining solution must be made fresh. Add the solution immediately after preparation to the osteogenic cultures.
6. For viability experiments with trypan blue solution, if the cells are exposed to trypan blue for an extended period of time the viable cells will also stain blue. Therefore, prolonged exposure to trypan blue can lead to significant error with the viability calculations.
7. For the cell rolling assay, the T25 flask should be at 50–80% confluency on the day of the rolling assays.
8. For the analysis of flow chamber assays, it is important to note that there are several other parameters that may be important to characterize including:
   (a) *Instantaneous velocity*: the velocity profile for the distribution of velocities with respect to time. This data should be plotted as velocity versus time. This velocity profile will characterize the transition from cell rolling to firm adhesion and vice versa and will explain the dual interactions. This type of quantification is also useful to investigate the transient rolling of the cells which rolls for less than 10 s (the criteria defined for rolling cells) and then join the free stream that undergo transition to firm adhesion or detachment, especially in the presence of multiple receptors on the surface (e.g., selectin with a coimmobilized antibody). The pause time during cell rolling provides direct insights into the kinetics of bond formation and dissociation; it has been observed that the pause time increases upon increasing the concentration of antibodies on the surface (against molecules on the cell membrane) (18, 19).
   (b) *Tethering*: defines the initial interaction of a cell as it transitions from the free stream to the substrate leading to cell rolling or firm adhesion (13, 20). This can be quantified as the number of cells that accumulate on the substrate during

a period of continuous flow. To compare tethering between different groups, one may maintain the same concentration of the cells between groups or normalize by dividing the number of cells tethered by the number of cells transported across the field of view.

9. Passage 3–6 MSCs are recommended for use in experimentation for modification of the cell surface.
10. Nonenzymatic cell dissociation solution (1×) is recommended to harvest the surface modified cells to prevent the cleavage of the conjugated ligands on the cell surface.
11. The factor 25 in the provided equation is specific for the surface area of theT25 flask and will change if different vessels with different surface areas are used. Plot the data as a proliferation curve with total number of cells (y-axis) versus time (x-axis).
12. Positive osteogenic staining is indicated by black foci (von Kossa staining) surrounded by cells that are red/pink staining (alkaline phosphatase). Nonspecific staining is indicated by black foci that are not surrounded by red/pink stained cells.
13. Lipid globules stain red and the nuclei stain purple for the adipogenic staining.
14. The flow cytometric analysis should also be performed following cell modification to examine any potential perturbations to the cell surface profile.

5. Conclusion

The methods presented here offer a simple approach to conjugate ligands on the cell surface for a variety of purposes including promoting cell rolling, adhesion, and cell–cell interactions. The biotin–streptavidin bridge permits conjugation of biotinylated ligands of which there are hundreds available. We have used the methods herein for imparting a rolling response through functionalization of primary human MSCs with SLeX through combined covalent and biotin–streptavidin chemistries. Moreover, this approach can be utilized without compromising cell viability, proliferation, adhesion, and differentiation of stem cells.

References

1. Picinich, S. C., Mishra, P. J., Mishra, P. J., Glod, J., and Banerjee, D. (2007) The therapeutic potential of mesenchymal stem cells, *Cell and Tissue-Based Therapy, Expert Opinion on Biological Therapy* 7, 965–973.
2. Burt, R. K., Loh, Y., Pearce, W., Beohar, N., Barr, W. G., Craig, R., Wen, Y., Rapp, J. A., and Kessler, J. (2008) Clinical applications of blood-derived and marrow-derived stem cells for nonmalignant diseases, *JAMA* 299, 925–936.
3. Abdel-Latif, A., Bolli, R., Tleyjeh, I. M., Montori, V. M., Perin, E. C., Hornung, C. A., Zuba-Surma, E. K., Al-Mallah, M., and Dawn, B. (2007) Adult bone marrow-derived cells for cardiac

repair: a systematic review and meta-analysis, *Archives of Internal Medicine* 167, 989–997.

4. Filho Cerruti, H., Kerkis, I., Kerkis, A., Tatsui, N. H., da Costa Neves, A., Bueno, D. F., and da Silva, M. C. (2007) Allogenous bone grafts improved by bone marrow stem cells and platelet growth factors: clinical case reports, *Artificial Organs* 31, 268–273.
5. Garcia-Olmo, D., Garcia-Arranz, M., Herreros, D., Pascual, I., Peiro, C., and Rodriguez-Montes, J. A. (2005) A phase I clinical trial of the treatment of Crohn's fistula by adipose mesenchymal stem cell transplantation, *Diseases of the Colon and Rectum* 48, 1416–1423.
6. Maitra, B., Szekely, E., Gjini, K., Laughlin, M. J., Dennis, J., Haynesworth, S. E., and Koc, O. N. (2004) Human mesenchymal stem cells support unrelated donor hematopoietic stem cells and suppress T-cell activation, *Bone Marrow Transplantation* 33, 597–604.
7. Karp, J. M., and Leng Teo, G. S. (2009) Mesenchymal stem cell homing: the devil is in the details, *Cell Stem Cell* 4, 206–216.
8. Phinney, D. G., and Prockop, D. J. (2007) Concise review: mesenchymal stem/multipotent stromal cells: the state of transdifferentiation and modes of tissue repair – current views, *Stem Cells (Dayton, Ohio)* 25, 2896–2902.
9. Ruster, B., Gottig, S., Ludwig, R. J., Bistrian, R., Muller, S., Seifried, E., Gille, J., and Henschler, R. (2006) Mesenchymal stem cells display coordinated rolling and adhesion behavior on endothelial cells, *Blood* 108, 3938–3944.
10. Sackstein, R., Merzaban, J. S., Cain, D. W., Dagia, N. M., Spencer, J. A., Lin, C. P., and Wohlgemuth, R. (2008) Ex vivo glycan engineering of CD44 programs human multipotent mesenchymal stromal cell trafficking to bone, *Nature Medicine* 14, 181–187.
11. Cheng, Z., Ou, L., Zhou, X., Li, F., Jia, X., Zhang, Y., Liu, X., Li, Y., Ward, C. A., Melo, L. G., and Kong, D. (2008) Targeted migration of mesenchymal stem cells modified with CXCR4 gene to infarcted myocardium improves cardiac performance, *Molecular Therapy* 16, 571–579.
12. Sarkar, D., Vemula, P. K., Teo, G. S., Spelke, D., Karnik, R., Wee le, Y., and Karp, J. M. (2008) Chemical engineering of mesenchymal stem cells to induce a cell rolling response, *Bioconjugate Chemistry* 19, 2105–2109.
13. Yago, T., Leppanen, A., Qiu, H., Marcus, W. D., Nollert, M. U., Zhu, C., Cummings, R. D., and McEver, R. P. (2002) Distinct molecular and cellular contributions to stabilizing selectin-mediated rolling under flow, *The Journal of Cell Biology* 158, 787–799.
14. Wiese, G., Barthel, S. R., and Dimitroff, C. J. (2009) Analysis of physiologic E-selectin-mediated leukocyte rolling on microvascular endothelium, *Journal of Visualized Experiments* **11**, 1009–1020.
15. Hong, S., Lee, D., Zhang, H., Zhang, J. Q., Resvick, J. N., Khademhosseini, A., King, M. R., Langer, R., and Karp, J. M. (2007) Covalent immobilization of p-selectin enhances cell rolling, *Langmuir* 23, 12261–12268.
16. Brunk, D. K., Goetz, D. J., and Hammer, D. A. (1996) Sialyl Lewis(x)/E-selectin-mediated rolling in a cell-free system, *Biophysical Journal* 71, 2902–2907.
17. Mazo, I. B., and von Andrian, U. H. (1999) Adhesion and homing of blood-borne cells in bone marrow microvessels, *Journal of Leukocyte Biology* 66, 25–32.
18. Eniola, A. O., Willcox, P. J., and Hammer, D. A. (2003) Interplay between rolling and firm adhesion elucidated with a cell-free system engineered with two distinct receptor-ligand pairs, *Biophysical Journal* 85, 2720–2731.
19. Rodgers, S. D., Camphausen, R. T., and Hammer, D. A. (2000) Sialyl Lewis(x)-mediated, PSGL-1-independent rolling adhesion on P-selectin, *Biophysical Journal* 79, 694–706.
20. Zou, X., Shinde Patil, V. R., Dagia, N. M., Smith, L. A., Wargo, M. J., Interliggi, K. A., Lloyd, C. M., Tees, D. F., Walcheck, B., Lawrence, M. B., and Goetz, D. J. (2005) PSGL-1 derived from human neutrophils is a high-efficiency ligand for endothelium-expressed E-selectin under flow, *American Journal of Physiology* 289, C415–C424.
21. Zhang, M., Mal, N., Kiedrowski, M., Chacko, M., Askari, A. T., Popovic, Z. B., Koc, O. N., and Penn, M. S. (2007) SDF-1 expression by mesenchymal stem cells results in trophic support of cardiac myocytes after myocardial infarction, *FASEB J* 21, 3197–3207.
22. Barbash, I. M., Chouraqui, P., Baron, J., Feinberg, M. S., Etzion, S., Tessone, A., Miller, L., Guetta, E., Zipori, D., Kedes, L. H., Kloner, R. A., and Leor, J. (2003) Systemic delivery of bone marrow-derived mesenchymal stem cells to the infarcted myocardium: feasibility, cell migration, and body distribution, *Circulation* 108, 863–868.
23. Chen, J., Li, Y., Wang, L., Zhang, Z., Lu, D., Lu, M., and Chopp, M. (2001) Therapeutic benefit of intravenous administration of bone marrow stromal cells after cerebral ischemia in

rats, *Stroke: A Journal of Cerebral Circulation* 32, 1005–1011.

24. Kawada, H., Fujita, J., Kinjo, K., Matsuzaki, Y., Tsuma, M., Miyatake, H., Muguruma, Y., Tsuboi, K., Itabashi, Y., Ikeda, Y., Ogawa, S., Okano, H., Hotta, T., Ando, K., and Fukuda, K. (2004) Nonhematopoietic mesenchymal stem cells can be mobilized and differentiate into cardiomyocytes after myocardial infarction, *Blood* 104, 3581–3587.

Index

A

B

Mohan C. Vemuri et al. (eds.), *Mesenchymal Stem Cell Assays and Applications*, Methods in Molecular Biology, vol. 698, DOI 10.1007/978-1-60761-999-4, © Springer Science+Business Media, LLC 2011

C

I

L

M

Q

R

S

T

U

V

W

LaVergne, TN USA
24 March 2011
221568LV00011B/20/P